MODERN DIESEL TECHNOLOGY: LIGHT DUTY DIESELS

Sean Bennett

DELMAR
CENGAGE Learning™

Australia • Brazil • Japan • Korea • Mexico • Singapore • Spain • United Kingdom • United States

DELMAR
CENGAGE Learning™

Modern Diesel Technology: Light Duty Diesels
Sean Bennett

Vice President, Editorial: Dave Garza

Director of Learning Solutions: Sandy Clark

Executive Editor: Dave Boelio

Managing Editor: Larry Main

Senior Product Manager: Sharon Chambliss

Editorial Assistant: Jillian Borden

Vice President, Marketing: Jennifer Baker

Marketing Director: Deborah Yarnell

Marketing Manager: Kathryn Hall

Production Director: Wendy Troeger

Production Manager: Mark Bernard

Senior Content Project Manager: Cheri Plasse

Senior Art Director: Joy Kocsis

Technology Project Manager: Cristopher Catalina

Production Technology Analyst: Joe Pliss

For product information and technology assistance, contact us at
Cengage Learning Customer & Sales Support, 1-800-354-9706
For permission to use material from this text or product,
submit all requests online at **cengage.com/permissions**
Further permissions questions can be e-mailed to
permissionrequest@cengage.com

Library of Congress Control Number: 2010942698

ISBN-13: 978-1-4354-8047-6

ISBN-10: 1-4354-8047-3

Delmar
5 Maxwell Drive
Clifton Park, NY 12065-2919
USA

Cengage Learning is a leading provider of customized learning solutions with office locations around the globe, including Singapore, the United Kingdom, Australia, Mexico, Brazil, and Japan. Locate your local office at: **international.cengage.com/region**

Cengage Learning products are represented in Canada by Nelson Education, Ltd.

To learn more about Delmar, visit **www.cengage.com/delmar**

Purchase any of our products at your local college store or at our preferred online store **www.cengagebrain.com**

Notice to the Reader
Publisher does not warrant or guarantee any of the products described herein or perform any independent analysis in connection with any of the product information contained herein. Publisher does not assume, and expressly disclaims, any obligation to obtain and include information other than that provided to it by the manufacturer. The reader is expressly warned to consider and adopt all safety precautions that might be indicated by the activities described herein and to avoid all potential hazards. By following the instructions contained herein, the reader willingly assumes all risks in connection with such instructions. The publisher makes no representations or warranties of any kind, including but not limited to, the warranties of fitness for particular purpose or merchantability, nor are any such representations implied with respect to the material set forth herein, and the publisher takes no responsibility with respect to such material. The publisher shall not be liable for any special, consequential, or exemplary damages resulting, in whole or part, from the readers' use of, or reliance upon, this material.

Printed in the United States of America
Print Number: 14   Print Year: 2019

Brief Contents

Contents

Preface

This ninth textbook in the Modern Diesel Technology series takes a look at the light duty diesel engines that are beginning to notch some sales success in North American markets. The American consumer has traditionally regarded the diesel engine as a workhorse powerplant, so most of the sales have been in the vocational pickup truck segment of the market. While the big three U.S. auto original equipment manufacturers (OEMs) have optioned diesel power in pickup trucks for more than two decades, sales have been sluggish until recently. One explanation was that the light duty diesel engines of the early 1990s were dirty, underpowered, and prone to costly breakdowns. Combine this with a poor infrastructure of diesel fuel supply outlets and it was no wonder that diesels struggled to achieve significant market share.

The technology of the American-built diesel engine changed more than a decade ago. A new generation of electronically controlled engines out powered and out torqued their gasoline-fueled rivals while providing greater longevity with lower maintenance. Farmers were probably the first to realize this, perhaps because as a consumer group they had the least reason to mistrust the diesel engine: for four decades, diesels have powered a large percentage of agricultural equipment. Although the growth of diesel engine sales is slower than I would have predicted five years ago, slowly but surely their market share is expanding, helped by the excellent offerings by Ford, GM, and Chrysler in their current small trucks. The 2011 Ford 6.7L is of particular note because this engine is engineered and manufactured from scratch within the Ford Motor Company.

More perplexing is the reluctance of the American consumer to consider diesel power in the family automobile. Auto diesels have been accepted in Europe for decades where they are purchased for fuel economy, lower maintenance, longevity, and better performance. The winner's circle in the grueling, high-speed LeMans 24-hour automobile race in France has been dominated by diesel power for years, and on-highway variants of these engines are a popular muscle car option. For some reason domestic OEMs have bet the barn on gasoline hybrid electric power (GHEP) and marketed this technology accordingly. While GHEP might make sense in a vehicle that operates in a stop-start city environment, it makes a lot less when that vehicle is run on the interstate, when it becomes exclusively gasoline powered. GHEP is a great solution for city vehicles such as pickup and delivery vans and transit buses, but it competes poorly with diesel power in a general purpose family automobile.

ORGANIZATION

While the focus of this book is primarily on the post-2010 crop of light duty vocational and automobile engines, some text is also devoted to some of the earlier electronic and hydromechanical diesels. Special attention is devoted to the Ford 2011 6.7L, the 2011 GM Duramax 6600, and the 2010 Cummins ISB that powers Dodge pickups. Today's stringent emissions requirements require an elaborate range of precombustion, incombustion, and postcombustion apparatus to be fitted on engines, and these are studied from a technician's perspective.

Multiplexing is a fact of life in any current vehicle. For this reason, the light duty generations of CAN technology have to be included. It is a challenge to make complex electronics technology understandable, so I have adopted a "have-to-know" approach in attempting to make CAN-A, CAN-B, J1850, and CAN-C easily understood.

Modern Diesel Technology: Light Duty Diesels covers the task fields required by the newly introduced ASE A9 certification test addressing light diesel technology. For this reason, there is a detailed maintenance and troubleshooting chapter designed to help readers undertake basic maintenance tasks and navigate OEM online service information and diagnostic systems, along with an introduction to some of the electronic service tools used in the industry. Throughout the book, the objective is to provide the key theory required to enable sound hands-on shop floor skills, presented in a user-friendly manner. I would like to especially thank John Murphy of Centennial College in Toronto, and Bob Starr of the Automotive Technical Training Center in New York City for their feedback in preparing the manuscript for this book.

Sean Bennett, January 2011

ABOUT THE SERIES

The Modern Diesel Technology (MDT) series has been developed to address a need for *modern*, system-specific textbooks in the field of truck and heavy equipment technology. This focused approach gives schools more flexibility in designing programs that target specific ASE certifications. Because each textbook in the series focuses exclusively on the competencies identified by its title, the series is an ideal review and study vehicle for technicians prepping for certification examinations.

Titles in the Modern Diesel Technology Series include:
MDT: Electricity and Electronics, by Joe Bell; ISBN: 1401880134
MDT: Heating, Ventilation, Air Conditioning, and Refrigeration, by John Dixon; ISBN:1401878490
MDT: Electronic Diesel Engine Diagnosis, by Sean Bennett; ISBN: 1401870791
MDT: Brakes, Suspension, and Steering Systems, by Sean Bennett; ISBN: 1418013722
MDT: Heavy Equipment Systems, by Robert Huzij, Sean Bennett, and Angelo Spano; ISBN: 1418009504
MDT: Preventive Maintenance and Inspection, by John Dixon; ISBN: 1418053910
MDT: Mobile Equipment Hydraulics: A Systems and Troubleshooting Approach, by Ben Watson; ISBN: 1418080438
MDT: Light Duty Diesels, by Sean Bennett; ISBN: 1435480473

ACKNOWLEDGMENTS

The author and publisher would like to thank the following for their comments, suggestions, and contributions during the development process.

Individuals

Sergio Hernandez, Palomar College, San Marcos, CA
Bob Huzij, Cambrian College, Sudbury, ON
Tony Martin, University of Alaska, Juneau, AK
John Murphy, Centennial College, Toronto
Ken Pickerill, Ivy Tech Community College, Indiana
Bob Starr, NYATTP, Brooklyn, NY

Companies

(Robert) Bosch gmbh
Caterpillar
Chrysler Motors
Cummins Engine
Ford Motor Company
General Motors Corporation
Navistar International
Volkswagen USA

SUPPLEMENT

An Instructor Resources CD is available with the textbook. Components of the CD include an electronic copy of the Instructor's Guide, PowerPoint® lecture slides that present the highlights of each chapter, an ExamView computerized test bank, and an Image Gallery that includes an electronic copy of the images in the book.

Shop and Personal Safety

Learning Objectives

After studying this chapter, you should be able to:

- Identify potential danger in the workplace.
- Describe the importance of maintaining a healthy personal lifestyle.
- Outline the personal safety clothing and equipment required when working in a service garage.
- Distinguish between different types of fire.
- Identify the fire extinguishers required to suppress small-scale fires.
- Describe how to use jacks and hoisting equipment safely.
- Explain the importance of using exhaust extraction piping.
- Identify what is required to work safely with chassis electrical systems and shop mains electrical systems.
- Outline the safety procedures required to work with oxyacetylene torches.

Key Terms

chain hoist

cherry picker

come-alongs

digital multimeter
 (DMM)

electronic service tool (EST)

ground strap

original equipment manufacturer
 (OEM)

single-phase mains

static charge

static discharge

three-phase mains

Underwriter's Laboratories (UL)

INTRODUCTION

The mechanical repair trades are physical by nature, and those employed as technicians probably have higher-than-average levels of personal fitness. While there are some heavy components on most vehicles, technicians in the modern workplace are never required to lift excessive weights. They are required to understand when and how to use shop jacks and hoisting equipment. Technicians should also make it

their business to safely handle materials that can be hazardous. It goes without saying that employers are required to ensure that the shop floor is a safe working environment. Employers who fail to ensure a safe working environment are both breaking the law and endangering their profitability.

Safety Rules

If you look at any service repair shop today, you will notice safety rules and regulations posted on walls and bulletin boards. Although these are posted for maximum exposure, the responsibility to work safely rests with the individual. A large part of safe working practice is common sense. But it is up to the individual to observe these common-sense rules and regulations. It simply does not make sense to take risks when safety is an issue. Those who do, sooner or later get burned by poor decisions, sometimes fatally.

Experience and Injuries

Most technicians do not want to get hurt in the workplace or anywhere else. But knowing something about potential danger minimizes the risk of injury. A major automotive **original equipment manufacturer (OEM)** monitored accidents in its assembly plants and came up with the following conclusion: line-production employees' risk of serious injury (defined as one that required some time off work) during the first year of employment was equal to that of years 2 through 6 combined. In simple terms, if you can survive your first year injury free, your risk thereafter diminishes.

Safety Awareness

Teachers of mechanical technology often complain that it is difficult to teach safe work practices to entry-level students. When students enroll in an automotive technology program, they appear to be well motivated to learn the technology but tend to tune out when it comes to learning the health and safety issues that accompany working life. The sad truth is that it is difficult to teach safe work practices to persons who have never been injured. On the other hand, a person who has been injured probably has acquired, along with the injury, a powerful motivation to avoid a repetition.

A Healthy Lifestyle

Repairing vehicles requires more physical strength than working at a desk all day, but it would be a mistake to say it was a healthy occupation. Lifting a 150 lb (70 kg) cylinder head or pulling a high load on a torque wrench requires some muscle power, but you cannot compare this with lifting weights in a gym. In the weight room, the repetitions, conditions, and movements are carefully coordinated to develop muscle power. Jerking on a torque wrench attempting to establish final torque on main caps on an engine job can tear muscle as easily as develop it.

It pays to think about how you use your body and use your surroundings to maximize leverage and minimize wear and tear. Make a practice of using hoists to move heavier components, even if you know you could manually lift the component. You may believe it is macho to lift a cylinder head off a block by hand, but it only takes a slight twist of the back while doing so to sustain an injury that can last a lifetime. There is nothing especially macho about hobbling around with chronic back pain for years.

Physical Fitness

Part of maintaining a healthy lifestyle means eating properly and making physical activity a part of your lifestyle. There are many different ways of achieving this. Team sports are not just for kids and teenagers. Whether your sport is hockey, baseball, basketball, or football, there are plenty of opportunities to compete through all ages and levels. If team sports are not your thing, there are many individual pursuits that you can explore. Working out in a gym, hiking, and canoeing are good for your mind as well as your body, and even golf gets you outside and walking. Because of the physical nature of repair technology, it makes sense to routinely practice some form of weight conditioning, especially as you get older.

PERSONAL SAFETY EQUIPMENT

Personal safety equipment refers to anything you wear on your body in the workplace. Some items of personal safety should be worn continually in the workplace. One of these essential items is safety shoes or footwear. Other personal safety equipment such as hearing protection may be worn only when required, for instance when noise levels are high.

Safety Boots

Safety boots or shoes are required footwear in a repair shop. Most jurisdictions require technicians to wear safety footwear. It is an employer responsibility to make sure this happens. Safety footwear is manufactured with steel shanks, steel toes, and UL (Underwriter's Laboratories: www.ul.com) certification. Keep in mind that safety is about you. If you lose a limb in the workplace, your whole life will be affected by the event. Even if the law did not require you to wear safety footwear, common sense should tell you that your feet should be protected in a shop environment. Given the choice, safety boots (see **Figure 1-1**) can generally be regarded as a better choice than safety shoes because of the additional support and protection for the ankle area.

Figure 1-1 UL-approved safety boots.

There is a range of options when it comes to selecting safety footwear. There is also a wide range of prices. If you are going to work on a car under a tree over a weekend, it may be that a low-cost pair of safety shoes is all you will require. However, for the professional technician who wears this footwear daily, it pays to invest a little more. Better-quality safety footwear will last longer and will tend to be more comfortable.

Safety Glasses

Many shops today require all their employees to wear safety glasses while on the shop floor. This is really just common sense. Eyes are sensitive to dust, metal shavings, grinding and machining particulates, fluids, and fumes. They are also more complex to repair than feet when injured. It also makes sense to wear safety glasses when working with chassis electrical equipment because of the potential danger represented by battery acid and arcing at terminals.

Perhaps the major problem when it comes to making a habit of using safety glasses is the poor quality of most shop-supplied eyewear. Shops supply safety glasses because in many cases they are legally liable if they do not. All too often, this means they provide low-cost, mass-produced, and easily scratched plastic safety glasses. If you have a pair of safety glasses that impairs your vision, you will probably want to wear them as little as possible. A pair of safety glasses in your pocket is not going to protect you from eye injuries.

Don't Be Cheap! The solution is not to depend on your employer to provide safety glasses. Get out of the mind-set that safety glasses should be provided to you at no cost. As we have said, "free" safety glasses are uncomfortable and may actually impair vision. Buy

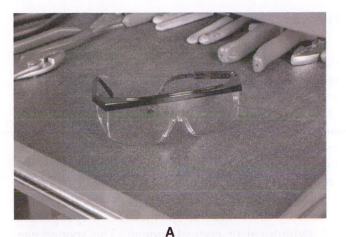

A

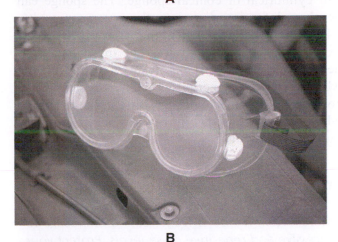

B

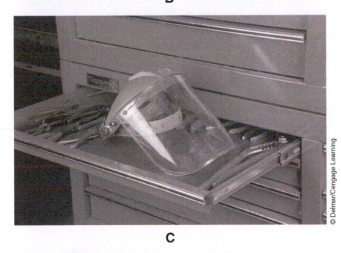

C

Figure 1-2 (A) Safety glasses, (B) splash goggles, and (C) face shield.

your own. Spend a little more and purchase a good quality pair of safety glasses. These will be optically sound and scratch proof. Even if you do not normally wear eyeglasses, after a couple of days, you will forget you are wearing them. **Figure 1-2** shows some eye protection options available to technicians.

Hearing Protection

Two types of hearing protection are used in shops. Hearing muffs are connected by a spring-loaded band and enclose the complete outer ear. This type of hearing protection is available in a range of qualities. Cheaper versions may be almost useless, but good-quality hearing muffs can be very effective when noise levels are extreme. But be careful. Hearing muffs that almost completely suppress sound can be dangerous because they disorientate the wearer.

A cheaper and generally effective alternative to hearing muffs are ear sponges. Each sponge is a soft cylindrical or conical sponge. The sponge can be shaped for insertion into the outer ear cavity. Almost immediately after insertion, the sponge expands to fit the ear cavity. The disadvantage of hearing sponges is that they can be uncomfortable when worn for long periods. Technicians should also consider using other types of soft earplugs, most of which are wax based. **Figure 1-3** shows some earmuffs and earplugs.

> **CAUTION** *Damage to hearing is seldom the result of a single exposure to a high level of noise. More often, it results from years of exposure to excessive and repetitive noise levels. Protect your hearing! And note that hearing can be damaged by listening to music at excessive volume as easily as exposure to buck riveting.*

Gloves

A wide range of gloves can be used in shop applications to protect the hands from exposure to dangerous or toxic materials and fluids. The following are some examples.

> **CAUTION** *Never wear any type of glove when using a bench-mounted, rotary grinding wheel: there have been cases where a glove has been snagged by the abrasive wheel, dragging the whole hand with it.*

Vinyl Disposable. Most shops today make vinyl disposable gloves available to service personnel. These protect the hands from direct exposure to fuel, oils, and grease. The disadvantage of vinyl gloves is that they do not breathe, and some find the sweating hands that result to be uncomfortable. Most shop-use vinyl gloves today are made of thin gossamer that allows some touch sensation.

Cloth and Leather Multipurpose. A typical pair of multipurpose work gloves consists of a rough leather palm and cotton back. They can be used for a variety of tasks ranging from lifting objects to general protection from cold when working outside. You should not use this type of glove after saturation with grease or oil.

Dangerous Materials Gloves. Gloves designed to handle acids or alkalines should be used for these tasks only. Gloves in this category are manufactured from unreactive, synthetic rubber compounds. Care should be taken when washing up after using this type of glove.

> **CAUTION** *Never wear leather gloves to handle refrigerants: leather gloves rapidly absorb refrigerant and can adhere to the skin.*

Back Care

Back injuries are said to affect 50 percent of repair technicians at some point in their careers seriously enough for them to have to take time off work. A bad back does not have to be an occupational hazard. Most of us begin our careers in our twenties when we have sufficient upper-body strength to handle plenty of abuse. As we age, this upper-body strength decreases, and bad lifting practices can take their toll.

Observe some simple rules for lifting heavy items:

- Keep your back vertical while lifting (do not bend).
- Keep the weight you are lifting close to your body.
- Bend your legs and lift using the leg muscles.

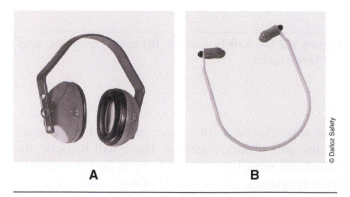

A   B

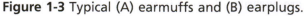

Figure 1-3 Typical (A) earmuffs and (B) earplugs.

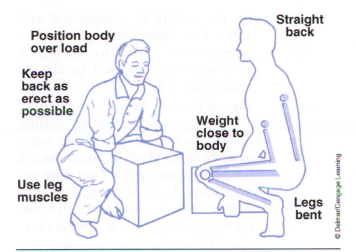

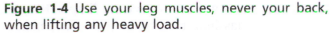

Figure 1-4 Use your leg muscles, never your back, when lifting any heavy load.

Figure 1-5 It is important to wear safety glasses and protective clothing when working on vehicles.

Figure 1-4 shows how to protect your back when lifting heavy objects: one of the keys is to hold the weight as close to your body as you can get it.

Back Braces. A back brace may help you avoid injuring your back. Wearing a back brace makes it more difficult to bend your back, so it "reminds" you to keep it straight when lifting. You may have noticed that the sales personnel in one national hardware and home goods chain are all required to wear back braces. As an automotive technician, you will be required to use your back for lifting, so you should consider the use of a back brace. Body shape plays a role when it comes to back injuries: if you are either taller than the average height or overweight, you will be more vulnerable to back injuries.

Coveralls and Shop Coats

Many shops today require their service employees to wear a uniform of some kind. This may be work shirts and pants, shop coats, or coveralls. Uniforms have a way of making service personnel look professional. Shop workwear should preferably be made out of cotton for reasons of comfort and safety. When ordering work clothing for personal use, remember to order at least a size larger than your usual nominal size: unless otherwise treated, cotton shrinks when washed. Shop coats can also be used, but because these come pretty close to our definition of loose clothing they are a second-best choice to coveralls. **Figure 1-5** shows a technician properly attired for working on a vehicle.

> **CAUTION**   *Avoid wearing any type of loose-fitting clothing when working with machinery. Shop coats, neckties, and not tucking a shirt into pants can all be classified as loose-fitting clothing.*

Artificial Fibers. When artificial fibers are used as material for coveralls, they should be treated with fire retardant. Cotton smolders when exposed to fire. That is, it smolders so long as it is not saturated with oil, fuel, or grease. When any material is saturated with petroleum products it becomes highly flammable. Cleanliness is essential: oily shop clothing not only looks unprofessional, it can be dangerous!

Artificial fibers can be especially dangerous. When not treated with fire retardant, artificial fibers melt when exposed to high temperatures. This can cause them to fuse to the skin. You should note that even when treated with fire retardant, some artificial fibers will burn vigorously when exposed to direct flame for a period of time.

> **CAUTION**   *Note that even when treated with fire retardant, some artificial fibers will burn vigorously when exposed to direct flame for a period of time.*

Butane Lighters

There are few more dangerous items routinely observed on the shop floor than the butane cigarette lighter. The explosive potential of the butane lighter is immense yet it is often stored in a pocket close to where it can do the most amount of damage. A chip of hot welding slag will almost instantly burn through the plastic fuel cell of a butane lighter. Owners of these devices will often compound the danger they represent by lighting torches with them. If you have to have a lighter on your person while working, purchase a Zippo!

Hair and Jewelry

Long hair and personal jewelry produce some of the same safety concerns as loose-fitting clothing. If it is your style to wear long hair, it should be secured behind the head, and you should consider wearing a cap. Because of the recent trend to wear more body jewelry, you should remove as much as possible of this while at work. Body jewelry is often made of conductive metals (such as gold, platinum, silver, and brass), so you should consider both the possibility of snagging jewelry and of creating some unwanted electrical short circuits.

FIRE SAFETY

Service and repair facilities are usually subject to regular inspections by fire departments. This means that obvious fire hazards are identified and neutralized. While it should be stressed that fire fighting is a job for trained professionals, any person working in a service shop environment should be able to appropriately respond to a fire in its early stages. This requires some knowledge of the four types of fire extinguishers in current use.

Fire Extinguishers

Fire extinguishers are classified by the types of fires they are designed to suppress. Using the wrong type of fire extinguisher on certain types of fires can be extremely dangerous and actually worsen the fire you are attempting to control. Every fire extinguisher clearly indicates the types of fires it is designed to extinguish. This is done by using class letters. This means that it is important to identify each of the four types of fires that could occur in the workplace. The role of the technician in suppressing a fire is to estimate the risk required. Intervention should only be considered if there is minimal risk.

Class A   A class A fire is one involving combustible materials such as wood, paper, natural fibers, biodegradable waste, and dry agricultural waste. A class A fire can usually be extinguished with water. Fire extinguishers designed to suppress class A fires use foam or a multipurpose dry chemical, usually sodium bicarbonate.

Class B   Class B fires are those involving fuels, oil, grease, paint, and other volatile liquids, flammable gases, and some petrochemical plastics. Water should not be used on class B fires. Fire extinguishers designed to suppress type B fires work by smothering: they use foam, dry chemicals, or carbon dioxide. Trained fire personnel may use extinguishers such as Purple K (potassium bicarbonate) or halogenated agents to control fuel and oil fires.

Class C   Class C fires are those involving electrical equipment. First intervention with this type of fire should be to attempt to shut off the power supply: assess the risk before handling any switching devices. When a class C fire occurs in a vehicle harness, combustible insulation and conduits can produce highly toxic fumes, so great care is required when making any kind of intervention with electrical fires in vehicle chassis or buildings. Fire extinguishers designed to suppress electrical fires use carbon dioxide, dry chemical powder, and Purple K.

Class D   Class D fires are those involving flammable metals: some metals when heated to their fire point begin to vaporize and combust. These metals include magnesium, aluminum, potassium, sodium, and zirconium. Dry powder extinguishers should be used to suppress class D fires.

Figure 1-6 shows the symbols used to categorize each type of fire and the types of fire suppressant required to put each out.

SHOP EQUIPMENT

There is an extensive assortment of shop equipment that technicians should become familiar with: some of this can be dangerous if you are not trained to use it. Make a practice of asking for help if you must operate any equipment with which you are not familiar.

	Class of Fire	Typical Fuel Involved	Type of Extinguisher
Class **A** Fires (green)	**For Ordinary Combustibles** Put out a class A fire by lowering its temperature or by coating the burning combustibles.	Wood Paper Cloth Rubber Plastics Rubbish Upholstery	Water*[1] Foam* Multipurpose dry chemical[4]
Class **B** Fires (red)	**For Flammable Liquids** Put out a class B fire by smothering it. Use an extinguisher that gives a blanketing, flame-interrupting effect; cover whole flaming liquid surface.	Gasoline Oil Grease Paint Lighter fluid	Foam* Carbon dioxide[5] Halogenated agent[6] Standard dry chemical[2] Purple K dry chemical[3] Multipurpose dry chemical[4]
Class **C** Fires (blue)	**For Electrical Equipment** Put out a class C fire by shutting off power as quickly as possible and by always using a nonconducting extinguishing agent to prevent electric shock.	Motors Appliances Wiring Fuse boxes Switchboards	Carbon dioxide[5] Halogenated agent[6] Standard dry chemical[2] Purple K dry chemical[3] Multipurpose dry chemical[4]
Class **D** Fires (yellow)	**For Combustible Metals** Put out a class D fire of metal chips, turnings, or shavings by smothering or coating with a specially designed extinguishing agent.	Aluminum Magnesium Potassium Sodium Titanium Zirconium	Dry powder extinguishers and agents only

Cartridge-operated water, foam, and soda-acid types of extinguishers are no longer manufactured. These extinguishers should be removed from service when they become due for their next hydrostatic pressure test.

Notes:
(1) Freezes in low temperatures unless treated with antifreeze solution, usually weighs over 20 pounds (9 kg), and is heavier than any other extinguisher mentioned.
(2) Also called ordinary or regular dry chemical (sodium bicarbonate).
(3) Has the greatest initial fire-stopping power of the extinguishers mentioned for class B fires. Be sure to clean residue immediately after using the extinguisher so sprayed surfaces will not be damaged (potassium bicarbonate).
(4) The only extinguishers that fight A, B, and C classes of fires. However, they should not be used on fires in liquefied fat or oil of appreciable depth. Be sure to clean residue immediately after using the extinguisher so sprayed surfaces will not be damaged (ammonium phosphates).
(5) Use with caution in unventilated, confined spaces.
(6) May cause injury to the operator if the extinguishing agent (a gas) or the gases produced when the agent is applied to a fire are inhaled.

© Delmar/Cengage Learning

Figure 1-6 Guide to fire extinguisher selection.

Lifting Devices

Many different types of hoists and jacks are used in automotive shops. These can range from simple pulley and chain hoists to hydraulically actuated hoists. Weight-bearing chains on hoists should be routinely inspected (this is usually required by law). Chain links with evidence of wear, bent links, and nicks are reason enough to place the equipment out of service. Hydraulic hoists should be inspected for external leaks before using. Any drop-off observed in hydraulic lifting equipment while in operation should be reason to take the equipment out of service. Never rely on the hydraulic circuit alone when working under equipment on a hoist: after lifting, support the equipment using a mechanical sprag or stands.

CAUTION *Never rely on a hydraulic circuit alone when working underneath raised equipment. Before going under anything raised by hydraulics, make sure it is mechanically supported by stands or a mechanical lock.*

Jacks. Many types of jacks are used in automotive service facilities. Before using a jack to raise a load, make sure that the weight rating of the jack exceeds the supposed weight of the load. Most jacks used in service repair shops are hydraulic, and most use air-over-hydraulic actuation because this is faster and requires less effort. Bottle jacks are usually hand-actuated and designed to lift loads up to 10 tons (1.02 tonnes): they are so named because they have the appearance of a bottle. Air-over-hydraulic jacks capable of lifting commercial vehicles are also available.

Using hydraulic piston jacks should be straight-forward. They are designed for a vertical uplift only. The jack base should be on level floor and the lift piston should be located on a flat surface on the equipment to be lifted. Never place the lift piston on the arc of a leaf spring or the radius of any suspension device on the vehicle. After lifting the equipment, it should be supported mechanically using steel stands. It is acceptable practice to use a hardwood spacer with a shop jack: it should be exactly level and placed under the jack. Whenever using a jack to raise one end of a vehicle, make sure that the vehicle being jacked can roll either forward or backward. After lifting, the parking brakes should be applied and wheel chocks should be used on the axles not being raised.

Cherry Pickers. **Cherry pickers** come in many shapes and sizes. Light duty cherry pickers can be used to raise a heavy component such as a cylinder head from an engine, while heavy duty cherry pickers (see **Figure 1-7**) can lift a complete diesel engine out of a chassis. Most cherry pickers have extendable booms. As the boom is lengthened, the weight that the device can lift reduces. Take care that the weight you are about to lift can be raised by the cherry picker without toppling.

> **CAUTION** *As the boom of a cherry picker is lengthened the weight it can lift reduces significantly. Make sure that the weight you are about to lift is appropriate for the boom length you have set: failure to do this can cause the cherry picker to topple.*

Chain Hoists. These are often called chain falls. **Chain hoists** can be suspended from a fixed rail or a beam that slides on rails, or they can be mounted on any of a number of different types of A-frames. Chain hoists in shops in most jurisdictions are required to be inspected periodically. An inspection on a mechanical chain hoist will include chain link integrity and the ratchet teeth and lock. Electromechanical units will require an inspection of the mechanical and electrical components. Where a chain hoist beam runs on rails, brake operation becomes critical: some caution is required when braking the beam because aggressive braking can cause a pendulum effect on the object being lifted.

Come-Alongs. **Come-alongs** describe a number of different types of cable and chain lifting devices that are hand-ratchet actuated. They are used both to lift objects and to apply linear force to them. When used as lift devices, come-alongs should be simple to use, providing the weight being lifted is within rated specification. However, come-alongs are more often used in automotive shops to apply straight-line force to a component, usually to separate flanges. Great care should be taken: make sure that the anchor and load are secure, and that the linear force does not exceed the weight rating of the device.

GENERAL SHOP PRECAUTIONS

Every service facility is different and because of that the potential dangers faced in each shop will differ. In this section, we will outline some general rules and safety strategies to be observed in truck and heavy equipment shops.

Exhaust Extraction

Engines should be run in a shop environment using an exhaust extraction system: in most cases this will be a flexible pipe or pipes that fit over the vehicle exhaust pipes(s). When moving vehicles in and out of service bays, park the unit in the bay and shut the engine off. Avoid running an engine without the extraction pipe(s)

Figure 1-7 Typical heavy duty cherry picker.

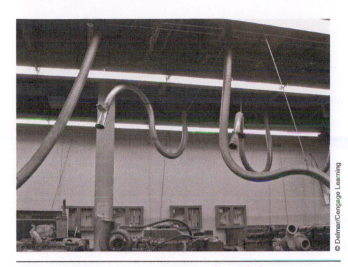

Figure 1-8 Shop exhaust extraction piping.

fitted to the stack(s). **Figure 1-8** shows exhaust extraction piping used in a shop.

> **WARNING** *The State of California has proven that diesel exhaust fumes can cause respiratory problems, cancer, birth defects, and other reproductive harm in humans. Avoid operating diesel engines unless in a well-ventilated area. When starting an engine outside a shop, warm the engine before driving it into the shop to reduce the contaminants emitted directly into the shop while parking the unit.*

> **WARNING** *When attempting to fit an exhaust extraction pipe to a vehicle exhaust system, check the temperature of the exhaust gas aftertreatment piping before attempting to handle it: diesel particulate filters can retain heat long after a regeneration cycle and cause severe burns.*

Workplace Housekeeping

Sloppy housekeeping can make your workplace dangerous. Clean up oil spills quickly. You can do this by applying absorbent grit: this not only absorbs oil but makes it less likely that a person will slip and fall on an oil slick. Try to organize parts in bins and on benches when you are disassembling components. This not only makes reassembly easier, it makes your work environment a lot safer.

Components under Tension

Even on light duty commercial vehicles, many components are under tension, sometimes under deadly tensional loads. Never try to disassemble a component that you suspect is under a high tensional load unless you are exactly sure of how to go about it. Refer to service literature and ask more experienced coworkers when you are unsure of a procedure.

Compressed Fluids

Fluids in both liquid and gaseous states can be extremely dangerous when proper safety precautions are not observed. Equipment does not necessarily have to be running to produce high-fluid pressures. Residual pressures in stationary circuits can represent a serious safety hazard. Technicians should also be aware of the potential danger represented by oxygen cylinders due to their high pressure and potential to aid in combustion and explosions. When shops receive fire safety inspections, fire personnel are more concerned about the storage location of compressed oxygen cylinders than compressed fuels such as acetylene and propane.

Pneumatics Safety

Compressed air is used extensively in automotive facilities. It can be dangerous if not handled with respect. Compressed air is used to drive both portable and nonportable shop tools and equipment.

> **WARNING** *Always wear safety glasses when coupling and uncoupling, or when using pneumatic tools.*

Some examples of shop equipment that use compressed air:

- pneumatic wrenches
- pneumatic drills
- shop air-over-hydraulic presses
- air-over-hydraulic jacks
- air-over-hydraulic cylinder hoists

Figure 1-9 shows a typical setup for a 1/2 inch drive impact gun used every day in diesel and truck repair shops.

Hydraulic System Safety. Vehicle and shop hydraulic systems use extremely high pressures that can be lethal when mishandled. Once again, never forget that idle circuits can hold residual pressures, and many circuits use accumulators. The rule when working with hydraulic circuits is to be absolutely sure about potential dangers before attempting to disassemble a circuit or component.

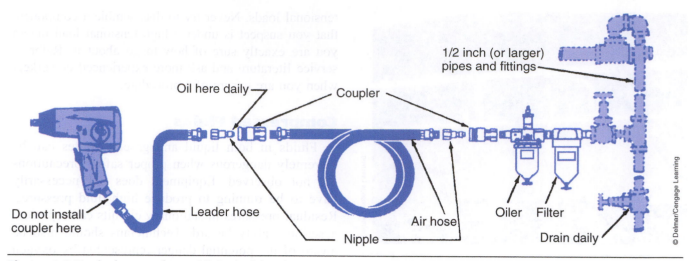

Figure 1-9 Typical setup for a 1/2 inch drive impact gun.

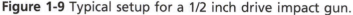

WARNING *Always wear safety glasses when working close to shop or vehicle hydraulic circuits, and check service literature before attempting a disassembly procedure. Ask someone if you are not sure rather than risk injury.*

Some examples of chassis systems that use hydraulic circuits:

- tailgate lift and auxiliary circuits
- high-pressure fuel management circuits
- automatic transmission control circuits
- air-over-hydraulic brake circuits
- clutch control circuits

Some examples of shop equipment using hydraulic circuits:

- jacks and hoists
- presses
- bearing and liner pullers
- suspension bushing presses

CHASSIS AND SHOP ELECTRICAL SAFETY

Vehicles today use numerous computers. These computers are all networked to a central data backbone using multiplexing technology. The chassis subsystems controlled by computers include:

- engine
- transmission
- brakes
- traction control
- vehicle directional stability
- lights
- dash electronics

- collision warning systems
- infotainment and communications systems

These systems function on low-voltage electrical signals and use thousands of solid-state components. While some of these electronic subcircuits are protected against voltage overload spikes, others are not. An unwanted high-voltage spike caused by static discharge or careless placement of electric welding grounds can cause thousands of dollars worth of damage.

Static Discharge

When you walk across a plush carpet, your shoes "steal" electrons from the floor. This charge of electrons accumulates in your body and when you go to grab a door handle, the excess of stolen electrons discharges itself into the door handle, creating an arc which we see as a spark. You will note that the accumulation of a **static charge** is influenced by factors such as relative humidity and the type of footwear you are wearing. Getting a little zap from the static charge that can accumulate in the human body is seldom going to produce any harmful effects to human health, but it can damage sensitive solid-state circuits.

Static Discharge and Computers. Static charge accumulation in vehicles running down a highway and in the human body can easily damage computer circuits. Because vehicles today use a wide range of computer-controlled circuits it is important for technicians to understand the effects of **static discharge**. The reason that static discharge has not caused more problems than it has in the service repair industry is due to:

- technicians' footwear of choice, usually rubber-soled boots
- shop floors that tend not to be carpeted

Rubber-soled footwear and concrete floors are not conducive to static charge accumulation. Having said this, technicians should remember that the flooring in vehicles is almost always carpeted. It is, therefore, good practice when troubleshooting requires you to access electronic circuits to use a **ground strap** before separating sealed connectors and before working on any vehicle electronic circuit, even if you are just connecting an **electronic service tool** (EST) to a data link. A ground strap electrically connects you to the device that you are working on, so an unwanted static discharge into a shielded circuit is unlikely. Special care should be taken when working with modules that require you to physically remove and replace solid-state components such as PROM chips from a motherboard.

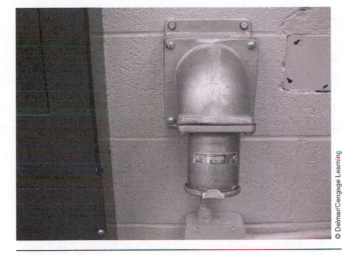

Figure 1-10 High-voltage three-phase electrical outlet.

Chassis Wiring and Connectors

Every year, millions of dollars worth of damage to vehicles is created by truck technicians who ignore OEM precautions about working with chassis wiring systems. Perhaps the most common abuse is puncturing wiring insulation with test lights and **digital multimeter** (DMM) leads. When you puncture the insulation on copper wiring, in an instant that wiring becomes exposed to both oxygen (in the air) and moisture (relative humidity!). The chemical reaction almost immediately produces copper oxides that then react with moisture to form corrosive cupric acid. The acid begins to eat away the wiring, first creating high resistance, and ultimately consuming the wire. The effect is accelerated when copper-stranded wiring is used because the surface area over which the corrosion can act is so much greater.

> **CAUTION** *Never puncture the insulation on chassis wiring. Read the section that immediately precedes this if you want to know why!*

The sad thing about this type of abuse is that it is so easily avoided. There are many ways that a vehicle technician can access wiring circuits using the correct tools. Use breakout tees, breakout boxes, and test lead spoons.

Mains Electrical Equipment

Mains electrical circuits, unlike standard vehicle electrical circuits, operate at pressures that can be lethal, so you have to be careful when working around any electrical equipment. Electrical pressures may be

single-phase mains operating at pressure values between 110 and 120 volts or **three-phase mains**, operating at pressures between 400 and 600 volts. In most jurisdictions, repairs to mains electrical equipment and circuits must be undertaken by qualified personnel. If you undertake the repair of electrical equipment, make sure you know what you are doing! **Figure 1-10** shows a three-phase outlet of the type commonly used in truck shops to power compressors, machine shop equipment, and arc welding stations. Learn how to identify these outlets with high electrical potential that can kill if mishandled.

Take extra care when using electrically powered equipment when the area in which you are working is wet. And remember that a discharge of AC voltage driven through a chassis data bus can knock out electronic equipment networked to it. Electrical equipment can also be dangerous around vehicles because of its potential to arc and initiate a fire or explosion.

> **CAUTION** *Do not undertake the repair of mains electrical circuit and equipment problems unless you are qualified to do so.*

Some examples of shop equipment using single-phase mains electricity:

- electric hand tools
- portable electric lights
- computer stations
- drill presses
- burnishing and broaching tools

Oxyacetylene Equipment

Technicians use oxyacetylene for heating and cutting on an almost daily basis. Less commonly this equipment is used for braising and welding. Some basic instruction in the techniques of oxyacetylene equipment safety and handling is required. The following information should be understood by anyone working with oxyacetylene equipment.

Acetylene Cylinders. Acetylene regulators and hose couplings use a left-hand thread. Left-hand threads tighten counterclockwise (CCW). An acetylene regulator gauge working pressure should *never* be set at a value exceeding 15 psi (100 kPa). At pressures higher than this, acetylene becomes dangerously unstable. An acetylene cylinder should always be used in the upright position. Using an acetylene cylinder in a horizontal position will result in the acetone draining into the hoses.

The quantity of acetylene in a cylinder cannot be accurately determined by the pressure gauge reading because it is in a dissolved condition. The only really accurate way of determining the quantity of gas in the cylinder is to weigh it and subtract this from the weight of the full cylinder, often stamped on the side of the cylinder.

Figure 1-11 High-pressure oxygen cylinder.

Oxygen Cylinders. Oxygen cylinders present more problems than acetylene when exposed to fire. For this reason they should be stored upright and in the same location in a service shop when not in use. This location should be identified to the fire department during an inspection. They should never be left randomly on the shop floor.

Oxygen regulator and hose fittings use a right-hand thread. A right-hand thread tightens clockwise (CW). An oxygen cylinder pressure gauge accurately indicates the oxygen quantity in the cylinder, meaning that the volume of oxygen in the cylinder is approximately proportional to the pressure.

Oxygen is stored in the cylinders at a pressure of 2,200 psi (15 MPa), and the handwheel-actuated valve forward-seats to close the flow from the cylinder and back-seats when the cylinder is opened. It is important to ensure that the valve is fully opened when in use. If the valve is only partially opened, oxygen will leak past the valve threads. **Figure 1-11** shows a typical oxygen cylinder.

Regulators and Gauges. A regulator is a device used to reduce the pressure at which gas is delivered. It sets the working pressure of the oxygen or fuel. Both oxygen and fuel regulators function similarly. They increase the working pressure when turned clockwise

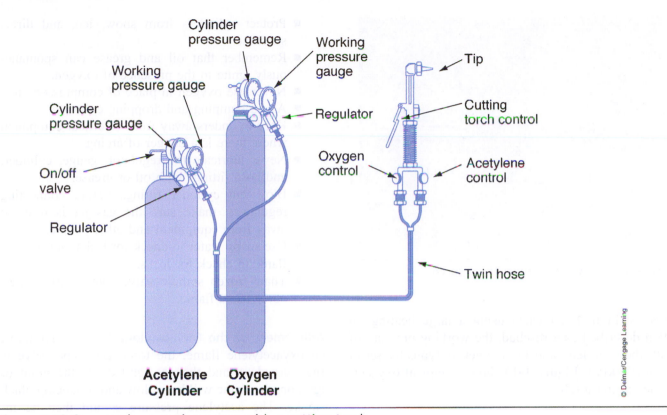

Figure 1-12 Oxyacetylene station setup with a cutting torch.

and close off the pressure when backed out counter-clockwise.

Pressure regulator assemblies are usually equipped with two gauges. The cylinder pressure gauge indicates the actual pressure in the cylinder. The working pressure gauge indicates the working pressure, and this should be trimmed using the regulator valve to the required value while under flow.

Hoses and Fittings. The hoses used with oxyacetylene equipment are usually color-coded. Green is used to identify the oxygen hose and red identifies the fuel hose. Each hose connects the cylinder regulator assembly with the torch. Hoses may be single or paired (Siamese). Hoses should be routinely inspected and replaced when defective. A leaking hose should never be repaired by wrapping it with tape. In fact, it is generally bad practice to consider repairing welding gas hoses by any method. They should be replaced when they fail.

Fittings couple the hoses to the regulators and the torch. Each fitting consists of a nut and gland. Oxygen fittings use a right-hand thread and fuel fittings use a left-hand thread. The fittings are machined out of brass that has a self-lubricating characteristic. Never lubricate the threads on oxyacetylene fittings. **Figure 1-12**

shows a typical oxyacetylene station setup with a cutting torch.

Torches and Tips. Torches should be ignited using the following sequence:

- Open the cylinder flow valve.
- Set the working pressure using the regulator valve for both gases under flow, then close.
- Next, open the fuel valve only and ignite the torch using a flint spark lighter.
- Set the acetylene flame to a clean burn (no soot) condition.
- Now open the oxygen valve to set the appropriate flame. When setting a cutting torch, set the cutting oxygen last.
- To extinguish a torch, close the fuel valve first, then the oxygen.
- Finally, the cylinders should be shut down using the main flow valve and the hoses purged.

Welding, cutting, and heating tips may be used with oxyacetylene equipment. Refer to a welder's manual to identify the specified working pressures for each type of tip. There is a tendency to set gas working

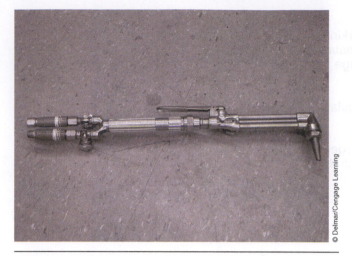

© Delmar/Cengage Learning

Figure 1-13 Oxyacetylene cutting torch.

pressure high. Even when using a large heating tip often described as a rosebud, the working pressure of both the acetylene and the oxygen is typically set at 7 psi (50 kPa). **Figure 1-13** shows a typical oxyacetylene cutting torch.

Backfire

Backfire is a condition where the fuel ignites within the nozzle of the torch, producing a popping or squealing noise: it often occurs when the torch nozzle overheats. Extinguish the torch and clean the nozzle with tip cleaners. Torches may be cooled by immersing in water briefly with the oxygen valve open.

Flashback. Flashback is a much more severe condition than backfire: it takes place when the flame travels backward into the torch to the gas-mixing chamber and beyond. Causes of flashback are inappropriate pressure settings (especially low-pressure settings) and leaking hoses/fittings. When a backfire or flashback condition is suspected, close the cylinder valves immediately, beginning with the fuel valve. Flashback arresters are usually fitted to the torch and will limit the extent of damage when a flashback occurs.

Eye Protection. Safety requires that a number 4- to number 6-grade filter be used when using an oxyacetylene torch. The flame radiates ultraviolet light that can damage eyesight. Sunglasses, even when UV rated, are not sufficient protection and can result in damage to eyesight with short exposure to an oxyacetylene flame.

Oxyacetylene Precautions. Things to do and not to do:

- Store oxygen and acetylene upright in a well-ventilated, fireproof room.

- Protect cylinders from snow, ice, and direct sunshine.
- Remember that oil and grease can spontaneously ignite in the presence of oxygen.
- Never use oxygen in place of compressed air.
- Avoid bumping and dropping cylinders.
- Keep cylinders away from electrical equipment where there is a danger of arcing.
- Never lubricate the regulator, gauge, cylinder, and hose fittings with oil or grease.
- Blow out cylinder fittings before connecting regulators: make sure the gas jet is directed away from equipment and other people.
- Use soapy water to check for leaks: *never* use a flame to check for leaks.
- Thaw frozen spindle valves with warm water; *never* use a flame.

Adjustment of the Oxyacetylene Flame. To adjust an oxyacetylene flame, the torch acetylene valve is first turned on and the gas ignited. At the point of ignition, the flame will be yellow and producing black smoke. The acetylene pressure should then be increased using the torch fuel valve. This increases the brightness of the flame and reduces the smoking. At the point at which the smoking disappears, the acetylene working pressure can be assumed to be correct for the nozzle jet size used. Next, the torch oxygen valve is turned on. This will cause the flame to become generally less luminous (bright) and an inner blue luminous cone surrounded by a white-colored plume should form at the tip of the nozzle. The white-colored plume indicates excess acetylene. As more oxygen is supplied, this plume reduces until there is a clearly defined blue cone with no white plume visible. This indicates the *neutral* flame used for most welding and cutting operations.

Electric Arc Welding

Electric arc welding and cutting processes are used extensively in truck and heavy equipment service garages. Arc welding stations work on one of two principles:

- Transformer: This receives a high-voltage feed (mains electrical) then reduces it to a lower-voltage, high-current circuit.
- Generator: This generates a high-voltage charge, then conditions it to lower-voltage, high-current circuits.

Just as with oxyacetylene welding, before attempting to use any type of arc welding equipment, make sure you receive some basic instruction and training.

Typical open-circuit voltages in industrial welding stations are usually around 70 volts while closed-circuit voltages are typically a little over 20 volts. The following types of welding stations are non-specialized in application and are found in many truck shops:

- Arc welding station: This uses a flux-coated, consumable electrode. Arc welding is often known as *stick welding*.
- MIG (metal inert gas): This uses a continuous reel of wire, which acts as the electrode around which inert gas is fed to shield the weld from air and ambient moisture. Closely related to MIG welding is flux-shielded reel welding.
- TIG (tungsten inert gas): This uses a non-consumable tungsten electrode surrounded by inert gas, and filler rods are dipped into the welding puddle created.
- Carbon-arc cutting: An arc is ignited using carbon electrodes to melt base metal while a jet

of compressed air blows through the puddle to make the cut.

Figure 1-14 shows a typical arc welding electrode holder used with an arc welding station.

Figure 1-14 Arc welding electrode holder.

Summary

- Auto service shops are safe working environments, but technicians must learn how to work safely.
- While all service shops play a role in ensuring a safe working environment, technicians should think of safety as a personal responsibility.
- Personal safety clothing and equipment such as safety footwear, eye protection, coveralls, hearing protection, and different types of gloves are required when working in a service garage.
- Technicians should learn to distinguish between the four different types of fires and identify the fire extinguishers required to suppress them.
- Jacks and hoists are used extensively in service facilities and should be used properly and inspected routinely.
- The danger of inhaling engine exhaust emissions should be recognized, and when engines are run

inside a garage, exhaust extraction piping must be fitted to the exhaust stacks.

- It is important to identify what is required to work safely with chassis electrical systems because of the costly damage made by simple errors.
- Shop mains electrical systems are used in portable power and stationary equipment and can be lethal if not handled properly.
- Oxyacetylene equipment is used for heating, cutting, and welding. Technicians should be taught how to work safely with oxyacetylene torches.
- Arc welding and cutting processes are also used in service facilities. This type of high-voltage equipment can be safely operated with some basic training.

Review Questions

1. When is a worker more likely to be injured?

    A. First day on the job

    B. During the first year of employment

    C. During the second to fourth years of employment

    D. During the year before retirement

2. When lifting a heavy object, which of the following should be true?

    A. Keep your back straight while lifting.

    B. Keep the weight you are lifting close to your body.

    C. Bend your legs and lift using the leg muscles.

    D. All of the above.

3. What is Purple K?

    A. A new type of stimulant

    B. A dry powder fire suppressant

    C. A toxic gas

    D. A type of paint

4. Which of the following is usually a requirement of a safety shoe or boot?

    A. UL certification

    B. Steel sole shank

    C. Steel toe

    D. All of the above

5. What type of gloves should never be worn when working with refrigerant?

    A. Synthetic rubber

    B. Vinyl disposable

    C. Leather welding gloves

    D. Latex rubber gloves

6. Which of the following is under the most pressure?

    A. Oxygen cylinders

    B. Acetylene cylinders

    C. Diesel fuel tanks

    D. Gasoline fuel tanks

7. Which type of fire can usually be safely extinguished with water?

    A. Class A

    B. Class B

    C. Class C

    D. Class D

8. When attempting to suppress a Class C fire in a chassis, which of the following is good practice?

    A. Disconnect the batteries.

    B. Use a carbon dioxide fire extinguisher.

    C. Avoid inhaling the fumes produced by burning conduit.

    D. All of the above.

9. What color is used to indicate the fuel hose in an oxyacetylene station?

    A. Green

    B. Red

    C. Yellow

    D. Blue

10. In which direction do you tighten an oxygen cylinder fitting?

    A. CW

    B. CCW

    C. Depends on the manufacturer

2 Introduction to Diesel Engines

Learning Objectives

After studying this chapter, you should be able to:

- Identify the scope of engines covered by this textbook.
- Interpret basic engine terminology.
- Identify the subsystems that make up a diesel engine.
- Calculate engine displacement.
- Define the term *mean effective pressure*.
- Describe the differences between *naturally aspirated* and *manifold-boosted* engines.
- Explain how volumetric efficiency affects cylinder breathing.
- Define *rejected heat* and explain thermal efficiency in diesel engines.
- Outline the operation of a diesel four-stroke cycle.
- Calculate engine displacement.
- Interpret the term *cetane number* and relate it to ignition temperature.

Key Terms

after top dead center (ATDC)

before top dead center (BTDC)

bore

bottom dead center (BDC)

clearance volume

combustion pressure

compression ignition (CI)

compression ratio

cylinder volume

diesel cycle

direct injection (DI)

engine displacement

fire point

friction

heat energy

ignition lag

indirect injection (IDI)

inertia

manifold boost

mean effective pressure (MEP)

naturally aspirated (NA)

oversquare engine

ratio

rejected heat

spark ignited (SI)

square engine

stroke

swept volume

thermal efficiency

top dead center (TDC)

torque

undersquare engine

vocational automotive diesels

volumetric efficiency

INTRODUCTION

We are going to begin this chapter by identifying the range of engines covered by this textbook. Light duty diesel engines are those with displacements of eight liters or less. The objective is to focus on automotive diesel engines. Unlike the Europeans, the American consumer has been slow to accept diesel engine power in passenger vehicles. The result is that the move toward adopting diesels has been led by the big three pickup truck manufacturers. Diesel engines suit pickup truck applications, especially when the truck is to be used as a work tool rather than a family vehicle. The term that we will use to describe this family of engines is **vocational automotive diesels**.

Vocational Automotive Diesels

The current line-up of Detroit's pickup truck diesel engine options is as follows:

- Dodge: 2010 Cummins ISB 6.7 liter, inline six-cylinder
- Ford: 2011 PowerStroke 6.7 liter, V8
- GM: 2010 Duramax 6600 (6.6 liter), V8

It should be noted here that engines specified for pickup truck applications are engineered as work engines. When these engines are used in a work environment they tend to produce much improved fuel economy over their gasoline-fueled counterparts. However, the pickup truck has always been more than merely a work vehicle, and the owner who specs diesel power for a family vehicle is probably going to be disappointed with the supposed fuel economy advantage. Included in the category of working automotive diesel engines is that used in Freightliner's popular Sprinter van application, along with some Japanese import vans. Freightliner uses Mercedes-Benz power. Mercedes-Benz offers inline, four-, five-, and six-cylinder engines. Vocational automotive diesel engines must meet U.S. EPA (Environmental Protection Agency) on-highway emissions standards with the result that all current engines are computer controlled with a full spectrum of exhaust aftertreatment devices.

Automobile Diesels

In some European countries, diesel engine power is preferred over gasoline power by a 60:40 ratio. The cost of fuel in Europe greatly exceeds ours, and the initial reason for the move to diesel engines was fuel economy. However, this has changed. Diesel engines engineered specifically for automobile applications can produce power, reliability, and quiet operation that compares with or exceeds the performance of gasoline engines, with the added advantage of fuel economy. For the last several years, the 24-hour Le Mans race has been dominated by diesels manufactured by companies such as Audi and Peugeot, and it will be only a matter of time before an American-built diesel takes the checkered flag at the Daytona 500. However, before this happens, the American consumer is going to have to be sold on accepting diesel power in the family vehicle. While automobile diesel engines are available in North America, they are usually imported. The following examples are offered by manufacturers who have led the way in providing a diesel engine option in their automobiles:

- Audi TDI, 3.0 liter, six-cylinder
- Mercedes-Benz CDI 2.0 liter, four-cylinder
- VM Motori (Jeep) RA620 3.0, six-cylinder
- Volkswagon TDI 2.0 liter, six-cylinder
- Volvo D3 2.0 liter, five-cylinder

Automobile diesel engines must meet the same EPA emissions standards as their truck counterparts so they have similar computer controls and exhaust after-treatment systems.

Off-Highway Light Duty Diesels

The diesel engine has fared better in light duty applications that are not intended for highway operations. This category of engines powers turf equipment, recreational marine equipment, farm equipment, and stationary applications such as generators and pumps. Manufacturers servicing this sector of the industry include Briggs & Stratton, Caterpillar, Deutz, Hatz, Isuzu, John Deere, Kohler, Kubota, Lister Petter, Mitsubishi, Perkins, and Yanmar. It should be noted that this class of small-bore engines (some may be single cylinder) is now regulated by U.S. EPA emissions standards. However, this is a recent development, so many of these engines still use hydromechanical (no computer) fuel management and no exhaust gas treatment. Some examples of recent versions of off-highway light duty diesel engines are:

- Caterpillar C 1.5 liter, three-cylinder
- Cummins QSB 4.5 liter, four-cylinder
- Deere 4024T 2.4 liter, four-cylinder
- Deere PowerTech PSX 6.8 liter, six-cylinder
- Deutz TCD 2.9 liter, four-cylinder
- Isuzu C-Series 1.5 liter, three-cylinder
- JCB Ecomax 4.4 liter, four-cylinder
- Kubota 1.12 liter, three-cylinder
- Lister Petter TR1, 0.773 liter, one-cylinder

- Lister Petter TR2, 1.55 liter, two-cylinder
- Yanmar 4JH 2.2 liter, four-cylinder

DIESEL ENGINE TERMS

Before you can properly understand how any engine functions you have to become familiar with some of the language we use to describe its operation. You will find that the basic engine terminology and principles introduced in this chapter are used repeatedly throughout the textbook. As you progress through it, make sure that you use the glossary to check the definitions of any words you are unfamiliar with. Mostly, we will use the technically correct term in this textbook, but remember that the terminology used on the shop floor might differ. When you reference manufacturer's service literature, you will be expected to have a basic understanding of the key terms introduced in this chapter. If you have studied spark-ignited (SI) gasoline-fueled engines, you may find you are familiar with many of these definitions, in which case you can just skip over them. **Figure 2-1** shows a Ford 6.7 liter engine that powers many pickup trucks.

Compression Ignition

A diesel engine is defined by the fact that it uses **compression ignition (CI)**. The key difference between what we know as a **spark-ignited (SI)**, gasoline-fueled engine and a diesel engine is in the way each ignites the fuel charge. In the SI engine, ignition of the fuel charge takes place when a spark is delivered to the engine

© Ford Motor Company

Figure 2-1 Typical direct-injected, turbocharged, V8 diesel engine.

cylinder. The CI engine relies on the heat of compression to ignite the fuel charge. For this reason, more heat must be generated in the engine cylinder of a CI engine, at least enough to ignite the fuel charge.

Engine Definitions

You probably are already familiar with some of the terms introduced in this section. Even if you are, read through them because the definitions here may differ in small ways from those you are familiar with.

Diesel Engine. A diesel engine is a type of internal combustion engine in which the fuel/air charge is ignited by the heat of compression. While some diesel engines are equipped with glow plugs, these are start-assist devices that do not play a role during normal engine operation.

Air. Air is a gaseous mixture of nitrogen and oxygen. We breathe this mixture we call air and so does a diesel engine. Air is composed of a little under 80 percent nitrogen and a little over 20 percent oxygen. The oxygen available in air is used as the reactant to combustion for the fuel in most internal combustion engines.

Fuel. The fuels we use in diesel engines are hydrocarbons (HCs). The source of most diesel fuel used today is petroleum. However, other hydrocarbon-based fuels (such as soy base, biodiesels) are used and under some conditions can work as efficiently as petroleum-based fuels.

Heat Energy. **Heat energy** is a rating of the available energy in any given fuel. As the heat energy of a fuel increases, so does the potential for converting this heat energy into useful energy at the flywheel. For instance, gasoline contains less heat energy by weight than a diesel fuel. Because of this, the power potential of diesel fuel per unit of fuel used is usually greater than that for an equivalent gasoline-fueled engine.

Naturally Aspirated (NA). The term **naturally aspirated** is used to describe an engine in which air (or air/fuel mixture) is induced into its cylinders by low cylinder pressure created by the downstroke of the piston. Naturally aspirated diesel engines today tend to be in nonmobile applications.

Turbo-Boost. Most diesel engines today are turbo-boosted like the example shown in **Figure 2-1**. Another way of saying turbo-boost is **manifold-boost**. Manifold-boosted describes any engine whose

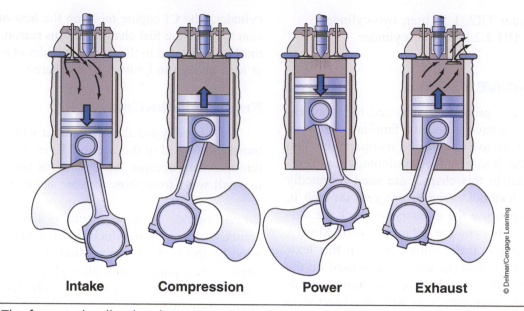

Intake   Compression   Power   Exhaust

© Delmar/Cengage Learning

Figure 2-2 The four-stroke diesel cycle.

cylinders are charged at pressures above atmospheric pressure. The diesel engines manufactured for applications such as Ford, General Motors, and Dodge pickup trucks and commercial vans are all turbo-boosted. The same is true of the diesel engine options in most Japanese and European import automobiles.

Volumetric Efficiency. **Volumetric efficiency** is defined as the measure of an engine's breathing efficiency. The best way to think of it is as the ratio between the volume of fresh air taken into the engine cylinder before the intake valve(s) close versus the cylinder swept volume. Volumetric efficiency is usually expressed as a percentage. In turbocharged engines, volumetric efficiency can often greatly exceed 100 percent. The best way of defining the term is to say that it is the amount of air charged to the engine cylinder in an actual cycle versus the amount it would contain if it were at atmospheric pressure.

These first few definitions should be sufficient to allow us to understand the operation of the diesel cycle, which we will introduce next. Toward the end of the chapter, we are going to introduce another group of terms to give us some of the building blocks required to understand the diesel engine.

THE DIESEL CYCLE

A cycle is a sequence of events. The **diesel cycle** is best introduced by outlining the four strokes of the pistons made as an engine is turned through two revolutions. A full cycle of a diesel engine requires two complete rotations. Each rotation requires turning the

engine through 360 degrees, so that a complete diesel cycle translates into 720 crankshaft degrees.

Each of the four strokes that make up the cycle involves moving a piston either from the top of its travel to its lowest point of travel or vice versa; each stroke of the cycle therefore translates into 180 crankshaft degrees. The four strokes that comprise the four-stroke cycle are:

- Intake
- Compression
- Power
- Exhaust

The four strokes of the diesel cycle are shown in **Figure 2-2**.

Direct Injection, Compression Ignition Engine

We can now take a closer look at what happens during the four-stroke diesel cycle. Make sure you refer **Figure 2-2** to help you understand the description provided here. In this description, we are going to describe what happens in a turbocharged diesel engine.

Intake Stroke. The piston is connected to the crankshaft throw by means of a wrist pin and connecting rod. The throw is an offset journal on the crankshaft. Therefore, as the crankshaft rotates, the piston is drawn from top dead center (TDC) to bottom dead center (BDC): while the piston moves through its downstroke, the cylinder head intake valve(s) are held open. As the intake valves open, turbo-boosted air is forced into the engine cylinder as the piston travels downward. This means that at the completion of the intake stroke when the intake valves close, the cylinder will be filled with a charge of filtered air. The actual quantity of air in the cylinder will depend on the extent of turbo-boost. Turbo-boost varies with how the engine is being operated.

The *air* that is taken into the cylinder is a mixture of approximately four-fifths nitrogen and one-fifth oxygen. The oxygen is required to combust the fuel. Note that no fuel is introduced into the engine cylinder during the intake stroke. When the air charged to the engine cylinder is pressurized using a turbocharger, more oxygen can be forced into the cylinder. All diesel engines are designed for lean burn operation; that is, the cylinder will be charged with much more air than that required to combust the fuel. Volumetric efficiency in most phases of engine operation will usually exceed 100 percent in turbocharged engines.

Compression Stroke. At the completion of the intake stroke, the intake valves close, sealing the engine cylinder. The piston is now driven upward from BDC to TDC with both the intake and exhaust valves closed. The quantity of air in the cylinder does not change, but compressing the charge of air in the cylinder gives it much less space and in doing so, heats it up considerably. Compression pressures in modern diesel engines vary from 400 psi (2750 kPa) to 700 psi (4822 kPa). The actual amount of heat generated from these compression pressures also varies, but it usually substantially exceeds the minimum ignition temperature values of the fuel. Compression ratios used to achieve the compression pressure required of diesel engines generally vary from a low of 14:1 to a high of 25:1. However, in modern turbocharged, highway diesel engines, compression ratios are typically around 16:1 to 17:1.

Power Stroke. Shortly before the completion of the compression stroke, atomized fuel is introduced directly into the engine cylinder by a multi-orifice (multiple hole) nozzle assembly. The fuel exits the nozzle orifices at very high pressures and in a liquid state. The liquid droplets emitted by the injector must be appropriately sized for ignition and combustion. These fuel droplets, once exposed to the heated air charge in the cylinder, are first vaporized (their state changes from that of a liquid to a gas), then ignited. The ignition point of the fuel is usually designed to occur just before the piston is positioned at TDC: you can compare the ignition point of diesel fuel in the cylinder with spark timing in a gasoline-fueled engine. However, in the diesel engine the cylinder pressures that result from combusting fuel in the cylinder can be

managed with more precision than in indirect-injected gasoline engines. This is because fuel can be injected at high pressures into the cylinder during combustion.

During the power stroke, cylinder pressure resulting from the combusting of fuel acts on the piston, driving it downward through its downstroke. The piston moves linearly, that is, in a straight line. It is connected to the crankshaft throw, which rotates. In this way, cylinder pressures are converted into a twisting force we know as **torque**. Because the crank throw is offset from the crankshaft centerline it acts as a lever. For this reason, in managing the power stroke it is desirable to have little pressure acting on the piston at TDC because it has zero leverage at this location. As the piston moves downward, the leverage will increase incrementally until the angle between the crank throw and connecting rod is 90 degrees (maximum leverage). By getting cylinder pressure (managed by the fuel system) and throw leverage (a hard value dependent on the rotational position of the engine) to work together, it is possible to deliver relatively constant torque from around 15 degrees after top dead center (ATDC) to 90 degrees ATDC. This relationship between pressure and throw leverage helps to transmit the energy produced in the engine cylinder as smoothly as possible to the flywheel. All current diesel engines use computers to precisely manage the relationship between cylinder pressure and crank throw angle. **Figure 2-3** shows the key events of the compression and power strokes.

Exhaust Stroke. Somewhere after 90 degrees ATDC during the expansion stroke, most of the heat energy

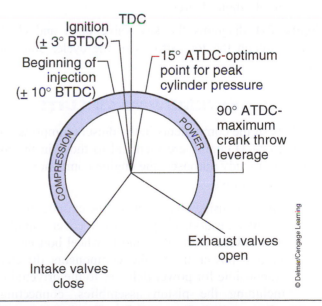

Figure 2-3 Events of the compression and power strokes.

that can be converted to useful mechanical energy has been converted, and the exhaust valve(s) open. The products of cylinder combustion are known as end gas. The exhausting of combustion end gases occurs in four distinct phases, and the process begins toward the end of the power stroke as the piston is traveling downward:

1. Pressure differential—At the moment the exhaust valves open during the latter portion of the power stroke, pressure is higher in the cylinder than in the exhaust manifold. High-pressure end gas in the cylinder will therefore flow to the lower pressure in the exhaust manifold. This phase is sometimes known as gas blowdown.

2. Inertial—Next the piston comes to a standstill at BDC at the completion of the power stroke. However, gas inertia established during the pressure differential phase will result in the end gases continuing to flow from the cylinder to the exhaust manifold while the piston is in a stationary and near-stationary state of motion.

3. Displacement—As the piston is forced upward through its stroke, it forces out (positively displaces) combustion end gases above it.

4. Scavenging—Toward the end of the exhaust stroke, as the exhaust valve(s) begin to close, the intake valve(s) begin to open with the piston near TDC. The scavenging phase takes place during valve overlap and can be highly effective in expelling end gases and providing some piston crown cooling. The efficiency of the scavenging process is greatest with turbocharged engines and because it is executed with air only, may be prolonged over a wide range of crank angle degrees.

Figure 2-4 diagrams the key valve open and close events through the four strokes of the cycle. Study it carefully.

ENGINE SYSTEMS AND CIRCUITS

Many of the components in the diesel (compression ignition or CI) engine are identical to those in the SI engine. For study purposes, the engine components are divided as follows:

1. Engine housing components—the cylinder block, cylinder head(s), oil pan, rocker covers, timing gear covers, manifolds, and flywheel housing.

2. Engine power train—the components directly responsible for power delivery to the drivetrain, including the piston assemblies, connecting rods, crankshaft assembly, vibration damper, and flywheel.

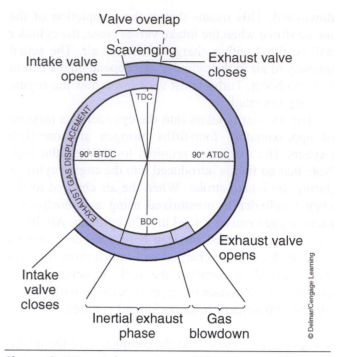

Figure 2-4 Key valve open and close events through the four-stroke diesel cycle.

3. Engine feedback assembly—the engine's self-management components also known as the valve train assembly. This term is used to describe the diesel engine timing gear train, camshaft, valve trains, valves, fueling apparatus, and accessory drive components.

4. Engine lubrication circuit—the oil pump, relief valve, lubrication circuitry, full flow filter(s), bypass filters, and heat exchangers.

5. Engine cooling circuit—the coolant pump, thermostat(s), water jacket, coolant manifold, filter, shutters, fan assembly, radiator, and other heat exchangers.

6. Engine breathing system—the engine intake and exhaust system components, including precleaners, air cleaners, ducting, turbocharger, charge air heat exchangers, intake and exhaust manifolds, pyrometers, exhaust gas recirculation (EGR) system, diesel particulate filter (DPF), exhaust piping, engine silencer, catalytic converter, and selective catalytic reduction (SCR) emission control apparatus.

7. Engine fuel management system—the fuel storage, pumping, metering, and quantity control apparatus, including a management computer, sensors and actuators, hydraulic injectors, electronic unit injectors (EUIs), hydraulically actuated electronic unit injectors (HEUIs), common rail (CR) injection, hydromechanical

injection pumps, fuel tanks, filters, and transfer pumps.

Diesel Fuel

It helps to know a little bit about the diesel fuel used to fuel compression-ignited engines. The fuel used in modern diesel engines on North American highways is composed of roughly 85 percent carbon and 12 to 15 percent hydrogen, not too much different from the chemical composition of gasoline. Unlike gasoline, diesel fuel does not vaporize easily at ambient temperatures, so it is less likely to form combustible mixtures of fuel and air. The heat required to ignite the fuel oil is defined by the most volatile fractions of the fuel: this is determined by the fuel's cetane number (CN). The ignition temperature of highway diesel fuel is usually around 250°C (482°F): this is equivalent to a CN of around 45. The ignition temperature would be higher, around 290°C (550°F), using fuel with the poorest ignition quality (rated with a CN of 40) that can be legally sold for use on North American highways.

Using diesel fuel oils with a fire point of around 250°C (482°F) means that at least this temperature must be achieved in the engine cylinder during the compression stroke of the piston if the fuel is to be ignited. In fact, actual cylinder temperatures generated on the compression stroke tend to be considerably higher than the minimum required to ignite the fuel. The greater the difference between these two temperature values, the shorter the ignition lag. **Ignition lag** is the time between the entry of the first droplets of fuel into the engine cylinder and actual ignition that begins combustion.

MORE ENGINE TERMS

Now that you have an understanding of some basic engine terms and know how the different engine cycles function, you should be ready for some more definitions. We have already used some of these terms in describing the engine cycles, so the objective here is to reinforce them. Refer to **Figure 2-5** as you read through the following definitions to make sure you understand each.

Top dead center (TDC): the highest point of piston travel in an engine cylinder.

Bottom dead center (BDC): the lowest point of piston travel in an engine cylinder.

Before top dead center (BTDC): a point of piston travel through its upstroke.

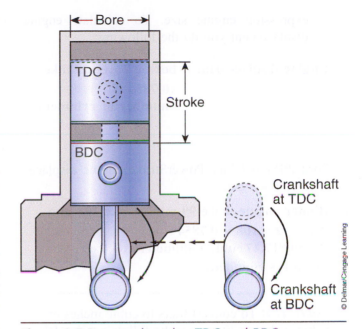

Figure 2-5 Bore and stroke, TDC and BDC.

After top dead center (ATDC): a point of piston travel through its downstroke.

Bore: cylinder diameter. Bore is how we express piston sectional area over which cylinder pressures act. **Figure 2-5** illustrates the bore dimension in a sectioned engine.

Stroke: the distance through which a piston travels from BDC to TDC. Stroke is established by the crank throw offset; that is, the distance from the crankshaft centerline to the throw centerline multiplied by 2 equals the stroke dimension. **Figure 2-5** illustrates the bore dimension in a sectioned engine.

Swept volume: the volume displaced by the piston in the cylinder as it moves from BDC to TDC. It can be calculated if both stroke and bore are known.

Clearance volume: the remaining volume in an engine cylinder when the piston is at the top of its travel or TDC. The clearance volume on older indirect-injection (IDI) diesel engines was considerable, but it is much less on today's direct-injected (DI) engines.

Cylinder volume: the total volume in the cylinder when the piston is at BDC: You can calculate cylinder volume by adding cylinder swept volume to the cylinder clearance volume.

Engine displacement: the *swept volume* of all the engine cylinders expressed in cubic inches or cubic centimeters/liters. It is one way of expressing engine size. To calculate engine displacement you do the following;

$$\text{Engine displacement} = \textbf{bore} \times \textbf{bore} \times \textbf{stroke}$$
$$\times \textbf{0.7854}$$
$$\times \textbf{number of cylinders}$$

Example

Ford 2011 6.7 liter PowerStroke engine displacement calculation data:

8-cylinder engine, bore 99 mm, stroke 108 mm
$$= 99 \times 99 \times 108 \times 0.7854 \times 8$$
$$= \textbf{6,650,817.47 cubic millimeters}$$
$$= \textbf{6.7 liters(rounded)}$$

....

Tech Tip: To convert liters to cubic inches or cubic inches to liters, use the following simple formula in which 61 is either multiplied or divided into the value to be converted:

6.7 liters $\times$ **61 = 408.7 cubic inches or**
rounded to 409 cubic inches

To change cubic inches into liters, you do the opposite:

408.7 cubic inches $\div$ **61 = 6.7 liters**

....

Square engine: an engine in which the cylinder bore diameter is exactly equal to the piston stroke dimension. For instance, an engine with a 4-inch piston diameter and a 4-inch stroke is classified as *square*. When bore and stroke values are expressed, bore always appears before stroke.

Oversquare engine: is the term used to describe an engine in which the cylinder bore diameter is larger than the stroke dimension. Most indirect-injected, gasoline-fueled, spark-ignited engines are oversquare.

Undersquare engine: the term used to describe an engine in which the cylinder bore diameter is smaller than the stroke dimension. Most high-compression diesel engines such as the one we based the earlier calculations on are undersquare.

Compression: when a gas is squeezed by driving a piston into a sealed cylinder, heat is created. For instance, when you use a hand-actuated bicycle pump to inflate tires, the pump cylinder heats up.

Compression ignition (CI): a high-compression engine in which the heat required to ignite the fuel in its cylinders is sourced from compression. More commonly known as a *diesel engine*. CI is an acronym commonly used to describe a diesel engine.

Spark ignited (SI): an engine in which the fuel/air charge is ignited by a timed electrical spark.

Direct injection (DI): either a CI or SI engine in which the fuel charge is injected directly into the engine cylinder rather than to a precombustion chamber or part of the intake manifold.

Indirect injection (IDI): a CI or SI engine in which the fuel charge is introduced outside of the engine cylinder to a precombustion chamber, cylinder head intake tract, or intake manifold. IDI engines are rare today, although they were common in light duty diesels a generation ago.

Ratio: the relationship between two values expressed by the number of times one contains the other. We use the term in automotive technology to describe the drive/driven relationships of two meshed gears, the mechanical advantage of levers, and cylinder compression ratio.

Compression ratio: a measure of the cylinder volume when the piston is at BDC versus cylinder volume when the piston is at TDC. Theoretically, compression ratios in diesel engines range between 14:1 and 24:1. In reality, most modern diesel engines have compression ratios typically between 16:1 and 17:1.

Compression pressure: the actual cylinder pressure developed on the compression stroke. Actual compression pressures in diesel engines range from 350 psi (2.40 MPa) to 700 psi (4.80 MPa) in CI engines. The higher the compression pressure, the more heat developed in the cylinder. Modern diesel engines typically produce compression pressures of ± 600 psi.

Combustion pressure: the highest pressure developed in an engine cylinder during the power stroke. In today's efficient, electronically controlled diesel engines, combustion pressures may peak at up to five times the compression pressure.

Fire point: the temperature at which a flammable liquid gives off sufficient vapor for continuous combustion to take place. Also known as **ignition temperature**. The fire point or ignition temperature of a diesel fuel is specified by cetane number (CN).

Friction: force is required to move an object over the surface of another. **Friction** is the resistance to motion between two objects in contact with each other. Smooth surfaces produce less friction than rough surfaces. Lubricants coat and separate two surfaces from each other and reduce friction.

Inertia: describes the tendency of an object in motion to stay in motion or conversely, the tendency for an object at rest to remain that way. For example, an engine piston moving in one direction must be stopped at its travel limit and its kinetic inertia must be absorbed by the crankshaft and connecting rod. The inertia principle is used by the engine harmonic balancer and the flywheel—the inertial mass (that is, weight) represented by the flywheel has to be greatest in a single-cylinder, four-stroke cycle engine. As the number of cylinders increases, the flywheel weight can be reduced due to the greater mass of rotating components and the higher frequency of power strokes.

Thermal efficiency: a measure of the combustion efficiency of an engine calculated by comparing the heat energy potential of a fuel (calorific value) with the amount of usable mechanical work produced. Electronically controlled CI engines can have thermal efficiency values exceeding 40 percent. A typical gasoline-fueled car engine has a thermal efficiency of just over 30 percent. A rocket engine produces thermal efficiencies of over 60 percent.

Rejected heat: that percentage of the heat potential of the fuel (see the thermal efficiency definition) that is not converted into useful work by an engine. If a CI engine operating at optimum efficiency can be said to have a thermal efficiency of 40 percent, then 60 percent of the heat energy of the fuel has to be dissipated as *rejected heat*. Half of the rejected heat is typically transferred to the engine hardware to be dissipated to the atmosphere by the engine cooling system, and the other half exits in the exhaust gas. A turbocharger makes use of rejected heat by compressing the intake air forced into the engine cylinders, thereby increasing the thermal efficiency of the engine. **Figure 2-6** shows how the potential heat energy of fuel is released when combusted in a diesel engine.

TWO KEY PRINCIPLES

Next we will take a look at a couple of principles that are key to understanding the operation of any engine.

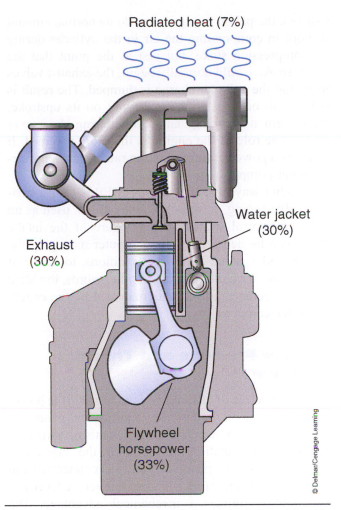

Figure 2-6 How the potential energy of diesel fuel is released in a diesel engine.

Mean Effective Pressure (MEP)

Mean effective pressure (MEP) is a way of describing the engine cylinder pressure that can actually be converted into useable torque. Simply, it is peak combustion pressure (produced during the power stroke) minus peak compression pressure (produced during the compression stroke). MEP describes the relationship between the *work performed* by the piston (in compressing the air charge) and the *work received* by the piston (through its downstroke on the power stroke). If the engine is going to continue to rotate, there has to be a net gain in terms of work. MEP is an important definition.

Modern, computer-controlled diesel engines can manipulate MEP by varying valve opening and closing. For instance, diesel engines using an internal compression brake open the exhaust valve(s) under engine braking just as the compression stoke is being completed. This means that when the engine brake is actuated, the piston is required to do its normal amount of work in compressing the air in the cylinder during the compression stoke, but just at the point that the power stroke would normally begin, the exhaust valves open, and the cylinder charge is dumped. The result is that the piston does its normal work on its upstroke, but is denied receiving any work during the power stroke. The role of the engine is therefore reversed. It becomes a power-absorbing pump rather than a power-receiving pump.

Another way of managing the MEP is to use variable valve timing. In diesel engines this is used as an emission control strategy. The closure of the intake valves can be delayed by the computer managing the engine under some running conditions to make it perform like a smaller engine. In other words, the MEP equation becomes a *soft* value managed by the engine controller computer.

Cylinder Pressure and Throw Leverage

The objective of any engine is to transfer the power developed in its cylinders as smoothly and evenly as possible to the power takeoff mechanism, usually a flywheel. The relationship between the crankshaft throw (to which the piston assembly is connected) and the crankshaft centerline is that of a lever. A lever is a device that provides a mechanical advantage. The amount of leverage (mechanical advantage) depends on the rotational position of the throw, which ranges from zero or no leverage when the throw is positioned at TDC to maximum leverage when the throw is positioned at a 90-degree angle with the connecting rod after TDC. This makes the relationship between cylinder pressure (gas pressure acting on the piston) and throw leverage (the position of the piston) critical in meeting the objective of smooth/even transfer of power from engine cylinders to drivetrain.

Let us take a look at this principle in operation. When the piston is at TDC beginning a power stroke, it is desirable to have minimum cylinder pressure, because in this position throw leverage is zero; therefore, no power transfer is possible. A properly set up fuel system attempts to manage cylinder pressure so that in any given performance mode it peaks somewhere between 10 degrees and 20 degrees ATDC when there is some but nevertheless a small amount of throw leverage. As the piston is forced down through the power stroke, gas pressure acting on the piston diminishes, but as it does so, throw leverage increases. Ideally, this relationship between cylinder pressure and crank throw leverage should be managed in a way that

Figure 2-7 2010 Cummins ISB engine used as optional power in Dodge pickup trucks.

Left Top of Engine

1. Engine coolant temperature (ECT) sensor
2. Waste-gate control vacuum solenoid
3. Fuel rail temperature (FRT)
4. Fuel pressure switch (FPS)
5. Engine-mounted fuel filter
6. Fuel supply line
7. Injector quantity adjustment (IQA) sticker
8. Turbocharger compressor outlet
9. Manifold absolute pressure (MAP) sensor

Figure 2-8 2011 Ford PowerStroke diesel engine.

results in consistent torque delivery from an engine cylinder through the power stroke until the throw forms a 90-degree angle with the connecting rod; this occurs a little before true 90 degrees ATDC.

MODERN DIESEL ENGINES

All highway diesel engines sold in the United States and Canada today are computer controlled. Computer controls are required to meet stringent EPA emissions standards. Every engine system is monitored and managed by computer. In addition, the engine management computer is networked to other chassis computers to optimize:

- engine power
- fuel economy
- emissions
- vehicle longevity

The clean looks of the diesels of a generation past have been replaced by engines that are cluttered with external emission control apparatus and complex exhaust aftertreatment hardware. **Figure 2-7** shows a side view of a 2010 Cummins ISB engine used as optional power in Dodge pickup trucks.

Because of the need to monitor the performance of every engine condition and adapt to every consequence, in addition to the emissions hardware, dozens of sensors and actuators are arranged over the engine. Sometimes it can be difficult to identify engine hardware beneath all the wiring harnesses. Working on the electronically controlled engines of today requires manufacturer training. Attempting to service and repair engines without the appropriate training can result in costly damage. **Figure 2-8** illustrates an overhead view of a 2011 Ford Power-Stroke engine.

Summary

- Most diesel engines are rated by their ability to produce power and torque. The tendency is to rate gasoline-fueled auto engines by their total displacement.

- Diesel engines have high compression ratios, so they tend to be undersquare.

- Almost every mobile highway diesel engine is manifold boosted; that is, it is turbocharged.

- The diesel cycle is a four-stroke cycle consisting of four separate strokes of the piston occurring over two revolutions; a complete engine cycle is therefore extended over 720 degrees.

- MEP is the average pressure acting on the piston through the four strokes of the cycle. Usually, the intake and exhaust strokes are discounted, so MEP is equal to the average pressure acting on the piston through the compression stroke subtracted from the average pressure acting on the piston through the power stroke.

- Ideally, engine fueling should be managed to produce peak cylinder pressures at somewhere around 10 to 20 degrees ATDC when the relative mechanical advantage provided by the crank throw position is low. This means that as cylinder pressure drops through the power stroke, throw mechanical advantage increases, peaking at 90 degrees ATDC, providing a smooth unloading of force to the engine flywheel.

- An engine attempts to convert the potential heat energy of a fuel into useful kinetic energy: the degree to which it succeeds is rated by its thermal efficiency.

- That portion of the heat energy of a fuel not converted to kinetic energy is known as *rejected heat*. Rejected heat must be dissipated to the atmosphere by means of the engine cooling and exhaust systems.

Review Questions

1. A diesel engine has a bore of 80 mm and a stroke of 100 mm. Which of the following correctly describes the engine?

   A. 10 inch displacement   C. Oversquare

   B. 10 liter displacement   D. Undersquare

2. Which of the following best describes the term *engine displacement*?

    A. Total piston swept volume   C. Peak horsepower

    B. Mean effective pressure   D. Peak torque

3. An eight-cylinder diesel engine has a bore of 80 mm and a stroke of 100 mm. Calculate its total displacement and round to one of the following values:

    A. 3.5 liters   C. 4.5 liters

    B. 4.0 liters   D. 6.7 liters

4. The tendency of an object in motion to stay in motion is known as:

    A. Kinetic energy   C. Inertia

    B. Dynamic friction   D. Mechanical force

5. What percentage of the potential heat energy of the fuel does the modern diesel engine convert to useful mechanical energy when it is operating at its best efficiency?

    A. 20 percent   C. 60 percent

    B. 40 percent   D. 80 percent

6. Where does scavenging take place on a four-stroke cycle, diesel engine?

    A. BDC after the power stroke   C. Valve overlap

    B. TDC after the compression   D. 10 to 20 degrees ATDC on the power stroke
       stroke

7. Ideally, where should peak cylinder pressure occur during the power stroke?

    A. TDC   C. 90 degrees ATDC

    B. 10 to 20 degrees ATDC   D. At gas blowdown

8. When running an engine at any speed or load, which of the following would be the best location to produce peak cylinder pressure?

    A. 20 degrees BTDC   C. TDC

    B. 10 degrees BTDC   D. 15 degrees ATDC

9. At which point in the engine cycle does the crank throw leverage peak?

    A. TDC   C. 90 degrees ATDC

    B. 15 degrees ATDC   D. BTDC

10. The energy of motion is known as:

    A. Inertia   C. Thermal

    B. Kinetic   D. Potential

CHAPTER

3 Cylinder Block Assemblies

Prerequisite

Chapter 2.

Learning Objectives

After studying this chapter, you should be able to:

- Define the roles of piston assemblies, crankshafts, flywheels, and dampers.
- Identify the different types of pistons used in current diesel engines.
- Explain the function of piston rings.
- Describe the role of connecting rod assemblies.
- Describe the role of crankshaft assemblies.
- Outline the forces a crankshaft is subjected to under normal operation.
- Identify some typical crankshaft failures and their causes.
- Define the term *hydrodynamic suspension*.
- Outline the roles played by vibration dampers and flywheel assemblies.
- Identify types of cylinder blocks.
- Outline the procedure required to inspect a cylinder block.
- Identify the types of cylinder heads used in diesel engines.
- Describe the function of the oil pan in the engine.

Key Terms

antithrust side	compressional load	direct injection (DI)
bearing shell	connecting rod	forged steel trunk-type pistons
big end	cracked rods	fractured rods
cam ground	crankcase	friction bearing
compacted graphite iron	crown	headland volume
compression ring	cylinder block	hone

29

hydrodynamic suspension	Ni-Resist™ insert	small end
indirect injection (IDI)	oil pan	sump
keystone ring	parent bore	template torque (TT)
keystone rod	pin boss	thrust bearing
lands	piston pin	thrust face
liners	Plastigage™	thrust washers
lugging	powertrain	torque
major thrust side	ring belt	torque-to-yield (TTY)
Mexican hat piston crown	ring groove	torsion
minor thrust side	rod eye	torsional stress
Monosteel™ pistons (Federal Mogul)	scraper ring	trunk-type piston
Monotherm™ pistons (Mahle)	short block	wrist pin
	sleeves	

INTRODUCTION

In an internal combustion pressure, the burning of fuel in an engine cylinder produces cylinder gas pressure. This pressure has to be converted into useful mechanical energy. This chapter addresses the engine components responsible for converting the gas pressures developed in engine cylinders to torque at the flywheel. We call this group of components the engine **powertrain**. The engine powertrain includes:

- pistons
- piston rings
- wrist pins
- connecting rods
- crankshafts
- friction bearings
- flywheels
- vibration dampers

Function of the Powertrain

The powertrain is driven by cylinder gas pressure. Cylinder pressure acts on a piston and drives it through its stroke. A piston moves straight up or down in the engine cylinder. This linear movement has to be converted into **torque** or twisting force. This is accomplished by connecting the piston assembly to an offset throw on the crankshaft. In this way, the linear force produced by the piston is converted to rotary force known as torque. As pistons travel up and down in the cylinder bore, the crankshaft rotates. Connecting rods pivot on both the piston and crankshaft throw. Torque from the crankshaft is transferred to a flywheel

bolted to the crankshaft. The flywheel acts as a coupling to transfer engine torque to the vehicle drivetrain.

Bicycle Powertrain

You can compare what is happening in a typical engine with another type of engine, a bicycle. A bicycle is an engine driven by muscle power. The powertrain of a bicycle consists of a pair of offset throws (we call them pedals), a crankshaft, and a bull gear: cogs on the bull gear allow torque to be transferred to a driven gear and wheel assembly. When a bicycle is ridden, linear force is applied to the pedals by the rider's legs. In just the same way as with a diesel engine, the crankshaft converts the linear force applied to the pedals into torque so that we can power the bicycle down a road.

Cylinder Block and Crankcase

The engine powertrain components are housed in a cylinder block assembly. All the other engine components are attached either directly or indirectly to the cylinder block. In a typically configured engine, an oil pan is bolted below the cylinder block. This lower region of the engine is known as the **crankcase**. The oil pan contains the lubricating oil. Lube oil is required to lubricate and help cool the powertrain components of the engine.

Short Block Assembly

The components of the powertrain and the cylinder block are often known as a **short block**. Although this might be considered a slang term, it is commonly used

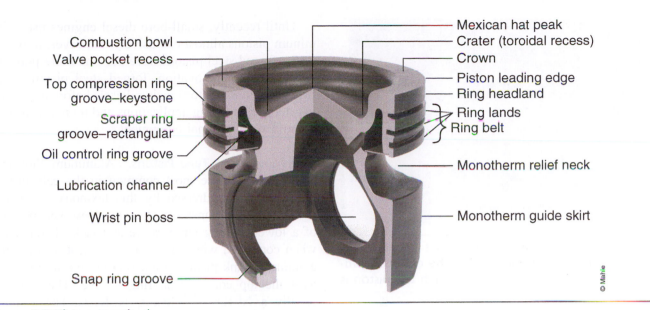

Figure 3-1 Piston terminology.

and is one means of partially reconditioning an engine. When a short block overhaul is undertaken, a fully reconditioned powertrain and cylinder block assembly is mated up to the existing engine's cylinder head(s) and peripheral components. In this chapter, we will begin with a study of the powertrain components in sequence and finish with a look at cylinder blocks.

PISTON ASSEMBLIES

A piston is a circular plug that seals the engine cylinder bore. It is connected to the crankshaft by means of a connecting rod and moves up and down in the cylinder bore. The role of the piston in the cylinder bore is to:

- Deliver force: It does this when traveling upward on its compression stroke.
- Receive force: It receives force when combustion pressures act on it during the power stroke.

A piston assembly consists of the piston, piston rings, a wrist pin, and connecting rods. Piston rings seal the cylinder and lubricate the cylinder walls. The wrist pin is installed through a boss in the piston; it connects the piston to the connecting rod.

Piston Terminology

A diesel engine piston is shown in **Figure 3-1** along with the terminology we use to describe it. The type of piston shown is a forged steel trunk piston used in late-model diesel engines. If you follow the callouts in **Figure 3-1** you should be able to make sense of the description that follows.

Piston Crown Geometry. In a diesel engine, the fuel is mixed with the air charge inside the engine cylinder. This means that the shape of the piston plays a role in determining the efficiency. The upper face of the piston is called the **crown**. The crown or top of the piston is exposed directly to the cylinder chamber and therefore the effects of combustion. Because of this, a piston should be capable of rapidly transferring the combustion heat it is exposed to. This ability to transfer heat is especially important when aluminum pistons are used. Aluminum has a much lower melting point than alloy steels. Many pistons have cooling jets that spray lubricating oil on the underside of the piston. This helps remove heat and keep piston crown temperatures lower.

Mexican Hat Crown. The shape of the piston crown is important. This shape determines how swirl and squish are generated in an engine. Swirl and squish determine how injected fuel mixes with the air in the cylinder. The **Mexican hat piston crown** features a recessed circular crater centered by a peaked crown. This crater may be concentrically located in the crown as shown in **Figure 3-1** or offset as shown in **Figure 3-2**. The Mexican hat design is commonly used on current diesel engines because it promotes good fuel/air mixing efficiency. The upper edge of the piston around the crown is known as its leading edge. Most modern diesel engines have low clearance volumes. This means that at top dead center (TDC), the piston rises in the cylinder bore until it almost contacts it. In **Figure 3-2** featuring an offset Mexican hat crown, the recesses beside the crater accommodate cylinder valve protrusion when the piston is at TDC.

Figure 3-2 Face-on view of an offset Mexican hat piston crown. The recesses beside the crater accommodate cylinder valve protrusion when the piston is at TDC.

Small-Bore Pistons.   While larger diesel engines may use two-piece and composite piston assemblies, small-bore diesels tend to use single-piece pistons. Single-piece pistons are known as **trunk-type pistons**.

Trunk-Type Pistons

Trunk-type pistons are single-piece pistons. Trunk pistons used in diesel engines can be divided into two categories based on what material they are made of:

- aluminum alloy
- forged steel

Until recently, small-bore diesel engines used aluminum pistons almost exclusively. However, in recent years, forged steel pistons have become more popular. As with much innovation, forged steel pistons were first adopted by factory racing diesels before migrating to aftermarket racing applications and more recently to manufacturer original equipment.

Aluminum Trunk-Type Pistons.   Aluminum alloy pistons are by far the most common in the category of diesel engines addressed by this textbook. They are lightweight and transfer heat easily. However, because of a lower melting temperature and lack of toughness when compared with forged steels or cast irons, many aluminum trunk pistons use a ring groove insert for at least the top compression ring groove. The insert is usually a **Ni-Resist™ insert**. Ni-Resist has a much higher melting temperature than aluminum, and in addition, it expands and contracts in the same way. **Figure 3-3** is an image of an aluminum trunk-type piston with callouts indicating many of the key terms used in this chapter; note the location of the Ni-Resist insert.

Ceramic Fiber Alumina.   Another method of increasing the toughness and high-temperature performance of an aluminum trunk-type piston is to reinforce it with ceramic fiber alumina (CFA). A CFA reinforced process is used on some smaller-bore diesels. The CFA process eliminates the requirement for a Ni-Resist insert because it reinforces the piston at the top of the

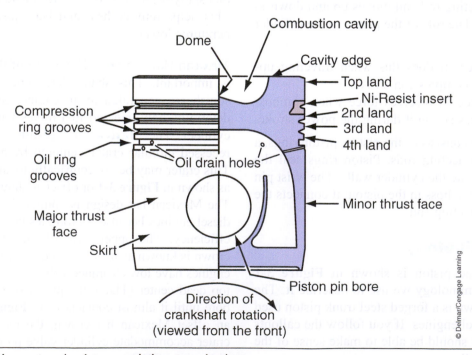

Figure 3-3 Aluminum trunk piston and ring terminology.

ring belt extending up into the top of the crown. This allows placement of the top compression ring close to the leading edge and reduces **headland volume** (see **Figure 3-1**). High location of the top compression ring is almost a requirement for today's low-emissions diesels because it reduces the dead gas zone between the upper compression ring and the piston leading edge. Another fiber reinforcement manufacturing practice known as squeeze cast, fiber reinforced (SCFR) is used by some manufacturers to toughen the crown area of the piston. This is also an alumina fiber manufacturing process.

CAM-GROUND PISTONS

Aluminum pistons are lightweight and transfer heat efficiently, but they also expand and contract much more than steels. This means that they have to be designed with much higher clearance in the cylinder bore when cold. It also means that they have to be **cam ground**. A cam-ground piston is slightly oval in shape when cold. As the piston warms up, the area around the **pin boss** expands relatively more than the thinner skirt area between the pin bosses. The idea behind a cam-ground piston is that at running temperatures, the piston should expand to an exactly circular shape with minimal bore clearance.

A big disadvantage of cam-ground pistons is that they should be warmed to operating temperature before being subjected to high cylinder pressures. Failing to warm up a cam-ground piston before loading an engine can overstress the piston rings and lands. Aluminum trunk pistons are shaped to beef up the piston where it is at its weakest. So, in addition to having reinforcement at the pin boss, they also have increased material around the crown, as you can see in **Figure 3-3**.

WRIST PINS AND RING BELT

Piston wrist pins when not full floating are usually press fit to the piston boss and float on the rod eye. Heating the piston to 95°C (200°F) in boiling water facilitates pin assembly. Because of their higher clearance when cold, aluminum trunk pistons are prone to piston slap during engine warmup. Piston slap is the tilting action of the piston when the piston is thrust loaded by cylinder combustion pressure. It can be minimized by tapering the piston so that the outside diameter at the lower skirt (where it does not get so hot) slightly exceeds the outside diameter over the ring belt region. The ring belt region is exposed to more heat and expands more as the piston heats to operating temperatures.

ADVANTAGES OF ALUMINUM TRUNK PISTONS

- Lightweight. This reduces the overall piston weight. Lightweight pistons reduce the inertia forces that the connecting rod and crankshaft are subject to. It also permits the use of lighter-weight components throughout the engine powertrain.
- Cooler piston crown temperatures. Because aluminum alloy pistons transfer heat so rapidly, they run cooler than equivalent steel-based pistons.
- Quieter. Engines using aluminum alloy, trunk-type pistons generally produce less noncombustion-related noise.

Forged Steel Trunk-Type Pistons. Although **forged steel trunk-type pistons** were used half a century ago in drag racing applications, they are a relatively new introduction to diesel engine technology. However, they have caught on quickly with diesel engine designers, and many of today's late-model diesel engines use them. The brand names of this new generation of pistons are **Monotherm**™ (manufactured by Mahle) and **Monosteel**™ (manufactured by Federal Mogul). In this text, we generally refer to them as *forged steel trunk-type pistons* to avoid using brand names. **Figure 3-4** shows two views of a Mahle Monotherm piston.

DESIGN AND CONSTRUCTION

The single-piece forged steel trunk piston has the appearance of an aluminum trunk piston with a large circumferential slot cut away between the pin boss and the ring belt. The skirt is designed to guide the piston over the thrust sides of the piston and is recessed across the pin boss. This permits the use of a shorter, lower-weight wrist pin. It also allows the piston pin

Figure 3-4 Two views of a forged steel trunk piston (Monotherm).

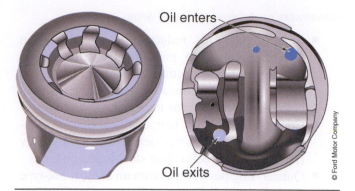

Oil enters

Oil exits

© Ford Motor Company

Figure 3-5 Forged steel trunk-type piston used in the Ford 2011, 6.7L PowerStroke.

boss support area to be increased, making the design well suited to engines producing very high cylinder pressures.

The problem of cold-start piston slap in aluminum trunk pistons has been overcome with the adoption of forged steel pistons. The steel alloy pistons expand much less than equivalent aluminum trunk pistons when heated from cold to operating temperatures. This allows them to be manufactured with a much tighter fit to the liner bore. Steel pistons run hotter than aluminum pistons because they dissipate heat more slowly. The lubricating oil cooling gallery used in a typical forged steel piston is closed to permit much higher oil feed and flow-back volumes as shown in **Figure 3-5.** The oil feed to the underside of the piston is delivered by a precisely targeted cooling jet. In some cases, the connecting rods used with steel forged pistons are not rifle-drilled with a lubrication passage. This means that the piston and wrist pin depend almost entirely on the oil cooling jet for lubrication and cooling.

The micro-alloyed steel construction of this class of trunk piston provides high strength, so much less material has to be used. The result is a tough steel piston of about the same weight as the older aluminum trunk pistons. In forged steel pistons featuring bushingless wrist pin bores, the pin boss is phosphate treated over the pin bearing bore.

ADVANTAGES OF FORGED STEEL TRUNK PISTONS

The growing adoption of forged steel trunk-type pistons in recent years is accounted for by the higher cylinder combustion pressures and temperatures required of today's engines. Some of the advantages of forged steel trunk pistons are:

- Reduction of headland volume. Headland volume (see **Figure 3-1**) is the volume in the cylinder above the top compression ring and below

the piston crown leading edge. This volume is less affected by cylinder turbulence and the effects of cylinder scavenging, so it collects dead end gas. Headland volume can be minimized by placing the top compression ring as close as possible to the crown leading edge. Because forged steel is much stronger than aluminum, steel trunk pistons require no groove insert, and the upper ring groove can be located close to the crown leading edge. Most forged steel trunk pistons place the top compression ring just slightly under the leading edge of the piston.

- Thermal expansion factors. Thermal expansion is simply how a material responds to heat. Forged steel expands and contracts as temperature changes much less than aluminum, so cam-ground designs are not required in steel trunk pistons, and piston-to-bore clearances are tighter.
- Long life. Micro-alloyed forged steel coated with phosphate provides much longer service life than equivalent aluminum trunk pistons.
- Lightweight. Forged steel trunk pistons are designed to be as lightweight as aluminum trunk pistons but to have much higher strength. Some forged steel trunk pistons are specified to sustain cylinder pressures that reach 3,500 psi (250 bar).

Piston Thrust Faces

When cylinder gas pressure acts on a piston, especially during the initial stage of combustion, it tends to cock (pivot off a vertical centerline) in the cylinder bore because it pivots on the wrist pin. This action creates thrust faces on either side of the piston. The major thrust face is on the inboard side of the piston as its throw rotates through the power stroke. The minor thrust face is on the outboard side of the piston as its throw rotates through its power stroke. Take a close look at **Figure 3-3** in which the piston thrust faces are identified. The major thrust face is sometimes simply called the **thrust face**, while the minor thrust face is called the anti-thrust face. You should learn to identify the thrust faces of a piston for purposes of failure analysis. On modern steel trunk-type pistons only the thrust faces are skirted. This allows the piston to be guided true in its bore while also reducing the weight of the piston.

Combustion Chamber Designs

In direct-injected diesel engines, the shape of the piston crown determines how the gas within the cylinder moves on the compression and power strokes. Up until around 1990, diesel engines tended to use piston crowns designed to produce developed high turbulence especially those using **indirect injection (IDI)**. IDI diesel engines are ones in which the fuel is injected into a cavity located remotely from the cylinder bore, usually in the cylinder head, less commonly within the piston. The cavity was known as a pre-combustion chamber, pre-chamber, or energy cell. Today, IDI engines are not common because they are not able to produce the combustion efficiency of **direct injection (DI)** diesel engines required to meet tough emissions standards.

Direct Injection. In a DI engine, the fuel charge is injected at high pressure directly into the engine cylinder above the piston. Most pickup truck and automobile diesel engines adopted DI diesel engines during the 1990s, and it is a requirement of any diesel engine meeting today's on-highway emissions standards. DI has always required high fuel injection pressures, but these have increased in recent years to improve combustion efficiency. As diesel fuel injection pressures have increased, the need for high cylinder turbulence has decreased. In fact, excessive turbulence tends to be avoided because it can propel fuel droplets away from the intended flame front.

Mexican Hat. We have already introduced the Mexican hat crown design because it is so common among recent diesel engines. While the Mexican hat crown produces desirable swirl characteristics, another reason diesel engine manufacturers use the design is that it allows injected diesel fuel droplets to be targeted into the crater. This allows atomized fuel droplets to vaporize and ignite before they physically contact the piston crown material. In this way, the Mexican hat piston crown provides a lower risk of fuel burnout scorching on the piston crown directly below the injector, and lengthens service life.

Mann Type (or "M" Type). The Mann-type piston crown is named after the German company responsible for its design. It tends to be used on older trunk-type pistons and consists of a radiused, recessed bowl located directly under the injector, though not necessarily in the center of the piston crown. Depending on the depth of this bowl, the Mann-type combustion chamber produces high turbulence and is most often seen in IDI diesel engines.

Dished. The dished piston crown has a slightly concave to almost flat design that produces low turbulence when compared with the previous types. You are most likely to see this design in some current small-bore and offshore-manufactured diesel engines.

Piston Cooling

Engine oil plays a major role in managing piston temperatures. Lube oil is routed to the pistons directly through rifled bores in connecting rods, or indirectly by piston cooling jets. Things such as the size of the piston, peak cylinder pressures, and whether the engine is turbocharged determine what type of piston cooling is required. Within the same engine family, you may find that engines with lower horsepower specifications do not use cooling jets, while those rated at higher horsepower do. Because combustion temperatures may sometimes be higher than the melting temperature of the piston materials (in cases where aluminum is used), it is essential to get rid of any heat that has not been converted to usable energy as quickly as possible. A percentage of cylinder heat is always transferred through the piston assembly. Three methods are used to cool pistons. Engines may use one or more combinations of these piston cooling methods.

1. Shaker—Oil is delivered through the connecting rod to a gallery machined into the underside of the piston crown. This oil is distributed by the motion of the piston after which it drains to the crankcase.
2. Circulation—Oil is delivered through the connecting rod rifling, through the wrist pin, and then circulated through a series of grooves machined into the underside of the piston crown. It then drains back into the crankcase.
3. Spray—A stationary cooling jet is located in the cylinder block just below the cylinder liner. This jet is fed by engine lube under pressure. The oil cooling jet is then aimed so that the spray is directed at the underside of the piston. This oil cools the piston crown and may also lubricate the wrist pin. Aiming cooling jets is usually done using a clear perspex template that fits over the fire ring groove on the cylinder block deck; an aim rod is inserted in the jet orifice. A target window is scribed in the perspex template and the aim rod has to be positioned within the window as shown in **Figure 3-6.** The spray cooling method is highly efficient and is used in many turbocharged diesel engines that run higher piston temperatures.

CAUTION *A piston cooling jet that is misaimed can destroy the piston it is supposed to cool. You have to especially take care when assembling an engine to avoid clunking a cooling jet when installing piston/rod assemblies.*

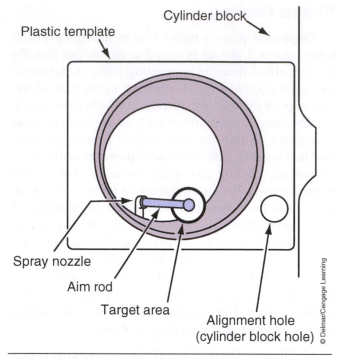

Cylinder block

Plastic template

Spray nozzle

Aim rod

Target area

Alignment hole
(cylinder block hole)

© Delmar/Cengage Learning

Figure 3-6 Spray nozzle targeting.

PISTON RINGS

The function of piston rings is to seal the piston in the cylinder bore. Most pistons require rings to effectively seal, and those that do not are usually found in automobile racing applications. Ringless pistons use special piston materials and are run at high rpms. Running engines at high rpms permits little *time* for cylinder leakage or blowby to occur. Rings have three important functions:

1. Sealing: They are designed to seal compression and combustion gases within the engine cylinder.
2. Lubrication: They are designed to apply and regulate a film of lubricant to the cylinder walls.
3. Cooling: Rings provide a path for heat to be transferred from the piston to the cylinder walls.

Piston rings are located in recesses in the piston known as **ring grooves**. Ring grooves are located between **lands**. Check out **Figure 3-1,** paying special attention to the ring area.

Roles of Piston Rings

Piston rings may be broadly categorized as:

- compression rings
- oil control rings

Compression rings are responsible for sealing the engine cylinder, and they play a role in helping to transfer piston heat to the cylinder walls. The term

scraper ring is used to describe rings below the top compression ring that play a role in sealing cylinder gas as well as managing the oil film on the cylinder wall. Oil control rings are responsible for lubricating the cylinder walls and also provide a path to dissipate piston heat to the cylinder walls.

Ring Materials

Piston rings are designed with an uninstalled diameter larger than the cylinder bore, so that when they are installed, radial pressure is applied to the cylinder wall. Some diesel engines use piston compression rings manufactured from cast-iron alloys that are similar to those used in production gasoline-fueled, spark-ignited engines. Cast iron rings are brittle and fracture easily. However, some modern workhorse diesel engines use steel alloy rings that have some flexibility and will not fracture like cast iron. These new generation piston rings are much tougher and more flexible than their cast iron counterparts. They are almost impossible to fracture and often show little wear at engine overhaul. The wall section of the piston in which the set of rings is located is known as the **ring belt**. You can identify the ring belt on **Figure 3-1.**

Ring Action

The major sealing force of piston rings is high-pressure gas. Piston rings have a small side clearance. The result of this minimal side clearance is that when cylinder pressure acts on the upper sectional area of the ring, three things happen in the following sequence to make it seal on the compression stroke:

1. Pressure forces the ring downward into the land.
2. Forcing the ring into the land allows developing cylinder pressure to get behind the ring.
3. When cylinder pressure gets behind the ring between it and the groove wall, it gets driven outward into the cylinder wall, creating the seal.

The lower right side of **Figure 3-7** explains ring action in diagram form. Make sure you understand this. What this tells us is that cylinder sealing efficiencies increase with cylinder pressure. The higher the cylinder pressure, the more effectively rings seal the cylinder.

Number of Rings

The number of rings used is determined by the engine manufacturer, and factors are:

- bore size
- engine speed
- engine configuration

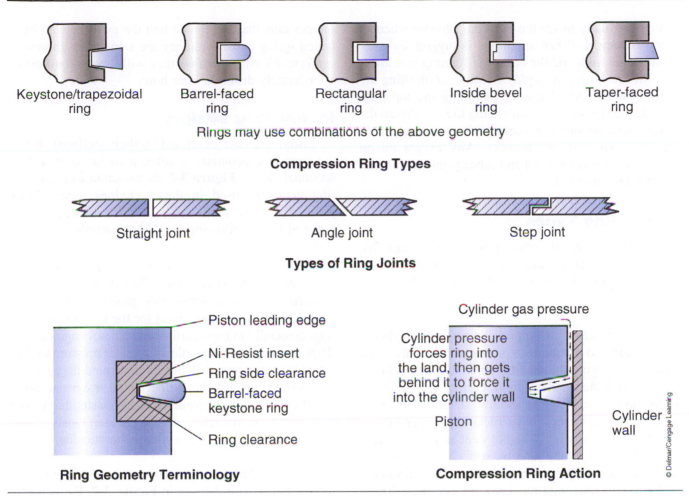

Figure 3-7 Piston ring geometry and action.

Time is probably the major factor in determining the number of compression rings. The slower the maximum running speed of the engine, the greater the total number of rings required because there is more *time* for gas blowby to occur. So, diesel engines with slower rated rpms usually require more rings. Higher-speed diesels require a lower number. Most automotive diesel engines run at peak speeds of 4,000 rpm or less, and this category of engines commonly use a three-ring configuration of two compression rings and a single oil control ring. **Figure 3-8** shows a close-up of the ring belt of a three-ring aluminum trunk piston. Note the Ni-Resist insert used to reinforce the top ring groove. On this piston, the top two rings are compression rings and the third ring is an oil control ring.

Gas Blowby. The top compression ring gets the greatest sealing assist from cylinder pressures. Gas blowby from the top compression ring passes downward to seal the second compression ring, and so on. Gas that blows by all the rings enters the crankcase, so high crankcase pressures are often an indication that an engine is wearing out. Because an engine cylinder is

Figure 3-8 Detail of a three-ring configuration on an aluminum trunk-type piston. Observe the Ni-Resist insert used on the upper compression ring.

sealed by rings, some cylinder leakage past the ring belt is inevitable. The limiting factor is time. When an engine is operated at 2,000 rpm, one full stroke of a piston takes place in 15 milliseconds (0.015), so there is quite simply insufficient time for significant cylinder leakage to take place.

You are likely to see the most gas blowby when an engine is lugged. When an engine is lugged, cylinder pressures are high and the engine is running at a lower rpm. All the piston rings play a role in controlling the oil film on the cylinder wall, including the top ring. The role the rings play in controlling the oil film on the cylinder wall becomes increasingly more important as emissions standards get tougher. Any excess oil on cylinder walls is combusted and subsequently exhausted as unwanted emissions.

Piston Ring Types

There are many different types of piston rings. We name them by function and shape. Make sure you take a look at **Figure 3-7** as we run through the following explanation.

Compression Rings. Compression rings seal cylinder compression and combustion pressures. They also play a role in managing the oil film applied to the cylinder wall. As indicated earlier, some current diesel engine piston rings can be substantially deformed without fracturing because they have metallurgical properties similar to stainless steel. Whether steel base or cast iron, compression rings may also be surface-coated by tinning, plasma, and chrome cladding to reduce friction.

Some ring coatings are break-in coatings. These temporary coatings wear off and end up in the crankcase lube, something that has to be taken into account when studying an oil sample analysis. Sometimes the upper compression ring is known as the fire ring, but more often this term is used to describe the cylinder seal at the top of the liner that is often integral with the cylinder head gasket.

Combination Compression and Scraper Rings. Combination compression and scraper rings assist in sealing combustion gases and controlling the oil film on the cylinder wall. Manufacturers who use this term are referring to a ring or rings located in the middle of the ring belt, between the top compression ring and the oil control ring(s).

Oil Control Rings. Oil control rings are designed to control the oil film on the cylinder wall. Too much oil on the cylinder wall will end up being combusted. Too little oil on the cylinder wall will result in scoring and scuffing of the cylinder wall. The action of first applying (piston traveling upward) and then wiping (piston traveling downward) lubricant from the cylinder wall helps remove heat from the cylinder by transferring it to the engine oil. Most oil control rings use circular

scraper rails that are forced into the cylinder wall by a coiled spring expander. They are sometimes known as *conformable* rings because they will flex to conform to a moderately distorted liner bore.

Piston Ring Designs

Piston rings are described by their sectional shape; the term *ring geometry* is often used to describe this sectional shape. **Figure 3-7** shows some examples of ring geometry used in diesel engines today. Most piston rings use combinations of the characteristics outlined here. Following are some examples.

Keystone Rings. **Keystone rings** (sometimes known as *trapezoidal rings*) are wedge shaped. A keystone ring is fitted to a wedge-shaped ring groove in the piston. This design is commonly used for the top compression ring, especially in high-performance diesel engines. The shape of a keystone ring allows cylinder pressure to first act on its upper sectional area to get behind the ring and then force it against the cylinder wall. Keystone rings are also less likely to form carbon deposits due to the scraping action that takes place as the ring twists within its groove. They are also less prone to sticking.

Rectangular Rings. The sectional shape of the ring is rectangular as you can see in **Figure 3-7**. Rectangular rings are loaded evenly (they do not twist so much) into the cylinder wall when cylinder pressure acts on them. Some pistons that use a keystone top compression ring use a rectangular second or third compression ring as you can see in **Figure 3-1**.

Barrel Faced. A barrel-faced piston ring is one in which the outer face is "barreled" with a radius (rounded). This increases the service life. The downside is that they do not seal quite as effectively because there is no sharp edge to bite into the cylinder wall when the ring twists within the groove. Keystone rings are often barrel faced. **Figure 3-7** shows an example of a barrel-faced keystone ring.

Inside Bevel. An inside bevel ring is one in which a recess is machined into the inner circumference of the ring as shown in **Figure 3-7**. This helps cylinder pressure to get behind the ring and causes it to twist in the groove. This twisting action produces high unit-sealing pressures because the ring bites into the cylinder wall, providing an effective seal.

Taper Faced. The design is similar to the rectangular ring but its outer face is angled, giving it a sharp

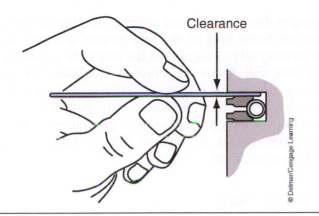

Figure 3-9 Oil ring side clearance.

lower edge. Once again, this enables the ring to achieve high unit-sealing pressures; that is, to bite into the cylinder wall when loaded with cylinder gas pressure. A taper-faced ring is shown in **Figure 3-7.**

Channel Section.   Channel section rings are used as oil control rings. They usually consist of a grooved ring with a number of slots to allow oil to be first applied, and then scraped from the cylinder walls. In most cases, an expander ring is used with channel section rings. This spring-loads the ring into the cylinder wall, allowing it to adapt to minor variations in the liner bore. The expander ring is a coiled or trussed spring installed into a groove behind the rails of the channel section. See the channel section oil control rings shown a little later in **Figure 3-9** when the procedure required to measure oil ring side clearance is shown.

Ring Joint Geometry

Piston rings must be designed so that when heated to operating temperatures, the ring does not expand so much that the joint edges come into contact. If this were to happen, the ring would buckle. A ring must also be capable of sealing with some efficiency when cold and cylinder pressures are low. **Figure 3-7** shows the three types of joint design described in the text:

1. Straight. The split edges of the ring abut. This design is more likely to leak blowby gas, especially during cold engine operation. It is, however, the most commonly used.
2. Stepped. This design uses an L-shaped step at the joint. It is least likely to leak blowby gas at the ring joint.
3. Angled. The ring is faced with complementary angles at the joint. It seals fairly efficiently at the ring joint.

Installing Piston Rings

Rings should be installed to the piston using the correct installation tool. Stretching rings over the piston by hand can fracture cast iron rings and crack the plating and cladding materials. Avoid using multipurpose ring expanders. This is one case where the engine original equipment manufacturer's (OEM's) special tool is usually the best bet: an example of an OEM ring expander is shown in **Figure 3-10.** You should not install a piston ring in which the coating appears cracked or chipped. Most rings must be correctly installed, and that means they usually have an up side, so check out the service literature before installation. Most rings are marked, typically by using a dot to indicate the up side.

Ring End Gap.   The ring end gap is checked by installing a new ring into the cylinder bore, making sure it is within the ring belt swept area. You then measure the gap using feeler gauges. The specification is usually in the region of 0.003 to 0.004 per inch of cylinder diameter (0.3 to 0.4 mm per 100 mm). When you check ring end gap you do so by installing a cold ring into a cold engine: the effect is to make the end gap appear larger than you might think. Remember that as with most engine specifications, the end gap is established for performance at engine operating temperature, and a ring at operating temperature is designed to expand to almost close the gap you measure. Never use a ring that measures out of specification. In certain cases where the OEM permits (this can only be done with cast iron rings), ring end gap can be adjusted by filing.

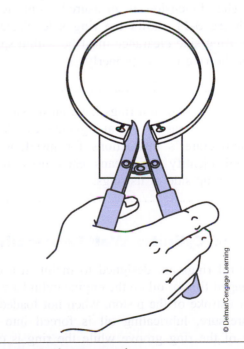

Figure 3-10 Ring expander.

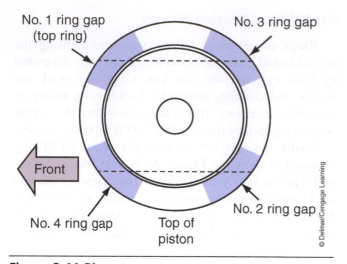

No. 1 ring gap
(top ring)

No. 3 ring gap

Front

No. 4 ring gap

Top of
piston

No. 2 ring gap

© Delmar/Cengage Learning

Figure 3-11 Ring stagger.

Ring Gap Spacing or Stagger. The manufacturer's instructions must be observed especially when forged steel trunk pistons are used. Ring stagger requires offsetting the ring gaps to disrupt the blowby gas route. The gaps are usually offset by dividing the number of rings into 360 degrees, so if there were three rings, the stagger would be 120 degrees offset. In the case of four rings, the stagger should be 90 degrees offset as shown in **Figure 3-11.** It is not recommended that ring gaps be placed directly over the thrust or antithrust faces of the piston.

Ring Side Clearance. Ring side clearance is the installed clearance between the ring and the groove it is fitted to. The dimension is measured using feeler gauges. **Figure 3-11** shows the ring side clearance dimension. Ring side clearance must be within specification for the ring to seal properly.

....

Tech Tip: Most piston rings have an up side that is often not easy to see at a glance. Check the manufacturer's instructions for installing rings and identify the means each uses to identify the up side of its rings.

....

Piston and Cylinder Wall Lubrication

Oil control rings are designed to maintain a precisely managed film of oil on the engine cylinder wall. On the downstroke of the piston, when not loaded by cylinder pressure, lubricating oil is forced into the lower part of the ring groove while the ring is contacting the upper ledge of the land. When the piston

changes direction to travel upward, the ring is forced into the lower land of the ring groove, allowing the lubricating oil to pass around the ring to be applied to the liner wall. While the action of simultaneously applying and scraping oil from the cylinder walls ensures that the applied film thickness is minimal, all engines will burn some oil. In the latest generation of low-emissions engines, the amount of burned oil must be held to a minimum.

WRIST PINS

The function of **wrist** or **piston pins** is to connect the piston assembly with the connecting rod eye or small end. The highest expected cylinder pressures determine whether the pin is solid or bored through. Because the weight of the wrist pin adds to the piston total weight, it is designed to be as light as possible while handling the forces it is subjected to. In applications that use forged steel trunk pistons, the wrist pin is reduced in length because the pin boss does not extend to the skirt.

The bearing surfaces of wrist pins are lubricated by engine oil in two ways:

- Directed upward through a rifle bore in the connecting rod: Method used in most engines until recently.
- Sprayed upward by a piston cooling jet targeted at a gallery entry port: This method is used on some forged steel trunk (Monotherm) pistons (see **Figure 3-5**).

Full-floating piston pins are fitted to both the rod eye and the piston boss with minimal clearance. Some newer piston bosses are bushingless.

Piston Pin Retention. All full-floating piston pins require a means of preventing the pin from exiting the pin boss and contacting the cylinder walls. Snap rings and plugs are used. When installing the internal snap rings used by most engine OEMs you must observe the installation instructions. In most cases, the split joint of the snap ring should be located downward, so in the absence of any other instruction do this.

Assembling Piston and Rings

When clamping a piston/connecting rod assembly in a vise, use brass jaws or a generous wrapping of rags around the connecting rod. The slightest nick or abrasion may cause a stress point from which a failure could develop in a con rod. It is also important to handle rings with the specified tools during reassembly. Overflexing of piston rings during as-sembly can damage the surface coatings of rings, while the rings' ends can score an aluminum piston during installation.

Reusing Piston Assemblies

Aluminum alloy trunk-type pistons must be in-spected with some precision before you can consider reusing them. Forged steel trunk-type pistons can often be reused, but these must also be carefully inspected. In all cases, the manufacturer's recommended practice should be observed. For engines under warranty, technicians should check to see whether piston re-placement is covered. If you are planning to reuse a piston, here are some common practices:

- Clean crystallized carbon out of the ring grooves using a correctly sized ring groove cleaner. If a used top compression ring can be broken (many cannot; use a grinder!), file it square and use this.
- The ring groove is correctly measured with a new ring installed square in the cylinder bore using feeler gauges and measuring to OEM specification.
- Ring end gap is measured by inserting the ring by itself into the cylinder bore and measuring the gap with feeler gauges. Remember, check OEM specifications and *always* measure new rings before installation.

Piston Thrust and Antithrust Side Identification. The piston thrust side is that half of the piston, when divided at the wrist pin pivot, that is on the inboard side of the crank throw during the downstroke. For a typical engine that rotates clockwise when viewed from the front, the major thrust side is the right side of the piston observed from the rear of the engine. The opposite side is known as the **antithrust side**. The terms **major thrust side** and **minor thrust side** are also used. Check out **Figure 3-2** and make sure you understand this; it could be the key to effective troubleshooting later.

CONNECTING RODS

Connecting rods are also known as con rods. They connect the piston with the throw on the crankshaft. The end of the connecting rod that links to the piston wrist pin is known as either the **rod eye** or the **small end**. The other end of the connecting rod links it to the crankshaft throw. This is known as the **big end**. Both the rod eye and big end have bearing surfaces. In this

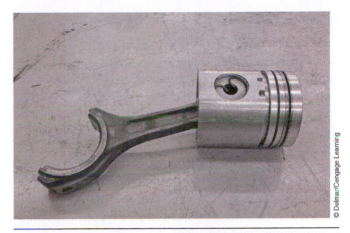

Figure 3-12 Aluminum piston and connecting rod assembly.

way the linear force that acts on the piston (cylinder pressure) can be converted to torque by the crank throw. **Figure 3-12** shows an aluminum trunk-type piston and its connecting rod. This design uses an offset big end, and the rod cap is not shown.

Cracked Rods

Cracked rod technology has been used in auto racing applications for many years, but it has now be-come common in diesel engines. Some diesel manu-facturers refer to cracked rods as **fractured rod** technology. Cracked rods have a big end that is ma-chined in one piece. After machining, the rod big end is fractured. Depending on the materials used, the fracture process may take place at room temperatures or when the rod is frozen to subzero temperatures. Fracturing requires a separation across the diameter of the rod big end. This produces rod and rod cap mating faces that appear rough but form a perfect final-fit alignment. Providing they are assembled properly, cracked rods make the procedure of checking rod sideplay after as-sembly unnecessary. **Figure 3-13** shows the cracked rod assembly used on a Ford 2011 6.7L PowerStroke engine.

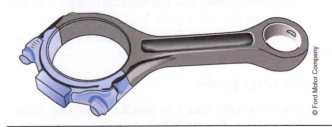

Figure 3-13 Cracked rod assembly used on a Ford 2011 PowerStroke engine. Note that the big end is offset.

Connecting Rod Construction

Most rods use an I-beam section design. The majority are rifle drilled from big to small ends to carry engine oil from the crank throw up to the wrist pin for purposes of lubrication and cooling. **Keystone rods** have become more common on new-generation engines that use high cylinder pressures. The keystone rod increases the area of the small end on which the greatest amount of force has to act. A keystone rod has a wedge-shaped rod eye or small end. The con rod shown in **Figure 3-13** has a keystone small end. Connecting rods are subjected to two types of loading: compressional and tensional.

Compressional Loading

During the compression and power strokes of the cycle, the connecting rod is subjected to **compressional loads**. When something is under compression, it is squeezed. Compressional loads on a rod can be calculated by knowing the cylinder pressure and the piston sectional area. Because of this, connecting rods seldom fail due to compressional overloading. When they do, it is usually coincidental with another failure such as hydraulic lock. Hydraulic lock is rare and usually results from cylinder head gasket failure that has allowed coolant to leak into the cylinder.

Tensional Loading

Tensional loading is stretching force. At the completion of each stroke, the piston has to stop each time it passes through TDC or BDC. This reversal of motion occurs nearly 140 times per second in each connecting rod when an engine is run at 4,000 rpm. The heavier the piston, the greater the tensile stress on the rod and crank throw. Tensile stresses on connecting rods increase with engine rpm because piston speeds increase. Any time an engine is overspeeded, the increased tensile loading on connecting rods can result in a tensile failure.

Offset Big End Caps. Many OEMs offset (from horizontal) the mating faces of the two halves of the big end. Doing this ensures that the rod cap fasteners do not have to sustain the full tensile loading of the rod. The con rod shown in the rod fixture in **Figure 3-14** uses an offset big end.

Inspecting Rods

Connecting rods should be handled with care. When assembling the rod to the piston, a brass jaw vise and light clamping pressure should be used. Slight nicks and scratches on connecting rods can turn into stress points

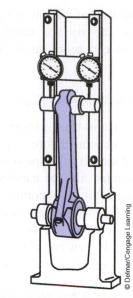

Figure 3-14 Connecting rod fixture.

that eventually cause a separation failure. Most diesel engine OEMs suggest that connecting rods be electromagnetic flux tested *every* time they are removed from the engine. The cost of magnetic flux examination is small compared to the damage caused by a rod failure. When a connecting rod fails in a running engine, the result is often a rod driven through the cylinder block casting. OEMs recommend that connecting rods that fail the inspection procedure be replaced rather than reconditioned. When rods are reconditioned, you should note that removing material from a con rod changes its weight. This changes the dynamic balance of the engine. **Figure 3-14** shows a rod inspection fixture.

Replacing Rods. Because reconditioning of rods is not widely practiced, connecting rods should be inspected using OEM guidelines, and if rejected, replaced. When replacing rods, they should be weight matched precisely, observing the manufacturer's recommendations. The consequence of replacing a connecting rod with one of either greater or lesser weight is an unbalanced engine. Manufacturers usually code connecting rods to a weight class, and typically each weight class has a window that varies with the size of the engine. This variability ranges from around 0.5 to 1.5 ounces (13 to 40 grams). When replacing defective connecting rods, always attempt to match the weight codes.

Rod Cap Fasteners. Most manufacturers prefer that the rod cap fasteners be replaced at each reassembly. However, it is a practice that is seldom observed. The fasteners should be replaced with the correct OEM fastener and not cross matched to an SAE-graded bolt.

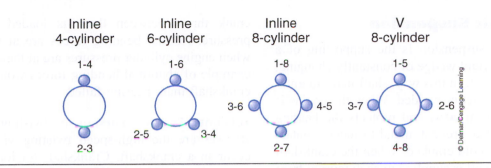

Figure 3-15 Crank throw configurations.

Deformed threads and stretching make these bolts a poor reuse risk, considering the costly consequences of fastener failure at the rod cap. The actual chances of failure are low, but remember technicians would do well to remind themselves that the consequences of a connecting rod separation are severe. If the work is being performed for a customer, recommend that they be replaced and leave the decision to reuse the fasteners to the customer. If you are a betting person, we should say that your odds are better when reusing fasteners with an offset big end cap.

Connecting Rod Bearings.   Most engine manufacturers use a single-piece bushing, press fit to the rod eye, and two-piece friction **bearing shells** at the big end. Rod eye bearings should be removed using an appropriately sized driver and arbor press. Using a hammer and any type of driver that is not sized to the rod eye bore is not recommended because the chances of damage are high. New one-piece bushings should be installed using a press and mandrel, ensuring that the oil hole is properly aligned, and then sized using a broach or **hone**. Split big end bearings should also be installed respecting the oil hole location. Bearing shells and rod eye bushings should both be installed to a clean, dry bore. Remove any packing protective coating from the bearings by washing them in solvent, followed by compressed air drying.

Tech Tip: Rod sideplay has to be checked after a rod cap has been torqued to the rod. Cocking of the rod cap on the big end can cause an engine to bind and damage the crankshaft by scoring the web cheeks. In applications in which cracked rods are used, this check is unnecessary as a perfect rod cap to rod fit can be assumed. Snapping the assembly fore and aft on the journal should produce a clacking noise indicating sufficient sideplay.

CRANKSHAFTS AND BEARINGS

A crankshaft is a shaft with offset throws or journals to which piston assemblies are connected by means of connecting rods. **Figure 3-15** shows some typical crank throw configurations. Note that in the case of V-configured engines, OEMs may use different numbering sequences than those shown here. **Figure 3-16** shows the cylinder numbering sequence preferred by Ford in its diesel engines. When a crankshaft is rotated, the offset crank throws convert the linear, back-and-forth movement of the pistons into rotary motion at the crankshaft. This works in the same way that the pedals function on a bicycle crankshaft. Crankshafts are supported by bearings at main journals. These main bearings require pressurized lubrication at all times the engine is run because crankshafts rely on **hydrodynamic suspension**.

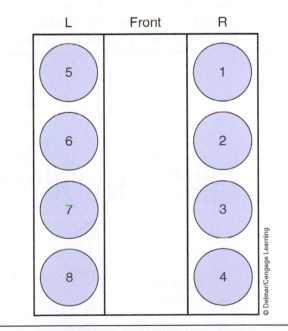

Figure 3-16 Cylinder numbering sequence used by Ford in their 6.7.

Hydrodynamic Suspension

Hydrodynamic suspension is the supporting of a rotating shaft on a fluid wedge of constantly changing, pressurized engine oil. In this way, shaft main journal-to-bearing bore contact is avoided. When the engine is stationary, an oil film coats and protects the bearing shell. It should also prevent metal-to-metal contact when the engine is cold cranked. When the crankshaft begins to turn, lube oil is supplied to each journal from oil passages in the cylinder block. This oil is pumped into the bearing and the main journals of the crankshaft. As the crankshaft rotates, its journals are rolled into this oil, creating the wedge required to hydrodynamically support the crankshaft. Angling the oil hole in the journal can help develop a thicker wedge of oil that better supports the crankshaft.

Dynamic Balance

Crankshafts are designed for dynamic balance and use counterweights to oppose the unbalancing forces generated by the pistons. These unbalancing forces reduce as the number of cylinders in an engine increases and companion throws (e.g., in an inline six-cylinder engine, 1 and 6, 5 and 2, 3 and 4) have a counterbalancing effect. Crankshafts are subjected to two types of force: bending forces and torsional forces.

Bending Forces. Bending stress occurs between the main journals between each power stroke. Crankshafts are designed to withstand the bending forces that result from the compression and combustion pressures developed in the cylinder. This normal bending stress takes place between the main journals at any time the crank throw between them is loaded by cylinder pressure. Normal bending forces are at their highest when engine cylinder pressures are at their highest. An example of abnormal bending forces would occur if a crankshaft main bearing failed.

Torsional Forces. Torsion is twisting. **Torsional stresses** are the high-speed twisting vibrations that occur in a crankshaft. Crankshaft torsional vibration occurs because when a crank throw is under compression (that is, driving the piston upward on the compression stroke) it slows to a speed just a little less than average crank speed. This same throw upon receiving the power stroke (from the piston) forces, accelerates to a speed just a little higher than average crank speed. These twisting vibrations take place at high frequencies and crankshaft design, materials, and hardening methods must take them into account.

Torsional stresses on a crankshaft tend to peak at crank journal oil holes at the flywheel end of the shaft. Torsional vibrations are amplified when an engine is run at slower speeds with high cylinder pressures, such as when an engine is lugged. This is because the real time interval between cylinder firing pulses is longer. Running an engine at lower speeds with high cylinder pressures is known as **lugging**. Torsional vibrations from the crankshaft can be amplified through the chassis drivetrain. This makes the proper matching transmissions and final drive carriers essential.

Crankshaft Construction

Figure 3-17 is a guide to crankshaft terminology. Understanding these terms is critical when measuring

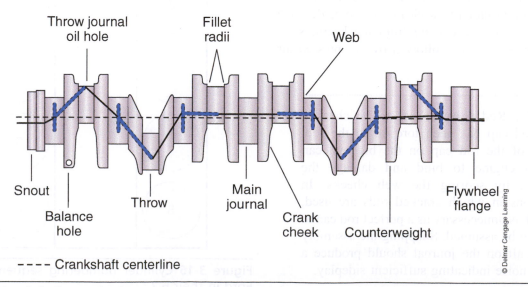

Figure 3-17 Crankshaft terminology.

and machining crankshafts. Diesel engine crankshafts today are made of cast iron, cast steel alloy, or drop-forged steel. In manufacture, crankshafts are usually tempered (heat treated) to provide a tough core with the ability to flex just enough to sustain the bending and torsional punishment they are subject to. Most diesel engine OEMs use special processes to forge crank-shafts. Most maintain some secrecy about the details of the process. An understanding of journal hardening procedures is important as the reconditionability of the crankshaft depends on this.

Journal Surface Hardening Methods.   There are basically three methods of surface hardening main and rod journals.

- Flame hardening. Used on plain carbon and sometimes middle alloy steels. Flame hardening consists of the direct application of heat followed by quenching in oil or water. This produces relatively shallow surface hardening. The actual hardness of the crankshaft depends on the amount of carbon and other alloys in the steel. Note: Flame is sometimes used as protection in induction hardening treatment ovens because it prevents air exposure to the crankshaft during the induction hardening process.

- Nitriding. Used on alloy steels. Involves higher temperatures than flame hardening and surface hardens to a greater depth—around 0.65 mm (0.025 inch). This provides a small margin for machining, but the process is mostly used on smaller-bore diesel crankshafts.

- Induction hardening. Area to be hardened is enclosed by an applicator coil through which alternating current (AC) is pulsed, heating the surface. Tempering is achieved by blast air or liquid quenching. This process results in hardening to depths of up to 1.75 mm (0.085 inch), providing a much wider wear and machinability margin than the previous two methods. Most current diesel crankshafts are surface hardened by this method. Sometimes during the process, flame is used in the induction oven to prevent air exposure to the crankshaft while it is being induction hardened. **Figure 3-18** shows the induction hardened, forged steel crankshaft used in the V-configured, Ford 6.7L engine.

Removing of Crankshaft from Cylinder Block.   The cylinder block should be in an engine stand and in an upside down position. After the main bearing caps and

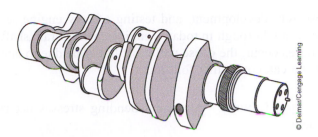

Crankshaft

The following improvements have been made to the crankshaft.

- Improved crank quality
- Fillet radius on each journal
- Forged modular steel
- Undercut rolled fillet radius
- Fully lightened crankshaft pins
- Two radiused counterweights
- One piece rear flange:
  – increase torque capabilities
  – improved sealing and balance
- Shrink fit assembled front drive gear
- Single mode torsional damper
- Direct accessory drive for improved NVH

Figure 3-18 Forged steel crankshaft used in the Ford 6.7L engine. (*Text courtesy of Ford Motor Company*)

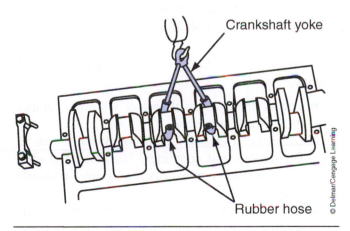

Figure 3-19 Crankshaft removal from cylinder.

any other obstructions have been removed, a rigid crankshaft yoke should be fitted to a hoist. The yoke prongs should be fitted with rubber hose (to prevent damage to the journals) and hooked to a pair of throws in the center of the crankshaft (for weight balance). **Figure 3-19** demonstrates how a yoke connects the crankshaft to a hoist.

Crankshaft Failures

Only a small percentage of crankshaft failures result from manufacturing and design problems. Because research, development, and testing of new engine series is so thorough in today's engines, when crankshaft failures occur, the cause will most likely be due to one of the categories that follows.

Bending Failures. Abnormal bending stresses occur when:

- Main bearing bores are misaligned. Any kind of cylinder block irregularity can cause abnormal stress.
- Main bearings fail or become unevenly worn.
- Main caps are broken or loose.
- Standard specification main bearing shells are installed where an oversize is required.
- Flywheel housing is eccentrically (relative to the crankshaft) positioned on the cylinder block. This produces a broken-back effect on the drivetrain. The condition usually results from a failure to (dial) indicate a flywheel housing on reinstallation.
- Crankshaft is not properly supported either out of the engine before installation or while replacing main bearings in-chassis. The latter problem is more likely to damage the block line bore, but it can also deform the crankshaft.

Bending failures tend to initiate at the main journal fillet and extend through to the throw journal fillet at 90 degrees to the crankshaft axis.

Torsional Failures. Causes of crankshaft torsional failures:

- Loose, damaged, or defective vibration damper or flywheel assembly.
- Unbalanced engine-driven components such as fan pulleys and couplings, fan assembly, idler components, compressors, and power take-offs (PTOs).
- Engine overspeed. Even a slight engine overspeed can create enough torsional stress to cause a crankshaft failure.
- Unbalanced cylinder loading. A dead cylinder or fuel injection malfunction of over- or underfueling a cylinder(s) can result in a torsional failure.
- Defective engine mounts. This can produce a "shock load" effect on the whole powertrain.

It should be remembered when performing failure analysis that the event that caused the failure and the actual failure may be separated by a considerable amount of time. High torsional stress produces fractures occurring from a point beginning in a journal oil hole extending through the fillet at a 45-degree angle or circular severing right through a fillet. In inline 6- and V8-cylinder engines, the rear journal oil holes are more likely to suffer from torsional failures when the crankshaft is torsionally overloaded.

Spun Bearing(s). Spun bearings are lubrication-related failures. They are caused by a lack of oil in one or all of the crank journals. The friction created in the bearing causes it to surface weld itself to the crank journal. It either rotates with the journal in the bore or continues to scuff the journal to destruction. When a crankshaft fractures as a result of bearing seizure, the surface of the journal is destroyed by excessive heat, and it fails because it is unable to handle the torsional loading. Causes of spun bearings and bearing seizure include:

- Misaligned bearing shell oil hole.
- Improper bearing-to-journal clearance. This produces excessive bearing oil throw off, which starves journals farthest from the supply of oil.
- Sludged lube oil causing restrictions in oil passages.
- Contaminated engine oil. Fuel or coolant in lube oil will destroy its lubricity.

Etched Main Bearings. Etched main bearings are caused by the chemical action of contaminated engine lubricant. Chemical contamination of engine oil by fuel, coolant, or sulfur compounds can result in high acidity levels that can corrode all metals. The condition is usually first noticed in engine main bearings. It may result from extending oil change intervals well beyond those recommended. Etching appears initially as uneven, erosion pock marks, or channels.

Crankshaft Inspection

Most diesel engine manufacturers recommend that a crankshaft be magnetic flux tested at every out-of-chassis overhaul. This process requires that the component to be tested be magnetized and then coated with minute iron filings. When AC current is pulsed through the shaft, the magnetic lines of force that result will bend into a crack or nick, causing the iron filings to collect in the flaw.

Tech Tip: Most small cracks observed when magnetic flux testing crankshafts are harmless, but beware of fillet cracks and cracks that extend into oil holes.

Visual Inspection. Following magnetic flux inspection, the crankshaft thrust surfaces should be checked for wear and roughness. It is usually only necessary to polish or dress these areas using the appropriate crank machining equipment. The front and rear main seal contact areas of the crankshaft should be checked for wear. Wear sleeves, interference fit to the seal race, are available for most engines and can be easily installed without special tools. It is a testament to the quality of modern manufacturing that most crankshafts in today's diesel engines require no special attention during the life of the engine. When checking a crankshaft, always use the manufacturer specifications.

Reconditioning Crankshafts. When a crankshaft does fail, most diesel engine manufacturers do not approve of any but their own in-house reconditioning practices. That said, crankshafts are routinely reconditioned in the aftermarket. In most cases, the reconditioning procedures try to preserve the original surface hardening, but where the damage is severe, such as in the example of a spun bearing, journal damage can penetrate deeper than the hard surfacing. While journal surfaces can be rehardened, this practice is not common. The reconditioning methods are simply outlined here, and technicians should be aware that most OEMs regard them as bad practice. There are four basic methods of crankshaft reconditioning:

1. Grinding to undersize. This may require oversize bearings: these are not always available from engine OEMs.
2. Metallizing journal surfaces followed by grinding to the original size.
3. Chroming surface to original size.
4. Submerged arc welding buildup followed by grinding to original size.

Front and Rear Main Seals. The crankshaft is sealed at both the front and rear of the engine by what are known as main seals. The mains seals are located at the snout (front) and flywheel flange (rear) of the crankshaft. Because hydrodynamic suspension requires a large volume of oil to be charged to the main bearings, the front and main seals must seal effectively. Front and main seals usually consist of a spiraled slinger (used to redirect oil away from the seal) and dynamic rubber seal. Main seals, especially the rear main seal, tend to be a not uncommon failure, especially in workhorse diesel engines. The procedure for replacing is simple enough, but in the case of the rear main, labor intensive because the transmission assembly has to be removed to access it. **Figure 3-20** shows the front main seal configuration on the Ford 6.7L engine.

Front crankshaft seal

© Ford Motor Company

Figure 3-20 Front main seal configuration used by Ford in their 6.7. The front seal and slinger are pressed into the casing bore together.

ROD AND MAIN BEARINGS

You should have a basic understanding of bearing construction because it can help you to diagnose failures. Most diesel engine manufacturers make available excellent bearing failure analysis charts and booklets, some of which are available on online service information systems (SIS).

Construction and Design

Two basic designs are used in current applications:

1. Concentric wall—uniform wall thickness
2. Eccentric wall—wall thickness is greater at crown than parting faces; also known as deltawall bearings

Materials. Rod and main **friction bearings** usually have a steel base or backing plate. Onto this is layered copper, lead, tin, aluminum, and other alloys. Friction bearings are designed to have embedability. This means that the outer face of the bearing must be soft enough to allow small abrasives to penetrate the outer shell, known as the overlay, to a depth at which they will cause a minimum amount of scoring to the crank journals that rotate in the bearing.

Bearing Clearance

The engine OEM's specification for bearing clearance must be observed. Bearing clearance is measured with Plastigage using the methods shown in **Figure 3-21** and **Figure 3-22**. Most manufacturers make several oversizes (of bearings) to accommodate a small amount of crankshaft journal wear or machining. It is important not to assume that a new engine will always have standard-size bearings. Typical bearing clearances in highway diesel engine

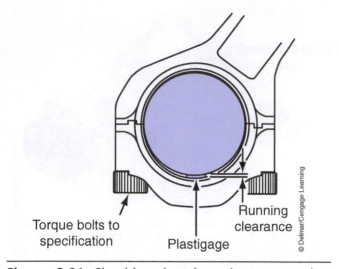

Figure 3-21 Checking bearing clearance using Plastigage.

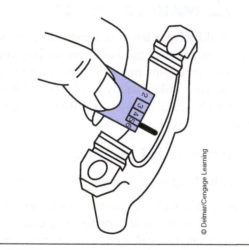

Figure 3-22 Correct method of locating Plastigage.

applications run from 0.035–0.100 mm (0.0015–0.004 inch). The ability to maintain hydrodynamic suspension of the crankshaft decreases as bearing clearances increase. When bearing clearance increases above specification, a drop in oil pressure results that may indicate the need for an in-chassis bearing rollover. Increased bearing clearance also increases oil throw off, and this may result in excessive lube being thrown up onto cylinder walls.

You should not attempt to measure bearing clearance when the engine is in-chassis because the results are not that accurate. This is because of the flexibility of crankshafts when not fully supported at each main bearing. The engine should be upside down and level so that there is no weight load acting on the retaining cap of the bearing being measured.

Interpreting Bearing Clearance Specs. Check the bearing clearance specifications first. Sometimes bearing clearances are expressed in service specs in ten thousandths of an inch because they are engineering specs. When this occurs, it is safe to round the value to the nearest half thousandth (0.0005 inch) and try to work in units of thousandths. For instance, an OEM bearing clearance spec of 0.0027 inch can be expressed as 2½ thou. Next select the Plastigage color code capable of measuring between the range of specifications. **Plastigage™** is soft plastic thread that easily squishes to conform to whatever clearance space is available when compressed between a bearing and journal. The crushed width can then be measured against a scale on the Plastigage packaging. The lower the clearance, the wider the Plastigage strip will be flattened. Check out the measuring process shown in **Figure 3-22.**

Using Plastigage™. A short strip of Plastigage should be cut and placed across the center of the bearing in line with the crankshaft. The bearing cap with the bearing shell in place can then be installed and torqued to specification. Do not rotate the engine with the Plastigage in place. Next the bearing cap and shell should be removed and the width of the flattened Plastigage checked against the dimensional gauge on the Plastigage packaging. If clearance is within specifications, carefully remove the Plastigage from the journal before reinstalling the cap and shell assembly.

Selecting the Correct Plastigage Strip. Plastigage is manufactured in four sizes, each color-coded for the range of clearances it is capable of measuring. **Table 3-1** identifies each Plastigage color-coded size window in

TABLE 3-1: PLASTIGAGE™ CODE INDENTIFICATION			
Part Number	**Color**	**Standard Dimension Window**	**Metric Dimension Window**
HPG1	Green	0.001–0.003 in.	0.025–0.76 mm
HPR1	Red	0.002–0.006 in.	0.051–0.152 mm
HPB1	Blue	0.004–0.009 in.	0.102–0.229 mm
HPY1	Yellow	0.009–0.020 in.	0.230–0.510 mm

both standard and metric dimensions. While most diesel OEMs still display their specifications using standard measurements, some display in both standard and metric, and one uses metric-only specifications.

Crankshaft Endplay

After installation of the crankshaft into the cylinder block, the amount of fore and aft movement of the crankshaft is known as endplay. Two methods are used to define crankshaft endplay:

- **Thrust bearings**: One of the main bearings is flanged.
- **Thrust washers**: A set of split rings is inserted around one set of main bearings.

Thrust bearings and thrust washers both accomplish the same thing. Depending on the engine, endplay specifications typically fall into the 0.06 to 0.15 mm (0.0025 to 0.012 inch) range.

Measuring Endplay. A dial indicator should be used to measure endplay. It should be positioned on a linear plane with the crankshaft, preferably at either the front or back but if it is inaccessible, place it on a crank cheek. A lever should be used that is not going to damage the crankshaft; a large screwdriver used gently should be okay. The crankshaft should be levered fore and aft while observing the indicator measurement.

Bearing Retention

Bearings are retained primarily by *crush*. The outside diameter of a pair of uninstalled bearing shells slightly exceeds the bore into which it is installed. This creates radial pressure that acts against the bearing halves and provides good heat transfer. The bearing halves may also be slightly elliptical: this helps hold the bearing in place during installation. It provides what is known as bearing spread. Some bearings use tangs. Tangs in bearings are inserted into notches in the bearing bore: the tangs reduce movement, prevent bearing rotation, and align oil holes. **Figure 3-23** shows the tangless friction bearing shells used in the Ford 6.7L engine.

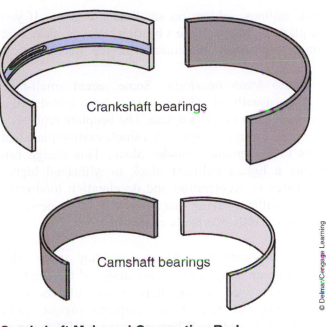

Crankshaft bearings

Camshaft bearings

© Delmar/Cengage Learning

Crankshaft Main and Connecting Rod Bearings

Both the crankshaft main and connecting rod bearings are color coded and a tangless design.

The lower half of the crankshaft main bearings are a dark gray color while the upper half is a bright metal with a lubrication groove and a slot for oil to flow through.

The upper half of the connecting rod bearings are dark gray while the lower half is a bright metal with no grooves.

Figure 3-23 Rod and main bearings used in the Ford 2011 PowerStroke: the bearings use a tangless design. (*Text courtesy of Ford Motor Company*)

Bearing Removal and Installation

Service literature should always be consulted and its procedures observed. This operation is very straightforward when performed out of chassis because the crankshaft is removed when the cylinder block side bearing shell is installed, and the technician is working above the engine with perfect visibility and accessibility for the rest of the procedure. Some manufacturers permit what is known as a *bearing rollover* to be performed on their engines. A bearing rollover consists of replacing a set of rod and main bearings while the engine is in-chassis and the crankshaft is in place. The usual reason for undertaking a bearing rollover is a drop-off in engine oil pressure, but the reality of today's friction bearings is that it is seldom necessary and is usually only performed on heavy duty truck engines, seldom on small-bore engines. If you are planning to undertake a bearing rollover, make sure you consult the manufacturer's service literature.

Bedplate Main Bearings. Some recent small-bore engines, usually of offshore origin, have introduced a lower cylinder block bedplate. The bedplate replaces a set of main bearing caps with a single casting plate that bolts to the engine cylinder block. This design can enable a lighter cylinder block to withstand higher resistance to acceleration and deceleration torsionals. It goes without saying that when a bedplate design is used, in-chassis overhauls cannot be done.

A Note on Fasteners. Many steel alloy and usually all aluminum alloy fasteners are designed for one-time usage. Many engine fasteners require the use of a **torque-to-yield (TTY)** or **template torque (TT)** tightening procedure. When installing a TTY fastener, it is initially tightened to a specified torque value using a torque wrench: then, final torquing is completed by turning the fastener through a set number of degrees. TTY and TT produce more consistent clamping pressures. The reason that many of these fasteners are specified as single use is that they are deformed when torqued to specification. The risk of reusing such fasteners is high, especially with aluminum fasteners.

VIBRATION DAMPERS

Vibration damper and harmonic balancer both refer to the same component, but the term *vibration damper* tends to be used by most diesel engine manufacturers. A vibration damper is mounted on the free end of the crankshaft, usually at the front of the engine. Its function is to reduce twisting vibrations and to add to flywheel mass in establishing rotary inertia. In other words, its main function is to reduce crankshaft torsional vibration (there is a full explanation of crankshaft torsional vibration earlier in this chapter under ''Crankshafts and Bearings'').

A typical vibration damper consists of a damper drive or housing and inertia ring (**Figure 3-24**). The housing is coupled to the crankshaft and using springs, rubber, or viscous medium, drives the inertia ring: the objective is to drive the inertia ring at *average* crankshaft speed. Vibration dampers therefore have three main components:

- Drive member: bolted to crankshaft
- Drive medium: either a fluid (silicone gel) or solid rubber
- Driven member: an inertia ring

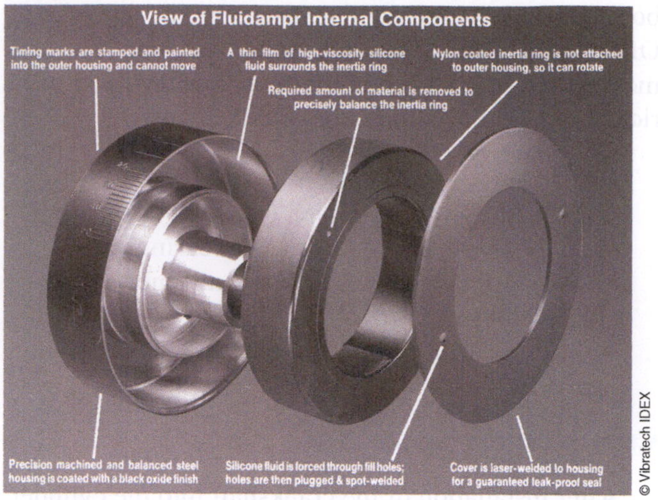

Figure 3-24 Exploded view of a viscous vibration damper.

Two types are used in current diesel engines:

- Viscous-drive type
- Solid rubber-drive type

Vibration Damper Construction

Viscous Drive Operating Principle. In a viscous-drive–type vibration damper, the drive member is bolted to the crankshaft and consists of a ring-shaped, hollow housing. Inside the hollow housing is a solid steel inertia ring. This inertia ring is suspended in and driven by silicone gel. At least it is gel when cold. But as it warms to operating temperature, the silicone drive medium becomes more fluid. Because the drive member is bolted to the front of the crankshaft, it moves with whatever torsional vibrations occur in this location. When the drive member rotates, it drives the drive medium (silicone gel) and inertia ring inside of it. Because the inertia ring contains most of the weight of the vibration damper, it rotates at *average* crankshaft speed. This means that it will be subject to the twisting effects occurring at the front of the crankshaft. Because the speeds of the drive member and inertia ring will differ, shear action will take place in the drive medium, the silicone gel. This shearing action of the viscous fluid film between the drive housing and the inertia ring will help smooth the twisting forces occurring at the front of the crankshaft. It would be correct to say that the inertia ring was hydrodynamically supported by the silicone medium.

Replacing Viscous Vibration Dampers. Most manufacturers recommend the replacement of a vibration damper at each major overhaul. In-house overhauls usually result in the automatic replacement of the vibration damper. The consequences of not replacing the damper when scheduled can be costly as it can result in a failed crankshaft. Viscous vibration dampers usually fail for one of the following two reasons:

- Damage to the outside housing, causing a leak or locking the inertia ring.
- Breakdown of the viscosity of the silicone drive medium.

CAUTION *Recommend that a viscous-type vibration damper be replaced at engine overhaul regardless of its external appearance. Explain that this is a manufacturer recommendation. If the customer declines, he has made the decision, not you. A failed vibration damper can cause crankshaft failure.*

Solid Rubber Vibration Dampers.   Solid-rubber–type vibration dampers are less effective at dampening torsionals through a wide rpm and load range on diesel engines, but they are widely used due to their lower cost. This type consists of a drive hub bolted to the crankshaft: a rubber compound ring is located between the drive hub and the outer inertia ring and bonded to each. The outer inertia ring contains most of the mass of the unit. The rubber compound ring therefore acts both as the drive and the damping medium. The elasticity of rubber enables it to function as a damping medium, but the internal friction generates heat, which eventually hardens the rubber and renders it less effective and vulnerable to shear failures. When cracks appear in the rubber it is usually an indication that it is time to replace it.

FLYWHEELS

The engine flywheel on a typical diesel engine is normally mounted at the rear of the engine. It has three basic functions:

1. Store kinetic energy (the energy of motion) in the form of inertia. This helps to smooth out the power pulses that occur in an engine as each cylinder fires and to establish an even crankshaft rotational speed.
2. Provide a mounting for engine output: It is the power takeoff device to which a clutch or torque converter is bolted.
3. Provide a means of rotating the engine with a starter motor.

Inertia

The inertia role of a flywheel relates to its ability to store energy. As an energy storage device, the flywheel plays a major role in smoothing the twisting vibrations that act on the rear of the crankshaft during engine operation. The weight of the flywheel helps rotate the engine between firing pulses. The actual weight of a flywheel depends primarily on two factors:

- number of engine cylinders
- engine operating rpm range

Because the number of crank angle degrees between power strokes on an eight-cylinder engine is twice the number on a four-cylinder engine, the eight-cylinder engine requires relatively less flywheel mass. Engines designed to be run at consistently high rpms also require less flywheel weight, while workhorse off-highway engines tend to use relatively heavier flywheels. Single-cylinder diesel engines used on farms and small boats for pumps, gensets, and propulsion often use two flywheels that comprise 60 percent of the total engine weight.

Flywheel Housings.   Flywheel housings are categorized by size, shape, and bolt configuration on diesel engines. They may comply with SAE standard dimensions, in which case some versatility in the selection of clutches, flexplates, and transmissions is provided. However, some are manufacturer specific, in which case they can only be mated to a manufacturer bell housing or power takeoff.

Ring Gear Replacement

A ring gear is shrunk fit around the outside of the flywheel. The ring gear is machined with external teeth. This allows the pinion on the starter motor to connect with the flywheel and rotate it to crank the engine. A worn or defective ring gear can be removed from the flywheel by first removing the flywheel from the engine and then using an oxyacetylene torch to partially cut through the ring gear, working from the outside on a single tooth. Cutting of a single tooth is usually sufficient to expand the ring gear so that the removal can be completed using a hammer and chisel. Care should be taken to avoid heating the flywheel any more than absolutely necessary or damaging the flywheel itself by careless use of the oxyacetylene flame. A hacksaw and chisel may also do the job, but this will require some care and a lot of sweat if no heat is to be used.

Installing a New Ring Gear.   To install a new ring gear, place the flywheel on a flat, level surface, and check that the ring gear seating surface is free from dirt, nicks, and burrs. Ensure that the new ring gear is the correct one. If its teeth are chamfered on one side, they will face the cranking motor pinion after installation. Next, the ring gear must be expanded using heat so that it can be shrunk to the flywheel. Most manufacturers specify a specific heat value because ring gears are heat treated, and overheating will damage the tempering and substantially reduce the hardness. A typical temperature specification would be around 200°C (400°F), but this may be higher in some cases.

Because of its size, the only practical method of heating a ring gear for installation is using a rosebud-type (high gas flow) oxyacetylene heating tip. To ensure that the ring gear is heated evenly to the specified temperature and especially to ensure that it is not overheated, the use of a temperature-indicating crayon such as a Tempilstick is recommended. When the ring gear has been heated evenly to the correct temperature, it will usually drop into position and almost instantly contract to the (cold) flywheel. Handle a hot ring gear with blacksmith tongs.

Reconditioning and Inspecting Flywheels

Flywheels are commonly removed from engines for reasons such as clutch damage, leaking rear main seals, or leaking cam plugs. Care should be taken when both inspecting and reinstalling the flywheel and the flywheel housing. Flywheels should be inspected for:

- face warpage (straight edge and thickness gauges)
- heat checks (visual)
- scoring (visual)
- axial and radial runout (dial indicator)

Damaged flywheel faces may be machined using a flywheel resurfacing lathe to manufacturer-specified tolerances. Typical maximum machining tolerances range from 0.030 inch to 0.090 inch (0.75 mm to 2.30 mm). It is important to note that when you are resurfacing pot-type flywheel faces, the pot face must have the same amount of material ground away as the flywheel face. If you do not do this, the clutch will not function.

ENGINE CYLINDER BLOCK

The engine **cylinder block** is the frame of the engine around which all the other components are assembled, in much the same way that subcomponents are assembled around a truck frame. The cylinder block houses the engine cylinders and the engine

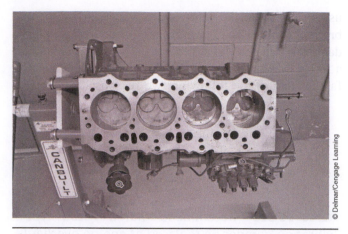

Figure 3-25 Overhead view of a four-cylinder Perkins engine.

crankcase. Diesel engine cylinder blocks differ according to the number of cylinders and whether the engine is inline or V-configured. Whether inline or V-configurations are used, in most cases a single crankshaft is required, and a small-bore cylinder block is usually cast as a single unit. **Figure 3-25** is an overhead view of a four-cylinder Perkins mounted in a stand in the process of being stripped down for reconditioning. This engine uses a simple gray cast iron cylinder block with parent bores.

Cylinder Block Forces

Cylinder blocks today have to be lighter and stronger than those of a generation ago because low-emission, fuel-efficient engines usually run higher cylinder pressures. Aluminum alloys are used in many imported diesel engines designed as automobile power plants, as are special alloy cast irons. Lighter cylinder blocks have the advantage of lower weight but are also more susceptible to torque twist when an engine is producing peak torque. Because of the undesirable flexibility of lightweight cylinder blocks, some are braced with plates to limit flexing. The forces that a cylinder block is subject to are:

- Torque twist: This occurs when twisting force from the crankshaft anchors through the cylinder block. A cylinder block can be subject to torque twist from either crankshaft input or crankshaft output. The condition can occur at high cylinder pressures or it can be generated through the drivetrain.
- Cylinder pressures: Excessively high cylinder pressures, especially in small parent bore engines, can generate failures. Today, combustion pressures tend to more than double those of the engines of a generation ago.

■ Sudden changes in temperature: This occurs when a hot cylinder block is cooled rapidly, either when immediately shut down after a hard run cycle or when splashed with cold water in a wash bay.

Cylinder Block Construction

A majority of current small-bore diesel engines use a parent block design, meaning that the cylinder bore is machined directly into the cylinder block. In some cases, the cylinder bore is toughened using a tempering process. Cylinder liners or sleeves are common in heavy duty diesels but are more likely to be a reconditioning option in light duty diesel engines. The advantage of using cylinder liners is ease of reconditioning because they can easily be replaced when the engine is overhauled.

Parent Bore. Engines that do not use liners or sleeves are usually known as parent bore engines. In a parent bore engine, the cylinder bore is machined directly into the cylinder block. Both aluminum alloy and cast iron parent bore blocks exist. The life of heavy duty parent bore engines can be increased by induction hardening the surface of the bore, as is done with many of the current Mercedes-Benz (MB) lineup of four-, five-, and six-cylinder engines. In the event that the cylinder bores have to be resurfaced in these engines, MB recommends that they be returned to them for reconditioning.

Parent bores in cylinder blocks can be bored to an oversize and fitted with sleeves, but this is obviously more costly than simply replacing a set of sleeves. The advantages of parent bore engines are:

■ Lower initial cost
■ No liner O-rings to fail
■ No liner protrusion specs to adhere to on reassembly

Crankcase. In addition to supporting the cylinders, a cylinder block must also support the crankshaft and flywheel. The crankshaft is located in a lengthwise bore in the cylinder block. It is supported in a cradle of main bearings in what is usually referred to as the crankcase. The flywheel housing is bolted to the rear of the cylinder block. The flywheel is bolted to the crankshaft and rotates within the flywheel housing.

Cylinder Block Design. Cylinder blocks may be bored to support the camshaft or camshafts and are cast with coolant passages and a water jacket. Even in a liquid-cooled engine, the cylinder block frame plays a major role in transferring engine heat to the atmosphere. Because all the other engine housing components are attached either directly or indirectly to the cylinder block, it must be designed to accommodate them.

Cylinder Block Materials. As indicated earlier, 20 years ago most diesel engines were manufactured from simple cast irons. Today's diesel engines are required to be lighter and many run at higher rpms and cylinder pressures. The materials used today are:

■ Gray cast iron (simple cast iron, still the most common cylinder block material)
■ Aluminum alloy (best suited to high-speed automobile applications)
■ Compacted graphite iron (used in some more recent engines such as the Ford 6.7 liter)

Compacted graphite iron (CGI) cylinder blocks have been a key to reducing engine weight. High-horsepower diesel engines today often weigh less than half as much as equivalent-power engines of just 10 years ago. **Figure 3-26** shows the CGI engine cylinder block used by the Ford PowerStroke 6.7 liter engine.

Cylinder Block Functions

Although there are some differences between manufacturers, a diesel engine cylinder block must perform some or all of the following:

■ House the piston cylinder bores
■ Support the crankshaft in main bearing bores
■ Support a cylinder block–located camshaft(s)
■ Incorporate coolant passages and a water jacket
■ Incorporate lubricant passages/drillings
■ Incorporate mounting locations for other engine components
■ Have sufficient surface area to dissipate heat if air cooled

Sleeves and Liners. The terms *sleeves* and *liners* are used interchangeably. Most heavy duty engines use sleeves and liners, but in light duty applications, their use tends to be limited to that of a repair option. However, you should be familiar with the terms wet and dry as they apply to liners. A wet liner is one with thicker walls that come into direct contact with the coolant in the cylinder block water jacket. Wet liners are not often found in light duty diesel applications. Dry sleeves are thinner walled than wet liners and are installed into the block bore. The dry sleeve does not

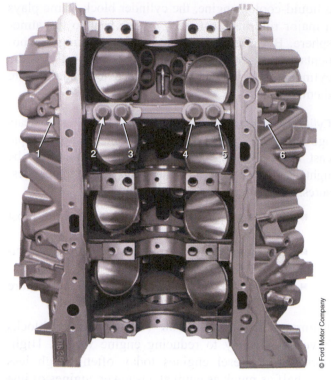

Engine Block

- The engine block is made from a compacted graphite iron (CGI), which enables best-in-class weight and improved noise, vibration, and harshness (NVH).
- Cam-in-block design with dry valley.
- Integrated direct mount for high pressure fuel pump.
- Six head bolts are used per cylinder and six bolts per main bearing cap.

Figure 3-26 Engine block used in a Ford PowerStroke 6.7 liter engine.

transfer heat as efficiently as the wet liner, but it is easily replaced and does not present coolant sealing problems.

When a parent bore fails and the repair strategy is to sleeve the engine, the original cylinder bore is machined to an oversize to permit the installation of a sleeve. The sleeve has an outside diameter that allows it to fit to the oversize bore machined to the cylinder block, and an inside diameter that is equivalent to the original specified bore of the engine cylinder.

Cylinder Sleeve Removal. Sleeves should be removed with a puller and adaptor plate or shoe. The shoe or adaptor plate must be fitted into the lower portion of the liner and must have an outside diameter just a little less than the inside diameter (ID) of the cylinder bore. Dry sleeves often require the use of mechanical, hydraulic, or air-over-hydraulic pullers.

Tech Tip: Removing pistons from cylinder liners can be made much easier by removing the carbon around the cylinder wear ridge that forms in use. The carbon in the wear ridge can be removed using a flexible knife blade followed by gentle use of emery cloth.

Checking a Cylinder Block

1. Strip block completely, including cup expansion and gallery plugs.
2. Soak in a hot or cold tank with the correct cleaning solution. This is important. A cylinder block hot soak solution designed for a cast iron block can destroy an aluminum block.
3. Check for scaling in the water jacket not removed by soaking. One OEM reports that 0.060 inch (1.5 mm) scale buildup has the insulating effect of 4 inch (100 mm) of cast iron.
4. Check for wear/erosion around the deck coolant ports and fire ring seats.
5. Electromagnetic flux-test the block for cracks at each out-of-chassis overhaul.

Final Inspection and Assembly

1. Check for deck warpage using a straightedge and thickness gauge. A typical maximum specification is approximately 0.004 inch, but refer to the manufacturer tolerances. The spec is usually less on CGI cylinder blocks.
2. Check the main bearing bore and alignment. Check the engine service history to ensure that the engine has not been previously line bored. If a master bar is available, use it. A master bar is a shaft with an outside diameter (OD) equivalent to that of the OD of the main bearings. The cylinder block should be inverted. Lube the master bar with engine oil. Then clamp to position by torquing down the main caps minus the main bearings. The master bar should rotate in the cylinder block main bearing line bore without binding. Should it bind, the cylinder block should be line bored.
3. If the engine uses a block-located camshaft, check the cam bore dimensions with a T-gauge. Then install the cam bushings with the correct cam bushing installation equipment. Use drivers with great care as the bushings may easily be damaged, and ensure that the oil holes are lined up before driving each bushing home.

4. Install gallery and expansion plugs. These are often interference fitted and sealed with silicone, thread sealants, and hydraulic dope.

Glaze Busting. When checked to be within serviceability specs, the liner should be deglazed. Deglazing involves the least amount of material removal. A power-driven (heavy duty electric drill with accurate low-rpm control) flex hone or rigid hone with 200 to 250 grit stones can be used for deglazing. The best type of glaze buster is the flex hone, typically a conical (Christmas tree) or cylindrically shaped shaft with flexible branches of carbon/abrasive balls.

PRESERVING CROSSHATCH

The objective of glaze busting is to machine away the cylinder ridge above the ring belt travel and re-establish the crosshatch. The drill should be set at 120 to 180 rpm. When glaze busting, the flex hone should be used rhythmically. Short in–out sequences work best; stopping frequently to inspect the finish produces the best results. A 60- to 70-degree horizontal crossover angle (some manufacturers express this as 120- to 130-vertical crosshatch) and a 15 to 20 micro-inch (3.8 to 5 microns)-deep crosshatch should be observed. Using a spring-loaded hone to deglaze liners produces faster results, but there is also more chance of damaging the cylinder liner. **Figure 3-27** shows what a properly executed crosshatch should look like after machining.

Honing. Honing is performed with a rigid hone, powered once again at low speeds by either a drill or overhead boring jig. The typical cylinder hone consists

of three legs. These are set before machining to produce a radial load into the liner wall. The abrasive grit rating of the stones determines the aggressiveness of the tool: 200 to 250 grit stones are typical. Overhead boring tools can be programmed to produce the required stroke rate for the specified crosshatch, but if using a hand-held tool, remember that a few short strokes with a moderate radial load tends to produce a better crosshatch pattern than many strokes with a light radial load.

CREATING CROSSHATCH

Once again, a 60- to 70-degree horizontal crossover angle, 15 to 20 micro-inch (3.8 to 5 microns) crosshatch should be observed. It should be clearly visible by eye as shown in **Figure 3-27**. Honing is designed to produce the specified crosshatch. This means that the liner contact surface retains oil required to allow the piston rings to seal. When installing wet liners, observe the OEM installation procedure.

OIL PANS OR SUMPS

The **sump** is a reservoir usually located at the crankcase flange below the engine cylinder block. It encloses the crankcase. **Oil pans** are manufactured from:

- cast aluminum
- stamped mild steels
- laminated steels
- plastics
- mineral and synthetic fibers

The oil pan is a reservoir. It collects engine lube oil that drains down into it by gravity. The oil pump pickup is positioned low in the oil pan. This allows the oil pump to pick up and pump oil through the lubrication circuit.

Oil Pan Functions

Oil pans can act as a sort of boom box and amplify engine noise, so they are usually designed to minimize noise. Laminated steels effectively reduce engine noise and also toughen an oil pan. The oil pan also plays a role in transferring lube oil heat to the atmosphere. The effectiveness of this role obviously has a lot to do with the material from which it is manufactured. Aluminum will transfer heat more effectively than a fiber-reinforced plastic. In more recent model diesel engines, lube oil plays a greater role in cooling the engine, so oil pan design and materials matter a lot. *Scavenging pumps* are secondary oil pumps designed to ensure that

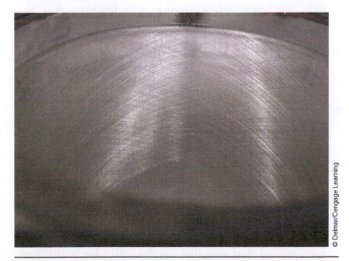

© Delmar/Cengage Learning

Figure 3-27 Detail of the crosshatch on a cylinder bore.

the engine is not starved for oil when the vehicle is operating on steep grades: you can expect to come across these on off-highway equipment designed to operate on grades.

Removing an Oil Pan. Oil pans are usually located in the air flow under the frame rails. They are therefore easily damaged by objects on the road such as rocks on rough pavement and small animals on the highway. Most highway diesel engines have oil pans that can be removed from the engine while it is in chassis. It is recommended that you drain the oil sump before removing it. Oil pans that seal to the engine block using fiber gaskets can be difficult to remove, especially where adhesives have been used to ensure a seal. A 4 lb (2 kg) rubber mallet may make removal easier. Avoid driving screwdrivers between the oil pan and crankcase flange because it is easy to damage the sealing faces.

Tech Tip: A pneumatic gasket scraper is a great way to remove gasket residue from the engine block oil pan flange face, but you must take care. It is easy to gouge the sealing surfaces. Avoid using pneumatic gasket scrapers on aluminum, glass fiber, or carbon fiber oil pans.

Tech Tip: Use the OEM recommended sealant on pan gaskets. With rubber compound gaskets, smeared engine oil does the job. Avoid using weatherstrip adhesive. Manufacturers do not recommend it, and it can be difficult to remove. It can also produce failures because it does not have great ability to allow creep between the oil pan and the crankcase flange.

Oil Pan Inspection. Oil pans that use rubber isolator seals tend not to present removal problems. In fact, sometimes rubber isolator seals can be reused if they are inspected and found to be in good condition. After removal, the oil pan should be cleaned of gasket residues and washed with a pressure washer. Inspect the oil pan sealing flanges, check for cracks, and test the drain plug threads. Cast aluminum oil pans that bolt both to the cylinder block and to the flywheel housing are prone to stress cracking in the rear due to cylinder block torque twist. Carefully inspect them before returning to service. When torquing aluminum oil pans, always meticulously observe torque sequences and values.

Summary

- For purposes of study, the engine powertrain refers to those engine components responsible for delivering power developed in the engine cylinders to the power takeoff mechanism, usually a flywheel.

- Aluminum trunk-type pistons are most commonly used in light duty diesels, but newly introduced, ultralight forged steel pistons are becoming more popular.

- Aluminum alloy trunk-type pistons used in diesel engines support the top compression ring with a Ni-Resist insert. They are also cam ground and tapered due to the way they expand in operation.

- Light duty forged steel pistons are known by the brand names Monotherm and Monosteel.

- In direct injection engines, the shape of the piston crown determines the type of combustion chamber. The Mexican hat piston crown, open combustion chamber is the most common in today's low-emission diesel engines.

- Engine oil is used to help cool pistons in three ways: shaker, circulation, and spray jet methods.

- Piston rings seal when cylinder pressure acts on the exposed sectional area of the ring: this pressure first forces the ring down into the land, then gets behind it to load the ring face into the cylinder wall. Because of this, we can say that piston rings seal with best efficiency when engine cylinder pressures are at their highest.

- Any cylinder gas that passes by the piston rings enters the crankcase. These are known as blowby gases. High crankcase pressure can indicate worn piston rings.

- A keystone is wedge shaped. Keystone rings are commonly used for the top compression ring in today's highway diesel engines. They may also be used for the rings below the top compression ring.

- Oil control rings are designed to apply a film of oil to the cylinder wall on the upstroke of the piston and "scrape" it on the downstroke. All the piston rings play a role in controlling the oil film on the cylinder because oil that remains on the cylinder walls during the power stroke is combusted.

- Piston pins may be full floating or semifloating. A full-floating wrist pin has a bearing surface with both the piston boss and the connecting rod eye. A semifloating wrist pin has a bearing surface only at the connecting rod eye.

- Full-floating wrist pins are retained in the piston boss by snap rings.

- Connecting rods are subjected to compressional and tensional loads. Most connecting rods will outlast the engine. That said, they should be carefully inspected at each overhaul.

- Crankshafts have to be flexible because they receive considerable bending and torsional stress.

- Most engine manufacturers do not approve of reconditioning failed crankshafts. However, the practice is widespread due to the high cost of new crankshafts.

- The term *friction bearings* is used to describe the half shell bearings used to support crankshafts and on rod big ends. The *friction* refers to the fluid friction created by the hydrodynamics of the lube oil.

- Friction bearings used in crankshaft throw and main journals are retained by crush.

- Vibration dampers consist of a drive member, drive medium, and inertia ring.

- The viscous-type damper is common on today's commercial diesels. A hollow drive ring bolts directly to the crankshaft: inside this a solid steel inertia ring is suspended in silicone gel. The hydrodynamic shear of the silicone drive medium between the drive ring and the inertia ring produces the damping effect.

- Some manufacturers use solid rubber vibration dampers. In these, the inertia ring is exposed and driven by solid rubber that is bonded to both the inertia ring and drive hub.

- The flywheel uses inertia to help smooth out the power pulses delivered to the engine powertrain.

- The engine cylinder block can be considered to be the main frame of an engine, the component to which all others are attached.

- Cylinder blocks are manufactured from cast irons, compacted graphite iron (CGI), or aluminum alloys.

- Most light duty diesel engines use a parent bore cylinder block.

- In a parent bore engine, the cylinder bore is machined into the engine cylinder block. Some parent bores are induction hardened to increase service life.

- A parent bore cylinder block may be reconditioned by installing liners.

- Engine cylinder blocks should be soaked in a tank, have every critical dimension measured, especially deck straightness and line bore, and be magnetic flux tested at every major engine overhaul.

Internet Exercises

1. Use a search engine to log onto the Mahle and Federal Mogul websites. Identify some of their products.

2. Log onto some diesel engine manufacturers such as Ford, GM, Cummins (Dodge), Caterpillar, Perkins, Volkswagen, Mercedes-Benz, and Volvo. Check out what each has to say about their diesel products.

3. Search out these key words: Monotherm, Monosteel.

4. Check out the following websites: Clevite, Perfect Circle, and Piomach.

5. Compare pricing of OEM and aftermarket diesel engine piston rings.

6. Use a search engine to research machine shops that will recondition diesel engine crankshafts. Check out the warranty offered with each type of repair.

Shop Exercises

1. Measure piston ring end gap in a liner using feeler gauges.

2. Research how one manufacturer identifies connecting rod weight class. Explain how other OEMs do this.

3. Assemble a set of rings onto a piston, identify the thrust faces of the piston, and set end gap stagger.

4. Install a piston assembly into a cylinder block. Torque the rod caps to spec and check rod sideplay (if cracked rods are not used).

5. Measure main bearing clearance using Plastigage.

6. Measure crankshaft endplay using a dial indicator.

Review Questions

1. Where would you likely find a Ni-Resist insert?

   A. Upper ring belt, aluminum trunk piston

   B. Lower ring belt, articulating piston

   C. Upper ring belt, crosshead piston

   D. Trunk piston, pin boss

2. Under which of the following conditions would a piston ring seal most effectively?

   A. Low engine temps

   B. High engine temps

   C. Low cylinder pressures

   D. High cylinder pressures

3. Where would you most likely find a *slinger*?

   A. Crankshaft main seals

   B. Piston leading edge

   C. Oil spray jet

   D. Big end journal

4. What should be used to measure rod and main journal bearing clearances?

   A. Tram gauges

   B. Dial indicators

   C. Plastigage

   D. Snap gauges

5. Technician A says that an advantage of forged steel trunk pistons such as Monotherm and Monosteel is that their weight is equivalent to that of aluminum and they are much tougher. Technician B says that steel trunk pistons are popular in recent engines because they have reduced headland volume. Who is correct?

   A. Technician A only

   B. Technician B only

   C. Both A and B

   D. Neither A nor B

6. Technician A says that keystone rings are common in high-performance diesel engines. Technician B says that keystone con rods have a wedge-shaped small end. Who is correct?

   A. Technician A only

   B. Technician B only

   C. Both A and B

   D. Neither A nor B

7. Which of the following would be most likely to occur if an engine was run under load with a failed vibration damper?

   A. Increased fuel consumption

   B. Increased cylinder blowby

   C. Camshaft failure

   D. Crankshaft failure

8. What is the upper face of a piston assembly known as?

   A. Crown

   B. Skirt

   C. Boss

   D. Trunk

9. Which term is used to describe the cylinder volume between the piston upper compression ring and its leading edge?

   A. Dead volume

   B. Headland volume

   C. Toroidal recess

   D. Clearance volume

10. Which of the following tools should be used to check a cylinder block deck for warpage?

    A. Master bar

    B. Dial indicator

    C. Laser

    D. Straightedge and feeler gauges

 Timing Geartrain and Cylinder Head Assemblies

Learning Objectives

After studying this chapter, you should be able to:

- Identify the engine timing geartrain and cylinder head components.
- Outline the procedure required to time an engine geartrain.
- Define the role of the camshaft in a typical diesel engine.
- Interpret camshaft terminology.
- Perform a camshaft inspection.
- Identify the role that valve train components play in running an engine.
- List the types of tappet/cam followers used in diesel engines.
- Inspect a set of push tubes or rods.
- Describe the function of rockers.
- Define the role played by cylinder head valves.
- Interpret valve terminology.
- Outline the procedure required to recondition cylinder head valves.
- Describe how valve rotators operate.
- Perform a valve lash adjustment.
- Outline the consequences of either too much or too little valve lash.
- Create a valve polar diagram.

Key Terms

base circle (BC)	creep	inner base circle (IBC)
cam geometry	cylinder heads	interference angle
cam profile	fire rings	interference fit
camshaft	followers	keepers
clevis	gasket	lifters
companion cylinders	helical gear	outer base circle (OBC)

overhead adjustment	tappets	valve float
pallet	Tempilstick™	valve margin
ramps	template torque	valve polar diagram
rockers	torque-to-yield	valve train
split locks	train	variable valve timing
spur gear	valve	yield point

INTRODUCTION

Timing geartrains, camshafts, tappets, rockers, and cylinder valves, anchored on or around the engine cylinder head(s), together form the engine's mechanical management train. This group of components includes:

- timing and accessory drive gearing
- the camshaft
- tappets
- valve and (if equipped) unit injector trains
- fuel pumping mechanisms

Its components are driven by and usually timed to the engine crankshaft. The camshaft(s) may either be mounted in the cylinder block or mounted overhead. The drive mechanism for most diesel engine valve train assemblies is a gearset. In some lighter duty engines, pulleys and belts, chains and sprockets may be used as drives. A typical timing gearset is illustrated in **Figure 4-1.** This is the timing gearset used on a Ford 6.7 liter PowerStroke engine, and it is noted for its simplicity. The PowerStroke uses a cylinder block–located camshaft. The camshaft bull gear acts as a drive for the fuel pump drive gear.

CYLINDER HEADS

Cylinder heads seal the engine cylinders. They also contain the cylinder **valves** that manage engine breathing. Diesel engines can use cast iron or aluminum alloy cylinder heads. Engines that use a cast iron cylinder block may use an aluminum cylinder head or heads, but an engine with an aluminum cylinder block always uses aluminum cylinder head(s). A cylinder head is machined with breathing tracts and ports, cooling and lubrication circuit manifolds, fuel manifolds, and injector bores. Cylinder heads may also support rocker assemblies and camshafts when overhead camshaft design is used. There are several configurations used in diesel engines.

Cylinder heads usually contain the valve assemblies, breathing ports, injector bores, and coolant and lubricant passages. **Figure 4-2** shows the underside of one of the cylinder heads used on a Ford PowerStroke 6.7 liter engine. This engine uses a pair of aluminum cylinder heads that mate to a compacted graphite iron (CGI) cylinder block. Note that there are four valves per cylinder. **Figure 4-3** shows a top view of the same cylinder head. The exhaust tracts are located on the inboard side to minimize heat losses upstream from the turbocharger.

Cylinder Head Disassembly, Inspection, and Reconditioning

Cylinder heads are often reconditioned in specialty machine shops. If you are required to disassemble and recondition a cylinder head, make sure you reference

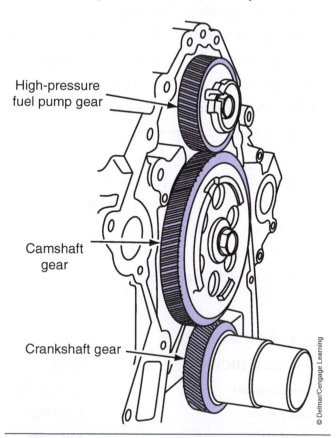

High-pressure fuel pump gear

Camshaft gear

Crankshaft gear

© Delmar/Cengage Learning

Figure 4-1 Ford PowerStroke 6.7L engine geartrain notable for its simplicity. The crankshaft drives the camshaft gear. The camshaft gear acts as the bull gear to impart drive to the fuel pump drive gear. All three gears must be timed.

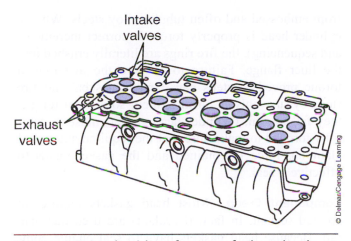

Figure 4-2 Underside of one of the aluminum cylinder heads used on a Ford PowerStroke 6.7L engine. Six bolts per cylinder are used to clamp the cylinder head to the engine block.

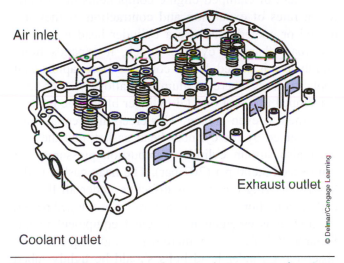
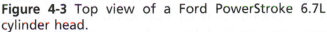

Figure 4-3 Top view of a Ford PowerStroke 6.7L cylinder head.

the manufacturer service literature. When removing variable valve timing and engine brake actuators there is usually a very specific procedure to observe. The following is a general procedure, and we will assume that the injectors have already been removed (it is required practice to remove injectors before removing the cylinder head from the engine):

1. Remove the valves with a C-type spring compressor and gently tap each valve with a nylon hammer to loosen the keepers and retainers.
2. Clean cylinder heads in a soak tank filled with a suitable cleanser. Remember that soak solutions suitable for CGI and cast iron can destroy aluminum alloy.
3. Check the cylinder head height dimension. Each manufacturer has its own preference on where to make this, so check with the service literature.

4. Clean up the head gasket surface with fine grit emery cloth.
5. Electromagnetic flux–test the head for cracks. Dye penetrant testing can also be used, but this is messy and usually inaccurate.
6. Hydrostatic pressure test. Cap and plug all the coolant ports. Place the cylinder head on a test bench and heat by running hot water through it. When hot to the touch, hydrostatically test using shop air at around 100 psi (7 bar). Areas to observe for leakage are the valve seats and injector sleeves. When brass injector sleeves are used (usually older engines), perform the hydrostatic test both cold and hot.
7. If injector sleeves have to be replaced, use appropriate removal and installation tools and repeat the pressure test after the operation. These are sometimes swaged and sometimes threaded into the injector bore. Some have internal threads that allow a removal tool to be threaded into them; these are pulled using a slide hammer.
8. Check for face warpage using a straightedge and feeler gauges. The specification will vary according to the size of the head.
9. In instances where valve guides are used, check the guide bores to spec using a ball gauge. If in need of replacement, the guide must be pressed or driven out with the correct driver. Damage to the guide bore may require reaming to fit an oversize guide. Integral guides can be repaired by machining for guide sleeves or knurling. Installation of new guides can be facilitated by freezing (dry ice) or use of press fit lubricant.
10. New guides may require reaming after installation, but some manufacturers use cladded guides that should *never* be reamed; check the service literature.
11. Check valve seats for looseness using a light and a ball peen hammer and listening—a loose seat resonates at a higher pitch in CGI and cast iron heads, and sounds flat in aluminum heads. To recondition, select the appropriate mandrel pilot and insert into the guide, then match valve seat to the mandrel grinding stone. Stellite-faced seats require special grinding stones. Dress the stone to achieve the required seat angle. Interference angles are used in some lighter duty engines but are less commonly used in late-model automotive diesels due to a lowered ability of the valve to transfer heat. An interference angle is usually not required when valve rotators are used.

New valve seats are installed with a marginal **interference fit** and require the use of the correct driver; after installation the seat is usually knurled or staked in position. Some diesels use alloy steel valve seats, but where cast iron seats are used, do not attempt to stake them because this could fracture the seat. After valve and seat reconditioning, check valve head height (valve protrusion) with a dial indicator.

12. Valve reconditioning. Clean valves with wire wheel or glass bead blaster, then check for stretching, cupping, burning, or pitting. When refacing (dressing) valves, ensure that the valve margin remains within manufacturer specifications. Check valve seating using Prussian blue (aka *machinist's blue*). Lapping is not required if the grinding has been done properly.

13. Check the valve springs for straightness, height, and tension using a right-angle square, tram gauge, and tension gauge.

14. Valve rotators. In engines using positive rotators (rotocoil), they can be checked after hand assembly by tapping the valve open with a nylon hammer to simulate valve train action—they should rotate.

Cylinder Head Installation

Fortunately, most current cylinder head gaskets are one piece. This means that cylinder pressures, coolant ports, and oil passages are all sealed by the cylinder head gasket. Grommets are used to seal coolant ports and oil passages. Grommets are commonly made of rubber: the rubber is thermally bonded into a one-piece cylinder head gasket. **Fire rings** are ring shaped and are used to seal cylinder gases at the top of the cylinder bore flange. Fire rings are manufactured from alloy steels: they have to effectively seal through a wide range of temperatures and cylinder pressures. Whether fire rings are incorporated into a one-piece head gasket or are separate components, they can be regarded as one-time-use components. This is because when a cylinder head is torqued onto the cylinder block, the fire rings are designed to deform so they create the best possible seal.

Gasket Yield Point.   All gaskets are designed to yield to conform to two clamped components. Some gaskets and grommets can be safely reused, but many cannot. A head **gasket** must be properly torqued to ensure that its **yield point** is achieved. Yield point means that the gasket is crushed to conform to the mating faces of two clamped components to produce the best possible seal. The fire rings used in most cylinder heads are made

from embossed and often tubular alloy steels. When a cylinder head is properly torqued (correct increments and sequencing), the fire rings are literally crushed into the liner flange. Failure to observe the incremental torque procedure steps can damage gaskets and fire rings by unevenly deforming them so at final torque, they fail to seal. Cylinder head gaskets must be carefully torqued to make sure that the required amount of clamping force is obtained and the gasket yields to effectively seal.

Component Creep.   Most head gaskets require no applied sealant. In fact, if sealants are used they can fail. Because head gaskets have to seal engine components where both the temperature and pressure are at their highest, they have to accommodate a large amount of component **creep**. Creep is the relative movement of clamped engine components due to different rates of expansion and contraction as they are heated or cooled. Because a cylinder head has much less mass than the cylinder block, it expands more rapidly as it is heated. It also contracts more rapidly as it cools. In cases where an aluminum alloy cylinder head is clamped to a cast or CGI cylinder block, the relative movement is even greater. This creep differential must be sustained by the cylinder head gasket, and the seal must also sustain cylinder combustion pressures that can exceed 3,000 psi (207 bar). The head gasket is the key to ensuring this seal is effective under all the operating conditions of the engine over a prolonged period of time. It is essential that original equipment manufacturer (OEM) torque increments and sequences be observed. Cylinder head bolts should be lightly lubed with engine oil before installation. Oil should never be poured into the block threads because a hydraulic lock may result. The torque sequence for a Mercedes-Benz MB-906 cylinder head is shown in **Figure 4-4.**

Incremental Torque Sequence.   Failure to observe torquing increments and sequencing can result in cracked cylinder heads, failed head gaskets, and fire rings that will not seal. Because of the large number of fasteners involved, a click-type torque wrench should be used. Some OEMs require that a **template torque** method be used because this produces more even clamping pressures. Typically, this requires setting a torque value first and then turning a set number of degrees beyond the torque spec using a template or protractor. We also know this as **torque-to-yield.**

Tech Tip: Installing a cylinder head onto older engines that use individual fire rings for each cylinder can be made easier by using four guide

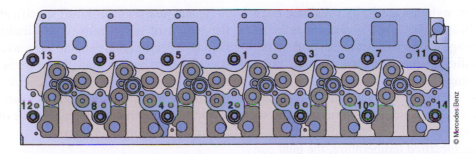

Figure 4-4 Cylinder head torque sequence on a © Mercedes-Benz MB-906 engine.

studs inserted into cylinder bolt holes. This reduces the chances of a fire ring misalignment occurring during head installation.

Rocker Housing Covers

Rocker housing or valve covers seal the upper portion of the engine above the cylinder head. Because valve and injector trains have become more complex over the years, the physical size of rocker housing covers has increased, especially on engines equipped with variable valve timing. Rocker housing covers are the most frequently removed engine component. They have to be removed to access the valves and in cases where injectors are mechanically actuated by the camshaft for purposes of adjustment. For this reason, some engine OEMs have adopted the use of multiuse, rubber compound sealing gaskets usually fitted to a captured groove in the rocker housing cover.

Rocker housing covers also reduce the noise produced by the rockers. Make sure you observe the OEM recommendations for installing these gaskets, especially if you really want to use them more than once. Usually, a light coating of engine oil is all that should be applied. Avoid using aggressive adhesives such as weatherstrip glue because it destroys these rubber gaskets, which are significantly more costly than the older fiber type. When removing the remains of fiber gaskets, a jackknife with a flexible blade does a good job. If you have to resort to using a pneumatic scraper, bear in mind that the sealing faces are easily damaged.

TIMING GEARS

Diesel engine timing gears can be located at either the front or rear of the engine. These gears are responsible for turning the camshaft and most of the engine accessories. The timing geartrain becomes necessarily more complex on engines that use overhead camshafts.

Timing Gear Construction

The timing gears are cast or forged alloys that are heat tempered and then surface hardened. The gear teeth are milled in manufacture to spur and helical designs. Combinations of both these gear designs are used in engine geartrains. The noise produced by the geartrain is a factor and for this reason, helical-cut gears tend to be more common because they are quieter. **Helical gears** provide increased tooth contact area: this lowers contact forces. The disadvantage of helical gears is that they have high thrust loads: this means that thrust forces have to be contained by thrust plates and bearings. The gearset shown in **Figure 4-1** earlier in this chapter uses helical-cut gear teeth.

The **spur gear** design offers much lower thrust loads but at the cost of greater noise and faster wear. Both helical and spur gear designs are used. The gears are commonly press fit to the shafts that they drive and are positioned on the shaft by means of keys and keyways.

Timing Gear Inspection and Removal

Visual inspection can usually determine the condition of timing gears. Indications of cracks, pitting, heat discoloration, or lipping of the gear teeth usually require the replacement of a gear. Press-fit gears have to be removed from gear shafts using mechanical, pneumatic, or hydraulic pullers. When the shaft and gear can be removed from the engine, a shop air-over-hydraulic press can be used. Make sure you take the usual safety precautions and support the components on separation. When a gear has to be separated from a shaft while on the engine, a portable hydraulic press is usually required; while using the press, ensure that it is mounted in such a way that it will not damage either the cylinder block or other gears.

Installation. Install the new gear to the shaft by heating to the manufacturer's specified temperature. The best way to heat the gear to the specified temperature is to use a thermostatically regulated oven or hot plate. If a bearing hot plate is used, make sure that it is large enough to fit the entire surface of the gear. Typical specified temperatures are around 150°C (300°F). Overheating a gear can destroy the heat treatment and possibly the surface hardening, resulting in premature failure. Heat-indicating crayon such as Tempilstick™ may be used to determine the exact temperature. At the specified temperature, the gear can be dropped over the shaft and allowed to air cool. The engine geartrain should be timed according to the OEM procedure.

A number of gears are used in the engine geartrain, especially when an OHC (overhead camshaft) is used. If the camshaft is to be rotated in the same direction as the crankshaft, an idler gear must be used between the crankshaft and camshaft gears. It is not necessary that the camshaft turn in the same direction as the crankshaft, and if you once again refer to **Figure 4-1** you will observe that the camshaft necessarily rotates in the opposite direction from the crankshaft.

Timing the Engine Geartrain. The timing procedure is simple but must be performed accurately. When timing the engine geartrain, you are phasing the crankshaft with the camshaft. Another way of saying this is that you are phasing the engine powertrain (crankshaft and the components that connect with it) with its feedback assembly (valve, injector, and pump drives). In most cases, you are required to locate the crankshaft in a specific position. The camshaft and other timing gears are then timed to the crankshaft gear. It is essential that you reference manufacturer service literature when performing this procedure.

Gears that have to be timed to each other are identified by stamped markings on the teeth that intermesh. Zeros and Xs are common markings. Engine timing geartrains may be designed with one or more idler gears.

....

Tech Tip: On a four-cylinder engine, a quick check for proper timing gear alignment can be made by locating number 1 piston at TDC (top dead center) on the compression stroke. In this position, the valves on number 1 cylinder should be closed, while those on its companion number 4 cylinder should be at overlap, that is, rocking.

....

Gear backlash should be checked after the engine geartrain has been assembled and timed. A dial indicator (preferred) or feeler gauges should be used.

A typical geartrain backlash specification would be in the region of 0.200 mm (0.008 inch), but the manufacturer specifications should be consulted. A backlash specification higher than the manufacturer specification is an indication of gear contact face wear and usually requires that a gear or gears be replaced. A lower-than-specified backlash factor often indicates an assembly or timing problem.

CAMSHAFTS

The **camshaft** in most diesel engines is gear driven by the crankshaft through one revolution per complete cycle of the engine. In a four-stroke cycle engine, to complete a full cycle the engine must be turned through two revolutions or 720 degrees. During this time, the camshaft would turn one revolution. Camshaft speed is therefore equal to one-half engine speed. Camshaft speed and crankshaft (engine) speed are therefore geared so each completes a full cycle in sync with each other.

Cam Profile

The camshaft in a diesel engine actuates the **valve trains**. The term **train** can describe any components that ride a cam profile and are actuated by it. In many diesel engines, the mechanical means of achieving injection pressures is provided by an injector train riding a cam profile. The camshaft is supported at its journals by bushings or bearings that are in most cases pressure lubricated. **Cam geometry** refers to the physical shape of the cams: the term **cam profile** is also used. The profile outside of a base circle will actuate the trains riding the cam and convert the rotary movement of the camshaft into reciprocating motion.

Overhead camshafts are becoming commonplace in current diesel engines, the GM/Isuzu Duramax being an example. Valve train timing as well as unit injection pump, or unit injector stroke in some engines, are determined by cam geometry, so this is dependent on the camshaft gear timing to the engine crankshaft. **Figure 4-5** shows the sequence of camshaft-actuated events during an engine cycle.

Rotary and Linear Motion. Camshafts rotate. The cams machined on the camshaft therefore have rotary motion. Positioned to ride the cam profiles are the followers. As a camshaft rotates, the followers that ride the cam profiles move linearly—that is, in a straight line. Followers therefore convert rotary motion into linear motion. Using the example of a cylinder block–mounted camshaft, the followers actuated by cam

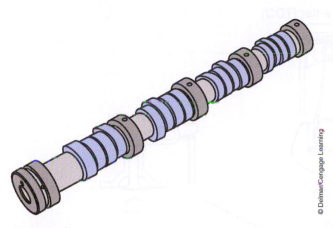

© Delmar/Cengage Learning

Camshaft

The camshaft is driven by the crankshaft. The camshaft has one exhaust and one intake lobe per cylinder.

The two exhaust valves are actuated by a single exhaust lobe and the two intake valves are actuated by a single intake lobe.

The camshaft bearings are lubricated by the rear cam groove. This groove receives oil from the block gallery and oil flows through the center of the camshaft to each cam bearing journal.

Figure 4-5 Camshaft used in the Ford PowerStroke 6.7L engine. (*Text courtesy of Ford Motor Company*)

profiles move up and down. Pushrods or tubes fitted to the followers also move linearly. These connect to rockers. Rockers pivot on a shaft. They are actuated linearly either directly by cam profiles or by push tubes. Rockers have rotary motion. Because of this, they have a linear input and produce a linear output. In doing this, they change the direction of the input. The valves or unit injectors actuated by rockers move linearly. The same thing happens in the case of an overhead camshaft: rotary motion is changed to linear motion.

Construction and Design

Camshafts are manufactured from middle alloy steels. They are then treated to provide surface-hardened journals and cams. Surface hardening is usually by nitriding or other hard-facing processes, followed by finish grinding. Diesel engine camshafts are not usually reconditioned. Resurfacing of the journals is possible, but most camshaft failures are cam lobe related. Camshafts are supported by bearing journals within a lengthwise bore in the cylinder block or on a pedestal arrangement on the cylinder head as in the case of an overhead cam.

Base and Outer Base Circle. Cams are designed to convert rotary motion to linear movement. The smallest radius of a cam is known as the **base circle (BC)** or **inner base circle (IBC)**. The largest radial dimension from the camshaft centerline is known as **outer base circle (OBC)**. The shaping of the profile that connects a cam's base circle with its outer base circle is described as ramping. A cam may be designed so that the larger percentage of its circumference is base circle: in this case, the train it actuates will be unloaded for most of the cycle. For instance, cams used to actuate engine cylinder valves use a mostly IBC design, such as that shown in **Figure 4-5.**

Alternatively, cams may be designed so that most of their circumference is outer base circle (OBC). In this case, the train it actuates will be loaded for most of the cycle. What a cam profile is required to do determines whether it uses a mostly IBC or mostly OBC design. In some cases, you will come across a design that is a mix of both, especially where the cam is required to actuate a unit pump or unit injector. **Figure 4-6** shows the camshaft timing events on an engine that uses cam-actuated unit injectors: in other words, the camshaft controls both valve and injector timing. **Figure 4-7** illustrates some cam terminology that you should understand in order to accurately interpret a manufacturer's service literature as it applies to measuring cam profiles.

Removing and Installing the Camshaft

The procedure for removing the camshaft from a cylinder head–mounted OHC is relatively straightforward. Usually, this requires the cam caps be removed from the cam pedestals. When a block-mounted camshaft has to be removed from a cylinder block, the tappet/follower assemblies will not permit the camshaft to be simply withdrawn while the engine is in an upright position. If the engine cannot be turned upside down, such as when removing the camshaft from an engine in-chassis, the tappet/follower mechanisms must be raised sufficiently so that they do not obstruct the camshaft as it is withdrawn. Most engine manufacturers provide special tools to perform this procedure—usually magnets on a shaft that lift the follower, after which the shaft is locked to the train bore in the cylinder head. All the followers must be raised in this fashion so that the cam lobes and journals do not interfere with them as the camshaft is withdrawn from the block.

When the correct tools are not available, the manufacturer's special tools may by improvised by

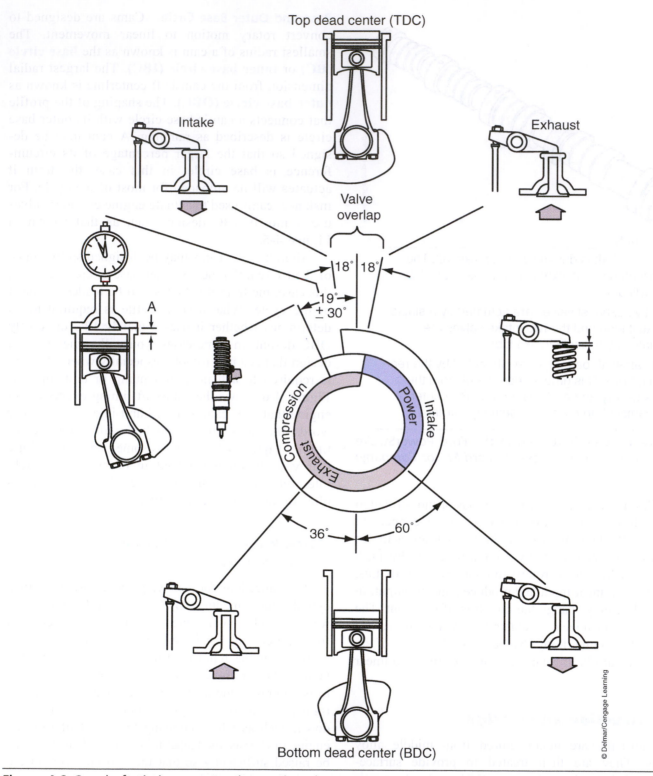

Figure 4-6 Camshaft timing events: the engine shown uses cam-actuated unit injectors, so the camshaft controls both valve and injector timing.

using coat hanger wire (mechanic's wire is usually not substantial enough) to first hook the follower and then lift it by bending the wire at the top. When withdrawing or installing the camshaft in a cylinder block bore, it is important that it never be forced because the result will almost certainly be damage to the camshaft or its support bearings. In certain engines, powertrain components such as the crank web and crank throws may interfere with the camshaft lobes. This means that the engine has to be rotated at intervals while the camshaft is being removed and installed.

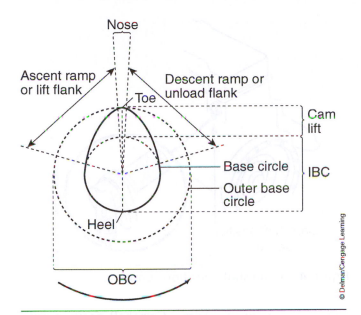

Figure 4-7 Cam terminology.

Camshaft Inspection

Camshaft inspection consists of the following steps:

1. Visual: Pitting, scoring, peeling of lobes and scoring, wear, blueing of journals. Check the drive gear keyway for distortion and key retention ability. Any visible deterioration of the hard surfacing on the journal or cam profile indicates a need to replace the camshaft. The inspection can be by touch. A fingernail stroked over a suspected hard surfacing failure can identify hard surfacing failure in its early stages better than the eye.

2. Mike the cam profile heel-to-toe dimension and check to manufacturer specifications. Mike the base circle dimension and check to specifications. Subtract the base circle dimension from the heel-to-toe dimension to calculate the cam lift dimension. Check to specifications. Cam lift can also be measured using a dial indicator with the camshaft mounted in V-blocks. Zero the dial indicator on the cam base circle and rotate the camshaft to record the lift. Ensure that a dial indicator with sufficient total travel to measure the specified lift is used. When checking the cam lift dimension with the camshaft in-engine using a dial indicator, the measurement may not exactly match the manufacturer spec, even when the cam profile is in good condition. This variance is usually due to the lie of the camshaft in the support journals.

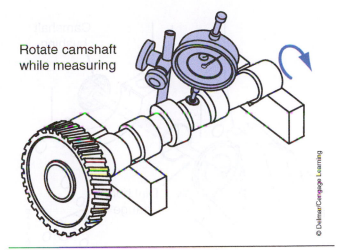

Figure 4-8 Checking a camshaft on V-blocks.

3. Check for cam lobe surface wear using a straightedge and thickness gauges sized to the maximum wear specification.

4. Place the camshaft in V-blocks (**Figure 4-8**) and check for a bent camshaft using a dial indicator. It is unusual for camshafts to bend in normal service. A bent camshaft is usually caused by the failure of another engine component.

Failure to meet the manufacturer specifications in any of the previous categories indicates that the camshaft should be replaced. You should be aware that the smallest indication of a hard surfacing failure on the cam profile can create a total and usually costly failure in very little time.

Probably the most important inspection is the visual inspection.

Camshaft Bushings/Bearings

The camshaft is supported by pressure-lubricated friction bearings at main journals. In other words, it is hydrodynamically supported. A camshaft is loaded whenever one of its cams is actuating the train that rides its profile. This loading can be considerable, especially when fuel injection pumping components are actuated by the camshaft, such as in many Volkswagen applications up to 2007. Cam bushings are normally replaced during engine overhaul, but if they are to be reused they must be measured with a dial bore gauge (**Figure 4-9**) or telescoping gauge, and micrometer to make sure that they are within the manufacturer's reuse specifications.

Interference-fit camshaft bushings located on the cylinder block should be removed in sequence starting usually at the front and working to the back of the

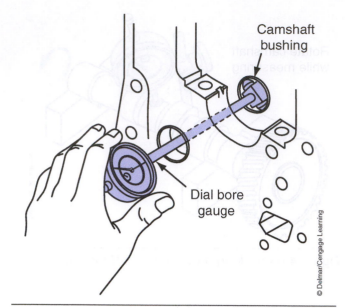

Figure 4-9 Measuring cam bushing bore.

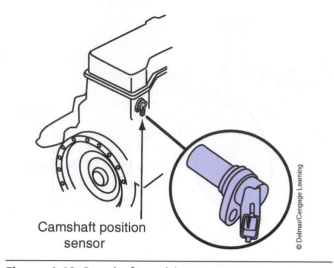

Figure 4-10 Camshaft position sensor.

cylinder block. The correct-sized bushing driver (mandrel) and slide hammer should be used. Cam bearing split shells are retained either by cam cap crush (overhead camshafts) or lock rings. Interference-fit cam bushings are installed using the same driver tools used to remove them. Care should be taken to properly align the oil holes. When installing bushings in a cylinder block in-chassis where access and visibility are restricted, painting the oil hole location on the bushing rim with a shop paintstick may help align the bushing before driving it. Ensure that the correct bushing is driven into each bore.

Bushings and their support bores vary in size, and bushings rarely survive being driven into a bore and then removed. Bearing clearance is the difference in size between the outside diameter of the camshaft journal and the inside diameter of the bushing. Prolonged use of certain types of engine compression brakes can be especially hard on camshaft bearings.

Camshaft Endplay

Camshaft endplay is its end-to-end or linear movement after installation. It is defined either by free or captured thrust washers/plates. Thrust loads are not normally excessive unless the camshaft is driven by a helical-toothed gear. In this case, the thrust is produced by the helical drive gear. When wear occurs, it is more likely to be observed at the thrust faces. Endplay is best measured with a dial indicator. The camshaft should be gently forced backward then forward and the total travel measured.

Camshaft Position Sensors. Camshaft position sensors (CPSs) are used in all current engines. They signal shaft speed and position data to the engine electronic control module (ECM). CPSs may use either an inductive pulse generator or Hall-effect electrical principle. The operating principles of camshaft position sensors are studied in some detail in **Chapter 12. Figure 4-10** shows a typical inductive pulse generator CPS. It signals a frequency value (for engine speed), and the irregular tooth at number 1 cylinder indicates TDC.

VALVE AND INJECTOR TRAINS

While a camshaft rotates, the trains it actuates move linearly. Followers **(Figure 4-11)** ride the cam profiles to actuate trains. The linear movement of the train that rides the cam profile is converted to rotary motion once again at the rocker. It is its rocking action that reverses the linear direction of the train to open and close valves. Cylinder valves move linearly. Rockers may be used in both cylinder block–mounted camshaft engines and those with overhead camshafts.

Followers

Tappet, lifter, and follower may all mean the same thing depending on which manufacturer is using the term. They describe components that usually are positioned to directly ride, or at least be actuated by, a cam profile. The term **tappet** has a broader definition and is sometimes used to describe what is more often referred to as a **rocker** lever. In this text, we will use the term cam **follower** or lifter to describe a component that rides the cam profile. The function of cam followers is to reduce friction and evenly distribute the

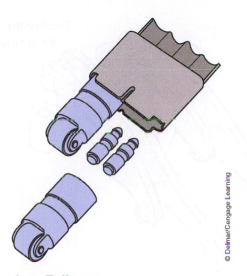

© Delmar/Cengage Learning

Cam Followers

- Patented design (Ford).
- Each roller lifter contains two hydraulic lash adjusters and each valve is individually actuated through its own pushrod and rocker arm.
- Simple stamped rocker arms provide robust quality and reliable motion.
- Eliminates the floating bridge used to open a pair of valves with a single rocker arm.

Figure 4-11 Hydraulic lifter assembly used in a Ford PowerStroke 6.7L engine equipped with a cylinder block–located camshaft. (*Text courtesy of Ford Motor Company*)

force imparted from the cam's profile to the train it is responsible for actuating. Light duty diesel engines using cylinder block–mounted camshafts use the following three categories of followers, while those using overhead camshafts use either direct-actuated rockers or roller-type cam followers:

- solid
- roller
- hydraulic

Solid Lifters. Solid **lifters** are manufactured from cast iron and middle alloy steels and are usually located in guide bores in the cylinder block so they ride the cam profile over which they are positioned. Pushrods or push tubes are fitted to lifter sockets. The critical surface of a solid lifter is the face that directly contacts the cam profile. This face must be durable and may either be chemically hardened, plated with a toughened

alloy, or have a disc of special alloy steel molecularly bonded to the face.

LIFTER INSPECTION

Solid lifters should be carefully inspected at engine overhaul. You should look for thrust face wear. Stem and socket wear should also be checked. The lifter guide bores in the cylinder block should be measured using digital calipers or a telescoping gauge and micrometer. When they wear to an oversized dimension they can be sleeved. Sleeving lifter guide bores is simple. The procedure involves boring out to the new sleeve outside diameter: this is installed with a slight interference fit. Check that the lifters do not drag or cock in newly sleeved, guide bores.

Roller-Type Cam Followers. Roller-type cam follower assemblies can handle higher mechanical forces, so they are a common choice for engines that use the camshaft to actuate injectors. They are used with both OHC and cylinder block–mounted camshafts. Roller-type cam followers usually consist of a roller supported by a pin mounted to a **clevis** (yoke). The clevis can either be cylindrical and mounted in a cylinder block guide bore or be a pivot arm fitted to either the cylinder block or the cylinder head.

ROLLER CAM FOLLOWER INSPECTION

In the case of some OHC designs, a roller-type cam follower assembly and the rocker arm are used. Roller-type cam followers distribute forces more evenly than solid followers. They also outlast them. The roller faces are usually chemically hard surfaced. Roller contact faces should be inspected for pitting, scoring, and other surface flaws. The roller assembly should also be checked for axial and radial runout.

Hydraulic Lifters. Hydraulic lifters are used to ride cam profiles in engines with cylinder block–mounted camshafts. They function using the same principles as those used in gasoline-fueled automobile engines. Engine oil under pressure is charged to the lifter assemblies and provides a damping effect to the trains they actuate. They are suitable for use in light duty diesel engines in which the camshaft is not responsible for actuating the pumping action of unit injectors. **Figure 4-11** shows the hydraulic lifter assembly used in a Ford PowerStroke.

Pushrods and Tubes

Engines using cylinder block–mounted camshafts require a means of transferring cam action to the rockers in the cylinder head. Push tubes and pushrods act as links in the train. They are located between the cam followers and the rocker assemblies. They are subject to shock loads that occur each time the cam ramps or deramps onto the follower. These shock loads increase proportionally with engine rpm. For this reason, they are manufactured from alloyed steels to handle these shock loads and also to keep their weight to a minimum.

Hollow push tubes are more commonly used than solid pushrods, especially in cases where the camshaft is mounted low in the cylinder block. A typical push tube is a cylindrical, hollow shaft fitted with a solid ball or socket at either end. Balls and sockets form bearing surfaces at the follower and rocker. The bearing contact surfaces are usually lubed by engine oil. This is especially important at the rocker end as the rocker moves through an arc while the push tube moves in linearly. Balls and sockets are usually hard-surfaced. Push tubes are preferred over solid pushrods because the tubular shape provides high strength along with less weight. Pushrods are cylindrical and solid. They are used mostly in applications that permit them to be short and relatively low in weight.

Inspecting Push Tubes. Both pushrods and push tubes seldom fail under normal operation, but when they do there are two main causes:

- Inaccurate valve lash or injector train adjustments
- Engine overspeeding

Ball and socket wear should first be checked visually. Reject if there is evidence of hard-surface flaking or disintegration. Next check:

1. Ball/socket-to-tube integrity: Test by dropping onto a concrete floor from a height of 2 inches. A separating ball will always ring flat. NEVER reuse a push tube with a separating ball or socket.
2. Straightness: Roll the push tube on a known true flat surface such as a (new) toolbox deck. The slightest bend (wobble as it is rolled) is reason to reject a push tube.

CAUTION *It does not make economic sense to ever straighten a push tube or push rod. The failure will recur—it is only a question of when. Even as a "temporary" repair, it makes little sense as the consequences of push tube failure can be much more expensive than the cost of replacing it.*

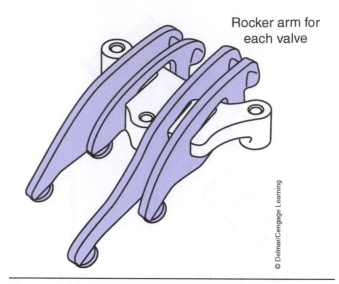

Figure 4-12 A rocker assembly in which individual rockers are used to actuate each of the four valves over one cylinder head.

Rocker Arms

Rockers are levers. They transfer camshaft motion to the valves and mechanically actuated injectors. They are used in both in-block camshaft and overhead camshaft configurations. The rocker pivots on a rocker shaft. When a cam **ramps** off its base circle, it acts on the train that rides the cam profile and "rocks" the rocker arm; this movement actuates the components on the opposite side of the rocker arm, either valves or unit injectors. Lubricating oil is supplied through the rocker shaft to each rocker arm. **Figure 4-12** shows a PowerStroke rocker assembly in which a four-valve cylinder head uses individual rockers for each valve. **Figure 4-13** shows a side view of the assembly.

Rocker Ratio. Rocker ratio may be used to increase the input cam lift to a greater amount of output travel. This requires that the distance from the centerline of the rocker pivot bore to the input (pushrod) side be less than its distance to the output side. Rocker ratio increases mechanical advantage. A rocker ratio of 1:2

Figure 4-13 Side view of the PowerStroke rocker assembly.

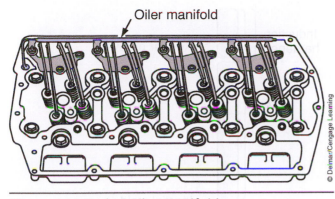

Figure 4-14 Rocker oiler manifold.

converts a cam lift dimension of 1 inch into 2 inches of valve-opening travel. In the case of a rocker with equal distance on either side of its centerline (center of pivot), the rocker ratio is 1:1, meaning equal cam and valve lift.

Lubing Rockers. Rockers can be lubed from the shaft on which they pivot, in which case the engine oil is routed through the rocker shaft pedestal assembly, or by using an oiler manifold assembly such as that illustrated in **Figure 4-14.** A breakdown in the lube delivery mechanism will result in a failure of the rocker assembly.

Inspecting Rockers. The rocker arms should be inspected for wear at the push tube end cup or ball socket, the pivot bore (usually a bushing), and the **pallet** end. The pallet end of a rocker is the bearing surface that contacts and actuates the valve stem or valve bridge. If the hard surfacing at either end of a rocker shows signs of deteriorating, the rocker arm should be replaced. The pivot bore bearing/bushing can be replaced if it shows signs of wear as can the adjusting screw ball. The rocker shafts should be checked for straightness and wear at the rocker bearing race.

Pallet Links. Because a rocker moves radially and whatever it actuates moves linearly, the "bearing" or pallet end of the rocker is subject to friction. This friction can be considerable when actuation forces are high, such as when a mechanically actuated fuel injector is used. A link provides a little forgiveness at the point at which radial motion is converted into linear motion. Links may clip onto the rocker pallet or be inserted into a bore in the injector tappet. It makes sense to use a link between the rocker pallet and the injector tappet. Links should be inspected for wear at overhaul.

CYLINDER HEAD VALVES

Cylinder head valves provide the means of admitting air into and exhausting end gas out of engine cylinders. Although any movement of cylinder valves in current engines is due to cam profile, the timing of engine valve movement depends on whether the engine is equipped with variable valve timing (VVT). With VVT, the engine management electronics can manage valve lift and alter valve timing.

Valve Design and Materials

Cylinder head valves are mushroom-shaped, poppet-type valves. The stems are fitted to cylindrical guides in the cylinder head and loaded by a spring or springs to seal to a seat. A disc-shaped spring retainer locks the spring(s) in position. **Split locks**, also known as **keepers**, are fitted to grooves at the top of the valve stem to hold the spring retainer in position. Valves rely on transferring heat through seat contact. This means that exhaust valves run at their highest temperatures when held open for the longest time. For this reason, exhaust valves are made of materials that do not distort when exposed to high temperatures, and often the valve head is manufactured using different alloys from the stem. The two sections are then inertia welded. Inertia welding is a type of friction welding that requires spinning the stem at high speed while also applying pressure. **Figure 4-15** identifies valve components and terminology.

Intake Valves. Intake valves do not have to sustain the high temperatures that exhaust valves are exposed to. They are actuated at high speeds and so must have some flexibility. This flexibility is required so that at high engine speeds when valve closing speeds are at their highest, the valves do not hammer out their seats. While most diesel engine designers use a valve seat angle of 45 degrees due to greater toughness, some choose to use 30-degree-cut seats on just the intake valves. This takes advantage of the better breathing efficiency of 30-degree over 45-degree seat angles.

Exhaust Valves. Running temperatures of exhaust valves are much higher than those of intake valves. When the exhaust valve is first opened at the end of the power stroke, cylinder temperatures are very high. Only about 20 percent of the heat that the valve is exposed to can be transferred through the stem: this means that the exhaust valves are exposed to high temperatures until they seat.

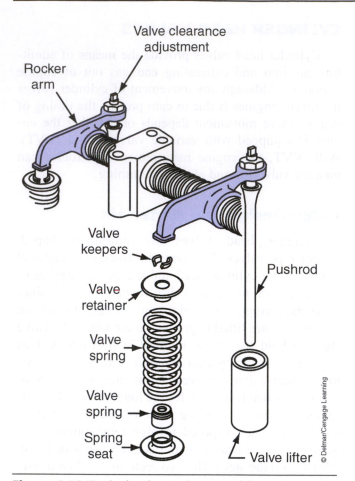

Figure 4-15 Typical valve train assembly.

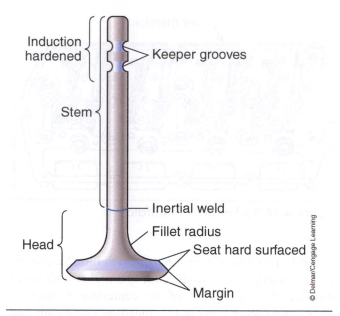

Figure 4-16 Valve terminology.

Exhaust valves are manufactured from special ferrous alloys. They may be cladded (coated) at the head and often use chromium, nickel, manganese, tungsten, cobalt, and molybdenum to improve flexibility, toughness, and heat resistance. Exhaust valves are cooled by:

- Transferring heat to the valve seats (when closed)
- Intake air (during valve overlap)
- Stem-to-guide contact

Inertia welding is commonly used to manufacture exhaust valves, using separate alloys for the stem and head. In this way, the stem can be made out of a tough, hard alloy while the head can be manufactured from an alloy with extreme heat resistance and flexibility. The separate stem and head are friction welded. If you take a look at **Figure 4-16** you can see the location where the stem and head are joined on a typical inertia-welded valve.

Valve Breathing Efficiency. Valves machined with 45-degree seats have lower breathing efficiencies than those with 30-degree seats given identical lift. Valves with 45-degree seats are widely used in

commercial diesel applications simply because they tend to last longer. Valves machined with 30-degree seats tend to run hotter than those with 45-degree seats because they have less material at the head. They also have less distortion resistance and lower unit seating force. To get around this problem, some engine manufacturers use a 30-degree valve seat angle on their intake valves and a 45-degree seat angle on the exhaust valves.

Valve Operation

Many diesel engine valves rely on maximizing the seat contact area for cooling purposes, and as a consequence, **interference angles** are not machined. An interference angle requires that the valve be machined at 0.5 degree to 1.0 degree more acute angle than the seat. This results in the valve biting into its seat with high unit forces. Interference angle also breaks up carbon formations on the valve seat. It is common in current diesel engines to use valve rotators to minimize carbon buildup on the seat and promote even wear. An interference angle is not machined to valves using valve rotators.

Valve Rotators. Some engines use valve rotators. Valve rotators use a ratchet principle or a ball and coaxial spring to fractionally rotate the valve each time it is actuated. Valve rotation should be checked after assembly by marking an edge of the stem and guide with a marker pen and then tapping the valve stem with a light nylon hammer a number of times. The valve should visibly rotate each time that the stem is struck with the nylon hammer.

Valve Harmonics. Cylinder head valves are seated by a spring or pair of springs. When a pair of springs is used, they are often wound in different directions (one clockwise [CW] the other counterclockwise [CCW]). This helps cancel the coincidence of vibration harmonies that may contribute to valve flutter or float. **Valve float** can occur at higher engine speeds when valve spring force is not sufficient to close the valve fast enough. This may result in out-of-time valve closing. Springs are a key component of the valve train. Their importance increases as engine rpm increases.

Inspecting Valve Springs and Retainers

To inspect valve springs and retainers, follow these steps:

1. Once the cylinder head has been removed from the cylinder block, the valves may be removed using a valve spring compressor. Check the keepers and the valve keeper grooves for wear. Inspect the retainers.
2. Measure the spring vertical height and compare to specifications. Minor differences do not matter.
3. Test valve spring tension using a tension gauge **(Figure 4-17).** The valve spring tension should be within 10 percent of the original tension specification. If it is not, replace the valve spring.
4. Check the spring(s) for abrasive wear at each end.
5. Check that the spring is not cocked using a straightedge.

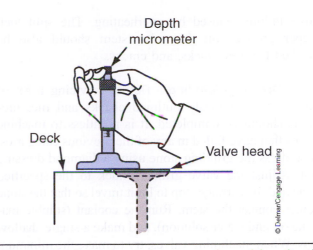

Figure 4-18 Measuring valve protrusion with a depth micrometer.

6. Check valve spring operation with the valve installed in the cylinder head. Remember that as the cylinder head valve seat material is ground away, the spring operating height is lengthened, reducing spring tension. Valve protrusion and recession must be checked to specification as shown in **Figure 4-18.**

Valve Servicing

Use a valve spring compressor to remove valves from the cylinder head. The valves should be tagged by location. Valves should first be cleaned on a buffer wheel, taking great care not to remove any metal. A glass bead blaster may be used to remove carbon deposits from the seat face, fillet, and head, but avoid blasting the stem.

Valve inspection should begin with visually checking for dishing, burning, cracks, and pits. Evidence of any of these conditions requires that the valve be replaced. Check the valve fillet for cracks and nicks—using a portable magnetic flux crack detector may help here—and again, reject the valve if cracks are evident in this critical area. Next, the valve should be measured. Using a micrometer, mark the valve stem at three points through the valve guide bushing sweep and check to specifications. Valve stem straightness should be checked with a vertical runout indicator.

Valve Margin. Measure the **valve margin** (see **Figure 4-16**). Valve margin is the dimension between the valve seat and the flat face of the valve head mushroom. This specification is critical when machining valves: it must exceed the minimum specified value after the grinding process has been completed. A valve margin that is lower than the specification will result in

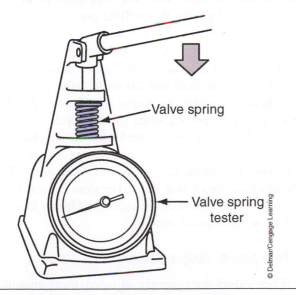

Figure 4-17 Valve spring tension tester.

valve failures caused by overheating. The split lock keeper grooves on the valve stem should also be checked for wear, nicks, and cracks.

Valve Dressing/Grinding. Before servicing a set of valves, the previously outlined checks and measurements should be completed. It is pointless to machine valves that have failed in any of the previous categories. First, dress the grinding stone using a diamond dressing tool. Adjust the valve grinder chuck to the specified angle and the carriage stop to limit travel so that the stone cannot contact the stem. Run the coolant (soluble machine oil and water solution), and make a single shallow pass. When machining valves, try to make the minimum number of passes to produce a valve seat face surface free of ridging and pitting. The final step should be to check that the valve margin is still within specification and that the installed valve protrusion measures within specification as shown in **Figure 4-18**.

In cases in which the valve has been loosely adjusted, the stem end may be slightly mushroomed due to hammering from the rocker. Grind a new chamfer, taking care to remove as little material as possible. Too much chamfer will reduce the rocker-to-stem contact area and may damage the rocker pallet. When valves must be replaced, the new valves must be inspected, measured, and sometimes ground using the same procedure as with used valves.

Valve Seat Inserts

Many diesel engines use valve seat inserts, especially when aluminum alloy cylinder heads are fitted. The advantage of valve seat inserts is that they can be manufactured from tough, temperature-resistant material and then easily replaced when the cylinder head is serviced. Valve seat inserts are press fit to a machined recess in the cylinder head. Sometimes they are staked to position (with a punch) after installation. Since most of the heat of the valve must be transferred from the valve to the seat, it is essential that the contact area of the seat and the cylinder head be maximized.

> **CAUTION** *It is generally inadvisable to stake aluminum. Consult the OEM service literature for the recommended retaining method for valve seats in aluminum heads.*

Valve Seat Removal and Installation. Valve guides must be removed using a removal tool. The removal tool is designed to expand into the valve seat, after which it can be either levered out or driven out with a

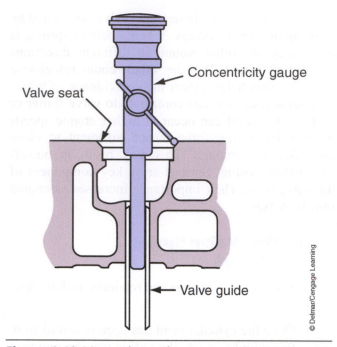

Figure 4-19 Measuring valve seat concentricity with a concentricity gauge.

slide hammer. When installing new valve seats, the seat counterbore must first be cleaned using low-abrasive emery cloth. The new insert should be inserted into the manufacturer-specified driver: this has a pilot shaft that fits tightly to the valve guide bore. A hammer should be used to drive the insert into its bore until it bottoms. Next check the concentricity of the valve seat with the valve guide bore using a dial gauge. This is a critical specification **(Figure 4-19)**. In most cases, the seat will have to be ground a small amount, even when new seats are finish ground. It is important that the valve grinders are serviced before fitting and machining the valve seats so that the new seats are machined to be concentric with the guides that will be put into service.

Valve Seat Grinding. Valve seats are ground using a specified grit abrasive stone, dressed to the appropriate angle, with a pilot shaft that fits to the valve guide. The valve head protrusion or recess dimension specifications must be respected when grinding valve seats. In cases in which the cylinder head deck surface has been machined, undersized valve seat inserts are available from most manufacturers. Whenever a cylinder head deck has been machined, the valve stem height, which is determined by the specific valve seat insert location, must be within specification.

Valve Lash Adjustment

When valves are properly adjusted, there should be clearance between the pallet end of the rocker arm and the top of the valve stem. Valve lash is required because as the moving parts heat up in an engine, they expand. If clearance were not factored somewhere in the valve train, the valves would remain constantly open by the time an engine reached operating temperature. Actual valve lash values depend on factors such as the length of the push tubes and the materials used in valve manufacture. Exhaust valves run hotter and as a consequence expand more than intake valves when at operating temperature. Valve lash specifications for exhaust valves are usually greater than the intake valve lash setting specification.

Maladjusted Valves. Loose valve adjustment retards valve opening and advances valve closing. This reduces cylinder breathing time and results in low power. Cam ramp shaping is designed to provide some "forgiveness" to the train at valve opening and valve closure to reduce the shock loading. When valves are set loose, the valve train is loaded at a point on the cam ramp beyond the intended point. The same occurs at valve closure as the valve is seated. High valve opening and closing speeds hammer the valve and its seat: this results in cracking, failure at the head-to-stem fillet, and scuffing to the cam and its follower.

Valve Adjustment Procedure

The following steps outline the valve adjustment procedure on a typical four-stroke cycle, inline six-cylinder diesel engine using solid lifters. Valves should always be adjusted using the manufacturer's specifications and procedure.

> **CAUTION** *Shortcutting the manufacturer-recommended valve (and injector) setting procedure can result in engine damage unless the technician knows the engine well. Cam profiles are not always symmetrical, and some engines may have camshaft profiles designed with ramps between base circle and outer base circle for purposes such as actuating engine compression brakes. Similarly, a valve rocker that shows what appears to be excessive lash when not in its setting position is not necessarily defective.*

Companion Cylinders. Cylinder throw pairings are called **companion cylinders**. Because of some variability in cylinder number, we will avoid using specific examples here other than that of a four-cylinder engine. In most four-cylinder engines, pistons 1 and 4 are

companions as are 2 and 3. In other words, when number 1 piston is at TDC completing its compression stroke, number 4 piston (its companion) is also at TDC having just completed its exhaust stroke. If the engine is viewed from overhead with the rocker covers removed, engine position can be identified by observing the valves over a pair of companion cylinders. For instance, when the engine timing marker indicates that the pistons in cylinders 1 and 4 are approaching TDC and the valves over number 4 are both closed (lash is evident), then the point at which the valves over number 1 cylinder rock (exhaust closing, intake opening) at valve overlap will indicate that number 1 is at TDC having completed its exhaust stroke and number 4 is at TDC having completed its compression stroke. This method of orienting engine location is commonly used for valve adjustment, but be sure you are familiar with the OEM cylinder numbering and the companion cylinders.

Adjustment. The valve adjustment procedure consists of the following steps:

1. Locate the valve lash specifications. These are often located on the engine ID plate on the rocker housing cover or cylinder block. The lash specification for the exhaust valve(s) is usually (but not always) greater than that for the inlet valve.
2. The valves on current diesel engines should usually be set engine off and with the engine coolant 100°F (37°C) or less. Locate the stationary engine timing indicator rotating indexes: these are set 120 degrees apart and may be located on a vibration damper, any pulley driven at engine speed, or the flywheel.
3. Ensure that the engine is prevented from starting by mechanically or electronically no-fueling it. The engine will have to be manually barred in its normal direction of rotation through two revolutions during the valve-setting procedure, requiring the engine to be no-fueled to avoid an unwanted startup, which can occur more easily than you might believe on older engines.
4. If the engine is equipped with valve bridges or yokes (some four-valve-per-cylinder engines) that require adjustment, this procedure should be performed before the valve adjustment **(Figure 4-20)**. To adjust a valve yoke, back off the rocker arm, then loosen the yoke adjusting screw locknut and back off the yoke adjusting screw. Using finger pressure on the rocker arm (or yoke), load the pallet end (opposite to the

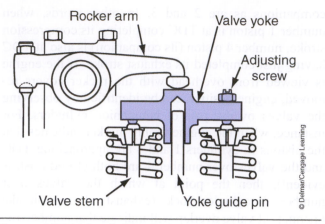

Figure 4-20 Valve bridge/yoke assembly.

adjusting screw) of the yoke until it just makes contact with the valve. Next, screw the yoke-adjusting screw clockwise until it bottoms on the other valve stem. Turn an additional one flat of a nut (60 degrees) as shown in **Figure 4-21,** then lock to position by torquing the jam nut.

> **CAUTION** *When loosening and tightening the valve yoke-adjusting screw locknut, the guide on the cylinder head can be easily bent. Most manufacturers recommend that the yoke be removed from the guide and placed in a vise to back off and final torque the adjusting screw locknut.*

5. To check that the yoke is properly adjusted, insert two similarly sized thickness gauges of 0.010 inch or less between each valve stem and the yoke. Load the yoke with finger pressure on the rocker arm and simultaneously withdraw both thickness gauges. They should produce equal drag as they are withdrawn. If the yokes are to be adjusted, do this in sequence as each

valve is adjusted. In some engines, valve yokes can only be adjusted after removing the rocker assemblies.

6. If the instructions in the manufacturer literature indicate that valves must be adjusted in a specific engine location, make sure you observe this. The cams that actuate the valves may only have a small percentage of base circle. Setting valves requires the lash between the rocker arm and the valve stem to be set. To do this, back off the valve adjusting screw jam nut first, then loosen the adjusting screw. Next, insert the specified size of feeler gauge between the rocker and the valve stem/yoke. Let go of the feeler gauge. Now turn the adjusting screw clockwise until it bottoms. Turn it an additional ½ flat of a hex nut (30 degrees). Hold the adjusting screw with a screwdriver and torque the jam nut to spec. Now for the first time since inserting the thickness gauge, handle it once again and withdraw the thickness gauge. A light drag indicates that the valve is properly set. If the valve lash setting is either too loose or too tight, repeat the setting procedure. Do NOT set valves too tight. Set all the valves in cylinder firing order sequence, rotating the engine 120 degrees between settings. It is preferable to begin at number 1 cylinder and proceed through the engine in firing order sequence. **Figure 4-22** shows this procedure.

Tech Tip: Never try to adjust valve lash while simultaneously checking it with a feeler gauge. Allow the gauge blade to be clamped by the rocker, then release it. Make the adjustment at the rocker screw, then check the gauge blade drag. This should reduce the time you spend performing this simple engine maintenance adjustment.

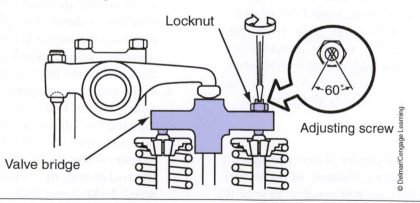

Figure 4-21 Adjusting a valve bridge/yoke assembly: the procedure on this engine is to bottom the adjusting screw, turn one flat, then torque the jam nut.

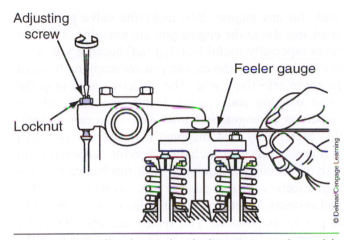

Figure 4-22 Adjusting valve lash on an engine with valve yokes/bridges.

Shortcuts. When you watch experienced technicians performing a diesel top end tune up, you will note that often they shortcut the service literature procedures. When you know a specific engine like the back of your hand, you will get a sense of which shortcuts involve zero risk. Until you do, avoid falling into the trap of thinking that a shortcut on one engine will work on another. The best kind of shortcut is one that the manufacturer approves of by outlining it in the service literature. Most engines have shortcuts, and it is fine to use them when approved by the engine manufacturer.

> **CAUTION** *Do not be persuaded that cylinder head valves should be set tighter than specified. This might have been OK in a different age when engine technology used wider specification margins, but not today. A feeler gauge used properly should define minimal drag; this is what you are trying to achieve. Setting valves tight can result in low power (valves not seating properly at operating temperature), burned valves, and physical engine damage.*

Barring Engines. When you are setting valves, you will have to rotate the engine. With older engines using symmetrical cam profiles, you can get away with bunting the engine over using a remote starter switch, providing the engine is no-fueled. With today's engines, you should always use the manufacturer-approved method described in the service literature. A valve adjustment may have to be performed using a precise location on the cam profile, and the consequence of not doing this could be to damage valves or pistons.

Variable Valve Timing

Variable valve timing (VVT) refers simply to any means used to vary either the opening or the closing of cylinder valves. Heavy duty highway diesel engines have used engine compression brakes for five decades, and this is a form of variable valve timing, but until recently, this was not offered in light duty diesels. Engine brakes are an option on the Cummins ISB. VVT can be applied to any engines capable of modulating valve timing, and to this date, VVT on diesel engines is mainly found in heavy duty engines. VVT today is electronically controlled and electrically or hydraulically (engine lube) actuated. VVT can be used to lower emissions. Retarding the moment of intake valve closure delays the beginning of the compression stroke. This means that the PCM can determine exactly how much intake gas (i.e., oxygen) is admitted to the engine cylinder. The net result is to enable a large engine to behave like a smaller engine when power output requirement is lower, along with a bonus of reducing NO_x emissions by limiting the mass of oxygen entering the engine cylinder. At the time of this writing, there are few examples of automotive diesels using VVT other than the 2010 Mitsubishi 4N13 engine with 1.8 liters of displacement.

Valves: Conclusion

Valves are normally but not always set when the piston is at TDC on the cylinder being set. There are some very critical exceptions to this rule, so when adjusting cylinder valves, make sure you always refer to the manufacturer's service literature. The procedure for setting valves in an engine is sometimes referred to as **overhead adjustment** or tune up. In most cases, the valve-setting procedure is accompanied by procedures such as injector timing and/or lash setting/train loading.

Other Feedback Assembly Functions

The engine feedback assembly in many modern engines is responsible for actuating the pumping of fuel to injection pressure values. A few current and many older engines are fueled by mechanically actuated injectors. These include different types of mechanical unit injectors (MUIs), electronic unit injectors (EUIs) (used by VW up to 2007), and electronic unit pumps (EUPs) (used by MB-900 series up to 2007). Pumping fuel pressures to the required injection pressures on all of these engines is accomplished by the engine camshaft. This requires considerable mechanical force to be delivered by the camshaft, so in these cases, camshafts have considerable bulk.

Every manufacturer has different requirements for setting pump/injector trains. This means you should assume nothing and always consult manufacturer service literature. Some injection pumping actuation trains are set at zero lash or even a slight load when the actuating cam profile is on its IBC. Injector train settings usually have to be precisely set because they help define injection timing. The consequences of failing to consult service literature can be to destroy the engine.

Creating a Valve Polar Diagram

If you refer back to **Figure 4-5,** you can see the key cylinder head valve opening and closing events during a typical diesel cycle. A valve polar diagram is a somewhat simplified version of this illustration that can be made for any engine. You make the valve polar diagram specific to the engine you are mapping. This can be an especially useful learning tool because you have to closely observe the engine you are mapping through an entire effective cycle. The objective is to map the valve opening and closing events by observing the tappets as an engine is barred over. You can perform this exercise by first mapping the valves only. Then you can advance to including injector actuation train activity on engines equipped with mechanically actuated injection systems such as those using EUI or EUP fuel systems. When you perform this exercise, try to be as precise as possible. You should end up with a diagram similar to that shown in **Figure 4-5** but with the exact specifications for the engine you are working on.

Summary

- Cylinder heads must be torqued in sequence to ensure even clamping pressure.

- Torquing cylinder heads in increments (steps) is designed to achieve the cylinder head gasket yield point by evenly achieving the required clamping force.

- Cylinder head bolts should be lightly lubed with engine oil before installation.

- Engine feedback components include the engine timing geartrain, the camshaft, valve and unit injector trains, and in some cases the injection pumping apparatus.

- Camshaft drive gears must be precisely timed to the crankshaft-driven engine geartrain.

- Timing the engine geartrain synchronizes the engine powertrain with the engine feedback assembly.

- The camshaft drive gear is usually interference fit to the camshaft. It may be positioned by a keyway.

- Camshaft gears are heat treated. To fit them to camshafts they must be heated up, so it is essential that they are heated evenly to a precise temperature. Overheating can destroy the gear.

- Camshafts may be rotated either CW or CCW.

- Camshaft gears may use spur or helical-cut gear teeth. Thrust loads are much higher when helical gears are used.

- Gear backlash should be measured using feeler gauges or dial indicators.

- Cam lift on block-located camshafts may be checked using a dial indicator mounted above the push tube or rod.

- Cam base circle or IBC is that portion of the cam circumference with the smallest radial dimension. Cam OBC is that portion of the cam periphery with the largest radial dimension.

- Critical cam dimensions can be checked on an overhead camshaft or an out-of-engine camshaft with a micrometer. Cam lift can be checked with a dial indicator in a block-mounted camshaft.

- Visual inspection of a camshaft should identify most cam failures. Cam profile wear may be checked to specification using straightedge and feeler gauges. The camshaft should be mounted in V-blocks to test for straightness.

- Out-of-engine camshafts should be supported on pedestals or on V-blocks, or hung vertically to prevent damage.

- A cam train consists of the series of components it is responsible for actuating.

- A diesel engine with a block-mounted camshaft uses trains consisting of a follower assembly, push tubes, and a rocker. The rocker actuates cylinder valves or an injector pump.

- Valve lash allows for expansion of the valve train materials as the engine heats to operating temperature.

- Some injection pumping actuation trains are set at zero lash or even a slight load when the actuating cam profile is on its IBC. Injector train settings may have to be precisely set because they help define injection timing.

■ Rocker assemblies provide a means of reversing the direction of linear movement of the push tube or follower, and in some cases they provide a mechanical advantage.

■ Cylinder head valves are used to enable the engine cylinders to breathe. They are actuated by cam profiles and time the gas (air and EGR) into, and end gases out of, the engine cylinders.

■ When reconditioning valves by regrinding, a critical specification is the valve margin.

■ Most diesel engines do not use an interference angle to seat valves because it reduces the seating contact surface area. In addition, valve rotators are widely used, and these cannot be used when valves are cut with an interference angle.

■ A 45-degree-cut valve seat has higher seating force but lower gas flow than a 30-degree-cut valve.

■ Valve seats are usually interference fit into the cylinder head. After installation they must be ground concentric to the valve guide bore.

■ When setting valve lash, the manufacturer specifications must be adhered to. The engine position for setting the valves over each cylinder must also be observed.

■ Valve lash should be set using feeler gauges. Valves in current engines are set cold.

■ Loosely set valves produce lower cylinder breathing efficiencies. They can also produce top end clatter and can damage cam profiles.

■ Creating a valve polar diagram is an effective way to map the valve opening and closing events in any engine you choose. You can also map how mechanically actuated injectors are phased into the cycle.

■ Gaskets used to seal the engine housing components must be able to accommodate creep without failing. For this reason, most gaskets can be regarded as single-use items.

Internet Exercises

1. Use a search engine to check out the QuickWay and Sioux Tools websites. Check out each company's valve servicing equipment.

2. Use a search engine and key in <tempilstick> to see what you come up with.

3. Use the Web to identify valve lash specifications for the intake and exhaust valves of a diesel engine of your choosing. Cross check this information with the manufacturer specs if they are available to you.

4. Use the Web to identify the forums for one of the Big Three diesels: Ford PowerStroke, GM Duramax, or Dodge Cummins ISB. Make a list of the key differences in the cylinder head design and timing geartrains.

5. Use the Web to research the specification data on the Dodge Cummins ISB inline six-cylinder engine versus the V8-configured GM Duramax and Ford PowerStroke engines. Make a note of the differences.

Shop Exercises

1. Check and set if necessary, valve lash on a diesel engine equipped with a cylinder block–located camshaft.

2. Check and set if necessary, valve lash on a diesel engine equipped with an overhead camshaft.

3. Set a complete top end on a diesel engine equipped with mechanically actuated injectors (such as EUIs on a VW). Note the method required to set injector height.

4. Use a dial indicator to check out cam lift on an engine equipped with a cylinder block–located camshaft.

5. Place a diesel engine camshaft in V-blocks: check cam profile dimensions to specification using the manufacturer-recommended methods.

6. Use the information in this chapter to create a valve polar diagram for a specific engine; include injector actuation if appropriate on the map you produce.

Review Questions

1. The location on a cam profile that is exactly opposite the toe is referred to as the:
   A. Sole
   B. Nose
   C. Ramp
   D. Heel

2. On a cam profile described as mostly inner base circle, the profiles between IBC and OBC are known as:
   A. Cam geometry
   B. Ridges
   C. Ramps
   D. Heels

3. In which direction must a camshaft be rotated on a diesel engine with a crankshaft that is rotated clockwise?
   A. Clockwise (CW) only
   B. Counterclockwise (CCW) only
   C. Either CW or CCW is OK depending on the engine

4. Which tool would be required to measure the cam lift of a cylinder block–mounted camshaft in position?
   A. Dial indicator
   B. Outside micrometer
   C. Depth micrometer
   D. Thickness gauges

5. A camshaft gear is usually precisely positioned on the camshaft using a(n):
   A. Dial indicator
   B. Interference fit
   C. Key and keyway
   D. Captured thrust washer

6. When grinding a valve face to remove pitting, the critical specification to monitor during machining would be the:
   A. Shank diameter
   B. Stem
   C. Poppet diameter
   D. Margin

7. Which of the following engine running conditions would be more likely to cause a valve float condition?
   A. Engine lug down
   B. Engine overspeed
   C. Operating in the torque rise profile
   D. Operating in the droop curve

8. When grinding a new set of cylinder valve seats, which of the following should be performed before the machining?
   A. Install the valves.
   B. Install the rockers.
   C. Adjust the valve yokes.
   D. Install the valve guides.

9. Which of the following methods is used to transfer drive torque from the crankshaft to the camshaft on most light duty diesel engines?
   A. Gears
   B. Belt and pulley
   C. Fluid coupling
   D. Timing chain and sprocket

10. You are about to set the valves over number 2 cylinder on a four-cylinder engine. Which of the following would likely be true on most engines?
    A. Valves over number 2 should be rocking.
    B. Valves over number 3 should be rocking.
    C. Valves over number 1 should show full lash.
    D. Valves over number 4 should show full lash.

CHAPTER

5 Intake and Exhaust Systems

Prerequisite

Chapter 2.

Learning Objectives

After studying this chapter, you should be able to:

- Identify the intake and exhaust system components.
- Describe how intake air is routed to the engine's cylinders.
- Describe how exhaust gases are routed out to aftertreatment devices.
- Define the term *positive filtration*.
- Outline the operating principle of an air precleaner.
- Service a dry, positive air cleaner.
- Perform an inlet restriction test.
- Identify the subcomponents on a diesel engine turbocharger.
- Define *constant* and *variable geometry turbochargers*.
- Outline the operating principles of turbochargers.
- Troubleshoot common turbocharger problems.
- Define the role of a charge air cooler in the intake circuit.
- Test a charge air heat exchanger for leaks.
- Relate valve configurations and seat angles to breathing efficiency.
- Outline the role of a diesel engine muffler device.
- Identify the different types of catalytic converters used on current diesels.
- Describe diesel engine exhaust aftertreatment devices, including EGR, SCR, and DPF systems.

Key Terms

catalytic converter

charge air cooler (CAC)

clean gas induction (CGI)

compressor housing

constant geometry (CG)

cross-flow valve configuration

diesel exhaust fluid (DEF)

Environmental Protection Agency (EPA)

exhaust gas recirculation (EGR)	parallel port valve configurations	sound absorption
heat exchanger	positive filtration	turbine
impeller	powertrain control module (PCM)	turbocharger
inlet restriction gauge	ram air	variable geometry (VG)
intake circuit	rejected heat	variable nozzle (VN)
manifold boost	resonation	variable valve actuator
naturally aspirated	selective catalytic reduction (SCR)	

INTRODUCTION

Most diesel engines today are turbocharged, even those used in small-bore applications. For this reason, diesel engine intake and exhaust systems share common components in the turbocharger and exhaust gas recirculation (EGR) system. It therefore makes sense to study intake and exhaust systems together. Turbochargers boost the intake manifold at pressures above atmospheric, and they use **rejected heat** from the exhaust system to do it. EGR systems reroute exhaust gas back into the intake circuit to reduce emissions, so they also have a role in both intake and exhaust systems.

The **Environmental Protection Agency (EPA)** defines and enforces diesel engine emissions standards in the United States and Canada. EPA Tier 4 or EPA 2010 sets the toughest emissions standards in the world. Euro VI standards are roughly equivalent to our 2010 standards, but these do not kick in until 2013 and appear to have many exemptions. Emission controls are the subject matter of **Chapter 13.** While this chapter will focus primarily on breathing gas flow, **Chapter 13** will examine the operating principles of the emission control systems.

OVERVIEW OF THE BREATHING CIRCUIT

Turbocharged diesel engine cylinders are *charged* with air. In a **naturally aspirated** diesel engine, air is drawn into the engine cylinder by the lower-than-atmospheric pressure created in the engine cylinder on the downstroke of the piston on its intake stroke. Almost all current light duty diesel engines meeting on-highway emissions standards are turbo-boosted, so when you see a naturally aspirated diesel it is usually older, low horsepower, and designed for off-highway use. For the most part during our study of diesel engines in this book, it will be assumed that an engine

is turbocharged. Turbocharging provides **manifold boost**. A manifold-boosted engine charges its cylinders with air at pressures above atmospheric in most phases of operation. Turbocharging has become a requirement for engines to meet the fuel efficiency standards expected by operators and the emissions standards mandated by government.

Gas Flow in Breathing Circuit

Figure 5-1 shows the components found in a typical diesel engine breathing circuit that meets post-2010 EPA emissions standards. The figure looks complicated, and it is. As we go through this chapter, the gas flow routing through these components will be explained and most of the components demystified. In addition, we will take a brief look at the exhaust gas aftertreatment devices: a more detailed study of exhaust aftertreatment appears in **Chapter 13.** Throughout this chapter we will follow the gas flow through the engine breathing circuit, beginning at the air cleaner and ending at the exhaust outlet.

Role of the Intake System

The role of the intake system in a diesel engine is to supply a charge of air or air/dead gas mixture to the engine cylinders. This gas mixture is used for combustion, cooling, and cylinder scavenging. Even when naturally aspirated and especially when turbocharged, diesel engines are designed for lean-burn operation. This means that under most operating phases, there is significantly more air present in the engine cylinder than that required to burn the fuel. A generation ago, the objective of diesel engines was to force as much air into the engine cylinders as possible. This was great for performance but bad news for emissions. Today, engines must be managed so that they achieve the best fuel economy while producing a minimum of harmful emissions, and that means precisely controlling the amount of oxygen that gets into the engine cylinders.

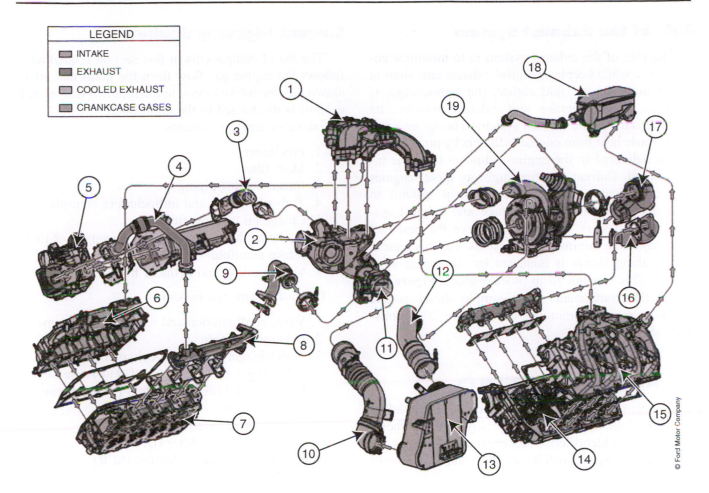

Figure 5-1 Components used in the 2011 Ford Powerstroke engine breathing circuit.

Number	Component	Number	Component
1	Upper intake manifold	11	Intake Throttle Body
2	Lower intake manifold	12	CAC inlet
3	EGR outlet tube	13	CAC
4	Exhaust manifold to EGR tube	14	Left cylinder head
5	EGR valve	15	Left valve cover
6	Right valve cover	16	Left turbo inlet pipe
7	Right cylinder head	17	Turbocharger down pipe
8	Right exhaust manifold	18	Crank Case Vent (CCV)
9	Right turbo inlet pipe	19	Turbocharger
10	CAC outlet		

Role of the Exhaust System

The role of the exhaust system is to minimize engine noise while keeping harmful exhaust emissions at a minimum. As we said earlier, the turbocharger is common to both intake and exhaust systems. Its function is to use the exhaust system to recapture some of the waste heat from engine cylinders by pressurizing the air delivered to the engine cylinders from the intake circuit. Current certified highway diesel engines have complicated exhaust aftertreatment systems to minimize emissions. Diesel intake and exhaust systems have to work together to achieve the best fuel economy while minimizing emissions. For this reason, the breathing circuit is managed by the engine electronics. This was not so in older engines. **Figure 5-2** shows the components of a simple diesel engine breathing circuit commonly used before 2004 emissions standards.

BREATHING COMPONENTS

To begin, the engine breathing components used on a current four-stroke cycle, highway-certified diesel engine will be identified. Older engines and small-bore off-highway engines will have fewer emission control devices.

Current Highway Engines

The list of components in this section sequentially follows the engine gas flow from the point that air is drawn into the intake circuit to the moment that treated exhaust is discharged to the atmosphere.

Intake system components:

1. Precleaner
2. Main filter
3. Intake ducting/piping
4. Turbocharger(s) and turbocharger controls
5. Charge air cooling circuit
6. Intake module for mixture management of EGR
7. Intake manifold
8. Valve porting and cylinder head design

Exhaust system components:

1. Valve configuration and exhaust tract geometry
2. Exhaust manifold
3. Turbocharger(s) and turbocharger controls
4. Exhaust gas recirculation (EGR) system
5. Positive crankcase ventilation (PCV) circuit
6. Exhaust piping
7. Catalytic converter(s)
8. Diesel particulate filter (DPF)
9. Selective catalytic reduction (SCR)
10. Muffler

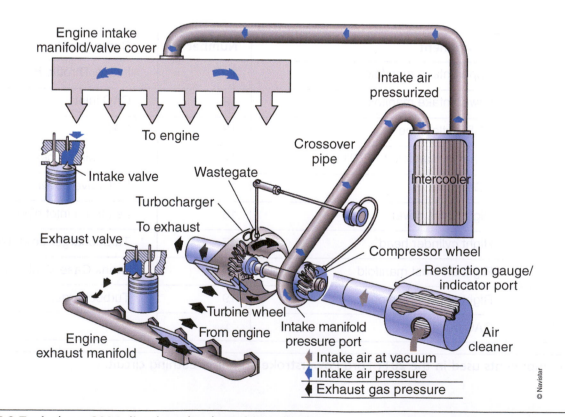

Figure 5-2 Typical pre-2004 diesel engine breathing circuit.

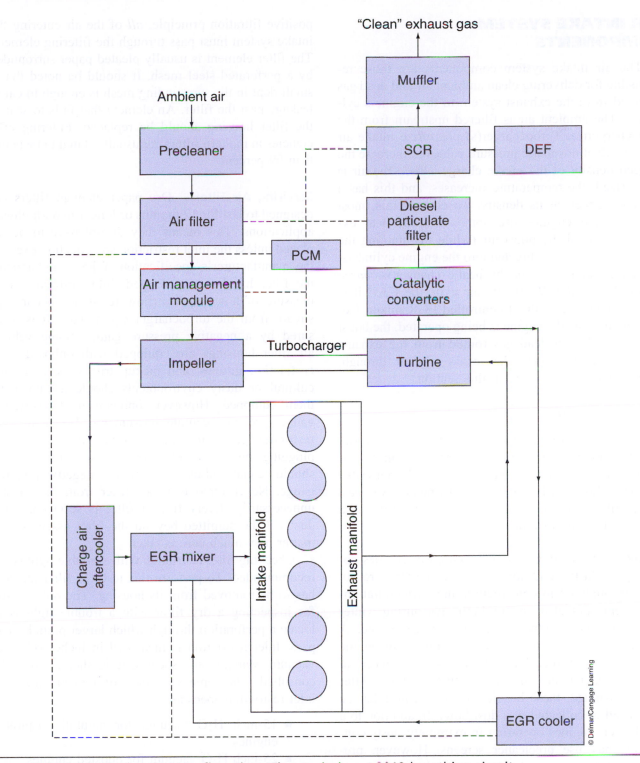

Figure 5-3 Schematic showing gas flow through a typical post-2010 breathing circuit.

Bear in mind that older engines may not be equipped with all of the components listed here, and some current engines can meet the EPA standard without using all of the components listed here. EGR became commonplace in 2004 and DPFs in 2007. These components are categorized by whether they are defined as intake or exhaust system components,

remembering that some play a role in both circuits. In most current diesel engines, the **powertrain control module (PCM)** manages engine breathing to optimize emissions. **Figure 5-3** shows the breathing circuit on a typical post-2010 diesel engine spec'd for on-highway use and equipped with a DPF and SCR.

AIR INTAKE SYSTEM COMPONENTS

The air intake system components are those responsible for delivering clean ambient air and dead gas rerouted from the exhaust system to the engine's cylinders. The ambient air is filtered upstream from the turbocharger(s). Turbocharger(s) pressurize intake air well above atmospheric pressure values to increase the oxygen density of the intake charge. When the air is pressurized, the temperature increases, and this has a reducing effect on its density. To counter this, most turbo-boosted engines use some form of heat exchanger to cool the boost air (while maintaining the pressure) before it is directed into the engine cylinders. The term used to describe this heat exchanger is *charge air cooler* (CAC). When the air exits the CAC it is routed to an engine PCM-controlled EGR mixer. Depending on how the engine is being operated, the boost air is mixed with dead gas routed from the exhaust system. We will trace the **intake circuit** through, component by component in this section.

Air Cleaners

The function of the air cleaner system on a highway diesel engine is to filter airborne particles from the air before it is routed to the engine cylinders. Airborne dirt can be highly abrasive, and when it finds its way past the air filtration system, it can destroy an engine in a very short period of time.

Precleaners. Precleaners are required in engines operating in dusty conditions. Dusty conditions are not only encountered in construction and agricultural environments but also on our highways during winter operation when highways are gritted and salted. A precleaner can triple the service life of the main engine air filter. Most diesel air filters use turbulence to separate heavier particulates by centrifugal force. After being separated, these heavier particles can either be exhausted or collected in a dust bowl. In addition to a precleaner, engines operating in extremely dusty conditions may use precleaner screens. However, precleaner screens can easily plug, so they have to be used in conjunction with inlet restriction gauges that signal the driver when a restriction threshold has been reached. In agricultural and mining applications, a precleaner can be connected to the exhaust system so that exhaust backpressure pulls particles into the exhaust gas stream.

Dry, Positive Filters. Dry, **positive filtration** air cleaners are commonly used today. Because they use a positive filtration principle, *all* of the air entering the intake system must pass through the filtering element. The filter element is usually pleated paper surrounded by a perforated steel mesh. It should be noted that a small dent in the surrounding mesh is enough to cause leakage past the filter. An element that fails to seal in the filter housing should be replaced. Filtering efficiencies in modern filters are usually stated to be better than 99 percent.

Servicing Air Filters. Dry paper element filters are designed to last for 12 months or longer in on-highway applications. This means they do not have to be replaced unless the inlet restriction specification exceeds the manufacturer's specification. When this happens, the filter has become plugged. Inlet restriction is a measure of how much airflow resistance occurs upstream from the turbocharger compressor. It is measured by a negative pressure gauge. Some vehicle air filter housings are equipped with onboard **inlet restriction gauges**, and pickup trucks used in agricultural or dusty environments should always have them equipped. However, onboard inlet restriction gauges are not accurate instruments. When an inlet restriction gauge indicates that the filter element is plugging before a scheduled service, the reading should be checked using an accurate negative pressure gauge. Never remove an air filter from its canister unnecessarily. Every time a filter is serviced, some dust will be admitted beyond the filter assembly, no matter how much care is taken.

Checking filter restriction with a trouble light is not recommended. To perform this test, the filter element has to be removed from its housing. The only reason for inspecting a dry filter with a trouble light is to locate a perforation through which larger particles can pass. Inlet restriction is measured in inches of water vacuum. Manufacturer specifications should always be consulted, but some examples of typical *maximum* inlet restriction specs are:

- 15-inch H_2O vacuum for naturally aspirated engines
- 25-inch H_2O vacuum for boosted engines

The practice of removing and attempting to clean a dirt-laden, dry filter on the shop floor should be avoided. A dirt-laden filter weighs several times as much as a new filter, and dropping it onto a concrete floor to shake dust free is usually sufficient to damage the filter element (by crumpling the perforated mesh) enough to prevent it from sealing in the housing. A worse practice is reverse blowing out of filters using compressed air. This may loosen sharp particles but leave enlarged openings through which other larger particles can pass.

Tech Tip: One of the first tests that should be performed when troubleshooting black smoke emission from an engine is to check inlet restriction. If the onboard restriction gauge is reading high, connect a negative pressure gauge to confirm the reading. The engine should ideally be tested under load, but a throttle snap to high idle should give a close indication. On some vehicles, inlet restriction is monitored with a negative pressure sensor; in the event of high inlet restriction, the driver is alerted to the condition by a dash display message.

Laundering Dry Element Filters. When a dry filter element becomes plugged, the ideal solution is to replace it. However, dry filter elements are expensive and in some operations, such as on construction sites, mining, and agricultural settings, a filter can become restricted in a couple of working days, even when precleaners are used. Professional laundering is a second-best option to replacement. The laundering process usually requires that the filter element be soaked in a detergent solution for a period of time, followed by reverse flushing with low-pressure clean water. The element is next dried with warm air and inspected for perforations. When professionally laundered filter elements are tested, filtering efficiencies are reduced with each successive laundering.

TURBOCHARGERS

Turbochargers are used on most domestic highway diesel engines. By definition, a **turbocharger** is an exhaust gas–driven, centrifugal pump that "recycles" some of the rejected heat from the engine's cylinders. Turbochargers may be driven to speeds exceeding 200,000 rpm in certain applications, and although typical maximum speeds are about 30 percent lower in most diesel engines, some post-2010 light duty diesels are capable of these speeds. Turbochargers are used to add to engine power in two ways:

- Deliver a pressurized charge of intake air to the engine's cylinders.
- Deliver drive torque to the engine drivetrain.

Principles of Operation

When a turbocharger is used to pressurize the intake air supplied to the engine cylinders, exhaust gas heat is directed onto a **turbine**. This exhaust gas heat rotates the turbine. Increase the exhaust gas heat dump (i.e., high-load operation) and the result is faster turbine rotation. Connected by means of a shaft to the other end of the turbine is a compressor wheel. When the compressor wheel is driven by the turbine, high-velocity intake air is developed. Place a restriction at the exit to the compressor housing and high-velocity air is converted into pressure. This increases the oxygen density in the air charge.

Construction. Another way of describing a turbocharger is to call it an exhaust gas–driven air pump consisting of a turbine and an **impeller**. The turbine and impeller are connected by a common shaft. The shaft is supported by bearings supplied with pressurized lube oil. The turbine wheel turns in the turbine housing. Engine exhaust gas is routed through the turbine housing. The impeller is enclosed in a separate **compressor housing**. When it is turned by the turbine, the impeller pumps intake air from the air filter, pressurizing it, before discharging it into the intake circuit. Note that the exhaust gas that drives the turbine and the intake air the impeller pressurizes, do not come into contact.

Compressor Operation. Filtered intake air is pulled into the compressor housing by the impeller. The turbine drives the impeller on the other side of the turbine shaft: this means that the actual speed of the impeller is determined by what is happening in the turbine housing. As the impeller rotates, the air in the intake system is accelerated to high speeds. High-speed air flows from the impeller to a restricted flow area. This restriction in the impeller housing converts the high-speed air into compressed air.

Turbine Operation. Exhaust gas is routed to the turbine housing. The greater the amount of heat in the exhaust gas the faster the turbine rotates. This means that turbine speeds are highest when engine loads are high. Exhaust gas enters the turbine housing radially. It is routed into a reduced-flow area known as a volute. The volute restricts flow. However, as the exhaust gas exits the volute it expands, acting on the turbine vanes before being routed into the exhaust. The amount that the exhaust gas expands depends on the amount of exhaust gas heat. The more exhaust gas heat, the more gas expansion and the faster the turbine rotates.

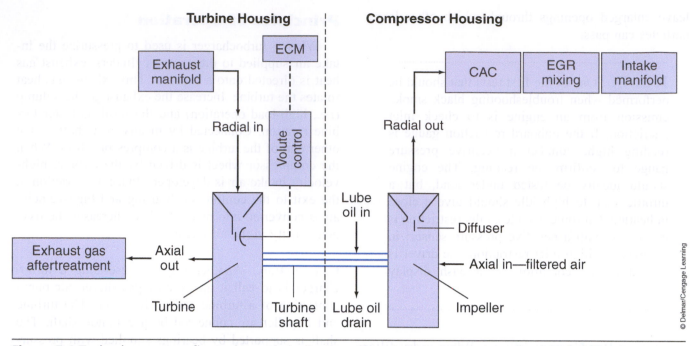

Figure 5-4 Turbocharger gas flow schematic indicating the roles played by the volute and diffuser.

Turbocharger speeds depend mainly on the exhaust gas heat and less on exhaust gas pressure. For this reason, it is not unusual to see insulated exhaust manifolds, usually with stainless steel, to maximize the heat delivered to the turbine housing.

The role of the volute should be understood. The smaller the volute size, the greater the restriction to exhaust gas flow. However, the smaller the volute, the greater the gas expansion as this gas exits to act on the turbine. The best arrangement is to be able to control the volute flow area, which we will look at next. **Figure 5-4** shows turbocharger gas flow. In the schematic you can see the key roles played by the volute and diffuser: note how gas flow enters and exits the turbine and compressor housings.

Types of Turbochargers

It is important to identify the two general categories of turbochargers. The definitions we will use in this textbook are as follows:

- **Constant geometry:** A turbocharger in which all of the exhaust flow is routed through the turbine housing regardless of how the engine is being operated.
- **Variable geometry:** A variable geometry turbo is one that uses either external or internal controls to determine how the exhaust gas passes through or around the turbine housing.

Constant geometry turbochargers were common a generation ago. Today, almost all turbochargers use one of a couple of types of variable geometry. In some cases, they use both in series with each other.

Constant Geometry Turbos. Constant geometry (CG) turbos are designed to produce their best performance at a specific rpm and engine load. They are not versatile. This limits output. Highway engines that used a single CG turbo were usually spec'd to produce peak efficiency under full load and at peak torque rpm. This meant that whenever the engine was operated outside that specified window, a performance shortfall would result. An advantage of CG turbos is that they were simple—there was little that could go wrong. **Figure 5-5** illustrates the gas flow through a typical CG turbocharger.

WARNING *Mismatching of CG turbochargers to an engine can result in damaging diesel engines either by producing high engine cylinder pressure or by causing low power, smoking, and high emissions.*

Variable Geometry Turbos. A variable geometry (VG) turbocharger is one that options exhaust gas flow either through (by varying flow area) or around (bypassing) the turbine housing using either pneumatic or electronic controls. For purposes of studying VG turbos they can be divided as follows:

- Wastegate controlled
- Volute controlled

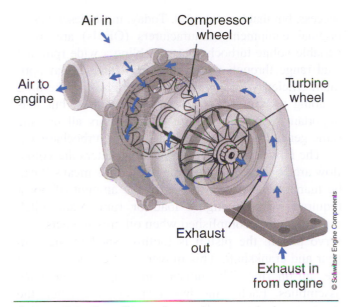

Figure 5-5 Gas flow through a typical constant geometry turbocharger.

The objective of VG turbos is to:

- Allow the turbo to act like a small turbocharger when engine loads are light.
- Allow the turbo to act like a large turbocharger when engine loading is high.

Today's PCM-controlled turbochargers can adapt to produce good performance however the engine is being operated. In addition, they can do this with almost no turbo lag (response to a change in conditions) and while keeping emissions at a minimum. **Figure 5-6** shows a cutaway of the Ford 6.7L turbocharger used on pickup trucks. This turbocharger has the following features:

- Single turbine wheel
- Dual impellers

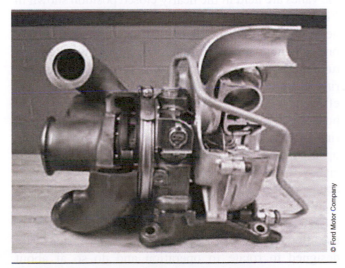

Figure 5-6 Ford's DualBoost turbocharger.

- Wastegate
- PCM-controlled, variable turbine vanes (oil-actuated servo)

The version used on non-pickup truck 6.7L engines is similar but is not equipped with a wastegate.

Wastegate-Controlled VG. Wastegated turbochargers have been around for many years. They can either allow all of the exhaust gas to be routed through the turbine housing, similar to a CG turbocharger, or they can allow a percentage of the exhaust gas to bypass the turbine. When this happens, the wastegate routes exhaust directly to the exhaust circuit. The *gate* in the wastegate functions as the exhaust gas control door. Two methods are used to control the gate:

- Manifold pressure: Wastegate movement depends on the amount of turbo boost.
- Electronic: Wastegate is positioned by the engine PCM.

Manifold Pressure Wastegate

A manifold pressure wastegate is often known as a pneumatically controlled wastegate. It uses a wastegate actuator, which has the appearance of a can with an actuator rod protruding from one end. The actuator rod moves the gate. The actuator rod connects to the wastegate door at one end. The other connects to a bellows inside the can. An internal spring within the can holds the actuator in its closed position. In the closed position, all of the exhaust gas is routed through the turbine housing. On the other side of the bellows, intake manifold boost acts against the spring. When boost pressure reaches a specified value, it begins to overcome the spring pressure, moving the actuator rod to allow some of the exhaust gas to bypass the turbine housing.

Electronic Wastegate

The engine PCM manages electronically controlled wastegates. This means that either exhaust gas could be full-flowed or a percentage could bypass the turbine housing. With a PCM-controlled wastegate, the wastegate door is controlled according to how the PCM is managing engine performance and emissions; it is not limited by how much intake manifold boost is being produced. In other words, an electronically managed wastegate provides much more flexibility than a pneumatic wastegate. **Figure 5-7** shows the wastegate actuator assembly on Ford's DualBoost turbocharger, and **Figure 5-8** shows the PCM-controlled wastegate solenoid. Vacuum is used as the servo medium in the

© Ford Motor Company

Wastegate Solenoid and Actuator

The wastegate on the DualBoost. VGT is vacuum controlled through a solenoid that is controlled by the PCM. The wastegate is only used on the DualBoost turbocharger.

The wastegate opens to relieve excessive exhaust pressure at higher RPMs.

When the VGT cannot react quickly enough to reduce MAP pressure, the wastegate will open to reduce EP to slow down the turbo.

Figure 5-7 Wastegate actuator on the DualBoost turbocharger.

© Ford Motor Company

Figure 5-8 Wastegate control solenoid on the DualBoost turbocharger.

actuator, and the PCM-driven solenoid is pulse-width modulated (PWM).

Volute-Controlled VG. Volute-controlled VG turbochargers have become more common on today's diesel engines. Race-car engines have used **variable nozzles (VNs)** to control volute flow area for many years. Diesel engines successfully adapted this technology about 20 years ago. The first volute-controlled diesel engine turbochargers appeared in the early 1990s, with limited success, but that has changed. Today, most diesel engine original equipment manufacturers (OEMs) are using variable volute turbochargers, enabling a wide rpm and load range through which the turbo functions at optimum efficiency. In this text, the Ford Powerstroke 6.7L is going to be used as a reference example, but it is important to note that Ford's competitors all use the same general principles to manage their turbochargers.

The Powerstroke VG turbocharger alters the volute flow area by moving the vane pitch. This means it can be managed to produce exactly the amount of boost required for any engine load or rpm. Vane pitch movement is accomplished when oil pressure acts on a servo piston: the piston is tooth-meshed to the cam gear and crankshaft. This movement "cranks" (turns) the unison ring. The unison ring supports the vane assemblies. Each vane has a spiral slot. When the unison ring rotational position changes, vane pitch angle changes. This is because each vane pivots on wheel pins. Whenever the crankshaft moves the unison ring, the vane pitch is altered simultaneously. In this way, the volute (inflow) flow area can be increased or decreased. As we have said previously, controlling the volute flow area onto the turbine has a lot to do with determining how fast the turbine rotates. Because this defines turbine speed, it is capable of managing boost pressure. Boost air is therefore managed by vane position because vane pitch determines how exhaust gas acts on the turbine. By examining **Figure 5-9**, **Figure 5-10**, and **Figure 5-11**, you can observe how vane pitch is altered by the VG actuator.

VG Actuator. The VG turbocharger is controlled by the VN control valve. The VN control valve is a pulse-width modulated (PWM), proportioning solenoid managed by the PCM. Maximum boost is produced when the vanes are held in the nearly closed position shown in **Figure 5-9**; it is important to note that the vanes are never fully closed. In **Figure 5-10**, oil pressure acting on the piston ring moves the cam and crank assembly, turning the unison ring clockwise. Moving the unison ring clockwise opens the vanes, reducing turbine gas efficiency. This has the effect of reducing manifold boost. As the VG actuator continues to move the unison ring clockwise, the vane pitch is changed to the fully open position (**Figure 5-11**), reducing turbine gas efficiency to its lowest level. **Figure 5-12** shows the VG actuator responsible for moving the unison ring.

Turbocharger Lubrication. All current turbochargers use floating friction bearings. Turbo bearings are designed to rotate in operation and do so at about one-third shaft speed floated with a film of oil on

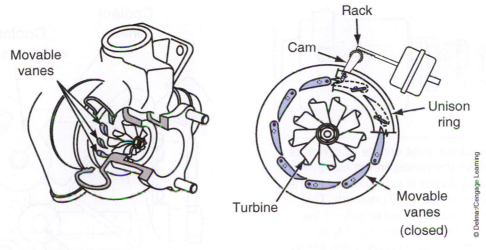

VGT Closed

When the VGT is closed it maximizes the use of the energy that is available at low speeds. Closing the VGT accelerates exhaust gas flow across the vanes of the turbine wheel. This allows the turbocharger to behave as a smaller turbocharger. Closing the vanes also increases the exhaust pressure in the exhaust manifold, which aids in pushing exhaust gas into the intake. This is also the position for cold ambient warm-up.

Figure 5-9 VGT closed. (*Text courtesy of Ford Motor Company*)

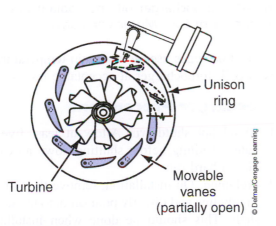

VGT Partially Open

During engine operation at moderate engine speeds and load, the vanes are commanded partially open. The vanes are set to this intermediate position to supply the correct amount of boost to the engine for optimal combustion, as well as providing the necessary exhaust pressure to drive exhaust gas.

Figure 5-10 VGT partially open. (*Text courtesy of Ford Motor Company*)

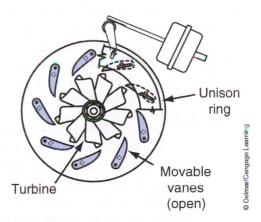

VGT Open

During engine operation at high engine speeds and load, there is a great deal of energy available in the exhaust. Excessive boost under high-speed, high-load conditions can negatively affect component durability. Therefore, the vanes are commanded open preventing turbocharger overspeed. Essentially, this allows the turbocharger to act as a large turbocharger.

Figure 5-11 VGT open. (*Text courtesy of Ford Motor Company*)

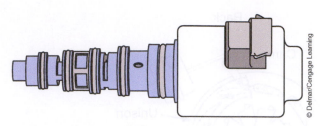

VGT Actuator

The 6.7L VGT actuator uses engine oil pressure on a piston to move the vanes on the exhaust turbine. The VGT actuator is commanded by the PCM, based on a desired MAP pressure. This actuator meters engine oil to either side of the piston. This design feature reacts quickly to changes in demand based on driving conditions. When one side of the piston is pressurized, the opposite side is vented.

Figure 5-12 VG actuator. (*Text courtesy of Ford Motor Company*)

either side of the bearing. The operating principle is that of friction bearings. This means that the turbine shaft is hydrodynamically supported during operation. Accordingly, the radial play specification is a critical determination of bearing and/or shaft wear. Turbine shaft radial play can be measured using a dial indicator. A thrust bearing, defines the endplay of the turbine shaft. Gas pressure on both the turbine and compressor housings is sealed by a pair of piston rings. The oil that pressure-feeds the bearings, spills into an oil drain cavity and then drains into the crankcase by gravity by means of a return hose. The lubricating oil required for the bearings also plays a major role in cooling the turbocharger components.

Pulse/Tuned Exhaust Manifold. Most current diesel engines use tuned exhaust manifolds. This means that the pipes from the cylinder heads that direct the exhaust gas from each cylinder are matched for minimum flow interference at most engine operating phases and are designed to compliment each other at peak power. Tuned exhaust manifolds help reduce turbo lag (turbocharger response time) and may increase low-engine load performance.

Turbocharger Cooling. Because heavy duty highway diesels carry 10 gallons (40 liters) and more of lube oil in the engine crankcase, high oil flow cooling is often sufficient to keep turbo temperatures at acceptable levels. This is not so with automobile and pickup truck diesel engines. For this reason, it is not unusual to cool turbochargers using a combination of both high oil flow and engine coolant. **Figure 5-13** shows the oil

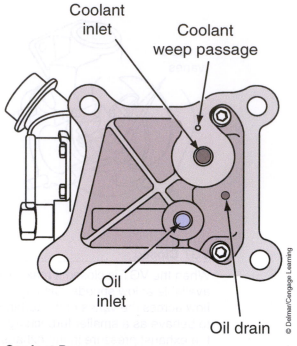

Coolant Passages

The turbocharger is cooled using coolant from the primary cooling system. Coolant enters the turbocharger from the block on the bottom of the turbocharger and flows through and out the top of the turbocharger through a line back to the coolant crossover tube.

Figure 5-13 Turbocharger oil and coolant passages. (*Text courtesy of Ford Motor Company*)

and coolant passages used to maintain temperatures in the Ford 6.7L turbocharger at acceptable levels.

Turbocharger Precautions

1. Avoid hot shutdown: Allow at least five minutes of idling before shutting down an engine after a hard run.
2. Prelubing: On installation, remove the turbo oil supply line and directly pour oil onto the turbine shaft. This should be done when installing a turbocharger before connecting the lube supply line or after an engine overhaul.
3. Prime oil filters: Most OEMs require that the oil filters be primed when replacing them at service intervals. This prevents oil starvation at the turbo shaft.

Turbocharger Failures

1. Hot shutdown: High-temperature failures resulting in warped shafts and bores.
2. Turbocharger overspeed: Caused by fuel rate tampering, high-altitude operation with a defective

altitude compensation fuel control, and mismatching of complete turbocharger units.

3. Air intake system leaks: Allows dirt to enter the compressor housing, causing the impeller vanes to be worn wafer thin.

4. Lubrication-related failure: Caused by abrasives in the engine oil, improper oil, broken-down oil, or restricted oil supply.

CHARGE AIR COOLERS

Compressing the intake air charge with a turbocharger can produce air temperatures of up to 300°F (150°C) when the ambient temperature is 70°F (21°C). These temperatures become proportionally more at higher ambient temperatures. The objective of **charge air coolers (CACs)** is to cool the air pressurized by the turbocharger as much as possible while maintaining the pressure. As intake air temperature increases, air density decreases. Less-dense air contains less of the oxygen that is required in the cylinder to burn fuel. Less-dense air results in lower power—along with higher cylinder temperatures. A charge air cooler is a **heat exchanger** and as such, its role is to reduce the temperature of boost air.

Charge air coolers use a couple of different principles to reduce boost air temperatures. These heat exchanger devices are a key component in a diesel engine management system for controlling emissions. Because the performance of charge air coolers influences combustion temperatures if they fail to perform at a specified efficiency, the result can be increased emissions, engine overheating, and lack of power. For this reason, most turbo-boosted diesel engines use some type of heat exchanger to cool intake air.

Types of Charge Air Coolers

For purposes of study, we will divide boost air coolers as follows:

- Air to air: Cooled by ram air
- Aftercoolers and intercoolers: Cooled by engine coolant

Air-to-Air Heat Exchangers. Air-to-air charge air coolers have the appearance of a coolant radiator. They are often chassis mounted in front of the radiator. As the vehicle moves down the highway, ambient air is forced through the fins and element tubing: we use the term **ram air** to describe this. Ram air cooling efficiencies are highest when the vehicle is moving at higher speeds. When the vehicle is not moving, they

have low efficiency. The engine fan may provide a little help, but generally air-to-air cooling is not going to suit vehicles that are not primarily run on the highway. When run down the highway, air-to-air coolers have better efficiency than any type of liquid-cooled heat exchangers: this is because the faster the vehicle is traveling, the greater the ram air effect.

TESTING AIR-TO-AIR COOLERS

When an engine is operating at close to rated speeds and loads, turbo-compressed air enters a charge air cooler at around 300°F (150°C). When an air-to-air heat exchanger is used, the boost air typically exits at around 110°F (44°C) when ambient temperatures are 75°F (24°C). CACs are designed so that they can leak small amounts of air without any performance drop-off. To test the air leak-off rate, plug the inlet and outlets of the charge air cooler and then pressurize it using a regulator to the OEM-recommended test value. Typically, OEMs require the heat exchanger to be charged with regulated compressed air at 30 psi (200 kPa) and then measure the drop-off over a 15-second period. The drop-off pressure should not exceed 5 to 7 psi (35 to 42 kPa). In other words, some leakage is permissible.

If the pressure drop-off exceeds specifications, drop the pressure down to 5 psi (35 kPa) and hold. Then try to locate the leak using a soap and water solution. **Figure 5-14** shows the equipment and method Caterpillar recommends to test its CACs.

Aftercoolers and Intercoolers. These normally refer to heat exchangers that use liquid engine coolant as the cooling medium. Boosted air is forced through an element that contains tubing through which coolant from the engine cooling system is pumped. Cooling efficiencies tend to be lower than with air-to-air heat exchangers due to the relatively high temperature of the liquid coolant. However, liquid media CACs are used in applications in which a good supply of ram air cannot be guaranteed. For this reason, it is used in some pickup truck applications. **Figure 5-15** shows the aftercooler assembly used in Ford pickup trucks.

Boost Circuit Troubleshooting

When manifold boost is either too high or too low, engine performance complaints result. Common causes of low manifold boost are:

- Intake circuit restriction
- Air leakage downstream from turbocharger impeller

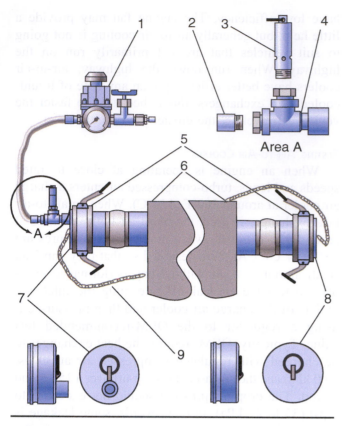

1. Regulator and valve assembly
2. Nipple
3. Relief valve
4. Tee
5. Coupler
6. Aftercooler
7. Dust plug
8. Dust plug
9. Chain

© Caterpillar

Figure 5-14 CAC test kit: Charge CAC to 30 psi (200 kPa). Pressure drop-off should not exceed 5 psi (35 kPa) after five minutes.

- Low fuel delivery
- Mismatched turbocharger

Common causes of high manifold boost are:

- Mismatched turbocharger
- Sticking unison ring on VGT turbos
- High inlet air temperatures
- Deposits on internal turbocharger components
- Overfueling
- Advanced fuel injection timing

CAUTION *Never regulate the air pressure at a value above 30 psi (200 kPa) when testing for boost side leaks because this may result in personal injury and damage to the components under pressure.*

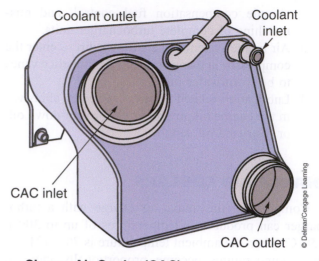

Charge Air Cooler (CAC)

The CAC is located on the left side of the engine, on top of the fender well.

The CAC is a air-to-coolant heat exchanger used to reduce the temperature of the compressed air from the turbocharger prior to entering the combustion chambers. Cooler air is denser (improving volumetric efficiency), resulting in increased power.

Figure 5-15 CAC that uses engine coolant as its medium. (*Text courtesy of Ford Motor Company*)

EXHAUST GAS RECIRCULATION

For a decade now, diesel engines spec'd for on-highway operation have used some form of **exhaust gas recirculation (EGR)** to limit NO_x emissions, though they may not call it EGR. It is not an ideal solution because it involves rerouting dirty end gas back into the engine cylinders, and this in itself presents problems. Because the role of EGR concerns emission control, we will take just a brief look at it in this chapter and address EGR in more depth in **Chapter 13**.

EGR Operation

Oxides of nitrogen, or NO_x, are produced when engine combustion temperatures are high, a condition that occurs during lean-burn combustion. Diesels run lean. Although EGR reduces engine power and fuel economy, diesel engine manufacturers had no option but to turn to EGR in order to meet 2004 emissions standards, and most have retained it with their current product even though SCR has helped. An EGR system redirects some of the exhaust gas back into the intake system. Exhaust gas is "dead" gas. That means it does not react with either the fuel or air in the engine cylinder.

Dead gas is routed back into engine cylinders to occupy space. In doing so, it reduces the lean-burn factor, and this makes NO$_x$ emission less likely. A better way of saying this is that EGR *dilutes* the intake charge of oxygen. It is not popular for the following reasons:

- Putting end gas back into the engine cylinder in place of fresh filtered air adds wear-inducing contaminants and increases engine oil acidity.
- EGR reduces engine power. However, the effect of this has been reduced by better control of combustion in modern engines.
- EGR reduces fuel economy. Again, the effect of EGR on fuel economy has been minimized due to improvements in engine technology.

Clean Gas Induction

Caterpillar adopted a type of EGR for some of its post-2007 generation engines. The Caterpillar system is called **clean gas induction (CGI)** because it sources the exhaust gas after it has exited the exhaust gas aftertreatment process. This explains the use of the term "clean." EGR and CGI are an effective means of reducing NO$_x$ emission because they function during combustion and not after NO$_x$ has been created. Because dead or unreactive gas is required to make EGR work, there is no better source (it has no cost) of dead gas than the exhaust system. It can be argued that any type of EGR defeats what took years of engineering to accomplish, the diesel engine with high engine breathing efficiency. But CGI is a way of routing dead gas back into the engine cylinders while avoiding some of the problems caused by soot-laden EGR systems.

EGR Components

In current commercial diesel engines, the control of EGR is by the PCM.

- PCM. Receives inputs from system sensors. The ECM output circuit drivers manage the operation of EGR flow gates and mixing.
- Sensors. Key sensors in the EGR circuit are ambient temperature, barometric pressure (altitude), boost pressure, mass air flow (MAF), coolant temperature, and oil temperature.
- Butterfly valve. Manages flow rate from the exhaust circuit to the EGR circuit.
- EGR mixer. Combines exhaust gas with charge air from intake system to be routed into engine cylinders. ECM controlled. In systems using a differential pressure–type MAF sensor, the venturi is usually built into the EGR mixer assembly.

The MAF sensors used in diesel engine EGR breathing systems can use either hot wire or differential pressure operating principles. These are explained in **Chapter 13.**

Intake Manifold Design

Because of the use of turbochargers on diesel engines, intake manifold design can be simple. This usually means that the runners that extend from the plenum can be of unequal lengths without affecting engine breathing efficiency. A single box manifold supplied with boost air and EGR gas is usually all that is required to meet the engine's breathing requirements. Intake manifolds can either be wet (coolant ports) or dry, but dry manifolds are used in highway diesel applications. Materials used are aluminum alloys, cast irons, poly-plastics, and carbon-based fibers.

VALVE DESIGN AND BREATHING

Cylinder head valves were studied in **Chapter 4**, so we will only take a look at them here insofar as they influence engine breathing. Most current diesel engines use multivalve configurations consisting of two inlet and two exhaust valves, but other arrangements are possible, such as two inlets and one exhaust. Two basic breathing configurations are used for the more common four-valve cylinder heads, cross-flow and parallel port.

Cross-Flow Configurations

Up until the electronic era, most diesel engines used **cross-flow valve configurations**. Cross-flow breathing locates both sets of valves transversely. This means that the inflow of intake air to the inboard valve interferes with that from the outboard valve. This is good if the idea is to create as much turbulence in the cylinder as possible. Back in the days when diesel engines produced lower fuel injection pressures, they were designed to maximize cylinder turbulence, and cross-flow, four-valve configurations did the job. **Figure 5-16** compares cross-flow with parallel port valve configurations.

Parallel Port Valve Configurations

Breathing efficiencies can be improved by using **parallel port valve configurations**. This allows each pair of cylinder valves to be responsible for an equal amount of gas flow without cross-flow interference. This generally results in lower cylinder turbulence.

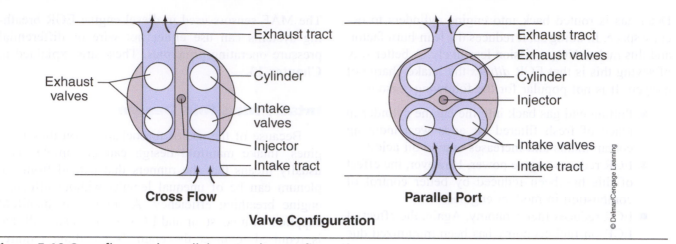

Figure 5-16 Cross-flow and parallel port valve configurations.

Parallel port valve configuration improves both cylinder charging and scavenging, but with the disadvantage of requiring a more complicated camshaft assembly (or dual camshafts). Reducing cylinder turbulence is desirable in some more recent engine designs that use high fuel injection pressures.

Valve Seat Angle

The valve seat angle also affects cylinder breathing. Valve seats in diesel engines are usually cut and machined at either 30 degrees or 45 degrees. The features of each include the following.

30-Degree Valve Seats. Gas flow both into and out of the engine cylinder is 20 percent greater using a 30-degree valve seat angle when compared with a 45-degree valve seat angle when each have equal lift. However, a 30-degree valve has less material around the seat area: this means that they do not last as long as valves cut with 45-degree valve seats.

45-Degree Valve Seats. Forty-five-degree seats tend to be used more in diesel engines due:

- Higher seating force
- Better distortion resistance

Some engines use a mixture of 30- and 45-degree valve seats. One popular engine uses a valve design in which the intake valves are machined at a 30-degree seat angle while the (hotter running) exhaust valves are machined at a 45-degree angle.

Variable Valve Timing

More recently, diesel engine manufacturers have introduced variable valve timing (VVT) managed by the PCM. At the moment, this technology is more likely to be seen in automotive high-performance and heavy duty diesel. VVT allows the PCM to reduce the compression ratio and therefore the actual air in the cylinder. Managing the compression ratio means that the engine breathes like a small engine when engine loads are light and like a large engine when more power is required. The electronically controlled, hydraulically actuated, **variable valve actuator (VVA)** assembly is mounted over the rocker assembly.

EXHAUST SYSTEM COMPONENTS

Diesel engine exhaust systems have become much more complicated in recent years. This is true because of exhaust aftertreatment devices required of diesel engines since 2007 and 2010. Once again, **Chapter 13** includes a more detailed study of emission control equipment, so the approach here will be to identify the hardware. Typically, an exhaust system is required to perform the following:

- Assist cylinder scavenging
- Minimize engine noise
- Minimize emissions
- Route heat, noise, and end gases safely to atmosphere

Exhaust Manifold

The exhaust manifold collects cylinder end gases and delivers them to the turbocharger. Exhaust manifolds are usually manufactured in single or multiple sections of cast iron, although some engines use cast aluminum alloy. Most diesel engine exhaust manifolds are "tuned." A tuned exhaust manifold is one that has been designed to efficiently route exhaust gas to the turbocharger without creating flow resistance. If the exhaust manifold and piping are properly designed, as each slug of cylinder exhaust gas is discharged, it will not "collide" with that from another cylinder but instead will be timed to unload into its tailstream. The pulsed manifolds discussed earlier in this section in the context of turbochargers are an example of tuned exhaust systems.

Exhaust backpressure factors in a diesel engine are the turbocharger turbine assembly, the exhaust aftertreatment apparatus, and the engine silencer. Exhaust manifold gaskets in truck/bus engine applications are usually of the embossed steel type, although occasionally fiber gaskets are used. Exhaust manifold gaskets are usually installed dry. Embossed steel exhaust manifold gaskets are designed for one-off usage because they yield to seal when they are clamped between a pair of mated components. Attempting to reuse them will almost certainly result in a leak, sooner or later.

Pyrometer

Pyrometers are used in emission control components, especially those equipped with DPFs. A pyrometer is a bimetal thermocouple. It is used to signal high temperatures. Two dissimilar metal wires are used: they are connected at the "hot" end, known as the sensing bulb. When the hot end is heated, a small voltage is produced: this voltage is measured using a minivoltmeter at the opposite end. Each voltage value indicates a specific temperature. The practice of locating one pyrometer in the exhaust tract on each engine cylinder to indicate cylinder balance is used on larger diesels but seldom on highway commercial diesel engines. Most diesel particulate filters use several pyrometers to signal temperature conditions to the ECM.

Exhaust Piping

Diesel engine exhaust systems are manufactured primarily from steel components. The steel is corrosion protected, usually by galvanizing (zinc coating) or using stainless steels, which effectively doubles the service life of exhaust components. Exhaust piping consists of:

- mild steel pipes
- stainless steel pipes
- flex steel pipes
- band clamps

Stainless steel band clamps have been used for many years and are an effective means of sealing flex pipe to straight pipe because they yield to shape at installation. They are not designed to be reused.

Figure 5-17 Stainless steel coupling used to clamp stainless steel piping to a Cummins ISB turbocharger.

The function of exhaust piping is to collect the engine exhaust gases discharged by the turbine housing and route them to the engine aftertreatment canister(s). **Figure 5-17** shows the stainless steel coupling clamp used at the turbocharger exhaust outlet on a 2010 Cummins ISB engine. Most diesel engines manufactured for North American highway usage after 2010 will use at least three aftertreatment canisters. After discharge from the aftertreatment canister, the exhaust should be discharged to atmosphere clear of the vehicle structure because intense heat can be generated by the aftertreatment process.

Engine Exhaust Aftertreatment

Today's diesel engines have complex exhaust gas aftertreatment canisters. A typical aftertreatment canister or set of canisters may incorporate:

- A muffler or engine silencer
- A diesel particulate filter (DPF)
- Oxidation **catalytic converter**
- Reduction catalytic converter
- Clean air induction piping (routes "dead" gas back to intake)
- **Selective catalytic reduction** (aqueous urea injection)
- Diffuser (redirects DPF heat away from trailer/ reefer, etc.)

These aftertreatment devices are usually contained within a single canister with many different sensors. The aftertreatment device is usually manufactured by the engine OEM and *must* be replaced with an OEM-approved equivalent. Because they are engine computer-controlled and part of the emission control circuit, we will address them in some detail in **Chapter 13.** In this chapter, we will address only the

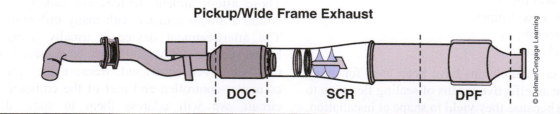

Figure 5-18 Coupling used to attach exhaust pipe to the DPF canister on a 2010 Cummins ISB.

principles of a muffler or engine silencer on a diesel engine. The design of the aftertreatment canister attempts to minimize restriction in routing "clean" exhaust gas to the atmosphere. **Figure 5-18** shows the coupling between the exhaust pipe and the DPF canister on a Cummins ISB engine.

> **CAUTION** *It is illegal to tamper with emission control hardware. Diesel exhaust aftertreatment canisters must be replaced with an OEM-approved equivalent that allows the engine to meet the emission controls that prevailed in the year of the engine's manufacture.*

Aftertreatment Canisters. Depending on the manufacturer and the type of chassis, the OEM may use separate canisters for each aftertreatment phase or may incorporate them all within a single housing. In most but not all automotive applications, the PCM is responsible for managing aftertreatment. In cases where a separate aftertreatment module is used, the module is assigned with an address on the J1850 data bus and is therefore networked directly with the PCM. Because almost all post-2010 diesel engines are required to use SCR, it is mandatory to have a method of alerting the operator to low aqueous urea levels. Aqueous urea is now known as **diesel exhaust fluid (DEF)** and is

available in refueling stations that supply diesel fuel. DEF is examined in some detail in **Chapter 13.**

Sonic Emission Control

Sound is an energy form and it is produced in an engine by the firing pulses in engine cylinders. The noise produced by a running engine is generally considered to be unpleasant to the human ear. The function of the silencer or muffler is to modulate the frequency of engine noise to legally acceptable levels.

Before the tougher sonic emissions standards of the 1990s, a turbocharger along with piping design was in some cases able to meet the minimum noise emissions standards, but this is no longer so. Engine silencers use two basic principles to achieve their objective of dampening sound:

1. **Resonation.** Resonation requires reflecting sound waves back toward the source; this process multiplies the number of sound emission points, changing the frequency. Separate chambers connected by offset pipes are used to achieve this. The idea is to reduce the noise while minimizing gas flow restriction. One of the reasons that the muffler section (usually the last section of the device) of a post-2010 aftertreatment canister looks simpler (or, is in some cases eliminated) is that DPFs, SCRs, and catalytic converters have a significant resonation effect that can sometimes render a muffler unnecessary.

2. **Sound absorption.** Exhaust gases pass through a perforated pipe enclosed in a canister filled with sound-absorbing material. Sound absorption involves converting sound energy into heat by friction. The efficiency of these devices depends on the packing density of the sound-absorbing material. Sound-absorption mufflers generally have improved flow resistance when compared with resonator mufflers but are usually insufficient to lower noise to legal levels without the help of a resonator.

Figure 5-19 shows the multistage exhaust canister arrangement used on post-2011 Ford pickup trucks powered by diesel engines.

Pickup/Wide Frame Exhaust

DOC   SCR   DPF

Figure 5-19 Components of the exhaust aftertreatment system used on a 2011 Ford pickup truck application.

Diffusers. Because of the amount of heat generated by DPFs during regeneration, it is sometimes necessary to fit a heat diffuser onto exhaust gas discharge pipes. The heat diffuser simply redirects heat away from chassis components that may be heat sensitive.

BREATHING CIRCUIT SENSORS

While we are not going to take a look at the operating principles of sensors until **Chapter 12,** we are going to identify those used in a typical system. Both intake and exhaust circuits are monitored in current diesel engines mainly because of the need to contain exhaust emissions within legal limits. The following is a list of sensors you may find on a diesel engine and in parentheses, the electrical principles under which they operate. Many sensors on today's engines are multifunctional. That means they perform multiple tasks, so what appears to be one sensor may read both temperature and pressure.

- Ambient temperature (thermistor)
- Barometric pressure (variable capacitance)
- CAC in temperature (thermistor)
- CAC out temperature (thermistor)
- Boost circuit temperature (thermistor)
- Manifold boost pressure (variable capacitance or piezo-resistive)
- Mass air flow (MAF) (hot-wire or delta pressure differential)
- Turbocharger shaft speed (inductive pulse generator)
- Exhaust gas backpressure (variable capacitance)
- DPF in temperature (pyrometer or PRT)
- DPF combustion temperature (pyrometer or PRT)
- DPF out temperature (pyrometer or PRT)
- NO_x sensor (SCR) (density or galvanic)
- Aftertreatment canister mass flow (delta pressure sensor)

Summary

- Most current diesel engines use a dry, positive filter system.

- Diesel engines operated in environments with airborne particulates such as grain chaff dust, construction, and road dust should use some kind of precleaner. Precleaners extend air cleaner element service life.

- Dry, positive-type filters have the highest efficiencies just before they become completely plugged.

- It is important not to over-service air filters because every time the housing is opened, some dust will find its way downstream from the filter assembly. Many air filters last for up to a year or more when operated in a clean atmosphere.

- Air inlet restriction should be tested with a water manometer or negative pressure gauge.

- Charge air heat exchangers cool the turbo-boosted air charge and therefore increase its oxygen density.

- Air-to-air charge air coolers have higher cooling efficiencies but must have adequate ram airflow; this makes them ideal for use in highway applications but not in high-load, low-road speed vocational applications.

- Most current diesel engines use some type of EGR system to dilute the intake charge with dead gas.

- Valve configuration affects both cylinder breathing efficiency and cylinder turbulence.

- Parallel port valve configurations generally produce better and more balanced cylinder breathing efficiency but also produce lower swirl.

- Valve seats cut at a 45-degree angle produce greater flow restriction and higher seating force than valves cut at a 30-degree angle, assuming identical lift. They are usually preferred in diesel engines.

- Turbochargers increase exhaust backpressure. Their objective is to recapture some of the engine rejected heat by using it to pressurize the intake charge to the cylinders.

- Turbochargers are driven by the *heat* in the exhaust gas, so the more heat, the faster the turbine rotational speeds.

- Turbochargers in diesel engine applications may wind out at up to 150,000 rpm with mean running speeds in the 70,000–80,000 rpm range.

- Diesel engine OEMs use variable geometry turbochargers to increase the efficient operating range and reduce the turbo lag duration. VG turbos may be managed internally by variable volute or externally by wastegates.

- Turbocharger radial and axial runouts should be routinely inspected at engine preventative maintenance (PM) intervals.

- Hot shutdowns are a major cause of turbocharger failures. After prolonged high-load operation, a cool-down period of at least five minutes is recommended.

- Engine silencers use resonation and sound absorption principles to alter the frequency of the sound emitted from the engine; most use a combination of both principles.

- The exhaust aftertreatment canister on today's diesel engines contains a DPF, SCR, oxidation and reduction catalytic converters, and a muffler device.

Internet Exercises

Use a search engine and information-sites such as Wikipedia and www.howstuffworks.com, and key in the following prompts:

1. Donaldson air filters
2. Donaldson exhaust aftertreatment devices
3. Borg Warner/Schwitzer turbochargers
4. Holset turbochargers
5. Exhaust gas recirculation
6. Diesel exhaust fluid (DEF)
7. Check out Ford Diesel, Duramax Diesel, and Dodge Cummins Diesel online
8. Research diesel engine power in racing applications: Begin with Audi Diesel and Peugeot Diesel

Lab Tasks

1. Use a manometer or negative pressure gauge to measure the inlet restriction of an air filter with the engine at idle and high idle.
2. Find manufacturer specifications for radial and axial turbine shaft play; measure both with a dial indicator.
3. Fit a negative pressure gauge to an exhaust port; measure exhaust backpressure.
4. Disassemble a cylinder head, removing the valves. Determine the seat angle on each. Note any differences between the intake and exhaust valves.
5. Examine a diesel engine "muffler." Determine exactly what components are contained within it: this will vary considerably depending on the age of the engine.

Review Questions

1. Which tool should be used to *accurately* check the inlet restriction of a dry air filter?

   A. Mercury manometer   C. Negative pressure gauge

   B. Trouble light   D. Restriction gauge

2. Which of the following would be a typical maximum specified inlet restriction for an air filter on a turbocharged diesel?

   A. 25-inch water   C. 25 psi

   B. 25-inch mercury   D. 25 kPa

3. Which of the following types of filters has the highest filtering efficiencies?

   A. Centrifugal precleaners   C. Dry, positive

   B. Oil bath

4. Which of the following should be performed *first* when checking an engine that produces black smoke under load?

   A. Injection timing

   B. Plugged particulate trap

   C. Fuel chemistry analysis

   D. Air filter restriction test

5. Which of the following best describes the function of an EGR system on a diesel engine?

   A. Increases engine breathing efficiency

   B. Dilutes intake charge with cooled dead gas

   C. Preheats the intake charge to the cylinder

   D. Assists the turbocharger in boosting intake pressure

6. Which of the following best describes the function of a wastegate in a current turbo-boosted diesel engine?

   A. Bleeds down intake boost air when excessively high

   B. Options exhaust gas to bypass the turbine

   C. Adjusts the volute flow area

   D. Holds the exhaust valves open when turbo-boost is low boost

7. What is the critical flow area managed by a VG turbo using a variable vane pitch operating principle?

   A. Compressor diffuser

   B. Impeller inlet throat

   C. Turbine volute

   D. EGR gate

8. Technician A says that some turbochargers use a combination of high-flow oil and engine coolant to help manage turbine operating temperatures. Technician B says that some turbochargers use both a wastegate and a variable pitch vane principle to manage gas flow through the turbine. Who is right?

   A. Technician A only

   B. Technician B only

   C. Both A and B

   D. Neither A nor B

9. Which of the following components would be located in a diesel exhaust gas aftertreatment canister?

   A. Diesel particulate filter

   B. Catalytic converter(s)

   C. Muffler

   D. Any or all of the above

10. Which of the following best describes the function of a VG turbocharger?

    A. Behaves like a large turbocharger when engine load is high and like a small turbocharger when engine loading is light

    B. Behaves like a small turbocharger when engine load is high and like a large turbocharger when engine loading is light

    C. Reroutes NO_x back into the engine cylinders

    D. Reroutes EGR gas into the DPF

CHAPTER 6

Cooling and Lubrication Circuits

Prerequisites

Chapters 2, 3, and 4.

Learning Objectives

After studying this chapter, you should be able to:

- Identify diesel engine cooling system components and their principles of operation.
- Define the terms *conduction*, *convection*, and *radiation*.
- Identify the three types of coolants used in current highway diesel engines.
- Outline the properties of a diesel engine antifreeze.
- Calculate the boil and freeze points of a coolant mixture.
- Mix coolant using the correct proportions of water, antifreeze, and SCAs.
- Perform standard SCA tests and measure antifreeze protection.
- Identify the problems scale buildup can create in an engine cooling system.
- List the advantages claimed for extended life coolants.
- Identify the role played by a diesel engine radiator.
- Test a radiator for external leakage using a standard cooling system pressure tester.
- Test a radiator cap.
- Identify the different types of thermostats in use and describe their principles of operation.
- Describe the role of the coolant pump.
- Define the role of the coolant filters and their servicing requirements.
- List the types of temperature gauges used in diesel engines.
- Describe how a coolant level warning indicator operates.
- Define the roles played by the engine fan in managing engine temperatures.
- Diagnose basic cooling system malfunctions.
- Identify the components of a typical diesel engine lubrication circuit.
- List the properties of diesel engine oils.

- Define the term *hydrodynamic suspension*.
- Describe the difference between thin film and thick film lubrication.
- Interpret API classifications and SAE viscosity grades.
- Replace and properly calibrate a lube oil dipstick.
- Describe the two types of oil pumps commonly used on diesel engines.
- Describe the operation of an oil pressure regulating valve.
- Define the term *positive filtration*.
- Outline the differences between *full flow* and *bypass* filters.
- Service a set of oil filters.
- Outline the role of an oil cooler in the lubrication circuit.
- Test an oil cooler core using vacuum or pressure testing.
- Identify the methods used to signal oil pressure in current diesel engines.

Key Terms

American Petroleum Institute (API)

antifreeze

boundary lubrication

bundle

bypass filter

bypass valve

centrifugal filter

conduction

convection

cross-flow radiator

diesel coolant additives (DCAs)

downflow radiator

ethylene glycol (EG)

extended life coolants (ELCs)

fire point

flash point

fluid friction

full-flow filter

gerotor

headers

heat exchanger

hydrodynamic suspension

hydrometer

inhibitors

lubricity

oil cooler

pH

positive filtration

pour point

propylene glycol (PG)

radiation

radiator

ram air

refractometer

rejected heat

relief valve

shear

single pass

supplemental cooling additives (SCAs)

synthetic oil

thermatic fan

thermostat

thick film lubrication

thin film lubrication

total dissolved solids (TDSs)

viscosity

waterless coolants (WCs)

INTRODUCTION

Because diesel engines produce combustion temperatures that exceed the melting point of some of the materials from which they are constructed, they depend on cooling and lubricating systems to manage maximum operating temperatures.

When a failure occurs in either one of these critical circuits, the result is usually rapid engine failure. For this reason, both cooling and lubricating system performance are monitored in real time by engine management computers: when the chance of imminent failure is detected, the engine is shut down.

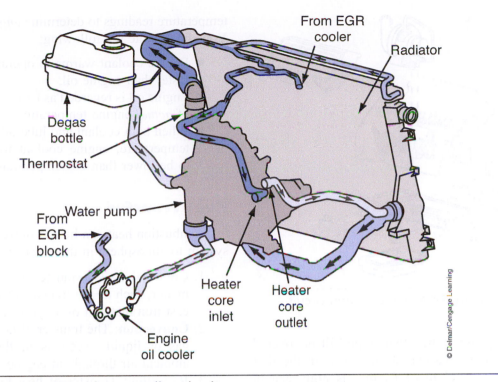

Figure 6-1 PowerStroke 6.7L primary cooling circuit.

Cooling System Objectives

Cooling systems have to transfer a percentage of engine **rejected heat**. Rejected heat (see **Chapter 2**) is that percentage of the potential heat energy of a fuel that the engine is unable to convert into useful mechanical energy. Engine rejected heat has to be transferred to the atmosphere, either in the exhaust gas or indirectly using the engine cooling system. If an engine is operating at 40 percent thermal efficiency, 60 percent of the potential heat energy is rejected. Approximately half of the rejected heat leaves the engine in the form of exhaust gas, which leaves the engine cooling system responsible for transferring the other half to atmosphere.

This task of safely transferring rejected heat is complicated by the extremes of our climate and the fact that it is necessary to manage a consistent engine operating temperature at all engine speeds and loads. Failure to manage engine operating temperatures has a negative effective on engine performance and emissions. Liquid cooling systems are universal in North American commercial diesel engine applications, and only they will be addressed in this section. The Deutz engine company of Germany manufactures air-cooled engines, but in North America, their engines are generally only found in agricultural and mining applications. **Figure 6-1** shows the primary cooling circuit used in a 2011 Ford PowerStroke engine.

Lube Circuit Objectives

The primary objective of the lubrication circuit is to minimize friction, but it also plays a major role in assisting the cooling system in managing engine temperatures. Most commercial diesel engines currently used on North American highways use a pressure lubrication system. The lube circuit supplies the bearings and other moving components with engine oil. Engine oil is formulated to fulfill the lubrication and service life requirements of diesel engines. In addition to lubricating an engine's moving parts, lube oil is also increasingly used to cool engine components. The basic components required of a modern diesel engine lubrication system and the flow routing are shown in **Figure 6-2.** Later in this chapter we will take a look at:

- Functions of a lubrication circuit
- Lubricants
- Oil pan
- Oil pump
- Filters
- Oil cooler
- Piston cooling requirements: spray nozzles

Friction

Friction is a common element in our lives that we simply take for granted. For instance, we can walk up a steep hill without slipping because of high friction between the soles of our feet and the ground surface.

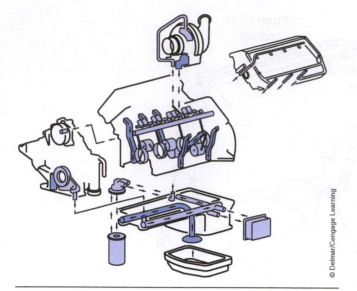

Figure 6-2 PowerStroke 6.7L lubrication circuit.

We also accept that when that same hill is covered with packed snow, we can ski down it. In the first instance, the coefficient of friction is high, and in the second it is low. Another way of thinking about the coefficient of friction is as a means of rating the aggressiveness of friction surfaces. Rubber soles have a higher coefficient of friction than skis. Lubricants are designed to reduce friction between surfaces that are, or could be, in contact with one another, creating wear at the contact points. When a lubricant fails to reduce friction, heat is generated. The heat generated can rapidly destroy components if not held in check.

COOLING SYSTEM

The functions of a diesel engine liquid cooling system are to:

- Absorb heat from engine components.
- Absorb heat from engine support systems such as EGR and oil coolers.
- Transfer the absorbed heat by circulating the coolant.
- Supply heat to heater cores.
- Transfer heat to atmosphere by means of heat exchangers.
- Manage engine operating temperatures.

The actual coolant temperature at any given moment of operation is one means that the PCM uses to determine engine temperature. In fact, in older hydromechanical engines the cooling system was entirely responsible for managing engine and under-hood temperatures. In today's engines, PCMs use both coolant and oil

temperature readings to determine *engine* temperature. You should note the following:

- Engine coolant warms to operating temperature faster than engine oil.
- Engine oil is regarded as being a better indicator of *actual* engine temperature.
- When both coolant and lube oil are at operating temperature, engine coolant temperatures tend to be lower than oil temperatures.

Heat Transfer

Combustion heat can be transferred by the cooling system to atmosphere in three ways:

1. **Conduction**: The transfer of heat through solid matter, such as the transfer of heat through the cast iron material of a cylinder block.
2. **Convection**: The transfer of heat by currents of gas or liquids, such as in the movement of ambient air through an engine compartment.
3. **Radiation**: Transfer of heat by means of heat rays not requiring matter, such as a fluid or solid. The turbine housing of a turbocharger radiates a considerable amount of heat.

Antiboil Properties

Cooling systems are sealed and maintained under pressure. By confining a liquid under pressure, you increase its boil point. Most cooling systems are designed to manage coolant temperatures at just below their boil points. The chemistry of the **antifreeze** and its concentration in the coolant will define the actual boil point of a coolant. Most antifreeze doubles as an antiboil agent. When an engine is approaching an overheat condition, the coolant will first boil at the location within the system where the pressure is lowest. In most cases, coolant boiling will occur first at the inlet (suction side) of the system water pump for the very reason that pressure is slightly lower in that location.

ENGINE COOLANT

Water-based coolant is the medium used to absorb engine rejected heat, transferring that heat to atmosphere via a **heat exchanger**. Engine coolant is a mixture of water, antifreeze, and **supplemental cooling additives (SCAs)** also known as **diesel coolant additives (DCAs)**. In most light duty diesel engine coolants, the SCAs are combined with the antifreeze mixture. It should be noted that if the only objective of diesel engine coolant was to transfer heat, pure water would accomplish this more efficiently than any currently used antifreeze mixture. However, water possesses inconvenient boil and freeze points, poor lubricating properties, and furthermore promotes oxidation and scaling activity.

Types of Antifreeze

Most light duty diesel engines use a mixture of ethylene glycol (EG), plus water as coolant, or a premix of carboxylate, EG-based extended life coolant (ELC). Propylene glycol (PG) solutions are not common and tend to be used only when an engine is operated at extremely low temperatures. Properly formulated diesel engine coolants are made up of the correct proportions of water, antifreeze, and SCAs. ELCs are low maintenance in that the coolant life can be up to six years, though the manufacturers' service literature should be consulted because ELCs are not all the same. Although not as yet approved by any of the light duty diesel engine manufacturers, waterless coolants (WCs) are widely used in heavy duty diesels and it is only a matter of time before their usage catches on elsewhere. WCs are EG based; they may last for the life of the engine with little or no maintenance provided and they are never contaminated with water or other coolants.

Coolant Expansion and Contraction

Water expands about 9 percent in volume as it freezes. This means that it can distort or fracture any container it is stored in, even when this container is a cast iron engine block. Water occupies the least volume when it is in the liquid state and close to its freezing point of 0°C or 32°F. As water is heated from a near freezing point to a near boiling point it expands approximately 3 percent. A 50/50 mixture of water and EG expands even more, approximately 4 percent through the same temperature range. This means that cooling systems must be designed to accommodate the expansion and contraction of the engine coolant while it is in the liquid state. Just as important to its operation is that antifreeze is also antiboil.

Properties of Antifreeze

The mixture of water, antifreeze, and SCAs that is referred to as engine coolant should perform the following:

1. Corrosion protection. Special anticorrosion ingredients in the antifreeze and the SCA package protect the metals, plastics, and rubber compounds in the engine cooling system.
2. Freeze protection. The freeze protection provided by any coolant relates to the proportion of antifreeze in the coolant mixture.
3. Antiboil protection. The antiboil protection provided by coolant relates to the proportion of antifreeze in the mixture.
4. Antiscale protection. Diesel engine antifreeze mixtures contain additives to prevent scale buildup in the engine. Scaling is caused by hard water mineral deposits.
5. Acidity protection. A pH buffer is used to control acid formation in the coolant, which would result in corrosion.
6. Antifoam protection. The action of pumping coolant through an engine can cause aeration or foaming of the coolant, so this must be controlled.
7. Antidispersant protection. This prevents insoluble matter from coagulating and plugging cooling system passages.

Toxicity of Coolants

While the antifreeze and antiboil characteristics of glycol-based coolant mixtures tend not to change much, the protection additives degrade with engine operation, especially at high temperatures. This requires that the manufacturer service requirements of coolants be observed. EG has been used as the standard antifreeze for some time, but the Federal Clean Air Act and the Occupational Safety and Health Administration (OSHA) regard EG as a toxic hazard. Leaks and spillage of antifreeze should be regarded as dangerous to mammals (including humans) and plant life. Engine coolant becomes more toxic as it ages. Although ELC uses an EG base, it is regarded as safer because it generally requires less handling (when premixed and not requiring routine testing) in service.

Antifreeze Protection

The temperature at which a glycol-based coolant solution stops protecting against freezing depends on the solution concentricity. **Table 6-1** compares the EG coolant mixture percentage with the freeze point temperature. Note that when the percentage of EG is increased above 60 percent, the freeze point protection starts to drop off.

Antifreeze Color

Antifreeze is colored with dye. The color of antifreeze is meaningless. Antifreeze may be dyed green,

TABLE 6-1: ANTIFREEZE CONCENTRATION AND FREEZE POINT

Concentration of EG Antifreeze by Percent Volume	Freeze Point of EG Coolant	
0% (water only)	32°F	0°C
20%	16°F	–10°C
30%	4°F	–16°C
40%	–12°F	–24°C
50%	–34°F	–37°C
60%	–62°F	–52°C
80%	–57°F	–49°C
100%	–5°F	–22°C

yellow, blue, orange, red, pink, or any other color in the spectrum. Chemically identical antifreezes can be sold in a number of different colors depending on which manufacturer's brand name is on the container. When green EG is mixed with orange EG, the result is a mud-colored solution. Nevertheless, it will perform as specified despite an unpleasant appearance.

Measuring Coolant Mixture Strength

Standard antifreeze **hydrometers** are calibrated to measure EG mixtures. However, even when measuring these, the readings are not accurate and require temperature correction. Most manufacturers prefer the use of a **refractometer** to test the antifreeze protection of a coolant. A refractometer produces an accuracy of within 4°C (7°F) of the actual freeze point of the coolant.

Tech Tip: Use a refractometer to test the freeze protection of coolant solution when one is available. If you use a hydrometer, make sure you calculate the temperature correction: this is based on coolant temperature, not ambient temperature.

Scaling. One engine manufacturer states in its service literature that a scale buildup of 1.5 mm (¹⁄₁₆") had the equivalent insulating effect of 100 mm (4") of cast iron. Scaling is caused by hard water mineral deposits (especially magnesium and calcium) forming on the surfaces of the cooling system where temperatures are highest. Scale acts as an insulator and severely limits the ability to transfer engine heat to the coolant. This condition can be especially serious in

engines with aluminum cylinder heads and cylinder blocks.

DESCALING ENGINES

There are commercially available descalants. These can sometimes remove minor scale buildup in the engine cooling system without removing the engine from a chassis. After descaling, the cooling system should be flushed. However, by the time scale buildup has progressed to the point that it is causing an engine to overheat, it is usually too late to rescue it with descalant. The engine usually has to be disassembled and the cylinder block and heads boiled in a soak tank. Note that the solution used to soak aluminum and cast iron engine components is quite different: check the manufacturer's service literature.

Testing Coolant pH. The **pH** level is a measure of the acidity of the coolant. Acids may form in engine coolant exposed to combustion gases, or in some cases, when cooling system metals start to corrode. High acidity is damaging to engine components because it is corrosive, especially to aluminum, but also because of its galvanic effect. The pH test is a litmus test in which a test strip is first inserted into a sample of the coolant, then removed, and the color of the test strip is compared to a color chart provided with the kit. The acceptable pH level is defined by each manufacturer: this is usually between 7.5 and 11.0 on the pH scale. A pH of 9.5 is typical. Higher acidity readings (below 8.0 on the pH scale) indicate that corrosion of engine copper and iron-based materials is taking place. Higher-than-normal alkalinity readings indicate aluminum corrosion or that low-silicate antifreeze is being used where a high-silicate antifreeze is required.

Testing TDS. Testing for **total dissolved solids (TDS)** requires using a TDS probe. A TDS probe measures the conductivity of the coolant by conducting a current between two electrodes. Distilled water does *not* conduct electricity. The ability of water to conduct electricity increases with its TDS content. The TDS test is performed by inserting the probe into the top radiator tank. A reading higher than the original equipment manufacturer (OEM)-specified TDS measured in parts per million is an indication that the condition of the coolant may be conducive to scale buildup.

Good-Quality Water. One of the reasons premixed ELC significantly outlasts EG and PG solutions is that distilled water is used in the solution. There are so many variables associated with tap water that it does not make sense to use anything but distilled water when mixing antifreeze solution. Distilled water is not costly, but it is not as readily available on the shop floor as tap water. Good-quality water should have the following specifications:

- Less than 40 ppm chloride
- Less than 100 ppm sulfate
- Less than 170 ppm calcium/magnesium (hardness)

Tech Tip: The main reason a premixed ELC antifreeze outlasts standard EG-based antifreeze is due not so much to the superior chemistry of the product but to the elimination of high-TDS tap water. To maximize coolant service life, spend a few extra pennies and mix EG with distilled water.

High-Silicate Antifreeze. High-silicate concentrations are used in antifreezes designed to protect aluminum components exposed to the engine coolant. However, most diesels using cast iron–based cylinder blocks and heads require that low- or no-silicate coolant be used in their engines. High-silicate and low-silicate antifreeze should not be mixed. Always adhere to the specific engine manufacturer's recommendations regarding the appropriate antifreeze. And remember, the recommendation for one manufacturer's 2000 model year may damage its 2010 model year engine.

WARNING When running a vehicle in severe winter conditions, the thermostat will close-cycle coolant through the engine. This means that the radiator, which is exposed to frigid ram air, can ice if there is insufficient freeze protection. Ensure that coolant freeze protection accounts for the lowest ambient temperature with a margin of at least another 10°F.

CAUTION After coming into contact with any antifreeze or coolant solution, wash the affected skin areas immediately and thoroughly.

COOLING SYSTEM COMPONENTS

The components used to store, pump, condition, and manage engine coolant flow and temperature are known as the cooling system components (**Figure 6-1**). These components vary little from one diesel engine manufacturer to the next. When servicing and repairing cooling system components, always consult the appropriate service literature.

Radiators

Radiators are heat exchangers. The engine power rating determines the size of a radiator required by any engine. In vehicle applications, coolant is circulated through the radiator, which transfers heat to **ram air**. Ram air is air forced through the radiator core as the vehicle moves down the highway. Vehicle speed and ambient temperatures determine the radiator's efficiency as a heat exchanger. Fan shrouds improve air flow through the radiator core and improve the efficiency of the fan.

Radiator Materials and Construction. Most vehicle engine radiators are fabricated mainly from copper, brass, aluminum, and plastic components. Radiators typically consist of bundled rows of round or elliptical tubes through which the coolant is flowed. Fins increase the sectional area to which ram air is exposed. Copper, brass (an alloy of copper and zinc), and aluminum all transfer heat efficiently and are ideal as materials for radiators. Aluminum corrodes more easily than copper and brass. The corrosion occurs from within (coolant breakdown, poor water quality) and outside (salt, both ambient and road salt). Plastics are being increasingly used in the construction of radiator tanks to replace metal tanks. Plastic tanks are usually crimped onto the main radiator core, enclosing the **headers**.

All radiators are equipped with a drain valve located at their lowest point, inlet and outlet piping, and a filler opening sealed with a radiator cap. Most use a **single pass**, downflow principle, which requires that the cooling tubes run vertically from the top tank to the bottom tank. Radiators are classified by their flow characteristics. The flow routing in a single pass, **downflow radiator** is shown in **Figure 6-3**.

In addition to single pass, downflow routing, it is common to use a single pass, **cross-flow** radiator as shown in **Figure 6-4**. Less common is the double pass radiator also shown in **Figure 6-4**; when it is used, it is usually in applications engineered for low ram air availability—that is, engines that have to operate with little or no ram air.

Aeration. Air in a cooling system can cause many problems. It reduces the coolant's ability to transfer

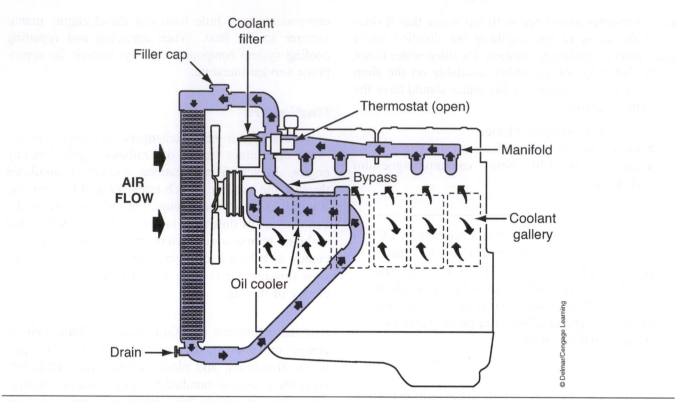

Figure 6-3 Single pass downflow radiator.

heat, may promote corrosion, and may cause a cooling system to become air bound. An air-bound system occurs when air becomes trapped in the inlet to the coolant pump. The result is that the coolant pump loses its prime. Most cooling systems are designed to limit aeration problems. Some radiators have a divided top tank in which coolant enters the lower section from the cylinder head and the upper section from a standpipe; the two sections are separated by a baffle. This design reduces cooling system aeration problems. Also, vent lines help to de-aerate the cooling system.

Split Circuit Cooling. As diesel engine emissions management becomes more complex, the overall cooling requirements of engines increase. In addition, some manufacturers have sectioned the cooling circuit with the objective of running differing coolant operating temperatures for the engine primary and secondary support apparatus coolant circuits. Now it is not uncommon to have a pair of heat exchangers for the engine systems, along with others for the transmission, power steering, and air conditioning condenser, all competing for whatever ram air can be driven through the grille. **Figure 6-5** shows the arrangement of heat exchangers used on a typical diesel-powered Ford pickup truck.

The management of each side within a split cooling circuit is usually separate. This means that separate water pumps and thermostat(s) are required for each side. When a split circuit is used in a vehicle the following is usually true:

- Primary circuit: Cools the main engine components, including the cylinder block and head(s).
- Secondary circuit: Cools the engine support circuits such as the EGR, CAC, fuel, transmission, and other support heat exchangers.

Figure 6-6 shows the secondary coolant flow routing used on a 2011 Ford pickup.

Servicing and Testing Radiators. Radiators in many vehicles tend to be left in the chassis with no maintenance until they fail, either because they leak or fail to adequately cool. An often overlooked maintenance practice is the external cleanliness of the radiator; buildup of road dirt and summer bugs can reduce a heat exchanger's ability to properly cool. Radiators should be cleaned externally using either a low-pressure washer or regular hose, detergent, and a soft nylon bristle brush. Never use a high-pressure washer because this will almost certainly result in damaging the cooling fins.

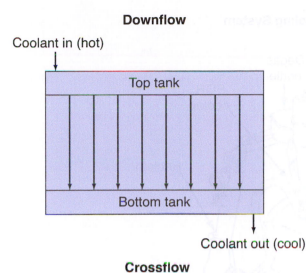

Downflow

Coolant in (hot)

Top tank

Bottom tank

Coolant out (cool)

Crossflow

Coolant in (hot)

Inlet tank

Outlet tank

Coolant out (cool)

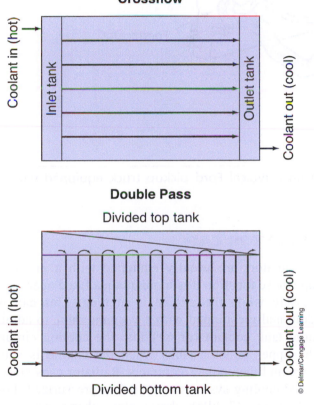

Double Pass

Divided top tank

Coolant in (hot)

Coolant out (cool)

Divided bottom tank

© Delmar/Cengage Learning

Figure 6-4 Single pass cross-flow radiator.

Repairing Leaks. Leaks are more often the result of external damage than corrosion failure. The location of a leak is often indicated by white or reddish streaks. Radiators are commonly pressure tested at around 10 percent above normal operating pressure, but be sure to consult the manufacturer's test specifications, especially where plastic tanks are used. Pressure testing may help identify the locations of leaks. It is important that radiator leaks are promptly repaired. If the leak has been caused by external damage and the radiator is in otherwise good condition, the radiator may be repaired by shorting out the affected tubes. This usually involves removing the top and bottom tanks and plugging the damaged tube(s) at the headers. If a leaking tube is accessible, a soldering repair may be possible.

Soldering radiators can be a risky job. Before beginning, assess how the heat will affect any nearby soldered joints. Low-melt point solder has little structural strength, and although it may be used to seal a hairline crack or small impact leak, it should not be used otherwise. Silver solder is a preferred solder repair medium, but much more heat is required to apply it than 50/50 lead-tin solder, so make sure you know what you are doing.

Radiator Flushing and Major Repairs. Most commercially available in-chassis radiator descaling solutions are a poor risk as they are seldom 100 percent effective and can dislodge scale without dissolving it. The loosened scale then plugs up somewhere else in the cooling circuit. For the same reasons, reverse flow flushing of the cooling system makes little sense. When manufacturers recommend radiator flushing, it is generally performed in the normal direction of flow, often aided by a cleaning solution.

Major radiator repairs should be referred to a radiator specialty shop. A scaled radiator falls into the classification of major repair. An ultrasonic bath will remove most scale rapidly and effectively. A properly equipped radiator repair shop will also be able to determine the extent of repairs required and whether recoring is necessary. One of the problems of performing radiator repairs without the proper test equipment is the inability to test the radiator until it is assembled and reinstalled in the chassis. Removing and installing radiators can be a labor-intensive operation in some chassis.

Auxiliary Heat Exchangers

The engine cooling system is required to cool other heat exchangers plumbed into the cooling system. Other heat exchangers that can be plumbed into the cooling circuit:

- CACs
- EGR heat exchangers
- Oil coolers
- Cabin heaters
- Transmission oil coolers

The principles required to diagnose and repair these heat exchangers are similar to those for radiators. Most manufacturers prefer that you replace rather than repair heat exchangers.

Powertrain Secondary Cooling System

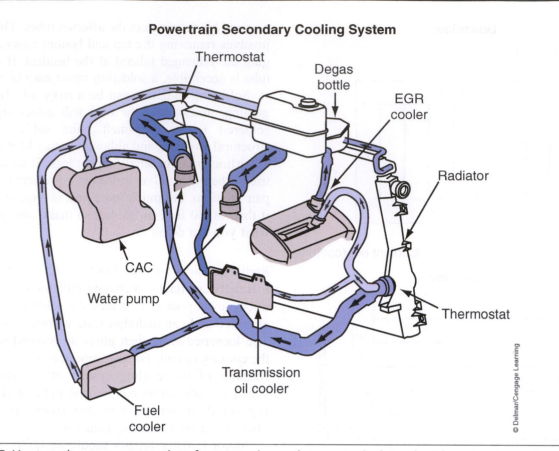

Figure 6-5 Heat exchangers competing for ram air used on a typical Ford pickup truck equipped with a PowerStroke 6.7L diesel engine.

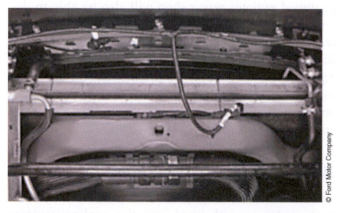

Radiators

The coolers that are located at the front of the grille opening include:

- primary radiator.
- secondary radiator.
- air conditioning condenser.
- power steering cooler.

The secondary radiator is located in front of the primary radiator to allow the powertrain secondary cooling system to operate at a lower temperature than the primary cooling system.

Figure 6-6 PowerStroke 6.7L secondary cooling circuit.

RADIATOR CAP

Radiators are usually equipped with a pressure cap whose function it is to maintain a fixed operating pressure while the engine is running. This cap is additionally equipped with a vacuum valve to admit surge tank coolant (or air) into the cooling circuit (the upper radiator tank) when the engine is shut down and the coolant contracts. Radiator caps allow sealed cooling systems to be safely pressurized. For each 1 psi (7 kPa) above atmospheric pressure, coolant boil point is raised by 3°F (1.67°C) at sea level; for every 1,000 feet of elevation, the boil point decreases by 1.25°F (0.5°C). System pressures are seldom designed to exceed 25 psi (172 kPa) and more typically they range between 7 psi (50 kPa) and 15 psi (100 kPa).

Radiator caps are identified by the pressure required to overcome the cap spring pressure and unseat the seal: when this occurs, the coolant is routed to a surge tank. The surge tank coolant level is always at its highest when the engine is running hottest. As the engine cools, the pressure within the cooling system drops, and when it falls to a "vacuum" value of about ¼ psi, the radiator cap vacuum valve is unseated: this

TABLE 6-2: SYSTEM PRESSURE AND BOIL POINT		
Cooling System Max Pressure	Increase in Boil Point	Coolant Boils at
7 psi (52 kPa)	21°F (12°C)	233°F (112°C)
10 psi (79 kPa)	30°F (17°C)	242°F (117°C)
15 psi (105 kPa)	40°F (22°C)	252°F (122°C)

allows coolant from the surge tank to be pulled back into the radiator. **Table 6-2** shows how much coolant boil point is raised correlated with rad cap trip pressure.

Tech Tip: When radiator hoses collapse as the engine cools, it indicates that the vacuum-relief valve in the radiator cap has failed. Collapsing hoses on diesel engines may destroy them internally, so they should be checked after this type of rad cap failure.

CAUTION *Great care should be exercised when removing a radiator cap from the radiator. If the system is pressurized, hot coolant may escape from the filler neck with great force. Most filler necks are fitted with double cap lock stops to prevent the radiator cap from being removed in a single counterclockwise motion; if the radiator is still pressurized, the cap will jam on the intermediate stops. Never attempt to remove a radiator cap until the cooling system pressure is equalized.*

Testing Radiator Caps. Radiator caps can be performance tested using a standard cooling system pressure testing kit. The radiator cap should first be installed on an appropriately sized adapter on the hand pump, then pumped to the seal crack value (this should exceed the cap rated value by 1 psi [7 kPa]). Next, release the pressure and once again recharge to the exact rated pressure value of the cap and observe the pump gauge: pressure drop-off should not exceed 2 psi over 60 seconds.

Water Manifold

Some diesel engines are equipped with a water manifold. A water manifold ensures an even distribution of coolant through the engine block, resulting in more consistent cylinder temperatures. In an engine with no water manifold, coolant flows from the front to the rear of the engine, resulting in higher temperatures at the rear of the engine. In the coolant system with a water manifold, coolant is routed through the cylinder block to cool, ensuring more even cylinder block temperatures.

Water Pumps/Coolant Pumps

Water pumps are usually nonpositive, centrifugal pumps driven directly by a gear or by belts. When the engine rotates the coolant pump, an impeller is driven within the housing, creating low pressure at its inlet, usually located at or close to the center of the impeller. The impeller vanes throw the coolant outward and centrifugal force accelerates it into the spiraled pump housing and out toward the pump outlet. Because the cooling system pressure is at its lowest at the inlet to the coolant pump (because of the low-pressure pull of the impeller at the inlet), system boiling always occurs at this location first. This will very rapidly accelerate the overheating condition as the pump impeller will be acting on vapor. Coolant pumps are the main reason that engine coolants should have some lubricating properties, because they are vulnerable to abrasion damage when the coolant TDS levels are high. A sectional view of a Navistar water pump is shown in **Figure 6-7**.

Pump Failures. Coolant pumps fail for the following reasons:

- Overloading of the bearings and seals caused by misalignment or tight drive belts.
- High TDS levels in the coolant that erode the impeller.
- Mineral scale buildup on the pump housing.
- Overheating—Boiling usually occurs first at the inlet to the water pump, so a system that is not properly sealed, or hot shutdowns, can cause vapor lock.

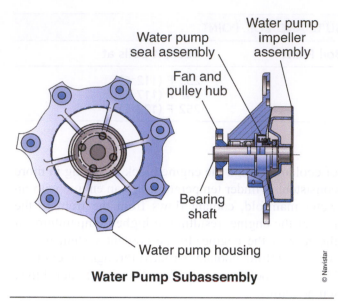

Water Pump Subassembly

© Navistar

Figure 6-7 Sectional view of a water pump and housing.

Inspecting, Replacing, Rebuilding Coolant Pumps. A defective coolant pump should first be removed from the engine. It should be carefully inspected to determine the cause of the failure to avoid a repetition. Water pumps are seldom reconditioned in the modern service garage. When defective, they are replaced as a unit with a rebuilt/exchange unit. Rebuilding of water pumps is usually performed at a rebuild center equipped with the proper equipment by persons who specialize in the process.

Although the technician who reconditions one or two coolant pumps a year may not be able to compete with the time of the specialist rebuilder, it is certainly possible to perform the work to the same standard. A slide hammer is usually required to remove the pulley from the impeller drive shaft, and an arbor press is generally preferred over power presses both for disassembly and reassembly. The pump is one of the simpler subcomponents of the engine. It consists of a housing, impeller, impeller shaft, bearings, and seals. When a pump is rebuilt, inspect the components thoroughly; in many cases, especially where plastic impellers are used, only the housing and shaft are reused. Examine the shaft seal contact surfaces for wear. When gear-driven coolant pumps are reconditioned, pay special attention to the drive gear teeth. The OEM instructions should be observed, and where ceramic seals are used, great care is required to avoid cracking them during installation. A critical specification is the impeller-to-housing clearance, and failure to meet this will reduce pumping efficiency. **Figure 6-8** shows a front view of one of the two water pumps used on a Ford PowerStroke 6.7L diesel engine.

© Ford Motor Company

Primary Coolant Pump

The primary coolant pump is located on the left front of the engine. The heater core transfers heat from the primary cooling system to the passenger compartment. Coolant is routed into the heater core from the coolant crossover pipe at the front of the engine. Coolant passes through the heater core and is routed to the lower radiator hose.

Figure 6-8 Primary water pump used on a Power-Stroke 6.7L engine.

Filters

Some diesel engines are equipped with coolant filters. When used, they are usually of the spin-on cartridge type and connected in parallel to coolant flow. Using a coolant filter reduces particulates pumped through the cooling circuit. Filters are usually plumbed in series with coolant flow, but they must be equipped with a bypass valve designed to trip if the filter plugs. **Figure 6-9**

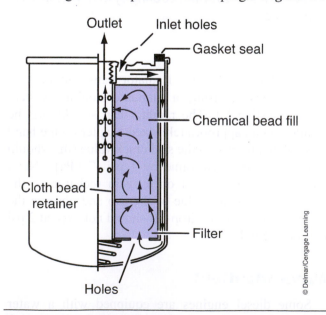

© Delmar/Cengage Learning

Figure 6-9 Cutaway view of a cartridge-type coolant filter showing the flow routing.

shows the flow routing through a typical coolant filter. Some coolant filters come charged with SCAs, so it is especially important to observe the manufacturer recommendations for filter type and service intervals.

When coolant filters have to be changed, check the type of shutoff mechanism used. Most current diesel engines use an automatic check valve, meaning that no spillage results when a filter is removed for servicing because a lock-off ball engages. Some engines have manual shutoff valves that must be turned off before removal. New coolant filters do not require priming. Some coolant filters are equipped with a zinc electrode to neutralize the electrolytic effect of the coolant, although these are more likely to be found in marine and off-highway applications.

COOLANT CIRCUIT MANAGEMENT

In modern low-emissions engines, the cooling circuit must be managed both indirectly and directly by the engine management electronics. This requires a complex monitoring circuit. A diesel engine cooling system is monitored in different ways according to the generation of the system. Coolant temperature is often a primary reference for the engine management software when it determines timing and air/fuel ratio parameters on electronically managed engine systems, so the temperature display to the operator is secondary in importance. All current electronically managed engines can be programmed to default to failure strategies, which may include engine shutdown based on the coolant temperature or level readings, or detection of malfunction in components that could result in high emissions. Thermistors are almost universally used to sense the temperature of the cooling system as well as other engine fluid temperatures, including ambient air, boost air, and lubricating oil. The following methods are used to sense coolant temperature.

Thermistors

Thermistors are solid-state semiconductor devices whose internal resistance varies with temperature change. They are supplied with a specific reference voltage, almost always around 5V DC, and they return a signal to the PCM based on temperature. Negative temperature coefficient (NTC) thermistors are commonly used. The internal resistance in an NTC thermistor decreases as temperature rises. Because the internal resistance in an NTC thermistor decreases as temperature increases, the signal voltage (returned to the PCM) goes up proportionally with temperature rise. The engine PCM broadcasts engine coolant

temperature information to the data bus so that it can be displayed on a dash digital display or gauge.

Electric

Electric sensors use a bimetal arm in conjunction with a resistor supplied by a modulated electrical signal from the temperature gauge: when the bimetal arm is heated, the greater linear expansion of one of the bimetal strips causes it to bend one way and as it cools and contracts, bend in the opposite direction. Connected to the bimetal strip is a wiper, which short-circuits the current flow through the resistor to ground, thereby altering the gauge value.

Expansion

An expansion-sensing gauge consists of a tube filled with a liquid that expands as it is heated and in expanding, activates the gauge indicator needle. This gauge tends not to be used much in today's highway diesel engines, but you will likely come across a few examples in older marine engines and off-highway, heavy-equipment applications.

Coolant Level Indicators

Almost all current electronically managed engines have radiators equipped with low coolant level warning systems. Most operate using the same principles. The PCM outputs a signal to a probe (or sensor), usually located in the top radiator tank, which grounds through the coolant. When the probe fails to ground through the coolant, a low coolant level warning is generated; the outcome depends on how the PCM has been programmed (this is usually a customer data program option). In most cases, a lag (somewhere around 5 to 12 seconds) is required before the PCM resorts to a programmed failure strategy. This may be simply to alert the operator, ramp down to a default rpm/engine load, or shut down the engine after a suitable warning period. Some radiators are equipped with two probes used to signal low and dangerously low coolant levels.

Thermostats

Thermostats function as a type of automatic valve that senses changes in engine temperature and regulates coolant flow to maintain an optimum engine operating temperature. To function effectively, a thermostat must:

- Start to open at a specified temperature
- Be fully open at a set number of degrees above the start-to-open temperature

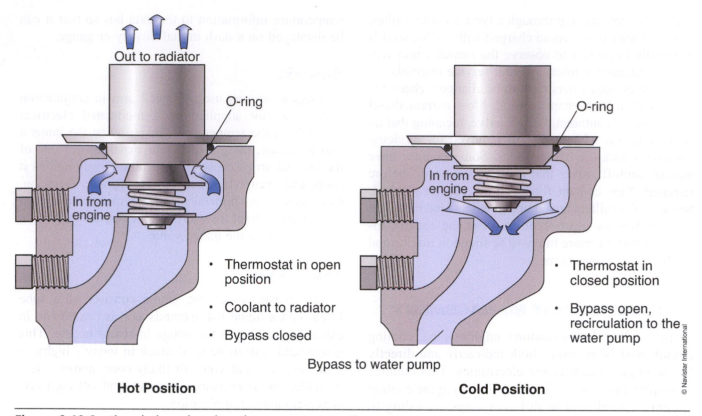

Figure 6-10 Sectional view showing thermostat operation.

- Define a flow area through the thermostat in the fully open position
- Permit zero coolant flow or a defined small volume of flow when in the fully closed position

The cooling system thermostat is normally located either in the coolant manifold or in a housing attached to the coolant manifold. It has two primary functions:

- Permit a rapid engine warmup by limiting flow to the in-engine cooling circuit
- Maintain a consistent temperature once the engine has attained its normal operating temperature by optioning flow to the radiator

Thermostats therefore function by optioning flow to either the in-engine circuit or the radiator circuit. They open and close based on temperature. During opening and closing, a flow area is defined by the extent the thermostat is open. For instance, a partially open thermostat limits the flow area of coolant routed to the radiator.

Operating Principles

Because a thermostat defines the flow area for circulating the coolant, there is often more than one in a single circuit. This means that a split-circuit engine could have four thermostats. The thermostat operates using a heat-sensing element that actuates a piston attached to the seal cylinder. When the engine is cold, coolant is routed back to the coolant pump to be recirculated through the engine. When the engine heats to operating temperature, the seal cylinder gates off the passage to the coolant pump and routes the coolant to the radiator circuit. The heat-sensing element consists of a hydrocarbon or wax pellet into which the actuating shaft of the thermostat is immersed. As the hydrocarbon or wax medium expands, the actuating shaft is forced outward in the pellet, opening the thermostat. Thermostats can be full blocking or partial blocking. **Figure 6-10** shows a typical engine thermostat in its engine warm and engine cold positions.

Bypass Circuit.   The term *bypass circuit* describes the routing of the coolant before the thermostat opens. When flow is confined to the bypass circuit, coolant flow is limited to the engine cylinder block and head. The flow of bypass coolant permits rapid engine warmup to the required operating temperature. **Figure 6-11** shows the flow routing through a Power Stroke 6.7L engine cylinder block when the engine is cold and the thermostat is closed. **Figure 6-12** shows the flow through the same engine when it is at operating temperature and the thermostat is open.

Running without a Thermostat.   This practice is not recommended, and engine manufacturers may void the

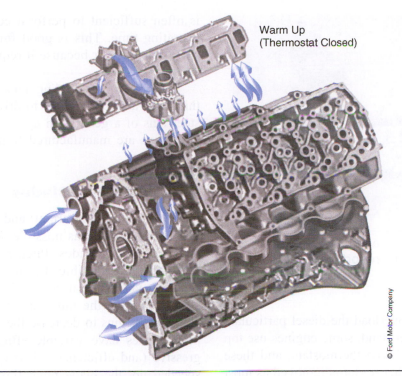

Warm Up
(Thermostat Closed)

© Ford Motor Company

Figure 6-11 Flow routing of coolant through a PowerStroke cylinder block during engine warmup: thermostat closed.

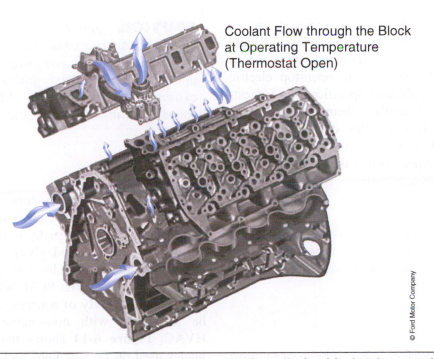

Coolant Flow through the Block
at Operating Temperature
(Thermostat Open)

© Ford Motor Company

Figure 6-12 Flow routing of coolant through a PowerStroke cylinder block when engine is at operating temperature: thermostat open.

warranty. It also violates the Environmental Protection Agency (EPA) requirements regarding tampering with emission control components. Removing the thermostat invariably results in the engine running too cool.

This can cause vaporized moisture in the crankcase to condense and can result in developing corrosive acids and sludge in the crankcase. Additionally, low engine running temperatures will increase the emission of HC

Figure 6-13 Water pump and thermostat housing with a thermistor used to report temperature data to the PCM.

particulates, which can overload the diesel particulate filter (DPF). On the other hand, some engines use top bypass or partial bypass–type thermostats, and these may overheat when the thermostat is removed as most of the coolant will be routed through the bypass circuit with little being routed through the radiator.

Testing Thermostats. Testing a thermostat can be performed using a specialized tool consisting of a tank, heating element, and accurate thermometer. You can make such a test device from an open-top electric kettle. Check the manufacturer specifications for testing thermostats and remember, there is a difference between start-to-open and fully open temperature values. **Figure 6-13** shows a water pump and thermostat housing equipped with a thermistor removed from the engine for bench testing.

> **CAUTION** *Exercise extreme care when handling close-to-boiling water in a thermostat test tank, and use eye protection, gloves, and tongs.*

Cooling Fans

Either suction or pusher fans can be used to move air through an engine compartment. Suction fans pull outside air into the engine compartment. Pusher fans do the opposite, so they push heated air out. Highway vehicles that receive ram air assist are best suited to using suction fans. Off-highway and stationary diesels that cannot take advantage of ram air use pusher fans to remove heated air from the engine compartment. In highway diesel engine applications, ram air

is often sufficient to perform cooling 95 percent of operating time. This is good for both fuel economy and engine power because it requires energy to drive a fan.

Modern engine cooling fans do not require anything like as much energy to drive them as the heavy steel fans of a generation ago. Lightweight, variable-pitch fans are manufactured from carbon composites and plastics.

Variable-Pitch Blades

Fan design is important, and fans should draw as little engine power as possible. Most current designs use flexible fan blades. Pitch means angle. Because the blades are flexible, the pitch is at its most aggressive when the fan assembly is being driven at a slow speed. As the fan's driven speed increases, air resistance begins to decrease the blade pitch. Flexible-pitch blades have variable efficiency. They are aggressive (and efficient) at low rpms, but as they are speeded up, the blade pitch flexes and reduces efficiency (decreasing the engine power required to drive them).

> **CAUTION** *Fan assemblies must be precisely balanced. An out-of-balance fan (a small fragment missing from one blade is sufficient) can unbalance the engine driving it. At its most extreme, this type of out-of-balance condition could result in a failure of the crankshaft.*

Fan Cycles

Because a fan assembly draws engine power, most diesel engines use lightweight, temperature-controlled fans. The objective is usually to have the fan running as little as possible. In today's electronically controlled diesel engines, fan cycles (on or off) are usually managed by the engine PCM, which can use sensor input from a variety of sources, some of which may be associated with non-engine systems such as HVAC. **Figure 6-14** shows the variable-pitch fan blades used on current Ford pickups. Fan cycling in this unit is PCM controlled, and fan speed is monitored by an inductive pulse generator–type sensor (see **Chapter 12**). Alternatively, fan cycling may be managed by a thermostat integrated into the fan hub. This type of temperature sensing is known as thermatic. **Thermatic fan** effective cycles are based on underhood temperatures rather than engine coolant or oil temperatures.

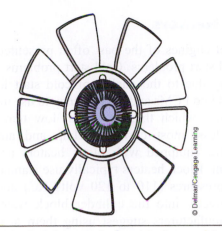

© Delmar/Cengage Learning

Figure 6-14 Variable-pitch fan blade used in a Ford pickup truck: fan cycling is PCM controlled and rotational speed is monitored by a sensor.

Fan Shrouds

Fan shrouds are usually molded fiber or plastic devices bolted to the inside of the radiator. The shroud usually partially encloses the fan. This provides some safety if the fan engages when the hood is open and the engine is running. Shrouds play an important role in directing air flow through the engine compartment. A missing or damaged shroud can result in temperature management problems. In hot weather conditions, fan efficiencies can be lowered by a defective or missing shroud, so they should be examined at each inspection.

Fan Belts and Pulleys

Fan pulleys use external V or poly-V grooves and internal bearings. Belts should be adjusted using a belt tensioner. If belts are not properly adjusted they will fail.

- Too tight: excessively loads the bearings and shortens bearing and belt life
- Too loose: causes slippage and destroys belts even more rapidly than a too-tight adjustment

Belts should be inspected as part of a PM routine. Replace belts when glazed, cracked, or nicked. Replacing belts with early indicators of failure costs much less in the long run than the breakdowns that may be caused by belts that fail in service.

COOLING SYSTEM PROBLEMS

When a problem occurs in the cooling system, it can result in an engine failure if not repaired. What initially appears as a minor leak can quickly progress to a condition that can destroy an engine. Cooling system performance problems can be grouped into the following categories:

- Overheating
- Overcooling
- Loss of coolant
- Defective radiator cap
- Defective thermostat

Leaks

Cooling system leakage is common, and most leaks are minor in nature and can be quickly repaired. A visual inspection of the cooling system is part of the driver or operator's daily inspection. Cold leaks may be caused by the contraction of components at joints as an engine cools. Cold leaks often cease to leak at operating temperatures because everything expands. Some fleet operators replace all the coolant hoses after a scheduled in-service period for no other reason than to avoid the costs of an over-the-road breakdown.

External Leaks. Silicone hoses are more expensive than the rubber compound type, but they usually have longer service life. Silicone hoses require special clamps. Take care to torque these hose clamps to specification and remember that overtightening can cause them to leak.

Pressure testing a cooling system will locate most external cooling system leaks. A typical cooling system pressure-testing kit consists of a hand pump and gauge assembly. The gauge is calibrated from 0 to around 25 psi (170 kPa). The kit contains various adapters for the different types of fill necks and radiator caps. Some are capable of vacuum testing.

Internal Leaks. Internal leaks can be more difficult to locate. When coolant is present in the engine oil, it forms a milky sludge that settles to the bottom of the oil pan. It is the first fluid to exit the oil pan when the drain plug is removed because it is heavier than oil. Internal leaks are not common in engines that do not use wet liners. When an internal leak occurs in a parent bore engine it can indicate a head gasket failure or a more serious engine condition such as a cracked block or head.

Bubbles in Coolant. The appearance of bubbles may indicate that combustion gases are leaking into the coolant or that the engine is being run without thermostat(s). There are a couple of ways of determining whether combustion gases are leaking into the cooling circuit, and the best method varies according to the engine you are testing. Pressure testing of the cooling system using the method outlined previously is the preferred method. On some engines, if you can disconnect the water pump drive, removing the piping to the upper radiator tank and the thermostat, filling the engine with coolant (water will do), and running the engine can identify cylinder leakage to the cooling system.

WARNING *The use of cooling system agents that claim to stop leaks should generally be avoided, even in a situation that might be described as an emergency. They may work temporarily but in doing so, they have been known to plug thermostats, radiator/heater cores, and oil cooler bundles. Generally, they cause more trouble than they cure.*

Stray Voltage Damage

Stray voltage grounding through engine coolant can result in electrolytic action that can cause considerable engine damage in unbelievably short periods of time. This electrolytic damage ranges from pinholes in heat exchangers to erosion pitting failures of cylinder blocks. Stray voltage damage has become more commonplace due to the increase in electrical and electronic components combined with the introduction of nonconducting components such as plastic radiator tanks. Chassis static voltage buildup can also discharge through engine coolant.

Testing for Stray Voltage. Stray voltage can be AC (alternator diode bridge leakage) or DC. It is more likely to be DC voltage. Use a digital multimeter (DMM) on autorange to perform the following tests, checking for DC first.

1. Run the engine and turn on all the vehicle electrical loads.
2. Place the negative DMM lead directly on the battery negative terminal and the other into the coolant at the neck of the radiator without touching metal.
3. Record voltage reading. 0.1V DC is OK; the maximum acceptable to most OEMs is 0.3V DC. If higher, you must locate the leakage source. Leakage of 0.5V DC is capable of eating out a cast iron engine block.
4. Shut down each electrical component in sequence while checking the voltage reading. When the leaking component or circuit has been identified, repair its ground. Attempting to ground the coolant at the radiator will not repair the problem.

Block Heaters

Diesel engines of the past often presented significant cold start problems. Cold start problems are more rare today due to the quality of cold start logic programmed into PCMs and the tendency to use multigrade engine oils, which thicken less at low temperatures. Nevertheless, most current pickup and automotive diesels are equipped with block heaters as original equipment. Block heaters typically use standard mains voltage pressures (110 to 120 volts AC) and are inserted directly into the cylinder block water jacket. Most manufacturers suggest using them at a recommended threshold temperature value: 0°F (−17°C) is typical, but as you can see in the example used in **Figure 6-15,** this might be lower.

Hoses and Couplings

The materials used in coolant hoses have improved significantly over the past two decades with the result that the incidence of cooling circuit leaks has been reduced. Rubber-based hose materials have given way to petro- and silicone-compound hoses that seldom fail. These hose materials usually require special hose clamps, and technicians should note that over-torquing these hose clamps can initiate leaks and destroy costly hoses. Tighter is not better when it comes to torquing coolant hoses. It is key when replacing them or reinstalling them. One way of circumventing this type of technician-generated problem is to use quick-couple

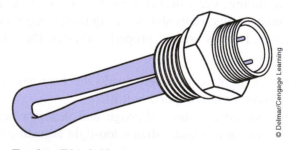

© Delmar/Cengage Learning

Engine Block Heater

The engine block heater is located on the right side of the engine block. The block heater uses 110V AC to heat the engine coolant in cold weather. The engine block heater should be used whenever ambient temperatures are at or below −23°C (−9°F).

The engine block heater is standard on the Ford 6.7 engine. The power cord is an optional accessory.

Figure 6-15 Engine block heater. (*Text courtesy of Ford Motor Company*)

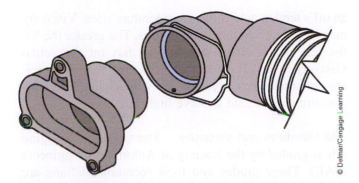

© Delmar/Cengage Learning

Figure 6-16 Quick-couple coolant hose clamps used on the larger coolant hoses on the Ford Power Stroke.

hose connections such as that shown in **Figure 6-16.** The seal within a quick-couple hose connector is an O-ring. The integrity of this O-ring should be inspected each time the coupling is separated.

ENGINE LUBRICATING CIRCUIT

The main functions of a diesel engine lubricating system are:

1. Lubrication. The primary task of an engine oil is to minimize friction and hydrodynamically support rotating shafts.
2. Act as a sealant to help the piston rings in the engine cylinder.
3. Coolant. Remove heat from hot engine components.
4. Cleaning agent. Neutralize acids and sludge formed by blowby gases.
5. Coat engine moving components with an oil film even when they are subjected to high thrust loads.

Lubricating oils are usually petroleum products that are complex mixtures made up of many different fractions. The fractioned compounds that make up an engine oil are refined from petroleum and asphalt bases. Synthetic oils are an option, but due to a reluctance of most engine manufacturers to increase service intervals when synthetic lubes are used, these have yet to prove themselves to be cost effective.

How Oil Works

Theoretically an engine oil is designed to form a film between moving surfaces. If it achieves this effectively, any friction that results occurs in the oil itself, preventing direct metal-to-metal contact. We call this **fluid friction**. Fluid friction generates a lot less

heat than dry friction. The lubricating requirements of an engine oil can be classified as:

- **Thick film lubrication.** Thick film lubrication occurs where the distances between two moving surfaces are wider, such as that between a rotating crankshaft and its main bearings.
- **Boundary lubrication (thin film lubrication).** Boundary lubrication is required where the distance between two metal surfaces is narrow, as would occur between a crankshaft and its main bearings when the engine is not running. A breakdown of boundary lubrication results in metal-to-metal contact.

Good-quality engine oil should be capable of performing both thick film and boundary lubrication. Because engine lube oil is usually petroleum based, it is flammable.

Principle of Hydrodynamic Suspension

When a shaft is rotated within friction bearings such as those used to support an engine crankshaft, and that shaft is stationary, a crescent-shaped gap is formed on either side of the line of direct contact due to the clearance between the journal and the inside diameter of the friction bearing. A static film of oil prevents shaft-to-bearing contact when stationary. When the shaft is rotated and the bearing is charged with engine oil under pressure, a crescent-shaped wedge of lubricant is formed between the journal and its bearing. The oil is introduced to the bearing where the shaft clearance is greatest, usually at the top. This wedge of oil is driven ahead of the direction of rotation in a manner that permits the shaft to be "floated" on a bed of constantly changing, pressurized oil. This principle is used to support crankshafts and camshafts in operation. We know this as **hydrodynamic suspension**.

In hydrodynamic suspension, the key is to maintain a liquid film between moving surfaces by pumping lubricant through a circuit. In the case of a rotating shaft and a stationary friction bearing, the shaft acts as a pump to maintain the lubricant film. The result is that the shaft journal floats on a film of oil the thickness of which depends on:

- Oil input: The rate at which oil is delivered to the bearing. This is why you will have a problem if oil pressure is low.
- Oil leakage rate: The oil that spills from a bearing during operation. This is why low oil pressure is an indication of worn engine main bearings.

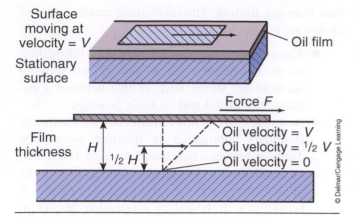

Figure 6-17 Concept of dynamic viscosity.

The thickness of the oil wedge created by hydrodynamic suspension depends on the following four factors:

1. Load increase: Causes oil to be squeezed out of the bearing at a faster rate.
2. Temperature increase: Causes oil leakage rate from bearing to increase.
3. Lower-viscosity oil: Flows with less resistance, creating higher leakage.
4. Changing shaft speed: Speed reduction (remember, the shaft is the "pump") results in a thinner film. An increase in speed provides a thicker film.

Figure 6-17 shows how hydrodynamic lubrication functions.

Engine Oil Classification and Terminology

Every technician should have a basic understanding of the codes and terms used to describe engine oils. This section introduces some of the basic language of lubricants.

Viscosity. The **viscosity** rating of an oil usually describes its resistance to flow. High-viscosity oils are thicker and have less fluidity than low-viscosity oils. However, the true definition of viscosity is that it is a measure of resistance to **shear**. When moving components are separated by engine oil, the oil film closest to each metal surface has the least fluid velocity while the fluid in the center has the greatest fluid velocity. Shear occurs when the fluid velocity is so high that it pulls the surface film away from the moving components.

Viscosity Index. Oils generally thin out as temperature increases. The viscosity index (VI) is a measure of

an oil's tendency to thin as temperature rises. Viscosity may be reduced as temperature rises. The greater the VI, the less of an effect temperature has on the actual viscosity of an oil. In other words, oils that show relatively small viscosity changes with changes in temperature can be said to have high VI.

SAE Numbers and Viscosity. The viscosity of engine oils is graded by the Society of Automotive Engineers (SAE). These grades and their recommendations are listed a little later in this chapter. SAE grades specify the temperature window within which the engine oil can function to provide adequate shear resistance under boundary lubrication conditions.

Multiviscosity Oils. Multiviscosity engine oils are the lubricating oil of choice in diesel engines today. The biggest advantage of multiviscosity oils over straight grade oils is that they provide proper lubrication to the engine over a much wider temperature range. In other words, they have a relatively flat viscosity-to-temperature curve.

Multiviscosity oils are produced by special refining processes and the addition of VI improvers. The objective is to provide them with good cold cranking features and show as little change in viscosity as is possible in a wide range of operating temperatures. Synthetic oils, when marketed with an SAE grade, usually greatly exceed the SAE grade specification.

Lubricity. Two oils with identical viscosity grades can possess different lubricity. The **lubricity** of an oil properly describes its flow characteristics. Lubricity is also affected by temperature: hotter oils flow more easily, colder oils less easily. In comparing two engine oils, the one that has the lower resistance to flow has the greater lubricity. Lubricity is an expression of the "oiliness" of an engine oil.

Flash Point. **Flash point** is the temperature at which a flammable liquid gives off enough vapor to ignite momentarily. The **fire point** of the same flammable liquid is usually about 10°C higher and is the temperature at which a flammable liquid gives off sufficient vapor for continuous combustion. The flash point specification has some significance for diesel engine lube oils because a large portion of the cylinder wall is swept by flame every other revolution. However, actual cylinder wall temperatures are lower than the temperatures of the combustion gases, and the oil is only exposed to these high temperatures for very short periods of time. Most diesel engine lubricating oils have flash points of 400°F (205°C) or higher.

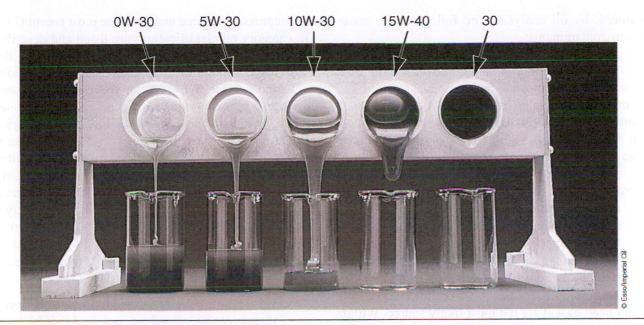

0W-30   5W-30   10W-30   15W-40   30

Figure 6-18 Demonstration of cold-weather flow ability of different oil viscosity grades.

Pour Point. The temperature at which a lubricant ceases to flow is known as the **pour point**. Engine lube oils suited for extreme cold weather operation have pour point depressants that act as ''antifreeze'' for lubrication oils. Pour point is an important engine oil specification, especially where operators use multigrade engine oil viscosities not suited for midwinter conditions in the northern part of the continent. **Figure 6-18** demonstrates the cold-weather flow ability of some common engine lubes.

Inhibitors. **Inhibitors** in an engine lubricant protect the oil itself against corrosion, oxidation, and acidity. They function to make the oil less likely to react with any contaminants that find their way into the crankcase, such as combustion by-products, moisture, and raw fuel. When lube oil inhibitors fail to work properly (usually due to extended service), the oil lacks protection and begins to degrade.

Ash. Ash in lubricant is mineral residue that results from oxide and sulfate incineration. Some ash content in diesel engine lube oils was desirable because it checked the formation of some acids. However, in the latest diesel engine lubricants, the objective is to keep ash levels at a minimum because they can plug emission control devices. When you observe higher-than-specified ash levels in oil analyses this is usually caused by high-temperature operation.

Film Strength. Most engine oils possess adequate film strength to prevent seizure and galling of contacting metal surfaces. However, those formulated for high-speed, high-output diesel engines usually contain additives to improve the film strength of the oil. The idea is that the film does not break down when pressure is applied. Synthetic oils generally provide superior film strength to mineral oils. You can test this by applying a couple of drops of each to a smooth surface: when you attempt to clean them off the surface, the mineral oil is usually easily wiped away, while the synthetic oil leaves a waxy film.

Detergents. Detergents are added to engine oils to prevent the formation of deposits on internal engine components. They also help reduce sludge formation. Some engine oils require a higher percentage of detergent, especially those recommended for newer diesels with EGR. Other engine oils may have almost no detergent.

Dispersants. Dispersants are added to engine oils to help keep insoluble contaminants in suspension and prevent them from forming into sludge and deposits. When sludge and deposits do form in the crankcase, the capability of the dispersants in the engine oil has been exceeded. This is usually an indication that the oil change interval should be reduced because the result is lube oil degradation.

Oil Contamination and Degradation

When oil becomes contaminated, it can lead to complete engine failure. Contaminated engine oil has some characteristic tattletales. These can be quickly identified by visual and odor testing, and confirmed, if required, by oil analysis. The following are some common contaminants:

Fuel. When fuel contaminates engine oil, the oil loses its lubricity and it appears thinner and blacker in color. The condition is usually easy to detect because small amounts of fuel in oil can be recognized by odor. When fuel is found in significant quantities in engine oil, the cylinder head(s) are the likely source.

Coolant. Coolant in the engine lubricant gives it a milky, cloudy appearance when churned into the oil. After settling, the coolant usually collects at the bottom of the oil sump. When the drain plug is removed, the heavier coolant exits first as long as sufficient time has passed for it to settle. When coolant is found in the engine oil, the cylinder head(s), wet liner seals, and cylinder head gasket are the most likely source. Within the cylinder head, the injector cup seals are usually the culprit.

Oil Aeration. Aerated engine oil occurs when engine oil is whipped into foam by moving engine components. Aerated engine oil cannot properly lubricate the engine. It can be caused either by the breakdown of the oil itself (such as detergent additives) or by contaminants such as water. The conditions that aerate oil in the crankcase are the churning action of the crankshaft, sucking of air into the oil pump inlet, and the free fall of oil into the crankcase from the oil pump **relief valve** and cylinder walls. Antifoaming additives reduce the tendency to oil aeration.

Cold Sludge. Cold sludge is caused by lube oil breakdown that occurs when an engine is operated under low loads at low temperatures for extended periods. Once formed, the sludge settles in the engine crankcase and can accelerate engine wear rates. When a diesel engine is operated with its coolant thermostat(s) removed, the result can be the formation of cold sludge.

API Classifications

The **American Petroleum Institute**, usually known as **API**, classifies all engine oil sold in North America. There are two main classifications designated by the prefix letters GF and C. The GF classes of engine oil designate those oils suitable for gasoline-fueled passenger cars and light trucks. The C classes of engine oils designate oils suitable for heavy duty trucks, buses, and industrial and agricultural equipment powered by diesel engines. The C represents CI

or compression-ignited engines. The most recent C and GF category oil classifications are listed and described here because many dealerships use engine oils that claim to be suitable for both C and GF classifications.

In most cases, OEMs have specific requirements for engine lubricants that should be observed because failure to do so could result in higher hydrocarbon (HC) emissions and plugging of costly emissions control devices such as diesel particulate filters (DPFs). What specific engine oil should be used in a diesel engine is the subject of many online chat threads and is outside the scope of this textbook. However, you should also note that using inappropriate oil in current diesel engines can create problems and unwanted repair costs. The following is a listing of the most recent engine oils and includes the API descriptor.

- CH-4—For use in high-speed, four-stroke cycle diesel engines used in on- and off-highway applications that are fueled with fuels containing less than 0.05 percent sulfur. Supercedes CD, CE, CF, and CG category oils.

- CI-4—Introduced in September 2002 for use in high-speed, highway diesel engines meeting 2004 exhaust emission standards, implemented in October 2002 by EPA agreement with engine OEMs. CI-4 oil is especially formulated for engines using fuels containing less than 0.05 percent sulfur and maximizes lube oil and engine longevity when exhaust gas recirculation (EGR) devices are used. CI-4 engine oil supercedes CD, CE, CF, CG, and CH category oils. It meets the requirements of heavy duty engine oil (HDEO) standard PC-9 but should not be used in any post-2007 diesel engines. It was available until 2010 and but it continues to be used by some operators due to stockpiling and its lower cost than CJ-4.

- CJ-4—Formulated for use in 2007 diesel engines equipped with cooled-EGR and DPFs, but it is also backwards compatible. The main difference between CJ-4 and CI-4 is a change in the additive package designed to reduce ash generated by the small amounts of oil combusted in the cylinder. This ash in the exhaust gas can produce a negative effect on the DPFs. This negative effect could be plugging or poisoning the catalyst.

Impact of Using CJ-4 Engine Oil. CJ-4 is a low-ash oil, and because ash has the effect of reducing acidity, oil analysis acidity counts will increase. Post-combustion acids are mostly produced when the sulfur

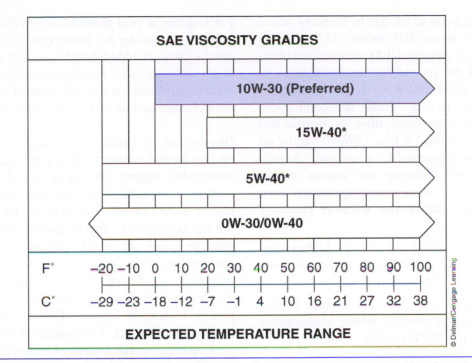

Figure 6-19 SAE viscosity grades correlated with expected ambient temperature. The chart indicates that Ford recommends 10W-30 for its PowerStroke engines.

component in diesel fuel is burned, and the use of ultra-low-sulfur (ULS) fuel should reduce this.

While CJ-4 is backward compatible, it costs more. This additional cost and the continued availability of CI-4 means that some may be tempted to continue to use CI-4 for pre-2007 engines until they are forced to change. Use of oil other than CJ-4 engine oil in post-2007 engines can result in voiding of engine and emission hardware warranty. CJ-4 engine oil will be around for a while.

Oils for Gasoline-Fueled Engines. Oils formulated for use in gasoline-fueled engines are categorized in much the same way, and they are no longer of much importance to diesel technicians because the days of dual fuel application engine lubes are past. The most recent engine oils formulated for service in gasoline engines are as follows:

SH 1993

GF-2 1996

GF-3 2001

GF-4 2004

There are engine oils supposedly formulated to meet the needs of both gasoline- and diesel-fueled engines. Most diesel engine OEMs recommend that multipurpose, multifuel-specified oils be avoided: this is because such oils tend to be loaded with excessive additives in an effort to meet too wide a range of lube specs.

SAE Viscosity Grades

Following is a list of the SAE engine oil grades and the recommended temperature operating ranges. **Figure 6-19** shows the ambient temperature operating range for some common SAE viscosity grades. The W denotes a winter grade lubricant. Note that 15W-40 diesel engine oil is seldom recommended by manufacturers of diesel engines in any part of the continent that experiences winter.

Multigrade engine oils:

0W-30 Recommended for use in arctic and sub-arctic winter conditions.

5W-30 Recommended for winter use where temperatures frequently fall below 0°F (−18°C) and seldom exceed 60°F (15°C).

5W-40 Recommended for severe-duty winter use where temperatures frequently fall below 0°F (−18°C).

5W-50 Recommended for severe-duty winter use in Arctic and sub-Arctic conditions where temperatures frequently fall below 0°F (−18°C). A synthetic oil viscosity grade.

10W-30 Recommended for winter use where temperatures never fall below 0°F (−18°C).

10W-40 Recommended for severe-duty winter use where temperatures never fall below 0°F (−18°C).

15W-40 Recommended for use in climates where temperatures never fall below 15°F (–9°C). Despite diesel engine OEM recommendations that support the use of lighter multigrades in winter conditions, this is by far the most commonly used viscosity grade for commercial diesel engines year round, often in climates that have severe winters. A true 15W-40 engine oil will freeze to a grease-like consistency in sub-zero conditions, making the engine almost impossible to crank; additionally, engine wear is accelerated during the warmup phase of operation.

20W-40 Recommended for use in high-performance engines in climates where temperatures never fall below 20°F (–6°C).

20W-50 Recommended for use in high-performance engines in climates where temperatures never fall below 20°F (–6°C).

Straight grades:

Straight diesel engine lubes should be avoided in modern automotive diesel engines. Though these oils continue to be available, they should be regarded as representing a distant past: the bottom line is that they have not been re-engineered for more than a generation. Those who insist on using them inevitably end up either damaging or reducing the longevity of their equipment.

Synthetic Oils

Most diesel engine OEMs approve of the use of **synthetic oils** and often recommend them for severe duty applications such as extreme cold weather operation, providing the oil used meets their specifications. The problem is that OEMs have been reluctant to sanction increased service intervals when synthetic lubricants are used, making them a high-cost option.

What Oil Should Be Used?

Each engine manufacturer has specific requirements for the engine oils it wants you to put in its engines. These requirements are outlined in tests that engine oils are subjected to prior to an approval. Oil refiners attempt to meet as many standards as possible when marketing engine oils for the obvious reason that they can service more potential buyers with a single product. The problem is, oil is like a brew, and you do not necessarily improve it by throwing more additives in.

For instance, if your preference is to drink cola, you are unlikely to consider it a better drink if some coffee and orange juice are added. For this reason, general purpose, heavy duty engine oil is not necessarily an improvement on the OEM-labeled oil, although of course it will be both cheaper and more heavily advertised.

Developing a Formula. In most cases, the best choice for the engine is to use the manufacturer-recommended engine oil. After all, the research and performance profiles have all been performed on the engine using this oil. This may not be the best choice for the pocketbook, and in fairness there are general purpose diesel engine oils proven to perform well in engines over time. You should also know that engine manufacturers do not "engineer" their own oils. This development is done by specialist oil research companies contracted by the OEM, such as Lubrizol (www.lubrizol.com). The primary business of a company such as Lubrizol is not to manufacture nor market oil. They are paid by a manufacturer to develop a formula (a "recipe" if you like). After the formula has been developed, the manufacturer takes it to an oil refiner, who then manufactures the product to specification. Developing the right additive package for an engine oil is a science in itself and is usually the result of extensive testing.

So, to directly answer the question of which engine oil to use, unless the manufacturer recommends a mass market, multiapplication engine oil, avoid using it. Use the manufacturer-branded engine oil for normal duty. However, research has proven synthetic engine lubes to substantially outperform mineral-based lubricants, so in severe-duty applications, consider the use of a well-known brand of synthetic engine oil.

LUBRICATION SYSTEM COMPONENTS

Engine lubricant must be stored and collected in a reservoir, pumped through the lubrication circuit, filtered, cooled, and have its pressure and temperature monitored. The group of components that performs these tasks is known as the lubricating circuit. The components in a diesel engine lube circuit vary little from one engine to another, but manufacturer service literature should be consulted before servicing and reconditioning any components.

Oil Pan

The oil pan or sump is a reservoir usually located at the base of the engine cylinder block enclosing the

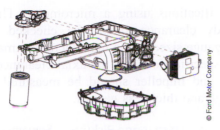

Oil Pan

The oil pan provides the mounting locations for the oil cooler and the oil filter adapter. The oil filter adapter is mounted to the right rear of the oil pan and contains the oil pressure switch and the Engine Oil Temperature (EOT) sensor. The oil cooler is mounted to the left rear of the oil pan. The oil cooler is a coolant-to-oil cooler.

Figure 6-20 A typical oil pan. Note the location of the pickup tube.

crankcase **(Figure 6-20)**. Oil pans are manufactured from cast aluminum, stamped mild steel, and various plastics and fibers. The oil pan acts as the reservoir to collect the lubrication oil, which gravity causes to drain to the crankcase, and from which the oil pump pickup can recycle it through the lubrication circuit.

> **Tech Tip:** Observe the torque sequence when fitting an oil pan, especially those designed to bolt both to the engine cylinder block and to the flywheel housing. The consequences of not doing so can be leaks at the pan gasket or stress cracks to the oil pan.

Dipsticks

The dipstick is a rigid band of hardened plastic or steel that is inserted into a round tube to extend into the oil sump. Checking the engine oil level should be performed daily by the vehicle operator, so its location is always accessible. It is crucial that the correct dipstick be used for an engine. Plastic dipsticks are engine-specific, so the serial number codes must be matched when one is replaced.

To replace a missing or defective steel dipstick, replace the engine oil and filters, installing the exact manufacturer-specified quantity. Run the engine for a couple of minutes, then shut down and leave for 10 full minutes. Dip the oil sump with the new dipstick and

scribe the high-level graduation with an electric pencil. Measure the distance from the high-level to low-level graduations on the old dipstick and then duplicate on the replacement. Remember, the consequences of low or high engine oil levels can be equally serious, so you have to perform this operation accurately. Some electronically managed engines have an oil level sensor that signals a low oil level condition to the PCM, which can then initiate whatever failure strategy it is programmed for.

> **Tech Tip:** To obtain an accurate oil level reading, ensure that an engine has been shut down for at least five minutes before reading the dipstick indicated level.

Oil Pump

Engine oil pumps are classified as positive displacement and have pumping capacities that greatly exceed the lube requirements of the engine. They are gear driven and are usually located in the crankcase close to the oil they pump, although in some Cummins and Caterpillar applications they are external. Oil pumps are driven either directly or indirectly by the engine geartrain. In cases where the oil pump is located in the crankcase, the drive source is a vertical shaft and pinion that engages with a drive gear on the camshaft. **Figure 6-21** shows the oil pump assembly along with the pickup circuit used on a typical diesel engine: this example is pinion driven.

External Gear. External gear pumps consist of two meshed gears, one driving the other within a housing machined with an inlet (suction) port and an outlet (charge) port. As the gears rotate, the teeth entrap inlet

Figure 6-21 A pinion-driven oil pump assembly.

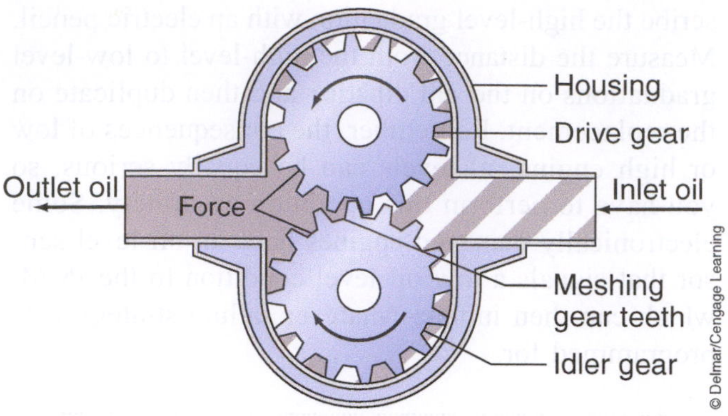

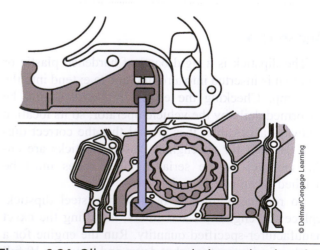

Figure 6-22 Operating principle of an external gear-type oil pump.

oil and force it outside between the gear teeth and the gear housing to the outlet port. Where the teeth mesh in the center, a seal is formed that prevents any backflow of oil to the inlet. **Figure 6-22** demonstrates the operating principle of a typical external gear–type oil pump.

Gerotor. Gerotor-type oil pumps use an internal crescent gear pumping principle. An internal impeller with external crescent vanes is rotated within an internal crescent gear also known as a rotor ring. The inner rotor or impeller has one less lobe than the rotor ring. The result is that as the inner rotor is driven within the outer rotor, only one lobe is engaged at any given moment of operation.

In this way, oil from the inlet port is picked up in the crescent that is formed between two lobes on the impeller, and as the impeller rotates, the oil is forced out through the outlet port as the lead lobe once again engages. The assembly is rotated within the gerotor pump body. **Figure 6-23** shows a cutaway view of a gerotor-type oil pump and its principal components.

Gerotor pumps tend to wear most between the lobes on the impeller and on the apex of the lobes on the rotor ring. These dimensions should be checked to

OEM specifications using a micrometer. The rotor ring-to-body clearance should be checked with a thickness gauge sized to the manufacturer's maximum clearance specification, and the axial clearance of the rotor ring and impeller should be measured with a straightedge and thickness gauges.

Scavenge Pumps/Scavenge Pickups. Scavenge pumps are used in the crankcase of some off-highway equipment required to work on grades that could cause the oil pump to suck air. They are designed with a pickup located at either end of the oil pan.

Pressure-Regulating Valves

Pressure-regulating valves are responsible for defining maximum system oil pressure. Most are adjustable. Typically, an oil pressure regulating valve consists of a valve body with an inlet sealed with a spring-loaded, ball check valve. Other types of poppet valves are also used, but the principle is the same. The regulating valve body is connected in parallel to the main oil pump discharge line. When oil pressure rise is sufficient to unseat the regulator spring-loaded check ball, it unseats. Unseating the regulator check ball permits oil to pass through the valve and spill into the oil sump, dropping the oil pressure. The regulating pressure value is adjusted by setting the spring tension, usually with shims or an adjusting screw. Some manufacturers use color-coded replacement springs to define the oil pressure values. **Figure 6-24** shows the location of the regulating pressure valve in a Ford PowerStroke engine.

Filters

Oil filters in a diesel engine lube system remove and hold contaminants. They must accomplish this

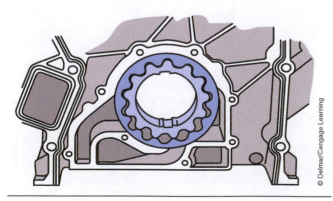

Figure 6-23 A gerotor-type oil pump.

Figure 6-24 Oil pressure regulating valve location downstream from oil pump outlet.

while providing the least amount of flow restriction. Filters use several different principles to accomplish this objective. In general, we can classify oil filters as:

- Full flow: In series between the oil pump and lube circuit.
- Bypass: Ported off a lube gallery in parallel.

The term **positive filtration** describes a filter that operates by forcing all the fluid to be filtered through the filtering medium. Most engine oil filters use a positive filtration principle. Filters function at higher efficiencies when the engine oil is at operating temperature. Filters work to clean the engine oil by using the following methods.

Mechanical Straining. Mechanical straining is accomplished by forcing the lubricant through a filtering medium, which if greatly enlarged would have the appearance of a grid. The size of the grid openings determines whether a solid particle passes through or is entrapped by the filtering medium. Most current diesel engine oil filters make use of mechanical straining. Straining media include rosin-impregnated paper, often pleated to increase the effective area, and cotton fibers.

Absorbent Filtration. These filters work by absorbing or sucking up engine contaminants as a sponge would. Effective absorbent filtering media include cotton pulp, mineral wools, wool yarn, and felt. These filters not only absorb coarse particles but may also remove moisture and acids.

Adsorbent Filtration. Filters adsorb by holding (by adhesion) dissolved liquids to the surface of the filtering medium. Adsorbent filtering media include charcoal, Fuller's earth, and chemically treated papers. Because adsorbent filters may remove oil additives, they are usually used only where low-additive engine oils are specified.

Filter Types and Efficiencies. Most current filters are spin-on disposable cartridges. Older engines may have permanent canisters enclosing a replaceable element; the canister was mounted to the filter pad with a long threaded shaft that extended through the length of the canister. They are seldom seen today as most OEMs made conversion adapters so that disposable spin-on cartridges could replace them. Oil filters are categorized by the manner in which they are plumbed into the lubrication circuit, so we divide them into:

- Full flow
- Bypass

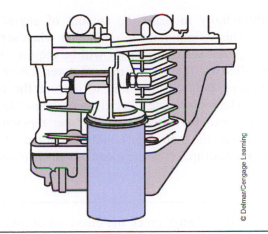

Figure 6-25 Oil filter assembly bolted to side of oil pan.

Full Flow. In this type, the filter mounting pad is usually plumbed into the lubrication circuit close to the oil pump outlet, and all of the oil exiting the pump is forced through the filter. This means it is in series. The filtering medium is usually rosin or otherwise treated paper or cotton fiber. The filters use a mechanical straining principle, so the particles entrapped by the element are those too large to pass through it. All **full flow filters** used on current commercial diesel engines use positive filtration and are rated to entrap any particles larger than their nominal specifications, which range between 25 and 60 microns. **Bypass valves** located on full-flow filter mounting pad(s) protect the engine should a filter become plugged. In this event, the oil exiting the oil pump would be routed around the plugged filter directly to the lubrication circuit. **Figure 6-25** shows a full-flow Power Stroke filter pad assembly.

Bypass. **Bypass filters** are used to complement the full-flow filters. While common on heavy duty engines, they are less common on current light duty highway diesel engines. They are often retrofit to smaller agricultural and off-highway engines with the objective of extending oil change intervals. Bypass filters are plumbed in parallel in the lubrication circuit, usually by porting them into the main engine oil gallery. They filter more slowly, but are rated to entrap particles down to 10 microns in size. Two types are used:

LUBERFINER FILTER

These older canister-type filters are designed to entrap much smaller particles than full-flow filters. Luberfiner filters are supplied from any point in the engine lubrication circuit, and gravity returns filtered lubricant directly to the oil pan. They also play a role as an oil cooler and for that reason they are often mounted in the airflow. Luberfiner filters are large-volume filters that substantially increase the amount of engine oil required. This is important to remember when servicing the engine because more oil will be required than that specified by the engine manufacturer. Drain the filter housing before attempting to remove the filter cartridge. The replaceable filter element is installed by dropping it into the filter housing. After replacing a filter element, the housing should be purged of air: after engine startup, crack the bleed nut until air ceases to exit.

Tech Tip: Always crack the bleed nut or pipe nut on the exit line of a bypass filter after each oil change to purge the air. Failure to do this can result in an air-locked filter.

CENTRIFUGAL FILTER

Centrifugal filtration is used to entrap smaller particles than most full-flow filters so they are usually, but not always, of the bypass type. The filter consists of a canister within which a cylindrical rotor is supported on bearings. It is plumbed into the lubrication circuit so that the rotor is charged with engine oil at lube system pressure. Oil exits the rotor via two thrust jets, angled to rotate the assembly at high velocity. The centrifuge forces the engine oil through a stationary cylindrical filtering medium wrapped outside the rotor. The filtering medium is usually a rosin-coated paper element. The filtered oil drains back to the oil sump.

Tech Tip: Bypass filters have high filtering efficiencies, and failure to observe scheduled maintenance can result in plugged filters.

CAUTION *The manufacturer-specified oil capacity of an engine seldom includes the volume of oil stored in the bypass filter housing. Purge bypass filters of air where required, and check oil level after running the engine following the oil change.*

Filter Bypass Valves. Filter mounting pad bypass valves operate in much the same way as the oil pressure regulating valves, except that their objective is to route the lubricant around a restricted full-flow filter to prevent engine damage by oil starvation. When a filter bypass valve is actuated and the check valve unseated,

instead of spilling the oil to the crankcase it reroutes the oil directly to the lubrication circuit. This shorts out the filter assembly. So, when a bypass valve trips, unfiltered oil is pumped through the lubrication circuit.

Replacing Filters. Filters are removed using a band, strap, or socket wrench. Ensure that the filter gasket and seal (if used) are removed with the filter. Precautions should be taken to capture oil that spills when the filter is removed. Disposable filters and elements are loaded with toxins and must be disposed of in accordance with federal and local regulations that apply to used engine oils and filters.

CAUTION *When performing lube service on a vehicle that comes in off the road, expect the engine oil you drain from the oil pan and filters to be HOT. In fact it is hot enough to cause serious burns, so take great care in dropping oil and filters from warm engines.*

Most manufacturers require that new oil filters be primed, especially on turbocharged engines. If this is not done, in some cases, the lag required to charge the oil filters on startup is sufficient to generate a fault code. Priming an oil filter requires that it be filled with new engine oil on the inlet side of the filter until it is just short of the top of the filter; this will take a little while as the oil must pass through the filtering media to fill the outlet area inside the element. The sealing gasket should be lightly coated with engine oil. Manufacturers usually caution against overtightening. In most cases, the filter should be tightened by rotating it one-half to a full turn after the gasket and filter pad mounting face contact.

Oil Coolers

Oil coolers are heat exchangers that consist of a housing and cooling core through which coolant is pumped and around which oil is circulated. Operating oil temperatures in diesel engines run higher than coolant temperatures, typically around $110°C$ ($230°F$). However, the engine coolant reaches its operating temperature more rapidly than the oil and plays a role in heating the oil to operating temperature in cold weather startup/warmup conditions. Two types of oil coolers are in current use:

Bundle. The **bundle**-type oil cooler is the most common design. It consists of a cylindrical "bundle" of tubes with headers at either end, enclosed in a

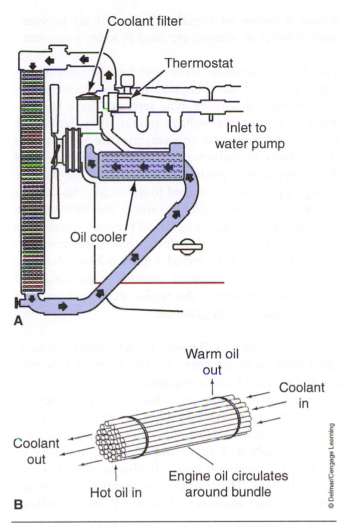

A

B

Figure 6-26 Oil flow through a bundle-type cooler.

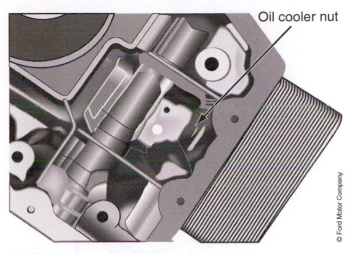

© Ford Motor Company

Oil Cooler

The oil cooler is mounted to the outside of the left side of the oil pan.

It is held onto the oil pan by bolts on the outside of the pan and a nut on the inside of the pan seals the center ports of the cooler.

Note: The lower oil pan must be removed to gain access to the oil cooler nut.

Figure 6-27 Plate-type oil cooler: heat transfer is from oil to coolant.

housing. Engine coolant is flowed through the tubes and the oil is spiral circulated around the tubes by helical baffles. The assembly is designed so that the oil inlet is at the opposite end from the coolant inlet. This arrangement means that the engine oil at its hottest is first exposed to the coolant at its coolest, slightly increasing cooling efficiency. **Figure 6-26** shows the routing flow through a typical bundle-type oil cooler.

The consequence of a failed cooler bundle or failed header O-rings is oil charged to the cooling circuit. Most manufacturers prefer that bundles be leak tested by vacuum because this tests the assembly in the direction of fluid flow in the event of a leak. One header should be capped with a dummy (blocking) plate and the other fitted with the vacuum adapter. Set the recommended vacuum value and leave for the required amount of time to observe any drop-off.

Alternatively, the bundle can be pressure tested using regulated shop air pressure and a bucket of water. This is known as reverse-flow testing. Whenever oil has leaked into the coolant circuit, the engine

cooling system must be flushed with an approved detergent and water with the engine run at its operating temperature for at least 15 minutes.

Plate-Type Coolers. In the plate-type oil cooler, the oil circulates within a series of flat plates, and the coolant flows around them within a housing assembly. They have lower cooling efficiencies than bundle element coolers, but they are usually easier to clean and repair. **Figure 6-27** shows a typical low-maintenance, plate-type oil cooler. Lower cost and lower maintenance, along with a more compact design, means that plate-type oil coolers tend to be more common in light duty diesel engines.

Oil Cooling Jets. Piston oil cooling jets have always been a critical component of the engine lubrication circuit. They are plumbed into a cylinder block oil gallery and spray engine lube onto the underside of each piston. They play a critical role in managing piston temperatures, especially in some more modern engines that may have no other means of delivering oil to the pistons. Bushingless piston bosses such as those used in Monotherm pistons and con rods with no delivery rifling have both added to the already important role played by piston cooling jets.

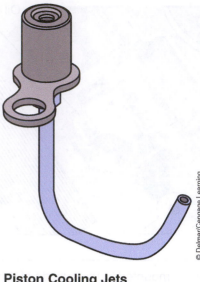

© Delmar/Cengage Learning

Piston Cooling Jets

The 6.7L engine incorporates piston cooling jets that spray oil into a hole in the bottom of the piston to cool the top of the piston.

The oil jets bolt into the bottom of the block and direct the oil into the piston.

Figure 6-28 Piston cooling jet. (*Text courtesy of Ford Motor Company*)

Targeting Piston Cooling Jets. Piston cooling jets have to be targeted to a specific area on the underside of the piston crown. A slightly misdirected cooling jet can result in rapid torching of a piston due to overheating. By targeting the cooling jet delivers oil to a circulation gallery machined into the underside of the piston. This enables engine lube to be routed to those areas of the piston that require lubrication and cooling. Be sure that you consult the manufacturer's service literature when targeting piston cooling jets and understand the consequences of an improperly targeted cooling jet. Just to make things more confusing, manufacturers have been known to alter the target window, so make sure you select the correct target template. **Figure 6-28** shows a piston cooling jet used on a Ford 6.7L engine.

Oil Pressure Measurement

Of all the engine monitoring devices used on an engine, the oil pressure is one of the most critical. Loss of engine oil pressure will, in most cases, cause a nearly immediate engine failure. Back in the days when few of an engine's operating conditions were monitored and displayed to the operator, there was

always a means of signaling a loss of oil pressure. Several types of sensors are used in today's engines.

Variable Capacitance (Pressure) Sensor. Most current commercial diesel engines managed electronically use variable capacitance–type sensors. These are supplied with reference voltage output from the engine PCM. Oil pressure acts on a ceramic disc and moves it either closer or farther away from a stationary steel disc, varying the capacitance of the device and thus the voltage signal value returned to the PCM. The PCM is responsible for outputting the signal that activates the dash display or gauge. Engine oil pressure usually has to fall to dangerously low levels before programmed failure strategies are triggered. **Figure 6-29** is a Navistar schematic showing how a variable capacitance oil pressure sensor signals the engine PCM and some of the fault codes it can generate.

Bourdon Gauge. A flexible, coiled, bourdon tube is filled with oil under pressure. The bourdon tube will attempt to uncoil and straighten incrementally as it is subjected to pressure rise. This action of somewhat bending the bourdon tube rotates a gear by means of a sector and pinion. A pointer is attached at the gear across a calibrated scale and provides a means of reading it. A bourdon gauge is also known as a mechanical gauge. They are the commonly used pressure gauges in older engines that are not PCM controlled.

Electrical. Engine oil pressure acts on a sending unit diaphragm, which in turn moves a sliding wiper arm across a variable resistor, which incrementally grounds out a feed from the electric gauge. The gauge is a simple armature and coil assembly that receives its electrical supply from the vehicle ignition switch.

Tech Tip: No single gauge should be used as the only means of diagnosing a low oil pressure complaint. Use a good-quality master gauge, usually a bourdon gauge with a fluid-filled display dial, to verify the dash gauge reading. Ensure that the engine oil is at operating temperature before recording a reading.

Oil Temperature Management. As EPA noxious emission standards become increasingly tougher, there is an ever greater requirement to manage the combustion temperatures to a tight operating window. Engine oil temperatures are among the most accurate indicators of engine temperature. It should also be

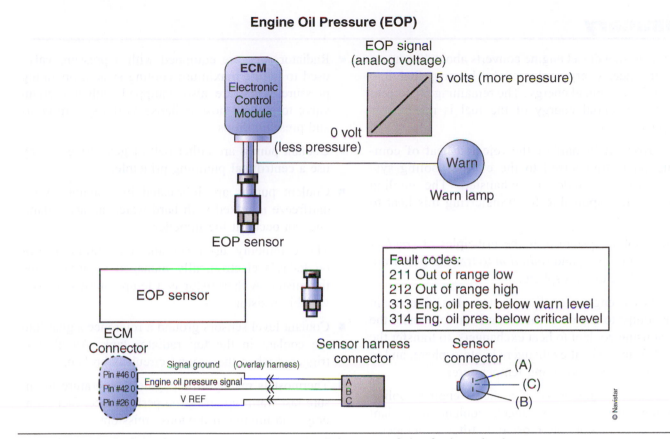

Figure 6-29 A typical oil pressure sensing circuit with some of the fault codes it can generate.

remembered that engine oil plays a major role as a coolant in engines, especially in managing piston temperatures. In addition, when engine oil is used as a hydraulic medium as it is in variable valve timing, engine brakes, and the HEUI fuel system, its performance is to some extent dependent on its operating temperature. On the other end of the scale, the time required to heat engine oil to operating temperature is a factor in the higher emissions output during a warm-up phase.

LUBRICATING CIRCUIT PROBLEMS

The following are some guidelines for troubleshooting lubrication circuit complaints. Be aware of the sequencing. You would be amazed at how many lubrication circuit problems are root-sourced by insufficient oil in the oil pan.

- When investigating a low oil pressure complaint, first check the oil pan level. Next, check the appearance and odor of the oil.
- Low oil pressure complaints must be verified with a master gauge.
- High oil pan levels aerate engine oil, causing low pressure and fluctuations or surging.

- After an engine rebuild, it is good practice to pressure-prime the engine lubrication system with an external pump.
- Most manufacturers recommend priming full-flow filters on all turbocharged engines. Oil filters should be primed by filling with new engine oil poured into the inlet side of the oil filter.
- It is not necessary to prime bypass filters, although it may be necessary to purge air from them as the engine is started up after an oil change.
- When taking an oil sample, draw off midway through the sump runoff, never at the beginning or end. If an oil sample is required outside of the normal oil change interval, use a syringe. Oil samples should be taken shortly after an engine has been run to obtain the most reliable results.
- Oil change intervals are usually determined primarily by mileage in on-highway engine applications and by engine hours in off-highway and stationary engines.

As with all troubleshooting on electronic engines, make sure that you connect to the chassis data bus: today, this is just part of sound shop practice.

Summary

- At its best, a diesel engine converts about 40 percent of the heat energy released by burning fuel into useful mechanical energy. The remaining 60 percent of the potential energy of the fuel is released as *rejected heat*.

- Approximately half of the rejected heat of combustion is transferred to the engine cooling system; the remainder is exhausted. The cooling system is responsible for transferring this heat to atmosphere.

- The cooling system uses the principles of *conduction*, *convection*, and *radiation* to transfer heat from the coolant to atmosphere.

- A diesel engine cooling system has four main functions: to absorb combustion heat, to transfer the heat using coolant to heat exchangers, to transfer the heat from the heat exchangers to atmosphere, and to manage engine operating temperatures.

- The main components of a diesel engine cooling system are the water jacket, coolant, a coolant pump, a radiator, thermostat(s), filter(s), temperature sensing circuit, and a fan assembly.

- Water expands in volume both as it freezes and as it approaches its boil point. The engine cooling system must accommodate this change in volume.

- Engine coolant is a mixture of water, antifreeze, and supplemental coolant additives (SCAs).

- Three types of diesel engine coolant are in current use: ethylene glycol (EG), propylene glycol (PG), and extended life coolant (ELC).

- Diesel engine coolant should protect against freezing, boiling, corrosion, scaling, and foaming. It should neutralize acid buildup.

- The most accurate instrument for reading the degree of antifreeze protection provided by a coolant is a refractometer.

- Mixing coolant solutions should be performed in a container outside of the engine cooling system and then added.

- Where possible, the radiator is located in the air flow at the front of the chassis to take advantage of ram air cooling.

- Most radiators are of the single pass, downflow type, but cross-flow and double pass routing are also used.

- Radiator caps are equipped with a pressure valve used to define maximum cooling system operating pressure. They are also equipped with a vacuum valve to prevent hose collapse as the system cools and pressure drops.

- Coolant pumps are either belt or gear driven. They use a centrifugal pumping principle.

- Coolant pumps are lubricated by coolant. When antifreeze is mixed with hard water, abrasive damage can occur at the impeller.

- The commonly used coolant temperature sensor on today's electronically managed engines is the thermistor. A thermistor is a temperature-sensitive, variable resistor.

- Coolant level sensors ground a reference signal into the coolant in the top radiator tank. An alert is triggered when the ground circuit is broken.

- Thermostats manage the engine temperature to ensure fast warm up and good performance, fuel economy, and minimum noxious emissions.

- Thermostats route the coolant through the bypass circuit to permit rapid engine warmup.

- Most engine compartment fans are temperature controlled, either directly on the basis of coolant temperature measured by a thermistor or fanstat, or indirectly based on under-hood temperature.

- Flex blade composite, plastic, or fiberglass fan blades alter their pitch based on engine rpm. This permits higher fan efficiencies at low rpm when coolant is being moved through the engine at a low rate.

- Fan assemblies are driven by V or poly-V belts. Belt tension should be adjusted using a belt tension gauge to avoid bearing and slippage problems.

- Testing a cooling system for external leaks is performed using a hand-actuated pressure testing kit. The kit consists of a pump, pressure gauge, and a variety of fill neck and radiator cap adapters.

- A diesel engine lubricant performs a number of roles, including minimizing friction, supporting hydrodynamic suspension, cooling, and cleaning.

- Viscosity describes a liquid fluid's resistance to shear.

- Lubricity describes the flow characteristics of a liquid fluid. Lubricity in engine oils is affected by temperature. Hot oils tend to flow more readily than cold oils.

- Diesel engine OEMs recommend the use of multigrade oils over straight grades and approve the use of synthetic engine oils, especially for operation in conditions of extreme cold.

- The pour point of engine oil is the temperature at which the oil starts to change into a solid state. Oils formulated for winter use have pour point depressant additives.

- Sludged oil is usually a result of oil degradation caused by prolonged low load, cold weather operation.

- Oil aeration can be caused by high oil sump levels.

- When interpreting API oil classifications, the GF prefix identifies oil formulated for gasoline-fueled engines and the C identifies oil formulated for compression-ignition engines.

- Post-2007 diesel engines with EGR and DPFs are required to use SAE CJ-4 rated engine lube.

- Most research indicates that synthetic oils outperform traditional oils.

- It is important to maintain the correct engine oil level. The consequences of excessively high oil level can be as severe as those for low oil levels.

- Positive displacement pumps of the external gear and gerotor types are used as oil pumps in diesel engines. Most current commercial diesel engines use the external gear pump design.

- Oil pumps are designed to pump much greater volumes of oil than that required to lubricate the engine. A pressure-regulating valve defines the peak system pressure.

- The filters used on a lubrication system may be classified as *full flow* and *bypass*.

- Full-flow filters are located in series between the oil pump and the lubrication circuit.

- Bypass filters are located in parallel, receiving oil ported from an oil gallery and returning it directly to the oil pan.

- Priming oil filters before installation helps prevent oil starvation to critical components in the lube circuit.

- Filters are usually rated by their mechanical straining specification in microns.

- Oil coolers are heat exchangers; the cooling medium used is engine coolant.

- There are two types of oil coolers: the bundle type and plate type.

- Engine oil pressure measurement is usually performed by a variable capacitance–type sensor. The sensor signals the PCM. The dash gauge or display is therefore a PCM output.

- Because oil pressure is rpm related, a low oil pressure reading at idle speed represents a less serious condition than the same oil pressure reading when the engine is run at rated rpm.

Shop Exercises

1. Identify the engine coolant used in a specific engine.

2. Use a refractometer or hydrometer to measure the antifreeze protection of a coolant.

3. Use a litmus paper to measure the acidity of an antifreeze. Color-code its pH.

4. Pressure-test a diesel engine cooling system and check for leaks.

5. Check the oil level using the dipstick. Research what the total oil capacity on that engine is.

6. Identify the type of oil cooler used on a diesel engine.

7. Outline the procedure required by the manufacturer to service engine fluids. Make a note of the recommended lube oil and viscosity for the season you are in.

Internet Exercises

Use a search engine and research what comes up when you enter the following key words:

1. ethylene glycol
2. propylene glycol
3. long life antifreeze
4. Texaco antifreeze products
5. Delo 400 and Amsol

6. "Big three" service chat lines: be a little skeptical of what you read here!

7. Want to know more? Try these websites:

   http:///www.api.org American Petroleum Institute

   http://www.dieselnet.com EPA diesel emissions news

   http://www.lubrizol.com Lubrizol home page

Review Questions

1. Which type of diesel engine coolant is regarded as potentially the most harmful to humans?

   A. EG
   B. PG
   C. Pure water
   D. ELC

2. What causes cooling system hoses on an engine to collapse when the unit is left parked overnight?

   A. This is normal
   B. Defective thermostat
   C. Improper coolant
   D. Defective radiator cap

3. When a radiator cap pressure valve fails to seal, which of the following is most likely to occur?

   A. Coolant boil-off
   B. Cooler operating temperatures
   C. Higher HC emissions
   D. Higher CO emissions

4. In a typical diesel-powered vehicle at operating temperature, which of the following should be true?

   A. Coolant temperatures run cooler than lube oil temperatures.
   B. Coolant temperatures run warmer than lube oil temperatures.
   C. Coolant temperatures should be equal to lube oil temperatures.

5. What operating principle is used by a typical diesel engine coolant pump?

   A. Centrifugal
   B. Constant volume
   C. Gear type
   D. Positive displacement

6. When the thermostat routes the coolant through the bypass circuit, what is happening?

   A. The coolant is cycled primarily through the radiator.
   B. The coolant is cycled primarily through the engine.
   C. The coolant is cycled primarily through auxiliary heat exchangers.

7. Fans with flexible-pitch blades are designed to drive air at greatest efficiency at:

   A. All speeds
   B. High speeds
   C. Low speeds

8. In the event of an engine overheating, where is the coolant likely to boil first?

    A. Engine water jacket   C. Inlet to the coolant pump

    B. Top radiator tank   D. Thermostat housing

9. What instrument do most engine manufacturers consider to be the most accurate for routinely checking the degree of antifreeze protection in a diesel engine coolant?

    A. Hydrometer   C. Spectrographic analyzer

    B. Refractometer   D. Color-coded test coupon

10. Which of the following API classifications would indicate that the oil was formulated for diesel engines meeting 2007 and later emission standards?

    A. CJ-4   C. CC

    B. CI-4   D. SF

11. Which of the following conditions could result from a high crankcase oil level?

    A. Lube oil aeration   C. Friction bearing damage

    B. Oil pressure gauge   D. All of the above
       fluctuations

12. A full-flow oil filter has become completely plugged. Which of the following is likely if the engine is running?

    A. Engine seizure.   C. Oil pump hydraulically locks.

    B. Engine lubrication ceases.   D. A bypass valve diverts the oil around the filter.

13. Oil that has a milky, clouded appearance is probably contaminated with:

    A. Fuel   C. Engine coolant

    B. Dust   D. Air

14. Which of the following statements correctly describes viscosity?

    A. Lubricity   C. Resistance to shear

    B. Resistance to flow   D. Breakdown resistance

Learning Objectives

After studying this chapter, you should be able to:

- Identify fuel subsystem components on a typical diesel engine.
- Describe the construction of a fuel tank.
- Explain the operation of and troubleshoot a fuel sending unit.
- Define the role of primary and secondary fuel filters.
- Service primary and secondary fuel filters.
- Explain how a water separator functions.
- Service a water separator.
- Define the operating principles of a transfer pump.
- Prime a fuel subsystem.
- Test the low-pressure side of the fuel subsystem for inlet restriction.
- Test the charge side of the fuel subsystem for charging pressure.
- Identify the some typical sensors used in diesel fuel subsystems.

Key Terms

canister

cartridge

centrifuge

charging circuit

charging pressure

charging pump

clockwise (CW)

common rail (CR)

crossover pipe

diesel fuel conditioning module (DFCM)

emulsify

fuel filter

fuel heater

fuel subsystem

fuel tank

gear pump

Hg manometer

inlet restriction

lift pump

low-pressure pump

micron (μ)

pickup tube

plunger pump

positive displacement

primary filter

secondary filter

sending unit

suction circuit

transfer pump

venting

water-in-fuel (WIF)

water separator

INTRODUCTION

The **fuel subsystem** on a diesel engine describes the group of components responsible for fuel storage and its transfer to the high-pressure injection circuit. While high-pressure injection circuits differ somewhat from manufacturer to manufacturer, the fuel subsystems that supply them tend to have much in common. It is important to understand fuel subsystems. Many of the problems encountered by diesel technicians are associated with the fuel subsystem. That makes this chapter one of the key chapters in this textbook.

A thorough knowledge of fuel subsystem components and how they can affect the performance of the high-pressure injection circuit is essential. First, we should say that the basic components used in a fuel subsystem have not changed much over the years. However, the way in which the fuel subsystem is monitored has changed. A generation ago, the fuel subsystem had one monitoring sensor called a sending unit. Today, depending on the original equipment manufacturer (OEM), system pressure and temperature are monitored at different locations in the circuit.

Fuel tanks, **fuel filters**, **water separators**, **transfer pumps**, **fuel heaters**, and all the plumbing that connects these components are studied in this chapter. **Figure 7-1** shows a fuel system schematic used in the Ford 6.7L engine. Note that the fuel system is divided into three circuits:

- High-pressure circuit (the injection circuit) **(Chapters 8** and **10)**
- Internal low-pressure circuit (supply circuit)
- Chassis low-pressure circuit (return circuit)

FUEL SUBSYSTEM OBJECTIVES

Study the fuel system shown in **Figure 7-1,** which is typical of a current vehicle fuel subsystem. Note the location of the transfer pump, identified as the **low-pressure pump** on this schematic. This pump divides the fuel subsystem into:

- the **suction circuit** (upstream from the transfer pump)
- the **charging circuit** (supplies the injection pump with low-pressure fuel: do not confuse with injection pressures, which are much higher)

A full explanation of how fuel is moved through the fuel subsystem appears a little later in this chapter under the heading of transfer pumps. **Figure 7-2** is a schematic of a 7L Caterpillar fuel subsystem: trace the fuel flow from the fuel tank up to the transfer pump (callout number 15) and try to differentiate between the primary and secondary circuits of the fuel subsystem.

A **primary filter** is most often located on the suction side of the transfer pump, while the **secondary filter** is located on its charge side, as illustrated in **Figure 7-2.** However, there are some fuel systems in which all movement of fuel through the fuel subsystem is under suction. Alternate arrangements include:

- gravity feed (stationary small-bore engines)
- fully charged (fuel pump is located within the fuel tank)

Most diesel engines use multiple filters, the exceptions being stationary small-bore applications.

In addition to ensuring that the high-pressure injection circuit receives a supply of clean fuel, the fuel system uses sensors to monitor conditions within its circuit such as pressure, temperature, and the presence of moisture. To summarize, the objectives of a typical fuel subsystem are as follows:

- Store fuel in a tank until required.
- Remove moisture from the fuel.
- Filter fuel to remove abrasive particulates.
- Deliver fuel to the injection components at the proper temperature.
- Monitor fuel tank level, fuel pressure, fuel temperature, and alert a water-in-fuel condition.

FUEL TANKS

Fuel is stored in fuel tanks. In highway vehicles, the fuel tank is securely mounted to the chassis. Large fuel tanks may hold a considerable weight in fuel. It is important that fuel weight is distributed evenly as it is consumed, and it is important to use internal baffles to minimize the effects of fuel sloshing. Many diesel fuel management systems are designed to pump much greater quantities of fuel through the system than that required to actually fuel the engine. The excess fuel is used to:

- lubricate high-pressure injection components.
- cool high-pressure injection components (especially those exposed to extreme temperatures).
- cool electronic components, such as engine control modules (ECMs) and injector drivers.

In its role as a cooling medium, the fuel transfers heat from engine components to the fuel tank. This means that one of the roles of the fuel tank(s) is to transfer heat from the fuel to atmosphere.

Fuel Tank Design

Because one of the roles of a fuel tank is to transfer heat from the fuel it circulates to atmosphere, it plays an important role as a heat exchanger. A vehicle fuel

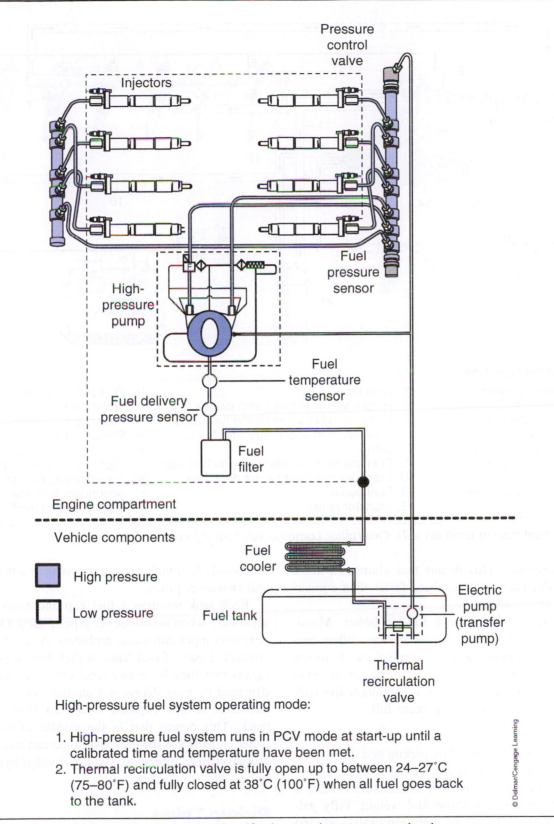

Figure 7-1 Ford 6.7L fuel system schematic. Identify the two low pressure circuits.

tank will function most effectively as a heat exchanger if the following are true:

1. Located in the air flow. This is not always practical, so in some instances, vehicle air

flow may be directed to flow around the fuel tank.

2. Aluminum construction. Aluminum transfers heat much more quickly than composite or steel

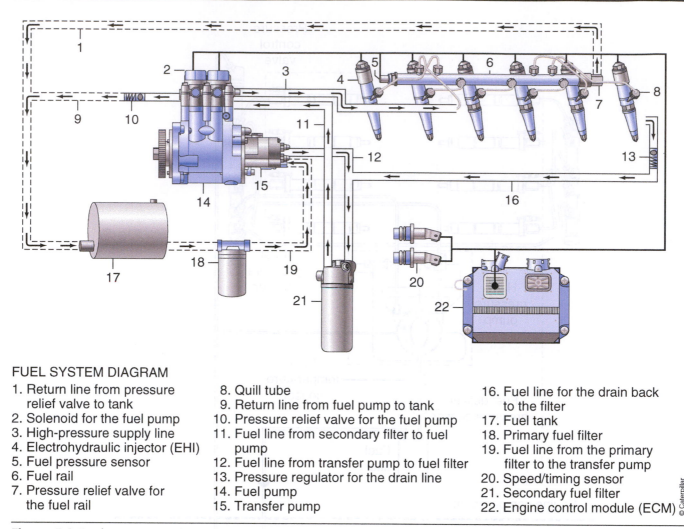

FUEL SYSTEM DIAGRAM

1. Return line from pressure relief valve to tank
2. Solenoid for the fuel pump
3. High-pressure supply line
4. Electrohydraulic injector (EHI)
5. Fuel pressure sensor
6. Fuel rail
7. Pressure relief valve for the fuel rail
8. Quill tube
9. Return line from fuel pump to tank
10. Pressure relief valve for the fuel pump
11. Fuel line from secondary filter to fuel pump
12. Fuel line from transfer pump to fuel filter
13. Pressure regulator for the drain line
14. Fuel pump
15. Transfer pump
16. Fuel line for the drain back to the filter
17. Fuel tank
18. Primary fuel filter
19. Fuel line from the primary filter to the transfer pump
20. Speed/timing sensor
21. Secondary fuel filter
22. Engine control module (ECM)

© Caterpillar

Figure 7-2 Fuel system used on a 7L Caterpillar common rail fuel injection system.

construction. This means that aluminum tanks transfer heat to atmosphere faster than equivalent steel tanks.

3. Maintained 25 percent full or better. Manufacturers of high-flow fuel systems often caution that running tanks low on fuel overheats the fuel, causing it to lose lubricity. It is good practice in such systems to maintain the tank level at better than 25 percent full.

Dual Tanks. Some heavy duty pickup and utility vans use multiple fuel tanks (usually two) to increase on-board fuel capacity. When two fuel tanks are used, it is important to evenly distribute fuel weight. Fifty gallons of diesel fuel weighs between 350 and 365 pounds. To maintain even weight distribution as fuel is consumed, a Y-type **pickup tube** pulls fuel from both tanks simultaneously. Providing the tanks are of equal volume, fuel level is automatically equalized. Most contemporary vehicles use a Y (sometimes known as a T-type) fuel pickup, but you still see plenty of

older-style fuel tank arrangements with single pickup and crossover pipes.

Each tank requires a fuel cap and must be filled separately when no crossover pipe is used. Older-style crossover pipes can cause problems. A **crossover pipe** connects a pair of fuel tanks at their lowest point. This means that they have low road clearance and can be damaged by road debris and animals. In addition, the crossover is also exposed to the air flow under the truck. This means that in the middle of winter, any water collected in the crossover pipe can freeze. When crossover lines freeze up, alcohol (methyl hydrate) has to be added to the fuel tanks.

Pickup Tubes

In instances where there is no transfer pump located inside the fuel tank, a fuel pickup tube is located inside the fuel tank. It is positioned so it can draw on fuel slightly above the base of the tank: in this way it avoids picking up water and sediment. Pickup tubes are quite often welded into the tank; in this case, if they fail, the tank may have to be replaced. Fuel pickup tubes seldom fail, but when they do it is usually by metal fatigue crack at the neck; this results in no fuel being drawn out of the tank by the transfer pump whenever the fuel level is below the location of the crack.

Lift Pumps

Some light duty diesels use an in-tank **lift pump** to supply fuel to the transfer pump. An in-tank lift pump is usually an electrically driven vane pump. It may cycle continuously when the ignition is key-on, or it may be cycled on powertrain control module (PCM) command. Lift pumps may be used to re-establish prime after servicing fuel filters(s), and this procedure usually requires multiple key-on, key-off cycles without cranking the engine; it is described in **Chapter 14.**

Fuel Tank Sending Units

In instances where the fuel transfer pump is located outside of the fuel tank, a stand-alone **sending unit** is required. This is usually flange-fitted to the top of a fuel tank. The function of a sending unit is to signal the fuel level to the chassis data bus, and from there to a dash-located gauge. A sending unit usually consists of a float and arm connected to a variable resistor. The float is suspended by whatever amount of fuel is in the tank. It moves a wiper over a variable resistor. The position of the wiper on the variable resistor determines how much control current flows back to a cab dash gauge. The gauge displays the fuel level in the tank(s).

Testing Sending Units. Fuel sending unit problems are easily diagnosed. On newer vehicles in which the sending unit messages the data bus, connect the appropriate scan tool equipped with the manufacturer's software. Then follow the appropriate diagnostic routines to source the problem. It is important to follow this procedure rather than resort to disconnecting the sending unit and testing this in isolation from its circuit.

With older units in which there is a direct connection between the sending unit and the gauge, first disconnect the terminals and use a DMM (digital multimeter) in resistance mode. Use service literature to find out the resistance of the sending unit. The procedure for checking a 90-ohm sending unit follows, but do not assume that this is the specified resistance of the unit you are working on. When the float arm is moved through its stroke, the readings observed should change as the arm angle changes.

DIAGNOSTIC PROCEDURE FOR A 90-OHM FUEL SENDING UNIT

1. Remove wire lead from sensor terminal.
2. Connect ohmmeter across the sensor terminal and mounting flange (ground).
3. Estimate the amount of fuel in the tank either visually or by using a stick.
4. Your fuel level estimate should connect with the following values:

Full tank	About 86 to 94 ohms
Half tank	About 40 to 50 ohms
Empty tank	About 0 to 4 ohms

5. Values that differ widely from those listed here indicate a defective sending unit.

Sealed Fuel Subsystem. At the time of this writing, most jurisdictions in North America permit **venting** of diesel fuel tanks to atmosphere. That said, manufacturers have tended to seal the fuel subsystem on recent vehicles for two reasons:

- Limit the ingestion of contaminants into the fuel tank
- Capture and retain the boil-off of the more volatile fuel fractions

In a sealed system, air is pulled into the tanks as fuel is consumed; this air passes through a filter. The air filter may be charcoal based to entrap (and store) fuel vapor boil-off. A plugged fuel tank filter may cause engine shutdown by fuel starvation, so checking it is an important maintenance procedure.

Vented Systems. In a vented system, a vent or breather permits gas (either air or fuel vapors) to enter and exit the tank without restriction. As fuel is pumped out of a tank to fuel the engine, it is replaced by air drawn in through a vent. The gas movement can be reversed if the fuel becomes heated. In hot weather conditions, some of the lighter fuel fractions evaporate. Evaporated fuel vapors are vented to the atmosphere.

Fuel tank vents or breathers should be routinely inspected for restrictions and should be protected from ice buildup. A plugged fuel tank vent or breather will rapidly shut down an engine, creating a suction side **inlet restriction**. The transfer pump cannot usually compensate for a plugged breather, so the result is an engine shutdown caused by fuel starvation.

WIF Checking. Diesel fuel tanks should be routinely checked for **water-in-fuel (WIF)**. To check for water in fuel tanks, first allow the fuel tanks to settle. Next, insert a probe (a clean aluminum welding rod will do the job) lightly coated with water-detection paste (you can usually get this from your fuel supplier) through the tank, fill neck until it bottoms in the base of the tank; withdraw the rod and examine the water-detection paste for a change in color. This test will give some idea of the quantity of water in the tank by indicating the height on the probe where the color has changed. Trace quantities (just the tip of the probe changes color) in fuel tanks are not unusual and are nothing to worry about.

FUEL FILTERS

Diesel fuel injection equipment is manufactured with very low tolerances. Dirt is a number one enemy of all diesel fuel systems. Impurities in fuel, if not removed by the fuel subsystem, can result in costly failures. Most dirt found in fuel is a result of conditions in stationary fuel storage tanks, refueling practices, and improper fuel filter priming techniques by service technicians. The function of a fuel filter is to prevent fine sediment in the diesel fuel from entering the fuel injection circuit. While some current secondary filters filter to the extent that water in its free state will not pass through the filter, a separator is often used to remove water before it gets to the secondary filter. All diesel fuel systems require clean fuel, and the function of the filters in a fuel system is to ensure that the fuel is as clean as possible before it is delivered to the injection pumping components.

A typical fuel subsystem with a primary circuit and a secondary circuit in most cases, uses a two-filter arrangement, one in each circuit. Two basic types of filters are used:

- disposable **cartridge** type (most common)
- permanent **canister** type, fitted with a disposable element

Spin-on filters are obviously easier to service and are the filter design of choice for most manufacturers. **Figure 7-3** shows the flow routing in some typical filters.

Primary Filters

A primary filter is the first-stage filter in a typical two-stage fuel filtering subsystem. Primary filters are therefore usually under lower-than-atmospheric pressure in operation. They are plumbed in series between the fuel tank and the fuel transfer pump. Filters are rated in **microns** (μ) that is, millionths of a meter; **Figure 7-4** illustrates the size of a micron relative to a

Fuel Filter
Two-stage box-type filter

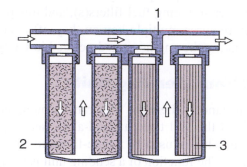

Multistage Filter
With spiral V-form filter element

1. Filter cover with mounting
2. Coarse filter
3. Fine filter

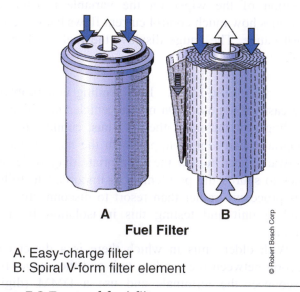

Fuel Filter

A. Easy-charge filter
B. Spiral V-form filter element

© Robert Bosch Corp

Figure 7-3 Types of fuel filters.

human hair. They are designed to entrap particles larger than 10–30 μ depending on the fuel system. They achieve this using pleated cotton threaded fibers and resin-impregnated paper. **Figure 7-5** shows the primary filter element used on a Ford 6.7L diesel

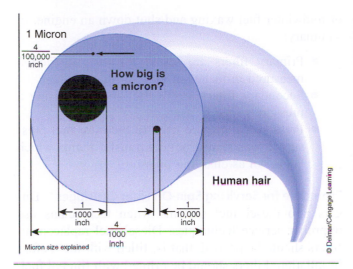
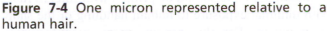

Figure 7-4 One micron represented relative to a human hair.

Figure 7-5 Pleated cartridge primary fuel filter used on a Ford 6.7L engine.

engine: it is a pleated cartridge located in series with fuel flow within the **diesel fuel conditioning module (DFCM)**. The entrapment capability of this filter is rated at 10 μ, which is typical for more recent diesel engines using **common rail (CR)** fuel systems that are especially sensitive to dirt intrusion.

Secondary Filters

Secondary filters (see **Figure 7-6**) represent the second filter in a typical two-stage filtering fuel subsystem. The secondary filter is charged by the transfer pump. Because this is at higher pressure, the filtering element used can be more restrictive. The secondary filter is located in series between the transfer or **charging pump**. The charging pump is responsible for pulling fuel from the fuel tank and charging the high-pressure fuel injection circuit. In some diesel fuel subsystems using two-stage filtering, a primary and secondary filter may be both located on the same circuit,

Secondary Fuel Filter Assembly

To provide additional fuel filtering, an engine mounted secondary fuel filter is located on the top of the left valve cover. The secondary fuel filter is a 4-micron cartridge style filter and is replaced as a complete unit.

Figure 7-6 Secondary fuel filter used on a Ford 6.7L engine.

usually the charge circuit. In such cases, both filters are mounted on the same base pad with the primary filter feeding the secondary filter in series. You are more likely to see this in off-highway diesel engines. Current secondary filters trap much smaller particulates than primary filters, sometimes as small as 2 μ. In the Ford example shown in **Figure 7-6,** the entrapment rating is 4 μ. Similar to primary filters, they use chemically treated pleated papers and cotton fibers.

Water and Secondary Filters. Water cannot be pumped through most current secondary fuel filters. This results in the filter plugging on water and shutting down the engine by starving it for fuel. Water-plugged filters should be replaced. As an emergency measure you can clean it out using methyl hydrate or other pure alcohol, then reprime it with fuel.

Single Circuit Subsystems. In a fuel subsystem that is entirely under suction, the terms primary and secondary are not used to describe multiple filters in the circuit. Every filtering device used in the fuel subsystem is therefore upstream from the transfer pump. This means that the filter or filters are held at a lower-than-atmospheric pressure (i.e., under "suction") any time the engine is running. In this type of fuel subsystem, the inlet restriction specification is critical. If the inlet restriction value is exceeded, the result is fuel starvation.

Uncommon Filter Arrangements. In some stationary and agricultural equipment, there exist some peculiar fuel filter arrangements that almost defy diagnostic common sense. Many of these fuel systems are older, and most are imported. When diagnosing such fuel systems without the benefit of service literature, take nothing for granted. The use of diagnostic pressure gauges can help.

Servicing Filters

Most fuel filters are routinely changed on preventive maintenance (PM) schedules. PM schedules are determined by highway miles, engine hours, or calendar months. Filters are seldom *tested* to measure serviceability. When filters are tested, it is usually to determine if they are restricted (plugged) to the extent they are reducing engine power by causing fuel starvation.

Primary filters should be tested for inlet restriction using a negative pressure gauge or a mercury-filled manometer (**Hg manometer**). A manometer is a clear tubular column formed in a U shape around a calibration scale marked off in inches. The U-shaped column is then filled with either mercury or water to a zero point on a measuring scale. When the manometer is connected to a fluid circuit, it produces a reading according to the pull (negative pressure) or push (positive pressure) acting on the fluid in the column. Actual inlet restriction values vary a lot according to which fuel system is being tested. Always refer to the OEM specifications.

The Hg manometer or low-pressure gauge should be connected into the circuit between the filter mounting pad and the transfer pump. Transfer pumps are **positive displacement**. This means they unload a constant slug volume of fluid per cycle. The faster these cycles go by, the more fuel they pump, meaning that accurate test results can be obtained without loading the engine. When testing circuit restriction on a fuel subsystem that is entirely under suction, the OEM specifications are usually pretty tight. Exceeding them by a margin as small as 1 inch of mercury may result in fuel starvation.

Secondary filters are usually charged by the transfer pump. Testing **charging pressure** (the pressure downstream from the charging/transfer pump) should be performed with an accurate, fluid-filled pressure gauge. This should be connected in series between the transfer pump and the high-pressure injection circuit. This is not generally used as a method of determining the serviceability of a secondary filter. Like primary filters, secondary filters tend to be changed according to preventive maintenance schedules rather than by testing—or they are changed when they plug on water

or midwinter fuel waxing and shut down an engine. In summary:

- Primary filters are tested for inlet restriction measured in inches of Hg.
- Secondary filters are restriction tested using a pressure gauge in psi.
- Pressure gauges are used to measure fuel pressure downstream from the transfer pump, which is known as charging pressure.

Procedure for Servicing Spin-On Filter Cartridges. Dirt gets into diesel fuel systems when technicians use improper service techniques. Diesel fuel replacement filters should be primed, that is, filled with fuel before installation. Filters should be primed with filtered fuel, with minimal exposure to human handling or the shop environment. For this reason, many manufacturers prefer that their systems be primed on chassis using the method described a little later in this section.

For chassis and equipment not equipped for on-chassis priming, the technician is required to prime the fuel subsystem. Shops performing regular engine services should have a reservoir of clean fuel. Any process that requires a technician to remove fuel from vehicle tanks will probably result in it becoming contaminated, no matter how much care is exercised. The container used to transport the fuel from the tank to the filter should be cleaned immediately before it is filled with fuel. Paint filters (the paper cone-shaped type) can be used to filter fuel. The inlet and outlet sections of the filter cartridge should be identified.

Priming

The filter being primed should be filled only through the inlet ports usually located in the outer annulus (ring) of the cartridge and never directly into the outlet port, usually located at the center. Most manufacturers prefer that only the primary filters be primed before installation during servicing. After the primary filter(s) has been primed and installed, the secondary filter should be installed dry and primed with a hand primer pump or inline electric primer pump if equipped. Many current diesel fuel subsystems are equipped with electric priming pumps. An electric priming pump when fitted should always be used to prime the secondary filter.

Replacement Procedure. The following sequence outlines a typical filter replacement routine:

1. Remove the old filter cartridge from the filter base pad using a filter wrench.
2. Drain the fuel to an oil disposal container.

3. Ensure that the old filter cartridge gasket(s) has been removed. Wipe the filter pad gasket face clean with a lint-free wiper.

4. Remove the new filter cartridge from the shipping wrapping. Fill the filter cartridge with clean, filtered fuel poured carefully into the inlet section. The inlet ports are usually located in the outer ring of the cartridge. Fuel poured into the filter inlet ports passes through the filtering media and fills the center or outlet section of the filter; this method will take a little longer because it requires some time for the fuel to seep through the filtration medium.

5. The fuel oil itself should provide the gasket and/or O-ring and mounting threads with adequate lubricant. It is neither necessary nor good practice to use grease or white lube on filter gaskets.

6. Screw the filter cartridge **CW (clockwise:** right-hand threads are used) onto the mounting pad; after the gasket contacts the pad face, a further rotation of the cartridge is usually required. In most cases, hand tightening is sufficient, but each filter manufacturer has its own specific recommendations on the tightening procedure, and these should be referenced.

Tech Tip: When a hand primer pump is fitted to a fuel subsystem, prime the primary filter by pouring fuel into it: make sure that all the fuel is poured through the filter inlet side only. Install the secondary filter dry and prime that using the hand primer pump. When an electric primer pump is fitted to the circuit, use it.

WARNING *When removing filter cartridges, ensure that the gasket is removed with the old filter. A common cause of air being sucked into the fuel subsystem is double gasketing of the primary filter. If you were to double-gasket a secondary filter, the result would be an external fuel leak, a condition that is likely to be identified more readily.*

Water Separators

Some current diesel engine–powered highway vehicles have fuel subsystems with fairly sophisticated water removal devices. Often these water removal devices are integrated into a filter assembly such as

Figure 7-7 Combined water separator/filter assembly used on a Cummins ISB engine.

that shown in **Figure 7-7,** which shows a combination filter water separator used on a Cummins ISB 6.7L engine. When water appears in diesel fuel it does so in three forms:

- free state
- emulsified
- semi-absorbed

Removing Free-State Water. Water in its free state appears in large globules, and because it is heavier than diesel fuel it collects in puddles at the bottom of fuel tanks or storage containers. Water separators can easily separate free-state water if it happens to be pulled into the fuel subsystem.

Removing Emulsified Water. Water emulsified in fuel appears in small droplets. Because of the small size of these droplets they may be suspended for some time in the fuel before gravity takes them to the bottom of the fuel tank. When free-state water collects at the bottom of a fuel tank, 3 miles (5 km) driving on a class B road is enough to **emulsify** it (finely dispersing it into the fuel), making it more of a problem.

Semi-Absorbed Water. Semi-absorbed water is usually water in solution, that is mixed, with alcohol. Semi-absorbed water in diesel fuel is a direct result of adding methyl hydrate to fuel tanks. Methyl hydrate is a type of alcohol added to fuel tanks as a deicer. Methyl hydrate either in pure form or as diesel fuel conditioner is added to fuel tanks to prevent winter freeze-up. Water that is semi-absorbed in diesel fuel is in its most dangerous form because it may emulsify in the fuel injection system, where it can seriously damage components.

Why Water Damages Fuel Systems. Generally, water damages fuel systems for three reasons.

- It has little ability to lubricate moving parts.
- It promotes corrosion.
- It flows less readily than diesel fuel.

Modern fuel injection systems pump diesel fuel at very high pressures. When water even in small quantities is pumped through the system, severe damage to fuel injectors may result. When you see a modern fuel injector with its tip blown off, the cause can often be traced to a water-in-fuel condition.

Water Separator Operating Principles. Water separators have been used in diesel fuel systems for many years. Often these were simple units that used gravity to separate the heavier water from the fuel. Today, a water separator will often combine a primary filter and water separator. Many of these combination primary filter/water separators are manufactured by aftermarket suppliers. These units use a variety of means to separate and remove water in free and emulsified states; they will not remove water from fuel in its semi-absorbed state.

Water separators use combinations of several principles to separate and remove water from fuel. The first is gravity. Water in its free state or emulsified water, if allowed to settle, will sink to the bottom of any container because it is heavier than diesel fuel. Some water separators use a **centrifuge** to help separate both larger globules of water and emulsified water from fuel. The centrifuge subjects fuel passing through it to centrifugal force: this throws the heavier water to the side walls of the separator, allowing gravity to pull it into the sump drain. A centrifuge acts to separate particulates from the fuel in the same manner, so sediment can also be removed in this way.

Positive filtration can also remove water from fuel, providing the medium has a sufficient entrapment rating. When fuel is forced through a fine resin-coated, pleated paper medium, it passes through it more easily than water. Water becomes trapped by this type of filtering medium. Once trapped, it collects in large enough droplets to allow gravity to pull it down into the sump drain. In many cases, aftermarket water separator/fuel filters are designed to replace the fuel system OEM's primary filter; in others this unit may work in conjunction with the primary filter. **Figure 7-8** shows an assortment of combination filter/water separators used on diesel engines.

All water separators are equipped with a drain valve. The drain valve may be manual or electrically operated. The purpose of this valve is to siphon water from the sump. Water should be routinely removed

|120|360|490|6120|6401|

Figure 7-8 An assortment of water separators used on diesel engines.

from the sump using the drain valve. A manual drain valve is located at the base of the DFCM on the Ford 6.7L engines as shown in **Figure 7-9**. The filter elements used in combination water separator/primary filter units should be replaced in most instances with the other engine and fuel filters at each full service. However, some manufacturers claim their filter elements have an in-service life that may exceed the recommended oil change interval by two or more times. Whenever a water separator is fully drained, it should be primed before attempting to start the engine.

Tech Tip: To troubleshoot the source of air admission to the fuel subsystem, a diagnostic

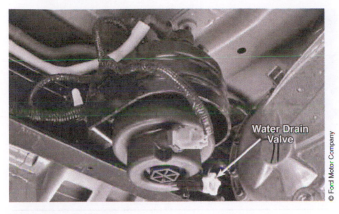

© Ford Motor Company

Water Drain Valve

The water drain valve is located on the bottom of the DFCM.

To drain water that has accumulated in the DFCM, turn the drain valve to the open position.

Figure 7-9 Location of the water drain valve on the Ford DFCM.

sight glass can be used; it consists of a clear section of tubing with hydraulic hose couplers at either end, and it is fitted in series with the fuel flow. However, the process of uncoupling the fuel hoses will always admit some air into the fuel subsystem, so the engine should be run for a while before reading the sight glass.

Fuel Heat Exchangers

In recent years, it is becoming more common to find fuel subsystems equipped with fuel heat exchangers. Fuel heat exchangers help to reduce fluctuations in the temperature of the fuel delivered to the high-pressure injection circuit. Fuel heat exchangers can be divided into:

- fuel heaters
- fuel coolers

Retrofitting either of these devices is generally not recommended. However, when a fuel heat exchanger is original equipment, this is not a problem because the functionality of the heat exchanger will be properly synergized with the fuel system.

Fuel Heaters. There are two types of fuel preheaters in current use:

1. Electric element type. An electric heating element uses battery current to heat fuel in the subsystem. This type offers a number of advantages, the most notable of which is that the

heater can be energized before start-up so that cranking fuel is warmed, facilitating start-up. Electric element fuel heaters may be thermostatically managed so that fuel is only heated as much as required and not to a point that compromises some of its lubricating properties.

2. Engine coolant heat exchanger type. This type of fuel heater consists of a housing within which coolant is circulated in a bundle (heat exchanger core) and over which the fuel is passed. A disadvantage of this type is that the engine cooling system must be at operating temperature before the fuel can be heated, and then it is often overheated during warm weather operation.

Fuel heaters exist that use both electric heating elements and coolant medium heat exchangers. This type of fuel heater can often manage the fuel temperature. Fuel temperatures should not exceed 90°F (32°C). Once fuel exceeds this temperature its lubricating properties start to diminish, and the result is reduced service life of fuel injection components.

Fuel Coolers. A more likely scenario with a modern diesel engine is a fuel cooler. The reason is that in a system where the fuel is circulated at a high rate and the onboard fuel storage volume is low, fuel draws heat away from the injectors and cylinder heads. **Figure 7-10** illustrates the fuel cooler used on a Ford 6.7L engine;

© Ford Motor Company

Fuel Cooler

A fuel cooler is located on the left-frame rail forward of the DFCM. The black fuel line is used for fuel return from the engine to the cooler. The gray fuel line returns fuel from the cooler to the DFCM. Depending on the temperature of the fuel from the injectors, the fuel cooler can be used to cool or heat the fuel going back to the DFCM. The powertrain secondary cooling system provides the coolant for the fuel cooler.

Figure 7-10 Fuel cooler used on a Ford 6.7L engine.

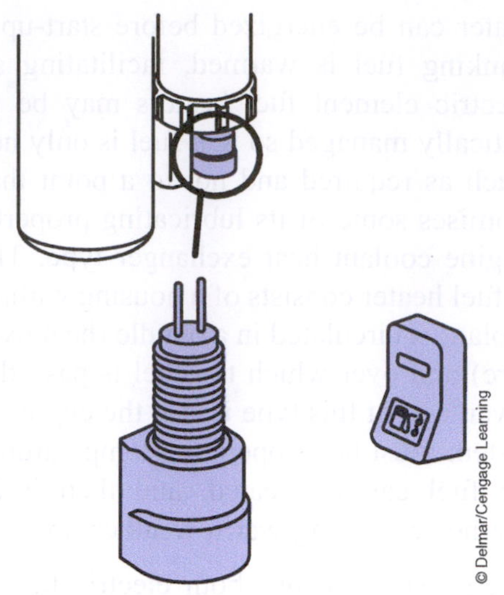

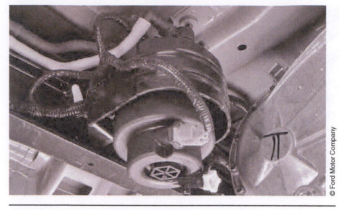

Figure 7-12 Location of WIF sensor in the Ford DFCM.

despite being called a fuel cooler, the Ford cooler acts as a comprehensive heat exchanger, allowing it to warm cold fuel and cool heated fuel. It uses a plate-type operating principle and engine coolant as a medium.

Figure 7-11 WIF sensor and its warning light.

Water-in-Fuel Sensors

Most current systems use a water-in-fuel (WIF) sensor to warn the operator of water contamination of fuel. A WIF sensor can be built into a replaceable filter cartridge or be integrated into a combination filter/water separator assembly. The sensor uses a couple of probes and a 12-volt supply. Because water has different electrical resistance than fuel, a signal from the WIF sensor can be triggered when the electrical path across the probes acts through water rather than fuel. At this point, the WIF broadcasts a service alert. Note that sometimes a WIF can produce a service alert immediately, after draining the water sump: the reason is that water-resident bacteria can coat the probes after draining and trigger a false signal. **Figure 7-11** shows a typical WIF sensor and its warning light, and **Figure 7-12** shows the location of the WIF sensor in the Ford DFCM.

FUEL CHARGING/TRANSFER PUMPS

Fuel charging or transfer pumps are positive displacement pumps driven directly or indirectly by the engine. A positive displacement pump displaces the same volume of fluid per cycle and therefore, fuel quantity pumped increases proportionally with rotational speed. Because of this, if a positive

displacement pump unloads to a defined flow area, pressure rise is proportional with rpm increase. On current small-bore diesel fuel systems, four types of transfer pumps are used:

- plunger-type pumps
- gear-type pumps
- vane-type pumps
- diaphragm pumps

Pumping Principles

In describing pump operation in the fuel subsystem, in common with most truck OEMs we use the terms *suction circuit* and *charge circuit*. Fuel movement through a fuel subsystem is created by a positive displacement pump. The way a transfer pump works is by creating flow that forces fuel out its discharge circuit. When this happens, lower-than-atmospheric pressure is created at the pump inlet. This allows atmospheric pressure acting on the fuel in the tank to exert "push" on the fuel, forcing it toward the pump inlet. In this way, fuel is moved through the fuel subsystem.

Plunger-Type Pumps

Plunger-type pumps (such as that shown in **Figure 7-13**) are usually flange-mounted to a housing and cam driven. Single-acting and double-acting plungers may be used. Double-acting plungers are often used in higher output engines that require more fuel.

Single-Acting Plunger Pumps. A single-acting **plunger pump** (with a plunger that pumps in one direction only) has a single pump chamber and an inlet and outlet valve. Fuel is drawn into the pump chamber on the inboard stroke. It is pressurized on the outboard or cam stroke. The principle of a single-acting plunger pump is shown in **Figure 7-14**.

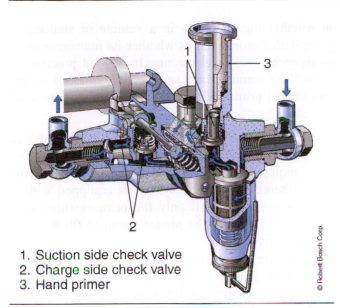

1. Suction side check valve
2. Charge side check valve
3. Hand primer

© Robert Bosch Corp.

Figure 7-13 Bosch charging pump with integral hand primer and primary filter.

Operating Principle (Single Acting)

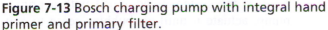

A. Cam stroke
B. Spring stroke

1. Drive eccentric
2. Camshaft
3. Pressure chamber
4. Suction chamber

Operating Principle (Double Acting)

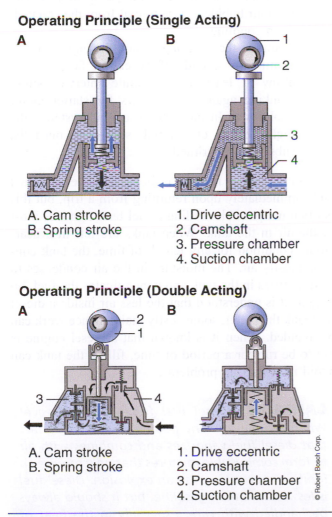

A. Cam stroke
B. Spring stroke

1. Drive eccentric
2. Camshaft
3. Pressure chamber
4. Suction chamber

© Robert Bosch Corp.

Figure 7-14 Action of single- and double-acting plunger pumps.

Double-Acting Plunger Pumps. A double-acting pump has twin chambers, each equipped with its own inlet and outlet valve. This permits the plunger to pump on both strokes. On the cam stroke, a two-way plunger pulls fuel in behind the plunger while discharging fuel in front of the plunger. The reverse occurs as the plunger is pulled back on its retraction stroke. The principle of a double-acting plunger pump is shown in **Figure 7-14.**

Gear-Type Pumps

Gear pumps are the most commonly used transfer pumps on large-bore diesel engines, but they are used less on small-bore diesels. These are normally driven from an engine accessory drive and are located wherever convenient. Gear pumps usually have a built-in relief valve that defines the system charging pressure, and they are often used where a high-flow fuel system is used. The more recent addition of diesel particulate filter (DPF) dosing systems to highway diesel engines has further increased the fuel flow requirement of the fuel subsystem and made gear pumps more common.

Vane and Diaphragm Pumps

Vane and diaphragm pumping principles are also used in small-bore diesel engine transfer pumps. A vane pump similar to that used in some gasoline-fueled automobile applications is used in some diesel engines, especially those that locate the pump inside the fuel tank. Vane pumps may be driven either mechanically using an engine accessory drive or by using an electric motor. Cam-actuated diaphragm transfer pumps may also be used.

The Ford 6.7L engine locates the fuel transfer pump within the DFCM assembly. The Ford CR system requires that the fuel subsystem deliver fuel to the high-pressure circuit at a pressure of 120 psi (830 kPa). In other words, the charging pressure of the Ford fuel subsystem is held at this specification. Pressure will not exceed this value due to a mechanical pressure regulator that trips at the specified pressure. **Figure 7-15** shows the location of the transfer pump on the Ford DFCM.

Hand Primer Pumps

A hand primer pump may be permanently fitted into the fuel subsystem. When it is, it is located either on the fuel transfer pump body or on a filter mounting pad (**Figure 7-13**). Some fuel subsystems are not equipped with a hand primer pump so the technician

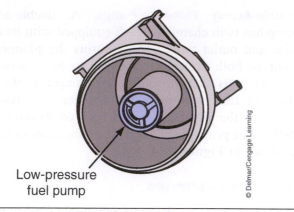

Low-pressure
fuel pump

© Delmar/Cengage Learning

Figure 7-15 Location of the transfer pump in the Ford DFCM.

may be required to temporarily fit one between the primary and secondary filters in order to prime the fuel subsystem.

A hand priming pump consists of a hand-actuated plunger and pump chamber. Most use a single-acting pumping principle. On the outward stroke, the plunger exerts suction on the inlet side, drawing in a charge of fuel to the pump chamber. On the downward stroke, the inlet valve closes and fuel is discharged to the outlet. When using a hand primer pump, it is important to purge air downstream from the pump on its charge side. Some fuel subsystems mount a hand primer to the transfer pump housing.

Electric Primer Pumps

More and more truck diesel engines today are equipped with self-contained, electric priming pumps. These electric pumps are used mainly to prime the fuel subsystem after servicing, but they can also be used when a loss of prime occurs. Electric primer pumps are a means of quickly and safely priming secondary fuel filters. Some more recent filters are designed to discourage any external priming methods, making it more likely that the electric prime feature will be used.

Priming a Fuel System

Although priming a fuel system is a simple procedure, you should still consult the OEM service literature. Most OEMs prefer that the technician avoid pressurizing air tanks with regulated air pressure to prime a diesel fuel system. Diesel fuel is easily vaporized, and fuel vapors may ignite. Especially avoid pressurizing fuel tanks in extremely hot weather conditions.

Recommended Procedure. The procedure required to prime a diesel engine varies considerably depending on whether the engine is in a vehicle or stationary, along with factors such as whether its management is hydromechanical or electronic. It is good practice to consult the manufacturer service literature before attempting to prime the engine. Following is a general procedure:

1. When an engine shuts down due to lack of fuel and it is determined that the fuel subsystem requires priming, remove the filters and fill with filtered fuel. If the system is equipped with a primer pump, fill only the primary filter with fuel. Then use the primer pump to fill the secondary filter.

2. Locate a bleed point in the system—this bleed point should be upstream from the injection circuit—and crack open the coupling.

3. Next, if the system is equipped with a primer pump, actuate it until air bubbles cease to exit from the cracked-open coupling. If the system is not equipped with a hand primer pump, fit one upstream from the transfer pump and actuate until air bubbles cease to exit from the cracked-open coupling.

4. Retorque the coupling. Crank the engine for 30 continuous seconds. If the engine fails to start, allow at least a two-minute interval before cranking again. This allows for starter motor cool-down. In most diesel engine systems, the high-pressure circuit will self-prime once the subsystem is primed.

Refueling. As a rule, it is good practice to refuel tanks immediately upon returning from a trip, but this is often not possible. Filling a fuel tank removes most of the air in the tank. When tanks are left in a near-empty condition for any length of time, the tank contains mostly air. The moisture in the air condenses to water, settles in the tank and contaminates the fuel. So long as it is understood that the less air inside a diesel fuel tank the better, some costly maintenance work can be avoided. When it is known that a diesel engine is not to be run for a period of time, filling the tank can avoid later start-up problems.

> **CAUTION** *Diesel fuel fumes are explosive! When refueling, many people overlook the fact that diesel fuel vaporizes and combines with air to form combustible mixtures that only require an ignition source to cause an explosion. Diesel fuel is less volatile than gasoline, but it should always be handled with care, especially in the heat of summer.*

MONITORING THE FUEL SUBSYSTEM

Almost all current fuel subsystems in electronically managed engines are monitored by the management system. We have already taken a look at a couple of monitoring devices, the WIF sensor and the sending unit, but in addition, temperature and pressure are monitored throughout the circuit. A typical fuel subsystem monitoring circuit comprises:

- Fuel sending unit
- Charging pressure sensor
- Primary circuit fuel temperature
- Secondary circuit fuel temperature sensor
- WIF sensor

The location of sensors in the fuel subsystem varies considerably by manufacturer, but the electrical operating principles are usually identical. The electrical principles used in vehicle temperature and pressure sensors are covered in **Chapter 12** of this book.

Ford monitors charging pressure by using a fuel pressure switch. This is a variable-capacitance device that feeds its signal to the PCM. Should charging

Figure 7-16 Location of the fuel pressure switch used to monitor Ford charging pressures.

pressure drop to a value below 50 psi, two events occur:

- Engine power is derated by 30 percent.
- An alert to the operator is broadcast to the driver display unit.

Figure 7-16 shows the location of a Ford fuel pressure switch located in a series between the secondary

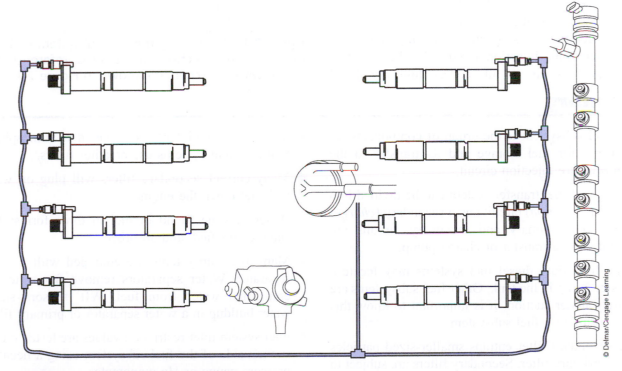

Injector Low-Pressure Connectors

The injector low-pressure connectors have a dual purpose. First, it is not really a return but a low-pressure feed to keep fuel pressure inside the hydraulic coupler. Without fuel pressure in the hydraulic coupler, the injector will not deliver fuel. Second, the fuel that passes through the injector during the injection process exits the injector through the low-pressure connectors.

Figure 7-17 The Ford low-pressure circuit used to hold the injectors under a moderate charge to prevent drainback.

filter and the high-pressure pump used in the system.

Return Circuit

Because of the need to cool injectors, most modern and many older fuel subsystems typically provided the injection circuit with up to three times the volume of fuel required to power the engine. This fuel is routed through the injector circuit, after which it has to be returned to the fuel tank. The return circuit may also be known as the:

- leak-off circuit
- low-pressure circuit

In some cases, fuel exiting the injectors and cylinder heads may be routed to a heat exchanger before being returned to the fuel tank. **Figure 7-17** shows the return circuit on a Ford 6.7L diesel engine, which they describe as the *low-pressure circuit*. In some cases, such as in the Ford system, the return circuit is held at a moderate charge to ensure that the injectors are never entirely drained of fuel.

Charging to Injection Circuit

The function of the fuel subsystem is to provide clean, filtered fuel at the specified charging pressure to the high-pressure injection circuit. In the CR fuel system used by almost all modern highway diesel

engines, the fuel subsystem fuels the high-pressure pump that manages injection pressures. **Figure 7-18** shows how a fuel subsystem connects with the high-pressure injection pump.

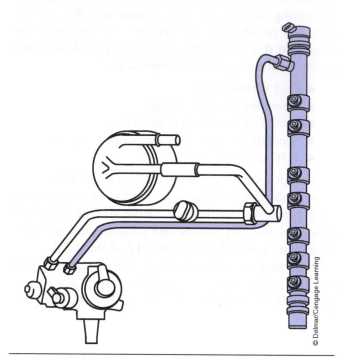

© Delmar/Cengage Learning

Figure 7-18 Connecting the fuel subsystem with the high-pressure injection circuit. Cross-reference this image with the schematic shown in Figure 7-1.

Summary

- The fuel subsystem is the group of components responsible for fuel storage and its transfer to the high-pressure injection circuit.

- The typical fuel transfer system can be divided into a primary (*suction*) circuit and a secondary (*charge*) circuit. The primary and secondary circuits are separated by a transfer or charge pump.

- Some light duty diesel fuel systems may locate a transfer pump in the tank. Other fuel subsystems are entirely under suction. It is important to know this when priming a fuel subsystem.

- The secondary filter entraps smaller-sized particles than a primary filter. Secondary filters are subject to charging pressure.

- Aluminum alloy, cylindrical fuel tanks located in the air flow on truck chassis can act as heat exchangers. Most trucks use dual fuel tanks mounted on either side of the chassis. This helps to evenly balance the weight of the fuel in the tanks.

- Most diesel fuel tanks are vented to atmosphere. Some current systems use breather filters.

- Many current secondary filters will plug on water and shut down the engine.

- Water may be found in fuel in three forms: free state, emulsified, and absorbed.

- Many fuel subsystems are equipped with a water separator. Water separators remove free-state and emulsified water from fuel. WIF sensors signal water buildup in a water separator or primary filter.

- Fuel system inlet restriction values are tested on the suction side of the fuel subsystem using a negative pressure gauge or Hg manometer.

- A common source of air in the fuel subsystem is double-gasketing of a filter under suction.

- Two types of fuel heaters are in current use: the electric element and coolant medium, heat exchanger types.

■ Diesel fuel systems commonly use one of two different fuel transfer or charge pumps: plunger pumps and gear pumps. Both these pumps use a positive displacement pumping principle.

■ Some fuel subsystems are equipped with a hand primer pump; the function of a hand primer pump is to purge air from the fuel subsystem.

■ It is good practice to prime secondary filters after installation using a hand or electric priming pump.

Review Questions

1. On the typical diesel fuel subsystem, which of the following would likely be at the lowest pressure when the engine is running?

   A. Outlet of the low-pressure pump
   B. Primary filter
   C. Secondary filter
   D. Charging circuit

2. Where is a secondary filter usually located?

   A. Upstream from the transfer pump
   B. On the charge side of the transfer pump
   C. In the fuel rail
   D. In the return gallery

3. What is the objective behind filling vehicle fuel tanks before overnight parking?

   A. To minimize moisture condensation in the tanks
   B. To minimize fuel evaporation
   C. To help cool down onboard fuel
   D. Drivers may forget the next morning

4. Besides fuel storage, the fuel tank may play an important role in a high-flow fuel system as a(n):

   A. Heat exchanger
   B. Fuel-heating device
   C. Ballast equalizer
   D. Aerodynamic aid

5. Which of the following could be used to test the low-pressure side of a typical diesel fuel subsystem for inlet restriction?

   A. Diagnostic sight glass
   B. H_2O manometer
   C. Negative-pressure gauge
   D. Accurate high-pressure gauge

6. Which of the following should fuel subsystem charging pressures be measured with?

   A. Diagnostic sight glass
   B. H_2O manometer
   C. Hg manometer
   D. Accurate pressure gauge

7. What should be used to check for air being pulled into a fuel subsystem?

   A. Diagnostic sight glass
   B. H_2O manometer
   C. Hg manometer
   D. Accurate pressure gauge

8. Which type of fuel transfer pump is more commonly used by electronically managed diesel engine/fuel systems?

    A. Plunger pumps   C. Diaphragm pumps

    B. Centrifugal pumps   D. Gear pumps

9. What type of pumping principle is used by a typical hand primer pump?

    A. Single-acting plunger   C. Rotary gear

    B. Double-acting plunger   D. Cam-actuated diaphragm

10. Charging pressure rise is usually directly related to which of the following?

    A. Throttle position   C. Peak power

    B. Engine load   D. Increased rpm

8 Injector Nozzles

Prerequisite

A good understanding of diesel engine operation.

Learning Objectives

After studying this chapter, you should be able to:

- Identify the subcomponents of a nozzle assembly.
- Describe the injector nozzle's role in system pressure management.
- Identify two types of injector nozzles.
- Describe the principles of operation of multiple-orifice and electrohydraulic nozzles.
- Define *nozzle differential ratio*.
- Describe a valve closes orifice (VCO) nozzle.
- Bench (pop) test a hydraulic injector nozzle for forward leakage and back leakage.
- Outline the procedure required to test an electrohydraulic injector (EHI).
- Explain the operation of a piezo-actuated EHI.
- Explain how to identify and program calibration codes on an EHI to the PCM.
- Outline the procedure required to remove, inspect, and reconnect high-pressure lines.

Key Terms

atomization	hard value	peak pressure
back leakage	hydraulic injectors	piezoelectric actuators
chatter	injector nozzle	piezo injectors
common rail (CR) injection	mechanical injectors	popping pressure
direct injection (DI)	multiple-orifice nozzle	pop tester
electrohydraulic injectors (EHIs)	nozzle differential ratio (NDR)	soft value
electronic unit injectors (EUIs)	nozzle opening pressure (NOP)	valve closes orifice (VCO) nozzle
forward leakage	nozzle seat	

INTRODUCTION

Almost all diesel engines manufactured today are direct injected. In a **direct injection (DI)** diesel engine, the fuel is injected into the cylinder immediately above the piston. Injected fuel must be atomized. **Atomization** of the fuel requires breaking it up into very small liquid droplets. These small droplets are produced by forcing very high-pressure fuel through minutely sized holes or orifices. The smaller the droplet exiting the injector, the faster it will vaporize and ignite when it is propelled into the engine cylinder. The actual size of droplets that exit the injector depends on:

■ Orifice size: The hole size in nozzles obviously does not change after it has been manufactured.
■ Pressure: The pressure a nozzle is subject to is managed by the injection pump. The higher the pressure, the smaller the fuel droplets exiting the nozzle.

The means used to inject fuel into the cylinder is an **injector nozzle**. An injector nozzle may be a standalone device or it may be a subcomponent of an electronically controlled pump and injector component. Although many different designs of injector nozzles have been used in the past, almost all of today's engines use one of the following two types:

■ Multiple-orifice (hole) hydraulic nozzles
■ Electrohydraulic nozzles

All of the injector nozzles covered in this textbook are closed-nozzle systems. Open-nozzle injectors use very different principles and are not discussed in this textbook.

Multiple-Orifice Nozzles

When manufacturers describe multiple-orifice injectors they are usually called:

■ **Hydraulic injectors**
■ **Mechanical injectors**

Perhaps the better of the terms is *hydraulic injector*. A **multiple-orifice nozzle** is opened and closed hydraulically. In fact, it really functions as a hydraulic switch. As a switch, the multiple-orifice nozzle is designed to open and close at a specific pressure. Once the pressure is set when the device is calibrated, it is unlikely to change unless it is due to old age or wear. The opening pressure of multiple-orifice nozzles is described as **nozzle opening pressure (NOP)**. Another way of saying NOP is to use the term **popping**

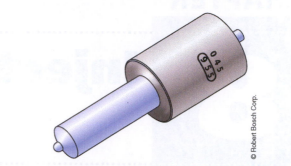

Figure 8-1 External view of a multi-orifice injector nozzle.

pressure. Because NOP in hydraulic injectors is a set value that cannot be controlled by the engine management electronics, they are described as having **hard value** NOPs. Hydraulic injectors with multiple-orifice nozzles can be divided into two general categories:

■ Integral injectors: A single-function nozzle device used to atomize fuel and also set the NOP. Integral injectors are connected to an injection pump by a high-pressure line.
■ Subcomponent nozzles: A subassembly built into a more complex injector assembly that also pumps and controls injection fuel quantity.

In general terms, we can describe hydraulic injector nozzles as devices with the following functions:

■ to open and close to begin or end fuel injection
■ to define *nozzle opening pressure (NOP)* (hydraulic nozzles only)
■ to atomize fuel to the correct size for combustion

Until the introduction of electrohydraulic injectors (EHIs), almost all the nozzles used in highway diesel engines were hydraulic nozzles. Today's diesel engines require injector nozzles that can be controlled by the engine management electronics, so hydraulic nozzles are rapidly becoming a thing of the past. **Figure 8-1** shows an external view of a hydraulic injector nozzle.

Electrohydraulic Nozzles

The key feature to electrohydraulic nozzles is that they do not have fixed-pressure opening and closing values. In other words, NOP becomes a **soft value**. We use the term *soft value* because both opening and closing of the injector valve are directly controlled by the powertrain control module (PCM) managing the engine. Electrohydraulic nozzles are found in two general types of current diesel injection systems:

Figure 8-2 External view of an EHI used in a modern CR-fuel system.

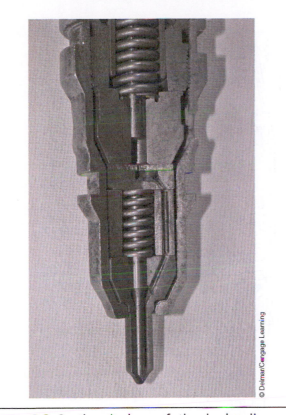

Figure 8-3 Sectional view of the hydraulic nozzle subassembly in an HEUI injector assembly: in this, the hydraulic nozzle is a subcomponent.

- **Common rail (CR) injection**: CR diesel fuel injection uses stand-alone **electrohydraulic injectors (EHIs)**. Each EHI is supplied by rail pressure fuel and is switched open and closed by the PCM.
- Dual actuator, **electronic unit injectors (EUIs)**: In a dual actuator EUI, the electrohydraulic nozzle is a PCM-controlled subcomponent. It operates almost identically to the EHI except that the pumping of fuel to injection pressures takes place immediately above the nozzle assembly.

We will study both the CR and EUI fuel systems in **Chapter 10. Figure 8-2** shows an external view of an EHI from a Cummins ISB engine used in Dodge pickup trucks.

MULTIPLE-ORIFICE NOZZLES

As we said in the introduction to this chapter, until 2007, most diesel injector nozzles could be classified as multiple orifice. They were used in:

- Pump-line-nozzle (PLN) diesel injection pumps
- Mechanical unit injectors (MUIs) as a subcomponent

- Single actuator, electronic unit injectors (as a subcomponent)
- Hydraulically actuated, electronic unit injectors (HEUI) (as a subcomponent). **Figure 8-3** shows a sectional view of an HEUI, focusing on the multiple-orifice nozzle assembly.
- Electronic unit pump (EUP) systems

A multiple-orifice nozzle is a hydraulically actuated switch machined with several tiny holes. The number of holes varies, but it is typically between four and eight. **Figure 8-4** shows a sectional view of a typical multi-orifice injector. It should be stressed that when EUIs or HEUIs use hydraulic nozzles as subassemblies, the operating principles are identical to those of the injector shown in **Figure 8-4,** which we will use as the basis for describing their operation.

Droplet Sizing

The size of droplets injected by a nozzle orifice depends on two factors:

- Flow area (size of nozzle holes)
- Pressure (managed by the injection pump)

Because flow area depends on the actual size of the holes machined into the nozzle, this remains constant.

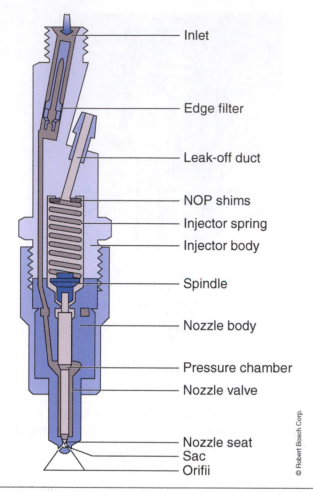

Figure 8-4 Sectional view of a multi-orifice injector nozzle.

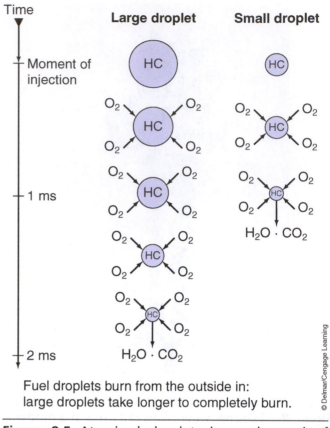

Fuel droplets burn from the outside in: large droplets take longer to completely burn.

Figure 8-5 Atomized droplet size and speed of combustion.

The pressure values vary. Actual pressure values range from values lower than NOP up to the **peak pressure** (the highest pressure a fuel injection system can produce) value. Because the flow area remains constant, as pressure increases, the droplet sizing decreases. Generally, the longer the injection pulse (more fuel injected to the cylinder), the higher the pressure peak. What this means is that the largest atomized fuel droplets are produced at the beginning of the injection pulse just after the nozzle opens. This is generally good for performance. Fuel droplets burn from the outside inward toward the center as shown in **Figure 8-5.** When large droplets are injected early on in combustion, there is plenty of time to combust them.

Because the injection pressure is designed to rise during an injection pulse, the droplet size reduces. The reduction in droplet sizing that occurs as the fueling pulse is extended generally favors the complete combustion of the fuel injected into the cylinder. Larger droplets require more burn time, smaller droplets shorter burn times. This means that as a fuel pulse is extended, the injected droplet size in the atomized fuel

reduces. This reduction in droplet size coincides with the reduction in real time available to burn the droplets injected late in the injection pulse. **Figure 8-5** shows how the size of an injected fuel droplet influences the speed of combustion: the fuel droplet is identified as hydrocarbon (HC) and the air it reacts with as oxygen (O_2) in this image.

Action

The multiple-orifice nozzle is located in the injector body by a dowel. The injector body is also usually positioned in its bore by a guide lug. This ensures that the atomized fuel spray pattern is directed at an exact location in the engine cylinder. In a typical diesel engine with a Mexican hat peak in each piston, the target location is in the crater that surrounds the Mexican hat peak. Fuel from the injection pump is delivered to the nozzle body. This fuel is then routed by a duct or ducts to the nozzle pressure chamber (see **Figure 8-4**) and then ducted to an annular recess in the upper nozzle valve body.

For most of the cycle, the nozzle valve is held closed by the injector spring. This spring pressure can be either directly or indirectly applied to the nozzle valve. When pressure is applied directly, the nozzle spring contacts the nozzle valve. When pressure is applied indirectly, spring pressure is relayed to the nozzle valve by a spindle. The spring tension that acts on the nozzle valve holds it closed. It sets the NOP value so it is adjustable either by a screw or by shims.

Opening Pressure. When line pressure is driven upward by the injection pump, pressure increases in the nozzle pressure chamber. As soon as this hydraulic pressure in the nozzle pressure chamber is greater than the mechanical pressure of the injector spring, the nozzle valve lifts. This instantly opens the nozzle. At this point, fuel flows around the **nozzle seat**. At the center of the nozzle valve seat, a single duct connects to the nozzle sac. The nozzle sac is a hollow circular cavity. Each nozzle orifice is machined into the sac. As the nozzle valve lifts, fuel passes around the nozzle seat and enters the sac. The function of the sac is to hydraulically balance the fuel about to be injected. The idea is that droplets start to exit from each orifice at the same time. When fuel begins to exit the nozzle, the fuel injection pulse begins.

Nozzle Closing. The injection pulse continues for as long as the nozzle valve remains open. Any time the nozzle valve is open, the engine is being fueled. A fueling or injection pulse ends when the pressure is collapsed by the injection pressure pumping element. The collapse phase takes time. Because this type of nozzle is controlled hydraulically, the collapse phase takes longer when peak pressures are high. For the nozzle valve to close, the pressure in the nozzle pressure chamber must collapse enough so that the mechanical force of the injector spring can take over once again. This pressure will be below NOP by a small margin. At the point the nozzle spring closes the nozzle valve, the nozzle valve is held on its seat by spring pressure.

Nozzle Differential Ratio

Nozzle differential ratio (NDR) tells us why it takes more pressure to open a hydraulic nozzle valve than the pressure required to hold it open. Nozzle valves are opened by hydraulic pressure acting on the sectional area represented by the pressure chamber. When this hydraulic pressure is sufficient to overcome the spring pressure holding the valve seated, it opens. The instant the nozzle valve opens, hydraulic pressure passes around the seat. This means that hydraulic pressure now acts over the whole sectional area of the nozzle valve. The whole sectional area of

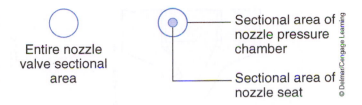

Figure 8-6 Nozzle differential ratio: sectional areas of a nozzle valve.

the nozzle valve consists of the sectional area of both:

- the seat
- the pressure chamber

The principle of nozzle differential ratio is shown in **Figure 8-6.**

Summary of NDR. NDR explains why more pressure is required to unseat a nozzle valve than is required to hold it off its seat. In older diesel injection systems, when a high-pressure pump element begins to increase pressure for an injection pulse, it does so by acting on a closed hydraulic circuit—closed by the nozzle valve seat, that is. However, the instant NOP is achieved, the hydraulic circuit opens. Opening the circuit results in a dip in pressure: without NDR, the nozzle valve would instantly close. Because of NDR, less pressure is required to hold the nozzle valve open when the pressure dips. Almost instantly after NOP is achieved, the restriction represented by the nozzle orifices comes into play, meaning that pressure rise continues to be controlled by the pump element.

Typical hydraulic injectors have nozzle closing pressures that are between 10 percent and 20 percent lower than their specified NOP. A real disadvantage of NDR is that the largest droplets are injected into the engine cylinder when they are least wanted, at the very end of the injection pulse. Because there is insufficient time to properly combust these last few droplets, these are exhausted as HC emission.

VCO Nozzles

Valve closes orifice (VCO) nozzles eliminate the sac (see **Figure 8-4**). The function of the sac in the nozzle is to provide balanced introduction of the fuel into the engine cylinder at the moment of NOP. This helps keep the interval between the delivery of the first fuel droplets and the moment they ignite, consistent. The disadvantage of the sac is at the completion of an injection pulse, the volume of fuel in the sac is wasted fuel that adds to HC emission. Most current injector nozzles have either reduced the sac volume or eliminated it

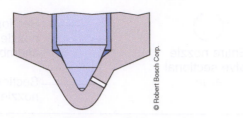

Figure 8-7 Valve closes orifice (VCO) nozzle.

entirely. They have been replaced by VCO nozzles. In a VCO nozzle, each orifice is bored directly into the nozzle seat. VCO nozzles are most often used on electronically managed injection systems that use high NOP values. Because of the much higher pressures used, hydraulic balance is achieved anyway. **Figure 8-7** shows a VCO nozzle design.

ELECTROHYDRAULIC NOZZLES

Electrohydraulic nozzles were introduced when EHIs were introduced on the first generation of common rail (CR), diesel fuel injection systems. As indicated earlier in this chapter electrohydraulic nozzles are used in:

- CR electrohydraulic injectors
- Dual actuator EUIs

Unlike hydraulic injector nozzles, EHIs are controlled electrically, that is, directly by the PCM. The first generation of EHIs was switched by solenoids. More recently, **piezoelectric actuators** have been introduced. The operating principles of solenoid-actuated and piezo-actuated injectors are almost identical. The **piezo injectors** (most manufacturers use this term) respond faster than solenoids to switching commands because they do not depend on creating and collapsing a magnetic field to open and close. They also require less current than solenoid-actuated EHIs. We will take a look at the general operating principles first, then examine piezoelectric actuators afterward. Although this description is based on an EHI, the way the electrohydraulic nozzle functions as a subcomponent of a dual actuator EUI is identical.

EHI Operation

Both solenoid-actuated and piezoelectric-actuated EHIs use hydraulic force from a rail or pump chamber to open and close the nozzle valve. The PCM precisely controls opening and closing events. An EHI (**Figure 8-8**) can be subdivided as follows:

- Nozzle assembly
- Hydraulic servo system

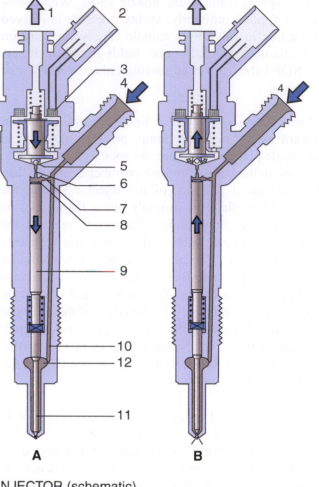

INJECTOR (schematic)

A. Injector closed (at-rest status)
B. Injector opened (injection)
1. Fuel return
2. Electrical connection
3. Triggering element (solenoid valve)
4. Fuel inlet (high pressure) from the rail

5. Valve ball
6. Bleed orifice
7. Feed orifice
8. Valve control chamber upper pressure field
9. Valve control plunger
10. Feed passage to the nozzle
11. Nozzle needle
12. Lower pressure field

Figure 8-8 Electrohydraulic injector (EHI).

- Actuator valve (either a solenoid or piezoelectric actuator)

EHI Injection Sequence. Referencing **Figure 8-8,** fuel at rail pressure is supplied to the high-pressure connection (4), to the nozzle through the fuel duct (10), and to the control chamber (8) through the feed orifice (7). The control chamber (upper pressure field) is connected to the fuel return (1) via a bleed orifice (6) that is opened by the actuator valve. With the bleed orifice closed, hydraulic force acts on the valve control plunger (9) exceeding that at the nozzle-needle pressure chamber (located between the shank and the needle of the nozzle valve). Although the actual hydraulic pressure values acting on the top of the nozzle valve (upper pressure field) (8) and that in the lower pressure field (12) are identical, the sectional area at the top of the nozzle valve is greater. As a result, the nozzle needle is loaded into its seated position, meaning that the injector is closed. The primary force holding the nozzle valve closed is hydraulic whenever the (common) rail is at operating pressure.

When a PCM signal triggers the actuator valve, the bleed orifice opens. This immediately collapses the control-chamber pressure (upper pressure field) and, as a result, the hydraulic pressure acting on the top of the valve control plunger (9) also drops. Because the hydraulic force acting on top of the valve control plunger collapses and the pressure in the lower pressure field remains, the nozzle needle lifts. Lifting the nozzle needle allows fuel to pass around its seat. This opens the orifices, allowing fuel to be injected into the combustion chamber. The nozzle needle remains open until de-energizing the actuator reestablishes the upper pressure field at rail pressure.

EHI Operating Phases.

When the engine is stopped, all injector nozzles are closed, meaning that their nozzle valves are loaded onto their seats by spring pressure. In a running engine, injector operation takes place in three phases.

Injector Closed

In the at-rest state, the actuator valve is not energized. In this state, the nozzle needle is seated by the injector spring combined with hydraulic pressure (from the rail). Although the hydraulic pressure in both the upper (control) and lower pressure fields is identical, the sectional area of the upper pressure field exceeds that of the lower pressure field by a considerable margin. With the bleed orifice closed (6), the actuator valve spring forces the armature ball check onto the bleed-orifice seat. This seals and holds intact the upper pressure field. The upper pressure field is the control field. Whenever the upper pressure field is held at rail pressure, the EHI is closed.

Nozzle Opening

When the actuator is energized by the PCM injector driver, the bleed orifice (6) opens. As the bleed orifice opens, fuel flows from the valve-control chamber into the return circuit. This collapses the hydraulic pressure in the upper pressure field (8). Because pressure in the lower pressure field (12) remains at rail pressure, the nozzle needle is forced open. The instant the EHI needle lifts, it exposes the injector orifices, beginning the injection pulse. Fuel continues to enter the valve control chamber (12) at rail pressure when the EHI opens but spills immediately to the return circuit. Note in **Figure 8-8** that the flow area at the bleed orifice is very small. This means that the actual volume of spill fuel from the control chamber is small. This spilled fuel is designed to be small enough not to affect the rail pressure.

The nozzle needle (11) opening speed is determined by the difference in the flow rate through the bleed and feed orifices. When the actuator control plunger reaches its upper stop, it is cushioned by fuel generated by flow between the bleed and feed orifices. When the injector nozzle needle has fully opened, fuel is injected into the combustion chamber at a pressure very close to that in the fuel rail.

Nozzle Closing

When the actuator valve is de-energized by the PCM, its spring forces the valve downward. This forces the check ball downward to close the bleed orifice. Closing the bleed orifice allows the upper pressure field to be reestablished. Almost instantly, the pressure in the upper pressure field is at rail pressure. This means that pressure acting in both the upper (8) and lower pressure fields (12) is equal. However, because the upper pressure field (8) has much greater sectional area, the valve control plunger (9) drives the nozzle needle downward until it seats. Seating the nozzle needle ends injection and seals the EHI. Once again, the primary seating force of the nozzle needle (11) is hydraulic relayed to it by the valve control plunger (9).

Piezoelectric Injectors

Piezoelectric actuators are based on reversibility of piezoelectric effect. When certain crystals are subjected to pressure, a voltage is produced in a process we know as piezoelectricity. These same crystals are capable of reversing the process, so when we subject them to a voltage, they almost instantly react, resulting in either expansion or contraction. This mechanical reaction produces a powerful force, making it possible for them to replace solenoids. They have an advantage over equivalent solenoid actuators. Solenoids use electromagnetism and respond more slowly due to the time required to build and collapse electromagnetic fields, whereas piezo actuators respond almost instantly when triggered by a voltage pulse.

Figure 8-9 Piezo-actuated, electrohydraulic injector.

Piezo actuators are therefore ideally suited to cyber-time requirements for an almost instant response to an output trigger. This makes them ideal to manage multipulse injection events. Multipulse injection means the breaking up of an injection pulse into as many as seven mini-injections during a single cycle. This results in giving the PCM much better control of combustion. In addition it results in quieter diesel

engine operation and lower emissions. **Figure 8-9** shows the piezo-actuator EHI used in the 2011 Ford Power Stroke. One EHI is used for each of the eight cylinders on this engine and is supplied by injection pressure fuel directly from the common rail. The injector is manufactured by Bosch and uses an eight-hole nozzle: it is retained in the cylinder head by a single clamp and bolt.

Piezo Actuator Operation. The actuator consists of a stack of wafer-thin piezo crystals as shown in **Figure 8-10.** When an electrical charge is applied to the piezo stack, the crystals expand, creating a mechanical kick. When the stack electrical charge is removed, the crystal wafer stack contracts. This can be observed by studying **Figure 8-10.** Drivers in the PCM are used to deliver a pulse-width modulated (PWM) electrical charge to the EHIs.

Hydraulic Coupler. The piezo stack in the EHI is directly linked to a control valve by means of a fuel-charged hydraulic coupler. The upper piston of the coupler has a larger diameter (larger hydraulic

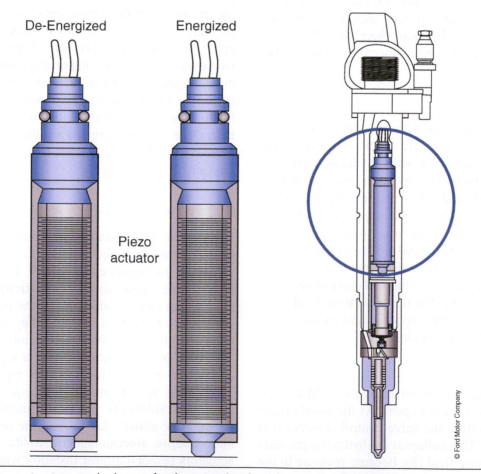

Figure 8-10 Piezo-actuator stack shown in de-energized and energized positions.

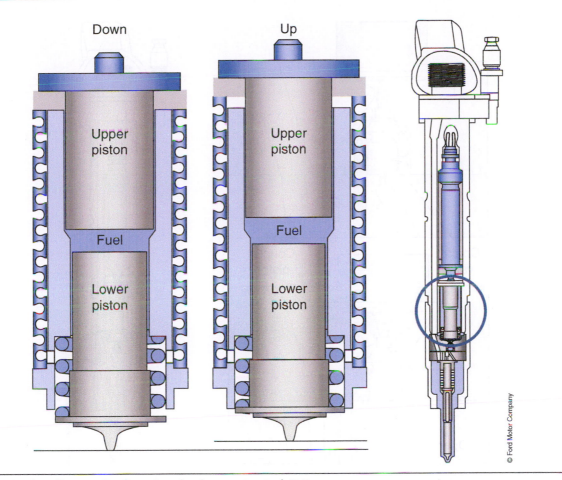

Figure 8-11 Hydraulic coupler in a Bosch piezo-actuated EHI.

sectional area) than the lower piston (see **Figure 8-11**); this difference in sectional area results in greater travel of the lower piston. The coupler piston at the base of the lower piston acts on and unseats the control valve. Fuel must be charged to the hydraulic coupler for the injector to function. Fuel is supplied to the hydraulic coupler by the low-pressure pump at key-on and through cranking, and by the return circuit fuel once the engine is running. This "hydraulic lock" fuel is required to relay actuator movement from the upper piston to the lower piston and thus move the control valve. With no fuel in the chamber between the upper and lower pistons, no injection can take place.

Control Valve. Movement of the control valve **(Figure 8-11)** is managed by the lower piston, which acts directly on it. When the control valve is moved downward, high pressure in the control chamber above the nozzle needle is collapsed. When the control valve is forced to its fully down position, it seals off an orifice in the intermediate plate: this action interrupts the fuel flow path of high-pressure fuel (from the rail) acting on the nozzle needle **(Figure 8-12).** Collapsing the high pressure acting on the top of the

nozzle needle while retaining it in the lower surface (see **Figure 8-13**) allows the needle to lift; the instant the nozzle needle is lifted, high-pressure fuel from the rail exits the nozzle assembly, beginning injection.

Nozzle Valve Operation. Nozzle position (open or closed) is determined by whether or not rail pressure is acting on its upper sectional area. For most of the cycle when the actuator is not energized, rail pressure acts on both the upper surface sectional area of the nozzle valve and the lower surface. We used the term *pressure chamber* to describe what is described as *lower surface* in **Figure 8-13,** and frankly, *pressure chamber* is a more accurate term. The upper surface comprises the entire sectional area of the nozzle needle. The lower surface comprises about half the sectional area of the needle valve. So, when rail pressure fuel acts on both upper and lower surfaces, hydraulic pressure holds the nozzle needle firmly seated. However, when the control valve is actuated, the pressure field acting on the upper surface immediately collapses: because full rail pressure continues to act on the lower surface, the nozzle needle triggers open at some velocity.

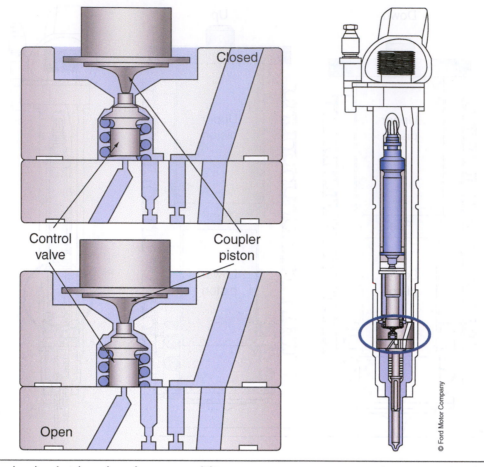

Figure 8-12 Control valve in closed and open positions.

Rail pressures are high, as we will see when we study common rail systems in **Chapter 10.**

When the piezo stack is de-energized, the resulting contraction causes the control valve to reestablish the pressure field acting on the top of the nozzle needle, slamming it closed.

The speed at which these events occur is measured in microseconds due to the extremely high hydraulic pressures (rail pressures can exceed 35,000 psi [2,400 bar]) and almost immediate response times of the piezo actuator. For this reason, up to seven injection events can take place within a real time period of three milliseconds.

Calibration Codes. Most contemporary injectors have a calibration code. This calibration code has to be programmed into the PCM whenever an injector change-out takes place. Failure to reprogram the correct calibration code into the PCM can result in an unbalanced engine. Full-load operation of an unbalanced engine could possibly result in complete engine failure. The calibration code is a bench test–derived fuel flow parameter that when programmed into the engine management computer, allows it to compensate

for the fuel flow variability of individual injectors. Calibration codes are known by different names depending on the manufacturer. Terms such as *trim code* and *quality code* are common. However, because we have been commonly using the Ford Power Stroke as a reference, we will use their term: *injector quantity adjustment (IQA) code.*

Ford's IQA is a 10-digit code printed on a factory label located on each injector cap. This data must be programmed into the PCM whenever a change-out takes place. If switching an injector from one cylinder to another, the reprogramming is required. **Figure 8-14** shows the IQA codes printed on the engine cylinder block for a set of eight injectors.

Summary of EHIs

Because of their soft opening and closing pressures and their ability to provide high injection pressures at low engine rpms, electrohydraulic nozzles are a requirement in most post-2007 diesel engines. Their advantages include:

- Ability to open and close almost instantly when signaled by the PCM.

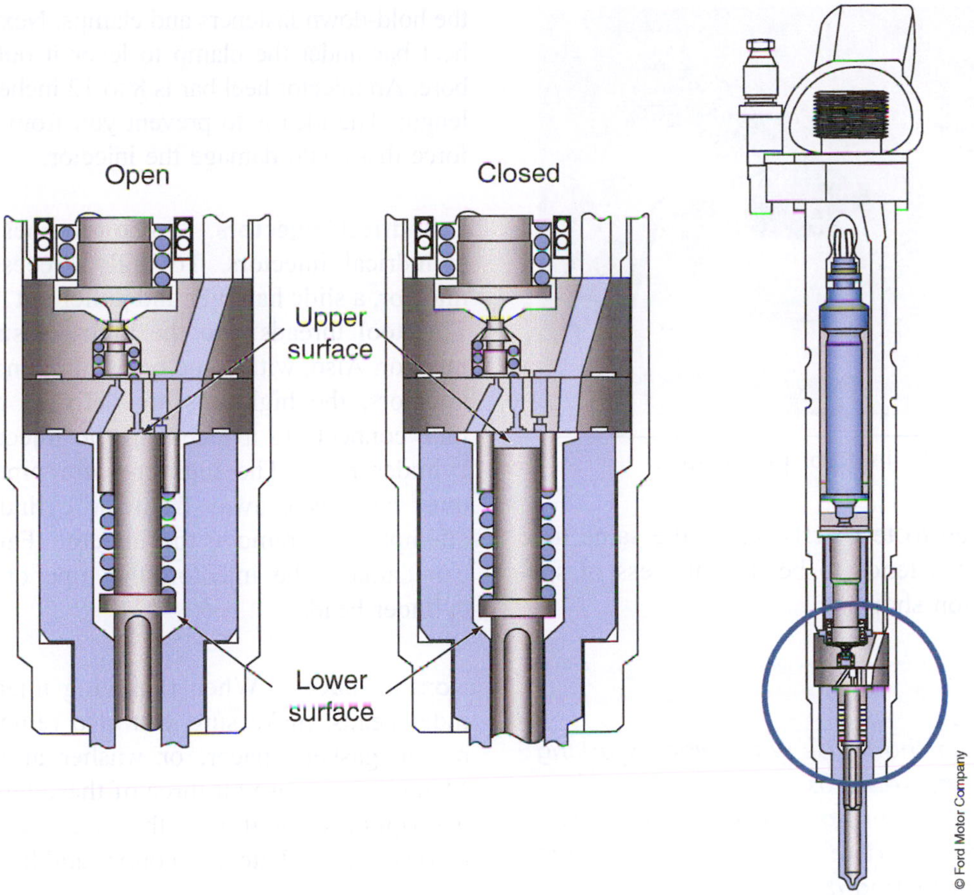

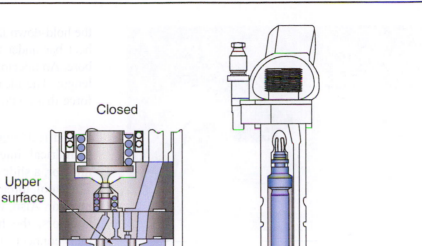

Open   Closed

Upper surface

Lower surface

© Ford Motor Company

Figure 8-13 Piezo-actuated EHI nozzle valve operation.

Inj. #1 FFFF01C028000000
Inj. #2 05EFC26028000000
Inj. #3 FC2EDD8028000000
Inj. #4 F7BDA08030000000
Inj. #5 03DF22E028000000
Inj. #6 01EF80F028000000
Inj. #7 FFBFC0D028000000
Inj. #8 F3CE9FC028000000

© Delmar/Cengage Learning

Figure 8-14 Fuel trim calibration: 10-digit IQA code located on each injector head that must be programmed to the PCM.

- EHI injection pressures remain consistent through a fueling pulse.
- Ability to close instantly eliminates the collapse phase of older hydraulic injectors.
- Suited to multipulse injection.
- May use either solenoid or piezoelectric actuators.
- May be integrated as a subcomponent of an EUI to provide it with soft opening and closing values.

Tech Tip: Practice special care when working with diesel fuel injectors. Use protective caps to seal the nozzle and the injector inout ports after removing an injector.

NOZZLE TESTING

Consult the manufacturer service literature before attempting to service injectors. Almost without exception, troubleshooting of engines with a suspected injector problem is performed with the injector in-engine using diagnostic software. Most manufacturers today do not include fuel injectors in preventive maintenance schedules. This usually means that when you are asked to service fuel injectors, you are working on older engines. You should know how to test hydraulic injectors because this is a procedure you will be tested on in certification exams. Testing hydraulic injector nozzles requires a simple bench test fixture (or **pop tester**—see **Figure 8-15**). Servicing of EHIs requires more costly test equipment because of the extreme accuracy required. We will go through the

© Delmar/Cengage Learning

Figure 8-15 Hydraulic injector pop tester.

procedure required to test EHIs but at the same time emphasize that this tends to be the business of specialty fuel injection shops.

CAUTION

- *Eye protection should be worn when working with high-pressure fluids.*
- *High-pressure atomized fuel is extremely dangerous, and no part of the body should be in danger of contacting it.*
- *Never touch an injector on the pop tester when it is at any pressure above atmospheric. Remember also that the entire injector assembly is under pressure.*
- *Diesel fuel oil is a known cancer-causing agent. Hands should always be washed after contact with it.*

Removal of Injectors from the Cylinder Head

You must follow the manufacturer-recommended method of removing an injector from a cylinder head. Before attempting to remove injectors, clean the surrounding area; any dirt around the injector bore can end up in the engine cylinder below it. It is also good practice to drain high-pressure lines when removing the pipe fittings. Use a line wrench or socket to loosen line nuts; they are easy to round off, and if you damage one you will end up replacing the line. Injector nozzles may be damaged if subjected to any side-load force. This means you should always use the correct pullers.

Flanged Injectors. Many hydraulic injector assemblies are flanged. This means you should first remove

the hold-down fasteners and clamps. Next use an injector heel bar under the clamp to lever it out of the injector bore. An injector heel bar is 8 to 12 inches (20–30 cm) in length. The idea is to prevent you from using excessive force that could damage the injector.

Cylindrical Injectors. Some engines use recessed cylindrical injectors. To pull a recessed cylindrical injector, a slide hammer and puller nut have to be used. This tool threads into the high-pressure inlet of the injector. Also, with some types of cylindrical hydraulic injectors, the high-pressure delivery pipe or a quill tube connects to a recess in the injector through the cylinder head. The high-pressure pipe or quill tube must be backed away from the cylinder head before attempting to remove the injector. Failure to do this can damage the injector, the pipe or quill, and the cylinder head.

Nozzle Spacer. When removing injectors from cylinder heads, make sure you also remove the injector nozzle gasket, spacer, or washer at the same time. Manufacturers use all three of these terms, but for this description we will go with spacer. A nozzle spacer is made out of soft steel or copper and has two functions:

- Define the exact distance the nozzle tip protrudes into the cylinder.
- Seal the injector cup from the engine cylinder.

The nozzle spacer has an id (inside diameter) just fractionally less than the cylinder head injector hole. This means that providing the piston is not at top dead center (TDC), it can usually be removed either by:

- inserting an O-ring pick, or
- inserting the tapered end of the injector heel bar and jamming it into the washer.

Do not reuse injector washers. Both steel and copper washers harden in service. They should be replaced because it is unlikely they will reseal properly.

Seal Lines and Injectors. After an injector has been removed, use the appropriate-sized plastic caps to seal all of the following components:

- cylinder head injector bores
- the inlet and return ports on the injector
- high-pressure injector lines

If you are removing a set of injectors, make sure you mark each by cylinder number. Use an injector tray if you have one because this protects the injectors. Wrap the injectors in shop rags if no injector tray is available.

Seized Injectors. Occasionally, an injector may seize in a cylinder head. Almost always this will result in damaging the injector; the trick is to get it removed without damaging anything else. Once the injector becomes damaged (for instance, the line threads that fit the injector to the puller are destroyed), you have no option but to remove the cylinder head. Never risk damaging a cylinder head while trying to remove an injector. With the cylinder head removed, a seized injector can usually be removed with a punch and hammer.

Testing

Manufacturer specifications and service literature should be referenced before testing a hydraulic nozzle. In using a nozzle test procedure such as the one outlined here, the first failed step fails the nozzle. Failing the nozzle means that it should either be replaced (best option) or reconditioned. A typical procedure for testing nozzle assemblies would be as follows:

1. Locate the manufacturer's test specifications and service literature.
2. Clean the injector externally with a brass wire brush.
3. Mount the injector in a pop test fixture. Check for leaks. Build pressure slowly using the pump arm, watching for external leakage.
4. Bench test the NOP value and record. Actuate to NOP three times. Record the average NOP spec. A variation of NOP greater than 10 bar (150 psi) fails the nozzle.
5. Test **forward leakage** by pumping the pop tester to 10 bar (150 psi) below the average NOP value. Hold this pressure by observing the gauge and using the pump handle just enough to prevent any drop-off. While doing this, observe the nozzle. It should be dry. Any leakage at the tip fails the nozzle because it is leaking at the seat.
6. Check **back leakage** by pumping the pop tester to 10 bar (150 psi) below the average NOP value. Now release the pump handle and observe the gauge. Pressure drop values should typically be in the range of 50–70 bar (700–1,000 psi) over a 10-second test period. The pressure drops because fuel leaks past the nozzle shank to the return circuit. A rapid pressure drop exceeding the original equipment manufacturer (OEM) specification suggests that the nozzle is worn out.
7. Check the nozzle spray pattern. Pump the pop tester to NOP, this time closely watching the spray pattern. It should be exactly even from each orifice. Ignore nozzle **chatter** (rapid pulsing of the nozzle valve). This sometimes happens when bench testing hydraulic nozzles and is nothing to worry about.

A hydraulic injector that passes the preceding test cycle can be regarded as OK. If the injector fails the test, then either the injector nozzle or the injector should be replaced. Injectors and injector nozzles can be reconditioned, but this is becoming an obsolete practice. The reason is that reconditioned nozzles seldom perform to the standards required of new injectors.

Reinstallation of Injectors. The injector bore should be cleaned before installation. A bore reamer can be used to remove carbon deposits. Blow out the injector bore using an air nozzle. Where injector sleeves are used, service/replace as per OEM instructions. In most cases, you will leave the injector sleeves in place when removing hydraulic injectors for purposes of testing them. Most diesel engine OEMs prefer that hydraulic injectors be installed dry. Pasting them with lubricants such as Never-Seez™ can limit the injector's ability to transfer heat to the cylinder head.

Reconnecting Lines. It is good practice to turn the engine over and pump a little fuel through the high-pressure pipe before reconnecting it to the injector. This eliminates air and flushes dirt from the pipe. Check the high-pressure line nut torque specs. Failure to torque to specification can choke down on the flow area at the line nipple. Always torque the nut on the high-pressure pipe with a line socket. Failure to use a line socket can result in damage to both the line nut and the nipple and seat. Ensure that each injector line is of identical length. A longer-than-specified injector line can produce retarded injection timing in the affected cylinder. A shorter-than-specified injector line would result in advanced timing in the affected cylinder.

WARNING *Whenever fuel has spilled over an engine, ensure that the engine is pressure washed afterward.*

Testing EHIs

As we indicated earlier in this chapter, the procedure for testing EHIs requires using manufacturer electronic service tools (ESTs) and software-driven diagnostic routines. The idea is to identify a defective injector and then remove and replace it. There

PHOTO SEQUENCE 1
Testing an EHI on a Computerized Bench Test Fixture

P1-1 Mount the injector into the test fixture. Plumb in the fuel lines. Connect the electrical terminals.

P1-2 Run the pretest. This verifies that the EHI is properly plumbed into the test fixture and responds electronically to PCM-driven triggers. Any anomaly will trigger a repair/remediation instruction.

P1-3 Test step one. This is the warmup phase and may take a little time. The test fuel temperature is warmed to specification, and the EHI is fired to ensure the actuator responds appropriately to trigger voltages.

P1-4 Test step two. Leak test. The EHI is charged with typical rail pressures. It is then checked for both internal and external leakage.

P1-5 Test step three. The first of two delivery tests. This verifies fuel flow at a specific test rail fuel pressure on the high side.

P1-6 Test step four. The second delivery test. This verifies fuel flow at a specific test rail fuel pressure on the low side.

P1-7 Test step five. Spray test. Verifies that there are no abnormalities in the spray geometry. An abnormality could be caused by a restricted or coked injector orifice.

P1-8 Test step six. Option to report. In this step, the technician will address a field in which he can choose to report the data gleaned from the test. This may be display only, printable, or networkable.

P1-9 View report fields. This will display the test data and provide an option to print it. The result is always binary: OK or Not OK.

are a number of aftermarket test fixtures designed exclusively for testing EHIs and because the procedure is software driven, it tends to be simple. That is, the test algorithm outcome can be one of the following:

- OK
- Not OK

When working with common rail (CR) fuel systems, NEVER crack the high-pressure pipe nuts in an effort to bleed them. It could prove to be dangerous, and you may have problems in getting them to reseal. Most manufacturers state that their CR high-pressure lines are single-use devices. The seats are torque-to-yield. This means that they yield to deform to create a perfect seal on initial torque. Like most modern high-pressure diesel fuel injection systems, CR fuel systems are designed to self-prime.

CAUTION *The high-pressure lines that connect EHIs to the rail in CR systems are expensive. Because these lines are designed for one-time use, avoid disconnecting them unnecessarily. Each time a line is disconnected it should be replaced with a new one.*

EHI Test Sequence. The photosequence on the previous page outlines the six-step procedure required to verify EHI performance. First, the EHI must be mechanically mounted and plumbed into the test fixture. After this, the test profile is driven by computer software, so the technician is required to do little more than input yes/no data. The result of the test profile is to determine that the EHI is either good or bad. The EHI used in the test profile is from a Cummins ISB and will be shown to test marginally bad: the LCD monitor displays are sequenced during the test routine.

Summary

- Injector nozzles act as switches to begin and end injection pulses.

- The injector nozzle is responsible for atomizing diesel fuel.

- Atomized diesel fuel is in a liquid state. After injection into the cylinder, it must be vaporized before it can be ignited.

- Today's diesel engines use multiple-orifice or electrohydraulic nozzles.

- Injector spring tension defines NOP value in hydraulic nozzles.

- Injector spring tension is set by adjusting either screws or shims.

- Dowels are used to position a nozzle in the injector body; this ensures correct spray dispersal into the engine cylinder.

- Electrohydraulic nozzles are required in current diesel engines. An electrohydraulic nozzle may be integral in an EHI or a subcomponent of an EUI.

- EHI nozzles are designed to open and close precisely when the PCM electrically switches them to do so.

- EHIs produce pressures that remain relatively consistent through a fueling pulse. The pressure depends on PCM-managed rail pressure.

- EHIs may use either solenoid or piezoelectric actuators.

- Piezoelectric actuators respond rapidly to control signals. They tend to be used in fuel systems requiring multipulse injection.

- Nozzle sac volumes have either decreased or been eliminated (VCO nozzle). This reduces emissions.

- Hydraulic injector nozzles are tested out of engine using a pop tester. Testing requires working through a simple test procedure. Nozzles tend not to be fully reconditioned in current trade practice.

- EHI injectors are diagnosed using the manufacturer EST and software-guided diagnostic routines.

- When diagnosed defective, EHIs can be bench tested, but if they are defective, they should be factory reconditioned.

- Torquing of high-pressure pipe union nuts at both the injection pump and injector ends is important to avoid damaging the nipple seat and hole size.

- A hydraulic injector nozzle test sequence checks NOP, forward leakage, back leakage, and the spray pattern.

- Most OEMs regard the high-pressure pipes that supply EHIs to be single-use devices. This is because when torqued, they deform to create a perfect seal.

Shop Exercises

1. Find a diesel engine in your shop that does NOT use CR fueling: referencing the information in this chapter, determine what type of injector nozzles are used in this engine.

2. Locate an engine in your shop that is CR fueled. Identify the rail and the means used to connect the EHIs.

3. Insert a multiple orifice, hydraulic injector into a pop tester and run the bench test sequence outlined in this chapter.

4. Place a CR-EHI into a computerized bench test fixture. Navigate the test fields and determine if the EHI classifies as OK or NOT OK.

5. Identify a CR-fueled engine in your shop: consult the manufacturer service literature and determine whether, and how, calibration codes have to be programmed to the PCM. Note: this is not a requirement of all CR fueled engines with EHIs.

Internet Exercises

Use a search engine and research what comes up when you enter the following key words:

1. Robert Bosch Diesel Common Rail

2. Bosch Diesel Common Rail animation

3. Diesel piezoelectric injectors

4. Testing diesel common rail injectors

5. Disadvantages of reconditioning diesel injectors

Review Questions

1. Which type of injection nozzle provides *soft* opening and closing pressures?

   A. Hydraulic multiple-orifice injector

   B. Electrohydraulic injector (EHI)

   C. Any hydraulically switched injector

   D. Any electronically switched injector

2. When injector *back leakage* is bench checked and is higher than specified, which of the following is the usual cause?

   A. Low NOP setting

   B. High NOP setting

   C. Sticking nozzle valve

   D. Wear

3. When nozzle *forward leakage* is bench checked, which of the following is being tested?

   A. NOP

   B. The seal at the nozzle seat

   C. Nozzle valve-to-body clearance

   D. Injector spring

4. Replacing a single high-pressure injection pipe on a multicylinder engine by one of shorter length would likely have what effect on injection timing in the affected cylinder?

   A. Advance

   B. Retard

   C. None

   D. Decrease the fuel pulse width

5. As injection pressure increases to a hydraulic injector nozzle, which of the following is likely to occur?

    A. Injected droplets decrease
       in size.

    B. Injected droplets increase
       in size.

    C. Ignition delay increases.

    D. Nozzle valve is more likely to unintentionally close.

6. Which type of injector nozzle would be used in most diesel engine fuel systems built after 2007?

    A. Pintle valve

    B. Hard opening value

    C. Multiple orifice

    D. Electrohydraulic

7. What should be used to remove the nut on a high-pressure injection line?

    A. Torque wrench

    B. Hex socket

    C. Line wrench

    D. Open-end wrench

8. Which of the following is another way of stating the NOP specification?

    A. Not operating properly

    B. Residual line pressure

    C. Peak pressure

    D. Popping pressure

9. Most diesel engine OEMs recommend that injectors be installed into the cylinder head injector bore:

    A. Dry

    B. Coated with engine oil

    C. Coated with Never-Seez

    D. Coated with Lubriplate

10. Which of the following types of injector would be best suited to multipulse injection?

    A. Hydraulic poppet

    B. Hydraulic multi-orifice

    C. EHI with solenoid actuator

    D. EHI with piezoelectric actuator

9 Pump-Line-Nozzle Injection Systems

Prerequisites

Chapters 2 and 7.

Learning Objectives

After studying this chapter, you should be able to:

- Identify the three types of pump-line-nozzle (PLN) fuel injection systems.

- Identify the major components in a typical port-helix metering injection pump.

- Explain the principles of operation of an inline, port-helix metering injection pump.

- Define the terms *effective stroke*, *port closure*, *port opening*, *NOP*, *residual line pressure*, and *peak pressure*.

- Explain how the pump element components create injection pressures.

- Identify the differences between hydromechanical and electronically controlled versions of port-helix metering injection pumps.

- Explain the operation of aneroid devices, altitude compensators, and variable timing/timing advance mechanisms.

- Outline basic governor operation.

- Time an injection pump to an engine.

- Relate event sequencing in the injection pump to that in the engine combustion chamber.

- Identify the two categories of rotary distributor pumps likely to be found on small-bore diesel engines.

- Identify the main components of an inlet-metering, distributor injection pump.

- Describe the operating principles of an opposed-plunger, inlet-metering syjection pump.

- Explain how fuel is routed, pumped, and metered from the fuel pump to the injector during an inlet-metering, opposed-plunger injection pump cycle.

- Identify the main components of a sleeve-metering, distributor injection pump.

- Describe the operating principles of a sleeve-metering, distributor injection pump.

- Explain how fuel is routed, pumped, and metered from the fuel pump to the injector during a sleeve-metering, distributor injection pump.

- Describe the operating principles of Stanadyne Roosa, Delphi Lucas CAV, and Bosch rotary injection pumps.

Key Terms

altitude compensator	distributor head	plunger
aneroid	distributor plunger	plunger geometry
bar	distributor rotor	plunger leading edge
barometric capsule	effective stroke	port closure (PC)
barrel	flutes	port opening
boosted engines	helix/helices	pump-line-nozzle (PLN)
calibration	high-pressure pipes	rack actuator
cambox	hydraulic head	register
cam plate	ignition lag	residual line pressure
camshaft	injection lag	retraction collar/piston
charging pressure	inlet-metering	sleeve metering
charging pump	internal cam ring	spill timing
control rack	lower helix	static timing
control sleeve	metering recesses	tappets
dead volume fuel	opposed plungers	thrust collar
delivery valve	phasing	transfer pump

INTRODUCTION

The first high-pressure liquid fuel injection to a high-speed diesel combustion chamber was developed in 1927 by Robert Bosch. This basic pump is still used today with surprisingly few modifications, although more recent versions are managed electronically. This chapter will be devoted to the study of three distinct **pump-line-nozzle (PLN)** fuel injection systems. The first is the inline injection pump invented by Robert Bosch. The other two are distributor pumps that each use very different operating principles. We will identify the three fuel injection systems as:

- Port-helix metering injection pumps
- Inlet-metering, rotary distributor pumps
- Sleeve-metering distributor pumps.

There is no doubt that PLN diesel fuel systems are a thing of the past, although electronic controls have allowed them some afterlife in our electronic age. But the inability of PLN fueling to meet current emissions standards is only one part of the story. PLN systems produce neither the fuel economy nor the precise control over combustion demanded by today's diesel engines.

PORT-HELIX METERING PUMPS

The typical port-helix metering pump (see **Figure 9-1**) used to fuel a diesel engine is inline configured and flange mounted to an engine accessory drive. It is driven through one complete rotation (360 degrees) per complete engine cycle (720 degrees). The internal pump components are housed in a frame constructed of cast aluminum, cast iron, or forged steel. The engine crankshaft drives the injection pump by means of timed reduction gearing.

The gear-driven pump drive plate is connected to the injection pump **camshaft** (the fuel pump shaft fitted with eccentrics designed to actuate the pump elements), so rotating the pump drive plate rotates the pump camshaft. The camshaft is supported by main bearings and rotates within the injection pump **cambox**. The cambox is the lower portion of the injection pump that houses the camshaft, **tappets**, and an integral oil sump. You can see how the cam actuates a pump element in the sectional view shown in **Figure 9-2**.

The camshaft has one cam profile dedicated to each engine cylinder. Riding each cam profile is a tappet assembly that drives a pump element consisting of a **plunger** and a **barrel**. The barrel is stationary and is drilled with two ports in its upper portion, which are exposed to the fuel-charging gallery. The tappet,

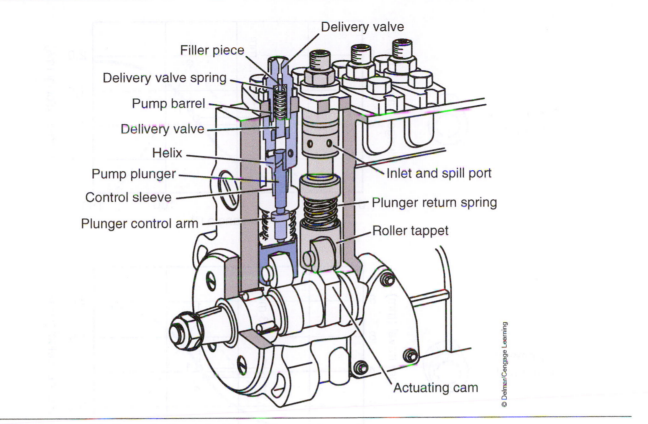

Figure 9-1 Inline, port-helix metering injection pump components.

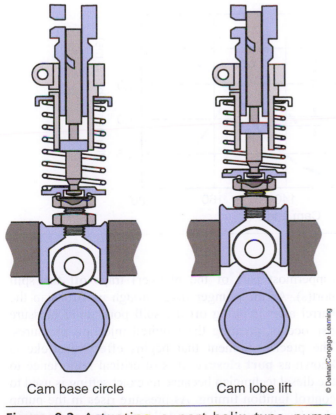

Cam base circle   Cam lobe lift

Figure 9-2 Actuating a port-helix–type pump element.

plunger, and barrel are shown in profile in **Figure 9-2.** The geometric shape of the cam profile determines how the plunger is actuated. The more common design is symmetrical, shown in the top image in **Figure 9-3.** However, asymmetrical (provides superior plunger cooling) and anti-back kick plunger geometry can also be used as shown in the lower two images in **Figure 9-3.** Anti-back kick cams make starting an engine in the reverse direction nearly impossible.

Fuel Routing

The fuel gallery is charged with low-pressure fuel, typically between 1 and 5 **bar** (15–75 psi). This charging pressure fuel flows into and through the barrel ports when they are not obstructed by the plunger. The plunger reciprocates within the barrel; it is loaded by spring pressure to ride its actuating cam profile. Therefore, the actual plunger stroke is constant. Plunger-to-barrel tolerance is close, the components being manufactured to a tolerance of 2 μ to 4 μ. Fuel quantity to be delivered in each stroke is controlled by managing plunger **effective stroke**. The plunger is milled with a vertical slot or cross and center drillings and helical recesses. The function of the vertical slot or cross and center drillings is to

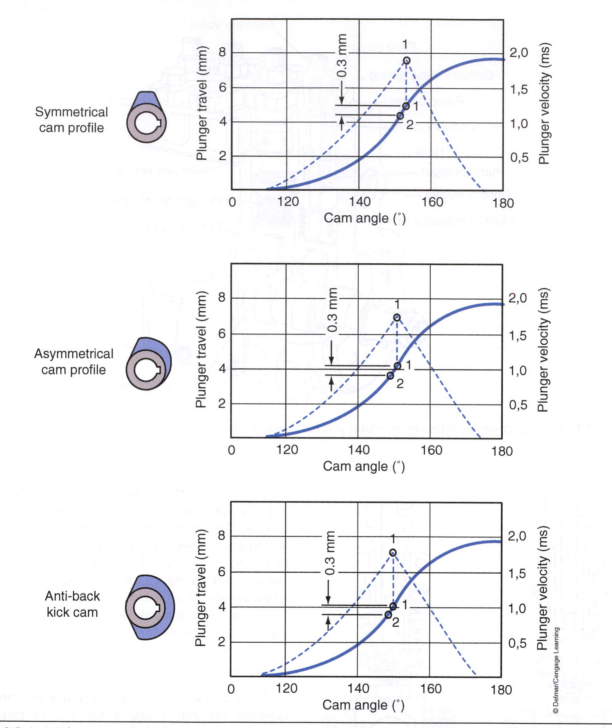

Figure 9-3 Actuating cam geometry.

maintain a constant conduit between the pumping chamber above the plunger and the helical recesses. In other words, whatever pressures exist in the pumping chamber must also exist in the helical recess.

Effective Stroke

Effective stroke describes the delivery stroke. The delivery stroke begins when the plunger is forced upward by the cam profile and the **plunger leading edge** (uppermost part of the plunger) traps off the spill port(s). As the plunger rises through its stroke in the barrel after trapping off the spill port, rapid pressure rise occurs, creating the required injection pressures. The precise moment that begins effective stroke is known as **port closure**. It is of critical importance to the diesel technician because its exact setting is used to control ignition timing. As pressure rises in the pump chamber, it acts first on a **delivery valve**, and next on the fuel confined in the high-pressure pipe that transmits fuel to the injector nozzle, finally delivering a fuel pulse to the engine cylinder. Effective stroke ends at **port opening**. This is the precise moment that the upward travel of the plunger exposes the helical recess(es) to the spill port. High-pressure fuel is spilled back to the charging gallery, causing a rapid collapse of pressure in the pump chamber, line, and nozzle.

Collapse Phase. The injection pulse ceases when there is no longer sufficient pressure to hold the delivery and nozzle valves open. Port opening always occurs while the plunger is moving in an upward direction, that is, not at plunger top dead center (TDC) or beyond. This is required because the pressure in a port-helix pump element is designed to rise through the delivery stroke, thereby producing smaller atomized droplets from the injector toward the end of effective stroke; however, at the point of port opening, regardless of the length of the effective stroke, pump pressure should collapse as rapidly as possible and minimize the larger droplets emitted from the injector as pump pressure falls to a value below nozzle opening pressure (NOP).

Control of Effective Stroke. The length of plunger effective stroke **(Figure 9-4)** depends on where the plunger helix **registers** (vertically aligns) with the spill port. **Control sleeves** lugged to the plunger permit the plunger to be rotated while reciprocating. Rotating the plunger in the bore of the barrel changes the location of register of the spill port with the helix. Therefore, plunger effective stroke depends entirely on the rotational position of the plunger. In multiple-cylinder engines, the plungers must be synchronized to move in unison to ensure balanced fueling at any given engine load.

The control sleeves are tooth-meshed to a governor **control rack**, which when moved linearly, rotates the plungers in unison **(Figure 9-4).** This is important. It means that in any linear position of the rack, all of the plungers will have identical points of register with their spill points, resulting in identical pump effective strokes. The consequence of not doing this would be to unbalance the fueling of the engine, that is, deliver different quantities of fuel to each cylinder.

No Fuel. Engine shutdown is achieved by moving the control rack to the no-fuel position. The rotational position of the plungers is now such that the vertical slot will be in register with the spill port for the entirety of plunger travel; the plunger will merely displace fuel as it travels upward, with no pumping action possible. In other words, as the plunger is driven into the pump chamber, the fuel in the chamber will be squeezed back

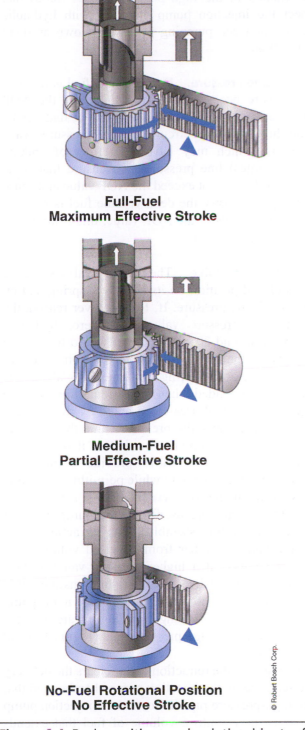

**Full-Fuel
Maximum Effective Stroke**

**Medium-Fuel
Partial Effective Stroke**

**No-Fuel Rotational Position
No Effective Stroke**

© Robert Bosch Corp.

Figure 9-4 Rack position and relationship to fuel delivery quantity.

down the vertical slot to exit through the spill port and return to the charging gallery **(Figure 9-4).**

Most port-helix metering injection pumps use delivery valves to reduce the amount of work required of each pump element per cycle. Delivery valves function to isolate the high-pressure circuit that extends from the injection pump chamber to the seat of the nozzle valve.

Fuel retained in the **high-pressure pipes** (pipes that connect the injection pump elements with hydraulic injectors) between pumping pulses is known as **dead volume fuel**.

Residual Line Pressure. Dead volume fuel is retained at a pressure necessarily somewhat below the NOP value; the high-speed hydraulic switching that occurs in the high-pressure circuit creates pressure wave reflections, which may have the effect of spiking (causing surges) line pressures. To ensure that these pressure spikes do not exceed the NOP value and cause secondary injections, the dead volume fuel is retained at about two-thirds of the NOP value; this is known as **residual line pressure**.

Role of Delivery Valve. The delivery valve is loaded into its closed position on its seat by a spring and by the residual line pressure. If, for whatever reason, the residual line pressure value were zero, hydraulic pressure of around 20 bar (300 psi) would have to be developed in the pump element to overcome the mechanical force of the spring. This mechanical force is therefore compounded when the residual line pressure acts on the sectional area represented by the delivery valve and establishes the pressure value that must be developed in the pump chamber before it is unseated.

The delivery valve **flutes** provide a means of guiding the valve in its bore while permitting hydraulic access between the **retraction collar** or **retraction piston** (both terms are used) and the pump chamber (plunger and barrel assembly). The retraction collar seals the pump chamber from the dead volume fuel, which is retained at a higher pressure value. Consequently, when the delivery valve is first unseated, it is driven upward in its bore by rising pressure in the pump chamber and acts as a plunger driving inward onto the fuel retained in the high-pressure pipe (**Figure 9-5** and **Figure 9-6**).

The moment the retraction collar clears the delivery valve seat, fuel in the injection pump chamber and that in the high-pressure pipe unite, and the injection pump plunger is driven into a volume of fuel that extends from the plunger to the nozzle valve seat. Rising pressure subsequently unseats the injector nozzle valve (NOP) and forces atomized fuel into the engine cylinder.

At port opening, the effective pump stroke ends, beginning a rapid pressure collapse as fuel spills from the barrel spill port. When there is insufficient pressure in the pump chamber to hold the nozzle valve in its open position, spring pressure overcomes hydraulic pressure and it seats, sealing the nozzle end of the

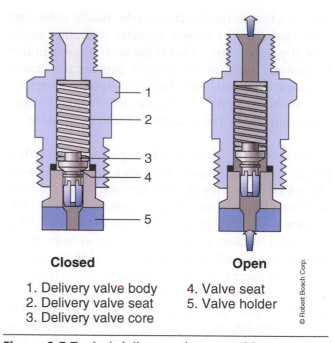

Closed   **Open**

1. Delivery valve body   4. Valve seat
2. Delivery valve seat   5. Valve holder
3. Delivery valve core

© Robert Bosch Corp.

Figure 9-5 Typical delivery valve assembly.

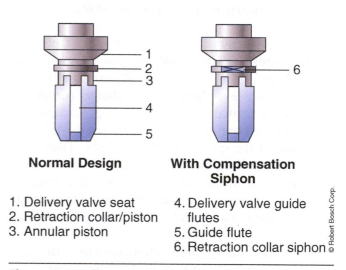

Normal Design   **With Compensation Siphon**

1. Delivery valve seat   4. Delivery valve guide
2. Retraction collar/piston   flutes
3. Annular piston   5. Guide flute
                             6. Retraction collar siphon

© Robert Bosch Corp.

Figure 9-6 Delivery valve core terminology.

high-pressure pipe. Almost simultaneously, the delivery valve begins to retract. The instant the retraction collar passes the delivery valve seat, it hydraulically seals the pump end of the high-pressure circuit. However, after sealing the pump end of the high-pressure circuit, the delivery valve must travel farther before it seats, increasing the volume available for dead volume fuel storage. This causes a drop in line pressure and defines the residual line pressure value.

In short, the volume available for fuel storage in the high-pressure pipe is increased by the swept volume of the retraction collar. This volume of fuel is known as dead volume fuel and is retained at residual line pressure. Retraction collar swept volume is matched to the length of the high-pressure pipe to achieve a precise residual line pressure value so that pipe length should not be altered.

INJECTION PUMP COMPONENTS

The following is a list of subcomponents that, once assembled, will form a typical port-helix metering injection pump.

Pump Housing

The pump housing is the frame that encases all the injection pump components and is a cast aluminum, cast iron, or forged steel enclosure. The pump housing is usually flange mounted by bolts to the engine cylinder block and driven by an accessory drive on the engine geartrain. In some applications of inline, port-helix metering injection pumps, the pump assembly is cradle mounted on its base, in which case it is driven by an external shaft from the timing gear train.

Cambox

The cambox is the lower portion of the pump housing that incorporates the lubricating oil sump and main mounting bores for the pump camshaft. Camshaft main bearings are usually pressure lubed by engine oil supplied from the engine crankcase, and the cambox sump level is determined by the positioning of a return port. In older injection pumps, the pump oil was isolated from the main engine lubricant, and the oil was subject to periodic checks and servicing.

Camshaft

The camshaft is designed with a cam profile for each engine cylinder and is supported by main bearings at the base of the pump housing. It is driven at one-half engine rotational speed in a four-stroke cycle engine by the pump drive plate, which is itself either coupled directly to the pump drive gear or to a variable timing device. Camshaft actuating profiles are usually, but not always, symmetrical (see **Figure 9-3**).

Tappets

The tappets are arranged to ride the cam profile and convert the rotary motion of the camshaft to the reciprocating action required of the plunger. A retraction spring is integral with the tappet assembly. This is required to load the tappet and plunger assembly to ride the cam profile, and it is necessarily large enough to overcome the low pressure (vacuum) established in the

pump chamber on the plunger return stroke. This low pressure can be considerable when plunger effective strokes are long, but it enables a rapid recharge of the pump chamber with fuel from the charging gallery. The time dimension within which the pump element must be recharged decreases proportionately with pump rpm increase.

Barrel

The barrel is the stationary member of the pump element, located in the pump housing so its upper portion is exposed to the charging gallery. This upper portion of the barrel is drilled with diametrically opposed ports known as inlet and spill ports that permit through flow of fuel to the barrel chamber to be charged. Because it contains the spill ports, both its height and rotational position in relation to the plunger are critical. Barrels are often manufactured with upper flanges so that their relative heights can be adjusted by means of shims, and fastener slots permit axial movement for purposes of **calibration** and **phasing** (the procedures of calibration and phasing are covered in more detail later in this chapter).

Plunger

Plungers are the reciprocating (something that reciprocates, moves backward and forward such as in the action of a piston in an engine cylinder) members of the pump elements, and they are spring-loaded to ride their actuating cam's profile. Plungers are lapped to the barrel in manufacture, to a clearance close to 2 μ, ensuring controlled back leakage directed toward a viscous seal that consists of an annular groove and return duct in the barrel. Each plunger is milled with a vertical slot, helical recess(es), and annular groove. In a majority of automotive diesel engine applications, a lower-helix design consistent with the images shown in **Figure 9-4** is used.

Plunger Geometry. The positioning and shape of the **helices** (plural of **helix**) on a plunger are often described as the **plunger geometry**. Plunger geometry describes the physical shape of the **metering recesses** machined into the plunger, and this defines the injection timing characteristics. The function of the vertical slot is to ensure a constant hydraulic connection between the pump chamber above the plunger and the plunger helical recess(es).

Timing and Plunger Geometry. A plunger with a **lower helix** will have a constant beginning, variable ending of delivery timing characteristic. Lower helix design is the most common and although other types exist, they are beyond the scope of this textbook. Some plungers are machined with identical, diametrically opposed helices to provide hydraulic balance to the pump element as it is driven through its stroke; specifically, they prevent the side loading of the plunger into the barrel wall that may occur at high-pressure spill off.

Rack and Control Sleeves

The rack and control sleeves permit the plungers in a multicylinder engine to be rotated in unison to ensure balanced fuel delivery to each cylinder. Plungers must therefore be timed either directly or indirectly to the control rack. The rack is a toothed rod that extends into the governor or **rack actuator** housing. The rack teeth mesh with teeth on plunger control sleeves, which are either lugged or clamped to the plunger. It must be possible to rotate the plungers while they reciprocate to permit changes in fuel requirements while the engine is running. Linear movement of the rack will rotate the plungers in unison, alter the point of register of the helices with their respective spill ports, and thereby control engine fueling. Timing of the plungers to the control rack is the means used to adjust the effective stroke in individual pump elements in a procedure known as calibration.

This timing procedure is effected either directly, by adjusting the relationship of the plunger with the rack, or indirectly, by adjusting the rotational position of the barrel and therefore its relationship with the plunger geometry. The means of calibrating an injection pump depends on the make and model.

Delivery Valves

Delivery valves isolate the high-pressure circuit that extends from the injection pump chamber to the seat of the injector nozzle valve. They act somewhat like check valves. Because they seal before they seat, they permit line pressure to drop to a residual value well below the NOP value, and this helps prevent secondary injections. The delivery valve is machined with a seat, retraction collar, and flutes to guide it in the bore of the delivery valve body **(Figure 9-6)**.

Charging Pumps

The terms **charging pump** and **transfer pump** are used interchangeably, depending on the manufacturer. The charging pump is responsible for all fuel movement in the fuel subsystem. In truck applications using port-helix metering injection, the charging pump is normally a plunger pump, flange mounted to the fuel injection pump and actuated by a dedicated eccentric on the injection pump camshaft. The charging or transfer pump is responsible for producing **charging pressure**.

Charging pressures range from 1 to 5 bar (15–75 psi) depending on the system. The role of charging/transfer pumps is dealt with in greater detail in the context of the fuel subsystem in **Chapter 7**.

Governor or Rack Actuator Housing

Either a governor or rack actuator housing must be incorporated with a port-helix metering injection pump. This acts as the control mechanism for managing fueling. The only factor that determines the output of a diesel engine is the amount of fuel metered into its cylinders. Given unlimited fuel, a diesel engine is capable of accelerating at a rate of 1000 rpm per second until self-destruction occurs. To prevent such an "engine runaway" and to provide the engine with a measure of protection from abuse to ensure that it meets the manufacturer's expectations of longevity, fuel quantity delivered to the engine cylinders must be precisely managed by a governor under all operating conditions. The mechanical fuel control mechanism on a port-helix metering injection pump is the rack. Therefore, the governor or rack actuator controls engine fueling by precise positioning of the rack.

Governors. Hydromechanical governors are seldom used today because only low-horsepower engines can meet emissions standards with them. In a simple mechanical governor, the vehicle accelerator linkage connects to a throttle arm or governor input lever located on the side of the governor housing. Enclosed within the governor housing, a weight carrier is mounted to the rear of the camshaft. Flyweights are mounted in the weight carrier, loaded inboard by spring force. When engine maximum speed is realized, the governor weights will produce sufficient centrifugal force to trim back fueling so it is not exceeded.

Rack Actuator Operation. When the port-helix metering injection pump is managed by a computer, engine governing depends on how the powertrain control module (PCM) is programmed. The governor housing attached to the rear of the injection pump is replaced by a rack actuator housing that consists of switched PCM output devices and sensors. The housing contains a rack actuator. In all light duty diesel engines, the rack actuator is a proportioning solenoid controlled directly by the PCM. Actual rack position is close loop signaled to the PCM.

The rack actuator housing may house other sensors to report rotational speed (pulse wheel fitted to the rear of the camshaft), engine position, and timing data to the PCM. Rack actuator housings are not governors. Where a rack actuator housing is fitted to an inline, port-helix metering injection pump, the engine governing functions are undertaken by the software programmed into the fuel management PCM.

Electronic governing offers infinitely more control over engine fueling, and it allows manufacturers to meet emissions requirements and achieve better fuel economy. It can also be easily programmed/reprogrammed with customer data to tailor the engine to varying applications.

Lubrication

The lower portion of the port-helix metering injection pump is lubricated by engine oil. In older and small-bore engines, the injection pump lubricating circuit was often isolated from that of the engine, and the lube level would be checked and replenished through a dipstick located in the cambox. Contemporary injection pumps tend to be plumbed into the main engine lubricating circuit. The camshaft main bearings are often pressure lubed directly from an engine oil gallery, and the remainder of the cambox components are splash lubricated from the oil held in the sump.

The upper portion of the injection pump is lubricated by the diesel fuel being pumped through the charging gallery, so the lubricity of the diesel fuel is critical. Compromising the lubricating qualities of the fuel may cause premature failure and/or leakage, and some manufacturers have reported such problems following the introduction of low-sulfur, low-lubricity fuels mandated by Environmental Protection Agency (EPA) standards. It is critical for the operation of both the fuel injection pump and the engine it manages that the fuel used to lubricate the upper portion of the pump never comes into contact with the engine oil in the cambox. Plunger-to-barrel clearance is lapped in manufacture to a minute tolerance, and fuel that leaks by the recessed metering areas bleeds to an annular belt in the barrel.

A duct connects the annular belt in the barrel to the charging gallery, permitting bleed-by fuel to be routed there. This is known as a viscous seal. Trace leakage of a viscous seal can rapidly cause engine oil contamination and lead to lubricant breakdown. Viscous seals fail due to plunger side loading (single-helix design), prolonged usage wear, and fuel contaminants—especially water. Fuel that bleeds by the metering recesses serves to guide the plunger true in the barrel bore, minimizing metal-to-metal contact.

Timing Advance and Variable Timing Mechanisms

Older port-helix metering injection pumps were usually directly driven by reduction gearing from the engine camshaft gear. Such a system would dictate that the static timing value (port closure) occur at the same number of crank angle degrees BTDC regardless of engine load or speed when lower-helix geometry was used. Fuel economy and noxious emission considerations led to the development of first, mechanical advance mechanisms; and more recently, electronically managed, variable timing.

Mechanical Variable Timing. Mechanical timing advance mechanisms (see **Figure 9-7**) are actuated using a set of flyweights and eccentrics to advance the drive angle of the pump camshaft in relation to the pump drive gear, using a spiral gear on a shaft. In other words, the position of the fuel injection pump relative to that of the engine is advanced in direct proportion to the centrifugal force generated by the weight carrier. Port closure was therefore advanced as engine rpm increased. The extent of the advance offered by mechanical advance mechanisms can be as little as 3 degrees crank angle and seldom more than 10 degrees crank angle.

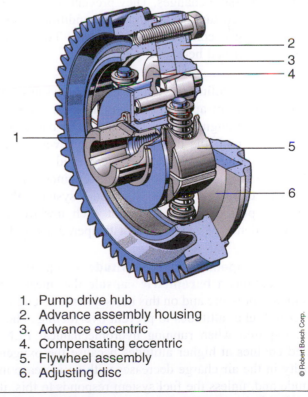

1. Pump drive hub
2. Advance assembly housing
3. Advance eccentric
4. Compensating eccentric
5. Flywheel assembly
6. Adjusting disc

© Robert Bosch Corp.

Figure 9-7 Mechanical timing advance mechanism.

Electronic Variable Timing. Electronically managed variable timing devices are used by partial authority PLN management systems and employ a variable timing coupling mechanism between the pump drive gear and the pump camshaft. This intermediary was designed to establish a variable timing window of up to 20 degrees crank angle, managed by the engine electronics.

The static timing value (that is, port closure) specified is usually the most retarded parameter in the variable timing window. For example, an injection pump timed at 7 degrees BTDC with a 20-degree variable timing mechanism would permit the PCM to select any port closure value between 7 degrees BTDC and 27 degrees BTDC. Injection pump static timing (port closure) specs of as little as 4 degrees BTDC are used to limit combustion heat and therefore NO_x emission; substantially retarded injection timing does not generally enhance either performance or fuel economy. In fact, the trade-off of retarded injection timing is an increase of hydrocarbon (HC) emission, but all emissions legislation is about keeping the noxious exhaust emissions within a window of acceptability.

Governor Trim Devices

Governor trim devices are used primarily on hydromechanical PLN pumps. Their objective is to modulate fueling as either ambient pressure (altitude) or manifold boost pressure changes. They prevent overfueling the engine when running or ambient conditions result in lower oxygen content in the cylinder. Two basic types are described here.

Aneroids. By definition, an **aneroid** is a low-pressure sensing device. In application, it is used on a turbocharged diesel engine to measure manifold boost and limit fueling until the boost pressure achieves a predetermined value. Such devices were used on most PLN **boosted engines** (turbocharged engines). They are easily tampered with and exhaust systems that produce a puff of black smoke at each upshift are usually evidence of a defective or tampered aneroid.

Altitude Compensator. An **altitude compensator** device contains a **barometric capsule** that measures barometric pressure and on this basis downrates engine power at higher altitudes to prevent overfueling. They were required when running hydromechanical PLN-fueled engines at higher altitudes because the oxygen density in the air charge decreases with an increase in altitude and, unless the fuel system responds to this, it effectively overfuels the engine. The critical altitude at which some measure of injected fuel quantity deration is required is 1,000 ft. (300 meters), but in older engines fueling corrections may not actually occur until 3,000 ft. (900 meters) in altitude. The performance consequence of an altitude compensator malfunction is engine smoking under load at altitude.

Governor Options

Port-helix metering pumps may be governed by either *limiting speed (LS)* or *variable speed (VS)* governors. The governor may be hydromechanical or an algorithm programmed to PCM memory, depending on the generation of pump. A short description of each category of governor follows, but this is not comprehensive.

Limiting Speed. A limiting speed (LS) governor is designed to make accelerator response similar to that used in a gasoline-fueled automobile. The automobile accelerator controls air flow into the engine, and the LS governor produces the same response by managing fuel-injected fuel quantity into the engine: it can do this because all diesels (whether boosted or not) run with excess air. If the operator increases pedal angle, the result is more fuel and faster rpm. LS governing is by far the most common type encountered in on-highway, light duty diesels.

Variable Speed. When a variable speed (VS) governor is used, accelerator input is a request for a specific rpm. If the accelerator angle is held constant by the driver, as load on the engine increases and decreases (as when going uphill then downhill), the governor will attempt to maintain the rpm request and will do so while within the limitations of the fuel system. When accelerator angle is increased, the request for rpm increases. Conversely, when accelerator angle is decreased, the rpm decreases.

It is important to understand the distinctions between LS and VS governing because both are programmable options in more recent fuel systems. However, VS governing tends to be popular with the drivers of commercial rigs and is seldom used in light duty highway vehicles.

Timing Injection Pumps to an Engine

Port-helix metering injection pumps are timed to the engine they manage by phasing port closure on the number 1 cylinder (usually in North American engineered engines—check the specification—for a variety of reasons, the pump may be required to be timed to the number 6 cylinder) to a specific engine position.

All injection pumps must be accurately timed to the engine they will fuel.

This usually means the phasing of pump port closure to a specific number of degrees BTDC on the cylinder to be timed to a specification that seldom can be outside of 1 degree crank angle and may have to be within ¼ degree crank angle. Methods used to time the injection pump to the engine vary by OEM. Actual spill timing of the pump to the engine is a procedure that has become obsolete on highway diesel engines; when you are required to spill-time an engine, it is most likely to be an off-highway, usually small-bore engine. However, most would argue that it continues to be a procedure that diesel technicians should understand because it can be used in the absence of OEM tooling.

Tech Tip: Most injection pumps on North American engines are timed to the number 1 engine cylinder, but not all, so watch for those that are not. The number 6 cylinder (on an inline six) is the next most common, but assume NOTHING. Always check the specifications in the service literature.

Spill Timing Procedure

Before beginning the **spill timing** procedure, check the engine (and OEM chassis) manual for positioning of the fuel control lever, stop fuel lever, brake valve, and gearshift lever position.

1. Check the injection pump data plate for port closure value. Manually bar the engine in the direction of rotation to position the piston within the number 1 cylinder on its compression stroke. Locate the engine calibration scale, usually to be found on the front pulley, vibration damper, or flywheel. Position the engine roughly 20 degrees before the port closure specification.

2. Remove the high-pressure pipe from the delivery valve on the injection pump number 1 cylinder. Unscrew the delivery valve body and remove the delivery valve core and spring. Replace the delivery valve body and install a spill tube. A discarded high-pressure pipe neatly cut and shaped to a gooseneck will suffice. When the hand primer pump is actuated, the charging gallery will be pressurized. The amount of pressure created by the hand primer pump will be insufficient to open the delivery valves, so it will

exit through the spill tube fitted to the number 1 cylinder. The fuel should exit in a steady stream, and it should be captured in a container held under the spill tube. Next, slowly and smoothly bar the engine in its direction of rotation, observing the stream of fuel exiting the spill tube.

When the plunger leading edge rises to trap off the spill port, the steady stream of fuel exiting the spill tube will break up first into droplets and then cease as the plunger passes the spill port.

The objective of this step is to locate the pump precisely at port closure—this means that the flow area at the spill port should exist but be minimal; two to six drops per 10 seconds should be within the specification window. Ensure that the pump has not been barred past port closure, cutting the fuel off altogether, as it is impossible to determine how much beyond port closure the plunger has traveled.

3. Next, check the engine calibration scale. The specification is typically required to be within 1 degree crank angle of the port closure (PC) specification. If this is so, the pump can be assumed to be correctly timed to the engine. If it is not, proceed as follows.

4. Remove the accessory drive cover plate. Loosen the fasteners that couple the pump drive gear to the pump drive plate. Bar the engine to the correct PC specification position. By uncoupling the pump drive plate from the pump drive gear, you hope to keep the pump stationary at PC on the number 1 cylinder while the engine is barred independently. When the engine is in the correct position, torque the fasteners that couple the pump drive gear to the pump drive plate. Back the engine up roughly 20 degrees before the PC specification and then repeat steps 1 through 3.

All diesel technicians should be acquainted with the foregoing procedure even though they may seldom practice it. Spill timing can also be performed using compressed air and manufacturer's service literature should be consulted before attempting this procedure. OEMs prefer other methods of static timing port-helix metering injection to be used that do not require the removal and disassembly of the delivery valve. Following are some of the alternatives:

1. Timing pin. This is probably the simplest method and least likely to present problems. It is also the least accurate. The engine is located at a specific position by inserting a timing pinion or bolt, usually in the cam gear but

sometimes in the flywheel. Similarly, the injection pump is pinned by a timing tool to a specific location. The injection pump is always removed and installed with the timing tools in position. It goes without saying that the timing tools must be removed before attempting to rotate or start the engine.

2. High-pressure pump. This involves connecting a high-pressure timing pump into the circuit. This portable electric pump charges the charging gallery at a pressure in excess of the 20 bar required to crack the delivery valves, which causes them all to open; consequently the injector leak-off circuit and pump gallery return must be plugged off. Next, the high-pressure pipe on the number 1 cylinder is removed at the pump, and a spill pipe discharging into the portable pump sump is fitted. The procedure then replicates that used to spill-time the injection pump—the engine must be correctly located well before the PC specification (to eliminate engine gear backlash variables) and then barred to spill cutoff.

    If the settings are out of specification, they are rectified by altering the coupling location of the pump drive gear on the pump drive plate.

3. Electronic. This method tends to be used with later generation hydromechanical and electronically managed, port-helix metering injection pumps. The static timing value is checked with an electronic timing tool that senses the positioning of a raised notch located on the pump camshaft.

4. Diesel timing light. Pump timing may be verified using a diesel engine timing light. This consists of a transducer that clamps to the high-pressure pipe and signals a strobe light when it senses the line pressure surge that occurs when the delivery valve opens. The test is not an accurate one, and the values read on the timing light should not be confused with the pump manufacturer's port closure specification because the pressure rise pulse used to trigger the light occurs after port closure.

Tech Tip: Although approved by manufacturers of older automotive diesel engines, using a timing light is not considered sufficiently accurate for current PLN systems. Limit the use of a timing light to verify the operation of timing advance mechanisms.

Removing a Port-Helix Metering Injection Pump from an Engine.
Probably close to half the port-helix metering injection pumps that are removed from engines have nothing wrong with them. This is due to poor field diagnostics caused by a low level of understanding of the operating principles of injection pumps combined with inaccurate interpretation of service literature. Another factor is a tendency of field technicians to black-box fuel pump technology as the responsibility of the pump room technician; as a result, they package up problems for someone else to repair and this can be unnecessarily costly.

The correct location to diagnose most fuel injection pump problems is with the pump on the truck engine. If the OEM-recommended onboard tests are performed before removing the fuel injection pump, then if the pump has to be removed from the engine, the job of the pump room technician has been made easier. When it has been determined that the injection pump is responsible for a fuel problem and it must be removed, the procedure should be as follows:

1. Power wash the injection pump and surrounding area of the engine.
2. If the fuel pump and engine are equipped with locking/timing pins, position the engine to install them.
3. Remove the fuel supply and return lines. Remove the lubricating oil supply and return lines.
4. Remove the accelerator and fuel shutoff linkages (if equipped) or the electronic connector terminals.
5. Disconnect the high-pressure pipes from the delivery valves. Ensure that the high-pressure pipes can be moved away from the delivery valves without forcing or bending them; this usually means releasing insulating and support clamps. Cap both the high-pressure pipe nipples and the delivery valves.
6. Unbolt the fasteners at the pump mounting flange. Depending on the pump, the pump may have to be removed separately from the variable timing/advance timing mechanism. Some pumps may require that the pump drive gear be separated from the pump drive plate or the variable timing device from the front of the pump drive gear, requiring the removal of the timing gear cover plate. Most pumps should slide back easily after this, but care should be taken not to support the weight of the injection pump on its drive gear.

Reinstallation. Essentially, the foregoing procedure is reversed, but the port closure timing should be confirmed using the manufacturer-recommended method (outlined earlier in this section). When installing the pump, ensure that it is installed with the pump drive gear/plate correctly registered with the engine timing gear. Resistance can be an indication that the pump is being installed a tooth out of register and contacting a dowel or key; never force an injection pump into position. When resistance is encountered, remove the pump and check both the pump position and its drive mechanism for problems.

CAUTION *If resistance is encountered while installing a fuel injection pump, remove and check for the cause. Forcing a pump home on its mounting flange using the fasteners can damage the pump drive and will almost certainly result in an out-of-time pump.*

DELIVERY, INJECTION, AND COMBUSTION

The function of any port-helix metering injection pump is to manage the fueling of an engine. This section attempts to match the critical events in the injection pump with the critical events in the engine; in other words, to describe how the fueling pulse affects combustion in the engine cylinders. **Figure 9-8** demonstrates the relationship between what is occurring in the fuel injection pump, the injector, and the engine

cylinder in real time. This graphic can actually be used to demonstrate fueling in most types of diesel fuel injection systems.

Injection Lag

Injection lag is the time measured in crank angle degrees between port closure in the injection pump and NOP, accounted for primarily by the time required to raise pump chamber pressure first to the residual line pressure value, then to the NOP value. Secondary causes of injection lag are:

1. The elasticity of the high-pressure pipe.
2. The compressibility of the fuel. This is approximately 0.5 percent per 1000 psi (67 bar), but it must be accounted for when the system is engineered.

Injection lag can be regarded by the diesel technician as a constant.

Ignition Lag

Ignition lag is the time period between the events of NOP and ignition of the fuel charge. It is usually measured in crank angle degrees at any given engine rpm and load. Liquid droplets exiting the nozzle orifices must be vaporized and ignited. This time period is variable and depends on the ignition quality of the fuel (CN value) and the actual compression temperature. Because it is variable, so will be the amount of fuel in the cylinder at ignition. In cold weather start-up

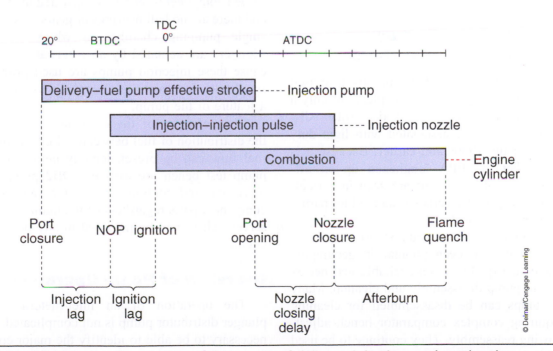

Figure 9-8 Graphic showing the phasing of the events of delivery, injection, and combustion.

conditions, this can be excessive and can result in a detonation condition known as diesel knock.

Combustion

The duration of combustion depends on the length of the injection pulse (that is, the total quantity of fuel injected to the cylinder and the point in crank angle degrees when injection ceases) and the engine rpm. The rate of injection is designed so that cylinder pressure and crank angle leverage are synchronized to deliver power to the flywheel as smoothly as possible.

Nozzle Closure Lag

Nozzle closure lag is the time period between the end of injection pump delivery and actual nozzle closure. This varies, increasing as the injection pressure pulse increases.

Afterburn

Afterburn is the normal combustion of fuel in the engine cylinder after injection nozzle closure. Its duration depends on the length of the injection pulse, the actual quantity of fuel in the cylinder, and many other factors including droplet sizing just before nozzle closure. Afterburn duration is managed to be as short as possible in later versions of hydromechanical, port-helix injection pumps by delivering higher peak pressures and using smaller injector nozzle orifices, the result of which are smaller droplets of fuel that oxidize more quickly.

OPPOSED-PLUNGER, INLET-METERING INJECTION PUMPS

Over the years, rotary distributor pumps have used a number of different operating principles, but only a couple of designs have survived. Today, rotary distributor pumps are mostly associated with light duty diesel engines of the 1990s and earlier. Because it is only recently that the EPA toughened up on off-highway emissions, these pumps are seen in agricultural equipment, especially versions adapted for partial authority, computer controls.

Because rotary distributor pumps are simple in construction, they have been popular in geographic regions where fuel quality is less reliable and necessitates frequent pump disassembly and cleanup. Many distributor pumps can be disassembled for cleaning without requiring complex comparator bench adjustments following reassembly. They continue to be used in Asia, Africa, and South America.

In this chapter, two distinct types of rotary distributor pumps are examined. It is important that you do not confuse them because they use very different operating principles. We begin by looking at what is known as a Roosa pump and its Delphi Lucas CAV sibling. This family of injection pumps can be generically described as **inlet-metering**, opposed-plunger, distributor injection pumps.

In terms of operating principles, the Roosa injection pump and its close relative, the CAV DPA, are nearly identical. Although the DPA significantly outsold Stanadyne/Roosa pumps worldwide, this was not true in North America, where both GM and Ford used the Stanadyne version of the pump into the 1990s. For purposes of describing the operation of the pump, the DB2 version of the Roosa pump is primarily referenced, but it should be remembered that there are a number of slightly different versions within the Stanadyne Roosa and Delphi/Lucas/CAV families of opposed-plunger pump.

ROOSA DB2

When the Roosa DB2 pump was first introduced, it was marketed as a Roosa-Master pump. Older technicians might often use this term today. The DB2 injection pump is an opposed-plunger, inlet-metering, distributor type, diesel injection pump. It was designed for low-cost production and simplicity. A typical DB2 pump has a total of around 100 component parts and only four main rotating members. There are no spring-loaded components, none are lap-fitted in manufacture, and there are no ball bearings or gears. The pump has a single pumping chamber in which two **opposed plungers** are actuated by an **internal cam ring**. Because these injection pumps are used primarily with small-bore, low-cost diesel engines, low initial cost is a feature of the pump.

The geometry of the **hydraulic head** determines the distribution of fuel between cylinders, and because fuel flow can be preset, lengthy periods on the fuel pump test bench are avoided. DB2 pumps are self-lubricated and contain roughly the same number of component parts regardless of the number of cylinders served. They may be mounted in any position on the engine.

Overview of Pump Operation

The operation of an inlet-metering, opposed-plunger distributor pump is not complicated. First, it is necessary to be able to identify the major components and circuits of the pump. A cutaway version of the

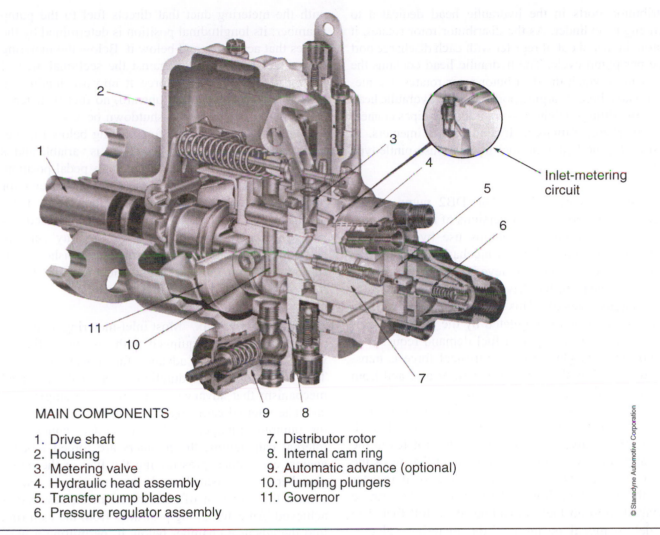

MAIN COMPONENTS

1. Drive shaft
2. Housing
3. Metering valve
4. Hydraulic head assembly
5. Transfer pump blades
6. Pressure regulator assembly
7. Distributor rotor
8. Internal cam ring
9. Automatic advance (optional)
10. Pumping plungers
11. Governor

© Stanadyne Automotive Corporation

Figure 9-9 Cutaway of a Stanadyne DB2 injection pump.

DB2 version of the Roosa pump which we use as our primary reference for study is shown in **Figure 9-9.**

In **Figure 9-9,** note that the drive shaft directly engages with the **distributor rotor** in the hydraulic head. In other words, the distributor rotor rotates within the stationary hydraulic head. It is difficult to see in **Figure 9-9,** but the drive end of the DB2 rotor incorporates two pumping plungers that oppose each other in a single chamber. These plungers are actuated by the inlet-metering circuit internal cam ring.

Operating Principles

The DB2 injection pump is driven through one complete rotation per complete engine cycle. This means that it is rotated at camshaft speed, that is, to turn the pump through one full 360-degree rotation, the engine has to be turned through two complete rotations or 720 degrees. The pump chamber is formed between the pump plungers, and they are actuated toward each

other simultaneously by an internal cam ring. The plungers ride on rollers and shoes that are carried in slots at the drive end of the rotor. The number of cam lobes normally equals the number of engine cylinders and in turn determines how many effective strokes will occur during each pump cycle.

Pump Components. A fuel transfer pump is located at the rear of the rotor. The pump is a positive displacement vane type, so it will displace fuel in proportion to rotational speed. The assembly is enclosed in the end cap, which also houses the fuel inlet strainer and the transfer pump pressure regulator. The face of the regulator assembly is compressed against the liner and distributor rotor and forms an end seal for the transfer pump. The injection pump is designed so that end thrust acts against the face of the transfer pump pressure regulator. The distributor rotor incorporates two charging ports and a single axial bore with one discharge port. This discharge port aligns with stationary distribution ports in the hydraulic head dedicated to each engine cylinder. As the distributor rotor rotates, it is brought in and out of register with each discharge port once per pump cycle. The hydraulic head contains the bore within which the distributor rotor rotates, the metering valve bore, charging ports, and the hydraulic head discharge fittings. High-pressure injection pipes connect these discharge fittings to hydraulic fuel injectors. In most cases, the hydraulic fuel injectors use pintle-type nozzles.

Governing and Metering.

The DB2 pump uses a mechanical governor. Other versions of inlet-metering, opposed-plunger injection pumps use hydraulic and electronic governors. When a mechanical governor is used, speed regulation is achieved by using centrifugal weights to measure the driven speed of the pump (to sense engine speed). This is opposed by variable spring force, which is regulated by the driver's accelerator position (this inputs a fuel demand request). In the governor weight carrier, centrifugal force is transmitted through a sleeve to the governor arm and from there uses a linkage to the metering valve.

All fuel that is allowed to enter the pump chamber formed by the two opposed plungers must pass through the metering valve. The metering valve rotates in the bore and defines a flow area to the duct that feeds the pump chamber. Depending on its rotational register with this duct, fuel flow to the pump chamber can be managed from no fuel graduating up to full fuel. The no-fuel rotational position of the metering valve is achieved by moving a dedicated shutoff solid linkage with an independently operated shutoff lever (outside of the governor housing) or an electrical solenoid.

Hydraulic Governing.

Many versions of the Delphi Lucas CAV pumps use hydraulic governing. Hydraulic governing simplifies the injection pump and reduces its size. In hydraulically governed versions of this pump, the metering valve moves longitudinally in its bore instead of rotationally.

Because the fuel transfer pump is a positive displacement, vane-type pump, it displaces fuel proportionately to engine rpm. If this fuel is unloaded into a defined (and unchanging) flow area, pump fuel pressure will rise in exact correlation with engine rpm. This means that transfer pump fuel pressure can be used as the indicator of rpm.

METERING VALVE OPERATION

The metering valve is cross and center drilled and milled with a graduated metering recess. This graduated metering recess can be brought into and out of register

with the metering duct that directs fuel to the pump chamber: its longitudinal position is determined by the forces that act above and below it. Below the metering valve, fuel pressure acts against the sectional area of the valve, attempting to force it upward, tending to reduce fuel delivery. Fully upward, no fuel is metered to the pump chamber, and shutdown occurs.

Opposing the hydraulic force acting below the metering valve is spring force. This force is variable and is determined primarily by the accelerator pedal position.

As the operator pushes harder on the accelerator pedal, increasing the pedal angle, the spring force acting against the hydraulic pressure is increased, resulting in more fuel delivery. So, in any running condition, the inlet metering valve establishes equilibrium between the spring and the hydraulic forces that position it to define a fuel quantity.

Advance Mechanism.

Most inlet-metering, opposed-plunger pumps are equipped with an automatic advance device that can advance fuel injection timing. This can be either a hydraulic or electrically actuated mechanism that advances or retards the pumping cycle. The internal cam ring is responsible for actuating the pumping plungers. When the rollers contact the internal cam ramps, the plungers are forced inboard against each other, pressurizing the fuel in the pump chambers in between them. If the cam ring is rotated toward the direction of rotor rotation, an advance is achieved. An external lug protrudes from the cam ring into the advance chamber below it, permitting a window of advance.

In the case of hydraulic advance, spring force acts on one side of the cam ring lug, positioning it in the most retarded location. An advance piston is located on the other side of the lug, and fuel pump pressure from the vane pump is allowed to act on this piston. Because a vane pump is positive displacement, when it unloads to a defined flow area, pressure rise will be proportional to pump rotational speed. When this fuel pressure acts on the advance piston, it overcomes the spring pressure opposing it to move the cam ring to advance fuel delivery timing. The result is that any advance achieved will be entirely speed sensitive. The same thing can be accomplished electrically in later versions of the pump by using a linear proportioning solenoid controlled by the engine PCM.

Fuel Flow

Now that you have some understanding of the pump operation, we can take a look at exactly how the fuel is routed through the pump during its cycle.

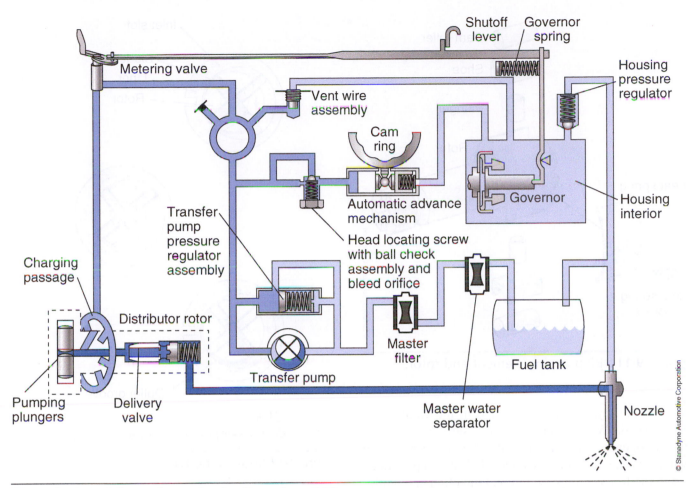

Figure 9-10 Fuel flow schematic of a Stanadyne DB2 pump.

It makes sense to use both the text and the figures to understand how this is achieved. Refer to **Figure 9-10,** a fuel flow schematic of the DB2 pump.

Movement of fuel through the fuel subsystem is the responsibility of the transfer pump, which is the vane pump integral with the injection pump. Fuel is pulled from the main fuel tank through a filter or filters into the pump inlet. At the pump inlet, fuel is pulled through the inlet filter screen by the transfer pump. Some fuel is bypassed through the pressure regulator and routed back to the suction side.

Fuel under transfer pump pressure flows through the center of the transfer pump rotor shaft, past the rotor retainers into an annular groove on the rotor. It next passes through a connecting passage in the head to the advance cylinder, up through a radial passage, and then through a duct to the metering valve. The rotational position of the metering valve, directly controlled by the governor, regulates fuel flow to the radial charging passage, which incorporates the hydraulic head charging ports.

As the rotor revolves, the two rotor inlet passages register with charging ports in the hydraulic head, allowing fuel to flow into the pumping chamber. Fuel flowing into the pump chamber spreads the pumping plungers outward. The extent to which they spread outward depends entirely on the fuel quantity metered into the pump chamber: this fact tends to mean injection timing is advanced proportionate with load. With further rotation, the inlet passages move out of register and complete the metering phase. Next, the discharge port of the rotor is brought into register with one of the hydraulic head distribution outlets that connect the pump with an individual fuel injector. During the register of the discharge port with the rotor, the pumping phase takes place. The plunger-actuating rollers contact the lobes of the internal cam ring, forcing the opposed plungers inboard into the pump chamber.

The pumping element is shown in **Figure 9-11.** The higher the fuel load in the pump chamber, the more advanced the timing of this moment that begins delivery. Fuel trapped in the pump chamber between the plungers is pressurized and unloaded through the discharge port in the hydraulic head during the delivery phase. The discharge port is connected by a high-pressure pipe to

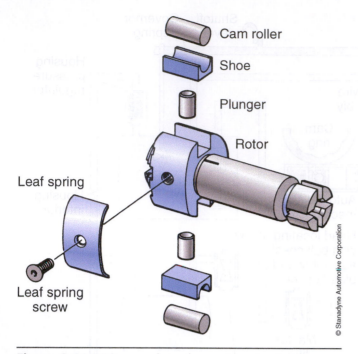

Figure 9-11 DB2 pumping element and rotor.

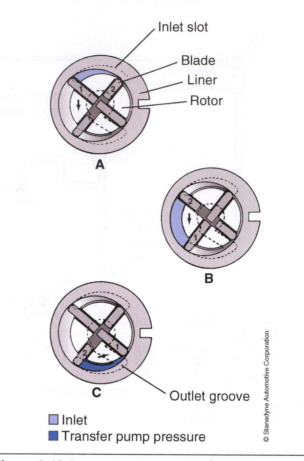

Figure 9-12 Roosa vane-type transfer pump.

a hydraulic injector nozzle located in the engine cylinder head. When fuel pump pressure exceeds the nozzle-opening pressure (NOP), the injector nozzle valve unseats and fuel is injected. Most applications using the Roosa pump use indirect injection and pintle-type nozzle assemblies.

Inlet-metering, opposed-plunger, distributor injection pumps are self-lubricating and self-priming. The lubricity of diesel fuel accomplishes the lubrication of the internal components of the pump. As fuel at transfer pump pressure reaches the charging ports, slots on the rotor shank allow fuel and any entrapped air to flow into the pump housing cavity. An air vent passage in the hydraulic head connects the outlet side of the transfer pump with the pump housing. This permits air and some fuel to bleed back to the fuel tank by means of a return line. Bypass fuel fills the pump housing, lubricates the internal components, acts as coolant, and purges the housing of small air bubbles. The pump is therefore designed to operate with the housing charged with fuel. During operation there should be no air within the pump.

Inlet-Metering Distributor Pump Circuits

There are nine subcircuits in an inlet-metering, opposed-plunger distributor injection pump. Technicians should understand the role of each circuit, as it will help isolate problems during troubleshooting.

Transfer Pump Circuit. The vane-type fuel transfer pump consists of a stationary liner and spring-loaded vanes or blades carried in slots in the rotor shaft. The inside diameter of the liner is eccentric to the rotor axis. Rotation creates centrifugal force that causes the blades to move outward in the rotor slots to hug the liner wall. As the pump rotates, the volume between the blade segments is varied, allowing the pump to discharge fuel to the outlet.

The transfer pump uses a positive-displacement operating principle, which means that transfer pump output volume and pressure increase in direct proportion to pump speed. Displacement volumes of the transfer pump are designed to exceed injection requirements by a margin, so some of the fuel is recirculated by the pump regulator that routes it back to the inlet side of the transfer pump. **Figure 9-12** shows the vane-type transfer pump.

Figure 9-12 illustrates the pumping principle. Radial movement causes a volume increase in the quadrant between blades 1 and 2 **(Figure 9-12A)**. In this position, the quadrant is in registry with a kidney-shaped inlet slot in the top portion of the regulator assembly. The increasing volume causes fuel to be pulled through the inlet fitting and filter screen into the transfer pump liner. Volume between the two blades continues to increase until blade 2 passes out of register with the regulator slot. At this point, the rotor is in a position in which there is little outward movement of blades 1 and 2 and the volume is not changing (**Figure 9-12B**). The slug of fuel between the blades is now carried to the bottom of the transfer pump liner.

As blade 1 passes the edge of the kidney-shaped groove in the lower region of the regulator assembly (**Figure 9-12C**), the eccentric liner wall compresses blades 1 and 2 in an inward direction (**Figure 9-12A**). This action reduces the volume available for fuel storage but not the quantity, so the fuel is pressurized. Pressurized fuel is unloaded through a groove in the regulator assembly and directed into a channel on the rotor leading to the hydraulic head. As the rotor continues with its rotation, volume between blades continues to decrease, pressurizing the fuel in the quadrant, until blade 2 passes the groove in the regulator assembly.

Regulator Assembly Circuit. The regulator prevents excessive vane pump pressures in the event of an engine overspeed condition. Fuel from the discharge side of the transfer pump forces the regulator piston against the regulator spring. As pump speed and therefore flow increases, the regulator spring is further compressed until the leading edge of the regulator piston starts to expose the pressure-regulating slot. Because fuel pressure acting on the regulator piston is opposed by the regulator spring, delivery pressure of the transfer pump is controlled by:

- regulator spring tension
- flow area defined by the regulating slot

Metering Circuit. Inlet metering refers to the fact that only fuel permitted to pass through the metering valve enters the pump chamber. The fuel delivered to the pump chamber during each pump cycle is the metered fuel quantity and defines the injection pulse duration. The metering valve rotates in the bore. It receives fuel from the charging passage. Because it is milled with a scrolled metering recess, it directs this fuel to the opposed plunger injection pump through a charging passage. In other words, the rotational position of the metering valve controlled by the governor defines the flow area to the duct that feeds the pump chamber. This flow area must accommodate all the fuel requirements of the engine from no fuel to full fuel. The no-fuel rotational position of the metering valve is achieved by moving a dedicated shutoff linkage or by an electrical solenoid.

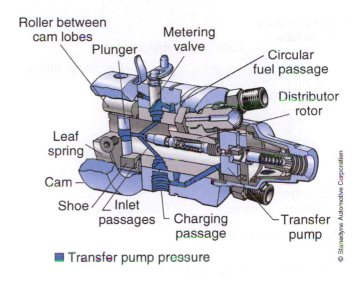

Figure 9-13 Charging cycle.

Charging Circuit. The charging circuit determines the exact quantity of fuel that enters the pumping chamber. As the rotor revolves (**Figure 9-13**), the two inlet passages in the rotor register with charging passage ports. Fuel from the transfer pump, controlled by the opening of the metering valve, flows into the pumping chamber, thereby forcing the plungers apart.

The plungers move outward only to the extent required to accommodate the fuel required for injection on the following stroke. If only a small quantity of fuel is charged to the pumping chamber, for instance that required to idle the engine, the plungers spread only a short distance. When engine fuel load is high, the plungers are forced outward further to accommodate the higher volume of fuel. Maximum plunger travel is limited by a leaf spring that contacts the edge of the roller shoes. Only when the engine is operating at full load will the plungers move to the most outward position.

Discharge Circuit. As the cycle continues, the inlet passages are moved out of register with the charging ports. Next, the rotor discharge port is brought into register with one of the outlets in the hydraulic head that will direct fuel to an injector. The plunger-actuating rollers then contact the cam profiles in the internal cam ring, driving the shoes against the plungers and forcing them inboard into the pump chamber. This action begins the high-pressure pumping phase called the discharge cycle shown in **Figure 9-14.**

The beginning of injection varies according to the required fuel load. Assuming that the internal cam ring remains stationary, the higher the fuel load admitted to the pump chamber, the more advanced the resultant

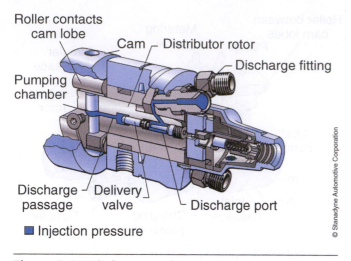

Roller contacts
cam lobe
Cam — Distributor rotor
Discharge fitting
Pumping
chamber
Discharge — Delivery
passage   valve   Discharge port

■ Injection pressure

© Stanadyne Automotive Corporation

Figure 9-14 Discharge cycle.

injection timing. During the discharge stroke, fuel trapped in the pump chamber between the plungers is forced through the axial passage of the rotor to the discharge port and out through the injection line to the injector nozzle. Fuel delivery continues until the plunger-actuating rollers are ramped beyond the highest point of the cam lobe and are allowed to move outward. Pressure in the axial passage is allowed to collapse, dropping line pressure and allowing the injector nozzle to close. Nozzle closure ends delivery.

Delivery Valve Circuit. Delivery valves are an option on most inlet-metering, opposed-plunger distributor pumps. The delivery valve accelerates injection line pressure drop after injection to a value approximately two-thirds of the specified nozzle closing pressure. This reduction in line pressure permits the nozzle valve to seat abruptly and minimizes the fuel injected during the collapse phase of injection. When larger droplets of fuel are injected to the engine cylinder during the collapse phase at the end of the injection pulse, there is insufficient time available to properly combust them, which causes higher HC emissions. Delivery valves are found on most highway applications.

The delivery valve is located in a bore in the center of the distributor rotor. This type of delivery valve requires no seat and uses just a mechanical stop to limit travel. Sealing is accomplished by the close-tolerance fit between the valve and bore. When injection begins, fuel pressure moves the delivery valve slightly out of its bore, adding its swept volume displacement to the spring chamber. Because at this moment the discharge port is already exposed to a hydraulic head outlet, the retraction volume and plunger displacement volume are discharged under high pressure to the nozzle.

Delivery ends when pressure on the plunger side of the delivery valve drops as the cam rollers ramp over the high point of the cam profile. Next, the rotor discharge port is closed off completely and a residual line pressure is sealed in the high-pressure injection pipe. The delivery valve seals only while the discharge port is open; once the port is closed, residual line pressure is maintained by the close-tolerance fit of the hydraulic head and rotor.

Fuel Return Circuit. Fuel under transfer pump pressure is discharged into a vent passage in the hydraulic head. Flow through the passage is restricted by a vent wire assembly to prevent excessive return fuel that could cause undue pressure loss. The actual amount of return fuel is controlled by the size of the wire used in the vent bore assembly. The smaller the wire, the greater the return flow, and vice versa. Vent wires are available in several size options to meet differing specifications. The vent wire assembly can be accessed by removing the governor cover. The vent passage is located behind the metering valve bore and connects with a short vertical passage containing the vent wire assembly, leading to the governor housing.

Any air entering the transfer pump is routed to the vent passage as shown. Both air and fuel then flow from the housing to return to the fuel tank via the return line. Housing pressure is maintained by a spring-loaded ballcheck return fitting in the governor cover of the pump.

Mechanical Governor. The governor used can be classified as a variable-speed governor, also known as an all-speed governor. This type of governor functions to maintain desired engine speed (requested rpm) and will do so within a certain window as load changes.

Speed sensing is performed by the flyweights as shown in **Figure 9-9** and indicated by arrow 11. Movement of the flyweights acting against the governor thrust sleeve rotates the metering valve in its bore by means of the governor arm and linkage hook. Rotation of the metering valve varies the register of the metering valve scroll to the passage from the transfer pump, thereby controlling how much fuel is metered to the pump element. Centrifugal force, which directly correlates with rotational speed, forces the flyweights outward, moving the governor thrust sleeve against the governor arm, actuating the linkages that rotate the metering valve. The force of the flyweights acting on the governor arm is balanced by the governor spring force. This spring force is variable and is controlled by the manually positioned throttle lever. The throttle lever connects to the vehicle accelerator linkage.

LOAD DECREASE

A variable-speed governor defines a specific rpm value. In the event of a load reduction on the engine, engine speed would tend to increase. This would in turn increase the centrifugal force produced by the flyweights, which would act to rotate the metering valve clockwise to reduce engine fueling.

LOAD INCREASE

When the load on the engine is increased, engine speed initially tends to drop. As engine speed falls, centrifugal force generated by the weights drops, permitting the spring forces that oppose it to rotate the metering valve in the counterclockwise direction, increasing fuel metered to the pumping element. Engine speed at any point in the operating range of the engine is dependent on the combination of forces that act on the governor thrust lever. As with any mechanical governor, centrifugal force acting on the thrust lever will result in speed reduction (less fuel) and spring forces acting on the thrust lever will result in increased speed (more fuel).

Governor Operation Summary. A light idle spring is provided for more sensitive regulation when flyweight centrifugal force is low, such as when the engine is operating close to idle speeds. The limits of throttle travel are set by adjusting screws to define low-idle and high-idle speeds. A light tension spring on the linkage assembly takes up slack in the linkage joints and permits the shutoff mechanism to close the metering valve without having to overcome the governor spring force, which means that only a little force is required to rotate the metering valve to the closed position to shut the engine down.

The function of any variable-speed governor is to attempt to hold engine rpm at a consistent value (based on a given throttle position) and let the governor adjust fueling to hold that rpm constant as engine load varies. Positioning the throttle lever simply defines a request for an rpm value, not a fuel quantity, which would be the case with a limiting-speed governor.

Advance Circuit. Opposed-plunger, inlet-metering distributor pumps permit the use of a simple, direct-acting hydraulic mechanism powered by fuel pressure from the transfer pump to rotate the internal cam within the pump housing. This changes the phasing of injection pump stroke with that of the engine that drives the injection pump and varies delivery timing. The advance mechanism defaults to the most retarded position when there is no or low pressure developed by the vane-type transfer pump. As vane pump pressure increases, fuel pressure acting on the advance piston overcomes the advance spring pressure and rotates the internal cam ring to advance the start of fuel delivery. Total movement of the cam is limited by the piston length. A trimmer screw is provided to adjust the advance spring tension, which essentially determines how much fuel pressure is required (that is, the specific rpm) to begin the advance. It can be incorporated at either side of the advance mechanism and may be adjusted on the test bench while the engine is running. Because the extent of advance depends on vane pump pressure, it is speed sensitive. **Figure 9-15** shows the advance circuit components and operating principles.

Advancing fuel injection timing compensates for inherent injection lag and greatly improves high-speed engine performance. Beginning the injection pulse earlier when the engine is operating at higher speeds ensures that peak cylinder combustion pressures are developed when the piston is ideally positioned in its downstroke to optimize torque transfer to the crankshaft. Advancing the beginning of the injection pulse results in the plungers completing their pumping stroke earlier.

Electronic Governing. When inlet metering, opposed plunger injection pumps use electronic governing, little changes within the pump or its operating principles. The system can be classified as a partial authority system. In addition to incorporating sensors to feedback status to the PCM, two PCM-controlled outputs are incorporated:

1. Metering valve actuator: Movement of the metering valve is controlled by the PCM.
2. Timing advance actuator: The timing piston is PCM-controlled.

Opposed-Plunger, Inlet-Metering Pump Summary

The worldwide success of Roosa's injection pump design is a testament to its low initial cost, ease of maintenance, and ability to manage a range of engines. Though not seen anymore on light duty trucks, opposed-plunger, inlet-metering, rotary distributor pumps are still found in many small diesel engine applications. They are especially prominent in agricultural equipment. As we indicated earlier in this chapter, the design lends itself to some limited enhancement with partial-authority computer controls, usually governing (PCM control of metering valve) and advance functions. When managed electronically,

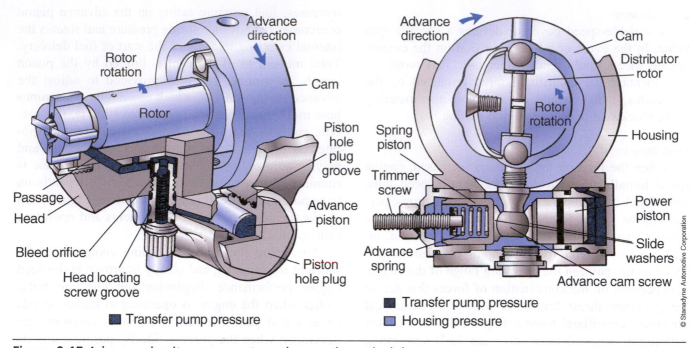

Figure 9-15 Advance circuit components and operating principles.

all of the critical components of the fuel pump are retained. However, with today's expectations for minimum emissions, better fuel economy, and higher longevity, this category of fuel injection pump has some limitations and can be regarded as a disappearing technology, at least in North America.

SLEEVE-METERING, SINGLE PLUNGER DISTRIBUTOR PUMPS

Bosch hydromechanical, **sleeve-metering** distributor injection pumps use a single plunger pumping element. In describing this technology, the popular VE pump is used as our primary reference. The VE pump has been used in a number of imported light duty highway diesel engines, including Volkswagen, and it was an option on some early versions of the Cummins B series engine. These pumps manage fueling for engines with up to six cylinders. A VE pump has four primary circuits:

1. Fuel supply pump
2. High-pressure pump
3. Governor
4. Variable timing

The VE distributor pump has been used in passenger cars, commercial vehicles, agricultural tractors, and stationary engines.

Subassemblies

The VE distributor pump uses only one pump cylinder and a single plunger. It is designed to fuel

multicylinder engines. The general layout of a typical VE fuel system is shown in **Figure 9-16.**

All movement through the fuel subsystem is the responsibility of a vane-type transfer pump integral with the VE pump assembly. Once fuel from the vehicle tank enters the VE pump, fuel is pressurized to injection pressures and routed to the engine cylinders by means of high-pressure pipes. Injection pressures

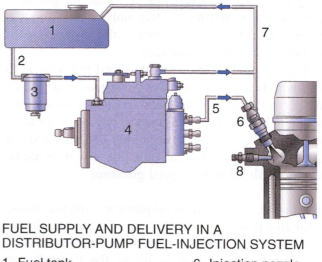

FUEL SUPPLY AND DELIVERY IN A DISTRIBUTOR-PUMP FUEL-INJECTION SYSTEM

1. Fuel tank
2. Fuel line (suction pressure)
3. Fuel filter
4. Distributor injection pump
5. High-pressure fuel injection line
6. Injection nozzle
7. Fuel return line (pressureless)
8. Sheathed-element glow plug

Figure 9-16 VE distributor pump fuel system layout.

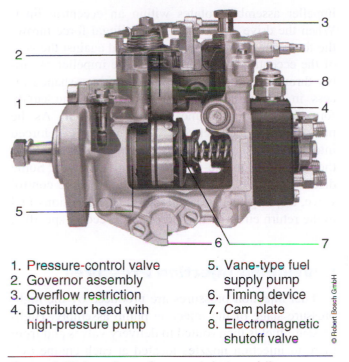

1. Pressure-control valve
2. Governor assembly
3. Overflow restriction
4. Distributor head with high-pressure pump
5. Vane-type fuel supply pump
6. Timing device
7. Cam plate
8. Electromagnetic shutoff valve

© Robert Bosch GmbH

Figure 9-17 Location of pump subassemblies.

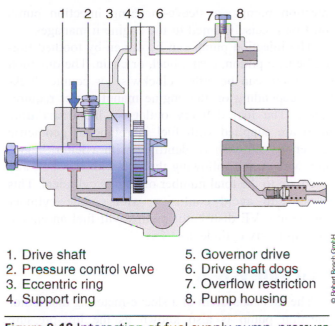

1. Drive shaft
2. Pressure control valve
3. Eccentric ring
4. Support ring
5. Governor drive
6. Drive shaft dogs
7. Overflow restriction
8. Pump housing

© Robert Bosch GmbH

Figure 9-18 Interaction of fuel supply pump, pressure control valve, and overflow restriction valve.

are created by a single plunger-type pump. Fuel delivered by the pump plunger is routed by a distributor groove to the outlet ports, which connect to hydraulic injectors located at each engine cylinder. The VE distributor pump housing contains the following subcircuits:

- High-pressure (injection) pump with distributor
- Mechanical (flyweight) governor
- Hydraulic timing device
- Vane-type, fuel supply pump
- Shutoff device
- Engine-specific add-on modules

Figure 9-17 shows a cutaway view of the subcircuits used on a typical VE pump. Add-on modules, some of which are described later in the chapter, allow the pump to be adapted to the requirements of specific diesel engines.

Design and Construction

The pump drive shaft is supported by bearings in the pump housing and drives the vane-type fuel supply pump. A roller ring is located inside the pump at the end of the drive shaft, although it is not connected to it. **Figure 9-18** shows the arrangement of the pump drive shaft and components in the front section of the pump. A rotating-reciprocating movement is imparted to the distributor plunger by means of a **cam plate** driven by the input shaft that rides on the rollers of the roller ring. The plunger moves inside the distributor head, which is itself bolted to the pump housing. Located in

the distributor head are the electric fuel shutdown, screw plug with vent screw, and the delivery valves. If the distributor pump is also equipped with a mechanical fuel shutdown, it is mounted to the governor cover.

The governor assembly, which includes flyweights and the control sleeve, is driven by the drive shaft. The governor linkage, made up of control, starting, and tensioning levers, pivots in the housing. The governor shifts the position of the control sleeve on the pump plunger and, in this way, defines plunger effective stroke. Located above the governor mechanism is the governor spring that connects with the external control lever by means of the control-lever shaft, itself held in bearings in the governor cover.

The control sleeve is used to control pump output. The governor assembly is located at the top of the pump, and it contains the full-load adjusting screw, the overflow restriction or the overflow valve, and the engine speed adjusting screw. The variable timing device is located under the pump assembly, and it functions to advance pump timing based on fuel pressure developed by the internal vane pump.

Pump Drive. The sleeve-metering, distributor injection pump is direct driven by the engine it manages, and it must be driven through one complete rotation per full engine cycle. This means that on a four-stroke cycle engine, the pump is driven through one complete revolution per two crankshaft revolutions, in other words at camshaft speed. In common with most other injection pumps, a sleeve-metering injection pump must be precisely timed to the engine it manages.

The injection pump can be driven by toothed timing belts, a pinion, gear wheel, or chain. The direction of rotation can be either clockwise or counterclockwise, depending on the engine manufacturer requirements. The fuel delivery outlets in the **distributor head** are supplied with fuel in rotational geometric sequence, and each is identified with a letter, beginning with A and following through with B, C, D, and so on, up to the total number of engine cylinders. This is done to avoid confusion with engine cylinder numbering. VE distributor pumps will fuel an engine with up to six cylinders.

Fuel Subsystem

The fuel subsystem of a sleeve-metering distributor injection pump is also known as the low-pressure circuit. It consists of a fuel tank, fuel lines, fuel filter, vane-type fuel supply pump, pressure control valve, and overflow restriction. The vane-type supply pump is responsible for all movement of fuel in the fuel subsystem. It pulls fuel from the fuel tank and routes it through a filter before it enters the injection pump. The vane pump operates on a positive displacement principle, so the volume of fuel it pumps is directly related to rotational speed. A pressure control valve ensures that injection pump internal pressure is managed as a function of vane pump speed. This valve sets a defined internal pressure for any given speed, meaning that pump's internal pressure rises directly in proportion to engine speed. In operation, some fuel flows through the pressure-regulating valve and is routed back to the suction side of the vane pump; some fuel also flows through the overflow restriction located at the top of the governor housing, and this fuel is routed back to the fuel tank. The excess fuel that flows through the injection pump cools and vents the injection pump housing.

In some applications, an overflow valve is fitted instead of the overflow restriction. The interaction between the supply pump, pressure control valve, and overflow restriction valve is shown in **Figure 9-18.** In some applications, the vane pump exerts insufficient suction to pull fuel through the fuel subsystem, and in these applications, a pre-supply pump is required. This pump is usually located in or close to the fuel tank.

Supply Pump Operation. The vane-type transfer pump is located on the injection pump drive shaft. The pump impeller assembly is concentric with the shaft and lugged to it by means of a Woodruff key. The

impeller assembly rotates within an eccentric liner. When the drive shaft rotates, centrifugal force throws the four vanes in the impeller outward against the wall of the eccentric liner. Fuel enters the impeller assembly through an inlet passage and a kidney-shaped recess in the pump housing and charges the cavity formed between the vanes and the liner wall. As the pump rotates, fuel between adjacent vanes is forced into the upper outlet, a kidney-shaped recess and from there, directed to the injection pump circuitry. Some discharge fuel is also directed to the pressure control valve, a spring-loaded spool valve, which options fuel to the return circuit when pressure exceeds a specified value.

Developing Injection Pressures

Fuel injection pressures are produced by the high-pressure stage of the injection pump assembly. High-pressure fuel is then routed to delivery valves and from there to injection nozzles located at each engine cylinder by means of high-pressure pipes.

Distributor Plunger Drive. Rotary movement of the drive shaft is transferred to the **distributor plunger** by a coupling unit as shown in **Figure 9-19.** Here, drive lugs or dogs on the cam plate engage with the recesses in a yoke, located between the end of the drive shaft and the cam plate. The cam plate is loaded onto the roller ring by a spring, so when it rotates, the cam lobes riding on the ring rollers convert the rotational movement of the drive shaft into a rotating-reciprocating movement of the cam plate.

The distributor plunger is locked into position relative to the cam plate by a pin. The distributor plunger is actuated upward through its stroke by cams on the cam plate, while a pair of symmetric return springs force it back downward. Because the plunger is actuated

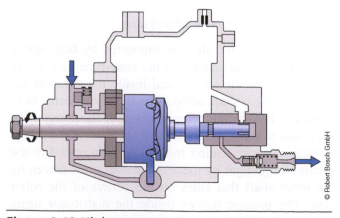

Figure 9-19 High-pressure pump components.

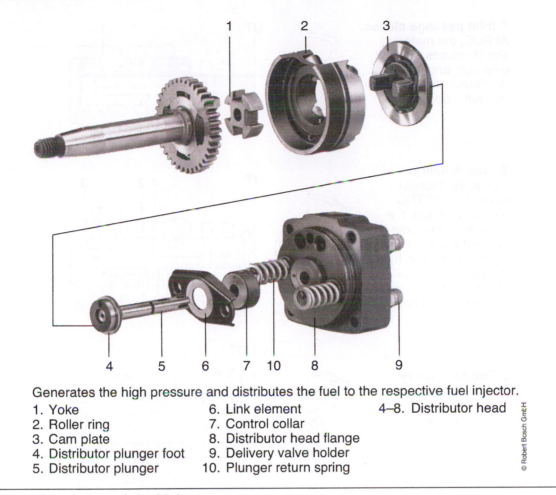

Generates the high pressure and distributes the fuel to the respective fuel injector.

1. Yoke
2. Roller ring
3. Cam plate
4. Distributor plunger foot
5. Distributor plunger

6. Link element
7. Control collar
8. Distributor head flange
9. Delivery valve holder
10. Plunger return spring

4–8. Distributor head

© Robert Bosch GmbH

Figure 9-20 Exploded view of the high-pressure pump components.

by cam profile and loaded to ride that profile by springs, its actual stroke does not vary.

The plunger return springs contact the distributor head at one end, and at the other, act on the plunger by a link element. These springs also have a dampening effect and can prevent the cam plate from jumping off the rollers during sudden speed change. Return spring length must be carefully matched so that the plunger is not side loaded in the pump bore. (See the lower portion of **Figure 9-20.)** An exploded view of the high-pressure pumping components is shown in **Figure 9-20.**

Cam Plates. The cam plate and its cam contour help define fuel injection pressure values and injection duration, along with pump-driven speed that determines plunger actuation velocity. Because of the different requirements of each type of engine, fuel injection factors produced by an injection pump are distinct to each engine. That means a specific cam plate profile is required for each engine type, and technicians should remember that cam plates are generally not interchangeable because of the engine-specific machining of each.

Distributor Head Assembly. The distributor plunger, the distributor head bushing, and the control collar are each precisely fitted by lapping into the distributor head. These components are required to seal at injection pressure values. Some internal leakage losses do occur and serve to lubricate the plunger. In common with other components that are lap finished in manufacture, the distributor head should always be replaced as a complete assembly.

Metering. Metering and the development of injection pressures takes place in four distinct phases shown in **Figure 9-21.** In a four-cylinder engine, the distributor plunger has to rotate through 90 degrees for a complete pumping stroke to occur. A complete pump stroke means that the plunger has to be stroked from BDC to TDC and back again. In the case of a six-cylinder engine, the plunger has to complete a pumping stroke in 60 degrees of pump rotation.

Use **Figure 9-21** and correlate the text with the four phases of a complete pumping stroke. As the distributor plunger is forced down from TDC to BDC, fuel flows through the open inlet passage and passes into

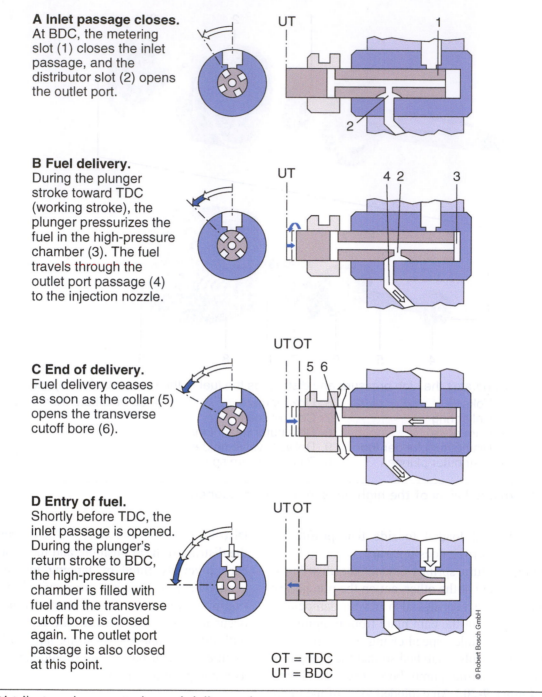

A Inlet passage closes.
At BDC, the metering slot (1) closes the inlet passage, and the distributor slot (2) opens the outlet port.

B Fuel delivery.
During the plunger stroke toward TDC (working stroke), the plunger pressurizes the fuel in the high-pressure chamber (3). The fuel travels through the outlet port passage (4) to the injection nozzle.

C End of delivery.
Fuel delivery ceases as soon as the collar (5) opens the transverse cutoff bore (6).

D Entry of fuel.
Shortly before TDC, the inlet passage is opened. During the plunger's return stroke to BDC, the high-pressure chamber is filled with fuel and the transverse cutoff bore is closed again. The outlet port passage is also closed at this point.

OT = TDC
UT = BDC

© Robert Bosch GmbH

Figure 9-21 Distributor plunger stroke and delivery phases.

the pumping chamber located above the plunger. At BDC, plunger rotational movement takes the plunger out of register with the inlet passage and exposes the distributor slot for one of the outlet ports as shown in **Figure 9-21A.** The plunger now reverses direction and is driven upward to begin the working stroke. Pressure rise is created in the pump chamber above the plunger and when sufficient, opens the delivery valve and forces fuel through the high-pressure pipe to the injector nozzle. The working (delivery) stroke is shown in **Figure 9-21B.** The delivery stroke is complete when

the plunger transverse cutoff bore (cross drilling) protrudes beyond the metering sleeve, collapsing the pressure. When collapse is initiated, pressure drops in the high-pressure line and pump chamber until there is no longer sufficient pressure to hold the nozzle valve open. Injection ceases the instant the nozzle valve closes.

As the plunger continues to move upward to TDC, fuel spills backward through the cutoff bore to the pump housing. During this collapse phase, the inlet passage is again exposed, ready for the next pump

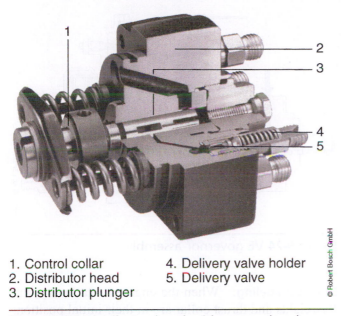

1. Control collar
2. Distributor head
3. Distributor plunger
4. Delivery valve holder
5. Delivery valve

© Robert Bosch GmbH

Figure 9-22 Distributor head with pump chamber.

working cycle as shown in **Figure 9-21C.** As the plunger is forced back down by the return springs from TDC to BDC, the transverse cutoff bore is taken out of register by the plunger rotational movement, and the pump chamber is again charged with fuel through the now exposed inlet passage as shown in **Figure 9-21D. Figure 9-22** shows a complete distributor head and high-pressure pumping assembly in cutaway view.

Delivery Valve. The delivery valve seals the high-pressure pipe from the injection pump chamber. It therefore retains dead volume fuel (static fuel in the pipe) at a value well above that in the pump chamber but comfortably below that required to open the injector nozzle. This static pressure is known as residual line pressure, and it ensures precise closure of the injector nozzle at the end of injection and ensures that a stable pressure is maintained in the high-pressure pipe between injection pulses, regardless of injected fuel quantity.

One delivery valve is used per engine cylinder. The delivery valve is a spring-loaded plunger. It is opened by delivery pressure developed in the injection pump chamber and closed by a return spring. Between injection pulses, the delivery valve remains closed. Its function can be most simply described by stating that it separates the high-pressure pipe from the distributor head outlet port for the larger portion of the cycle when no fuel is being pumped to a given engine cylinder.

DELIVERY VALVE OPERATION

The construction of each delivery valve is identical to that used on port-helix metering pumps, so you may

want to review this section and **Figure 9-5** earlier in this chapter. The valve core consists of a stem, seat, retraction piston, and flutes.

DELIVERY VALVE WITH RETURN-FLOW RESTRICTION

Pressure drop-off to a precise value in the high-pressure pipes is desirable at the end of injection. However, the high-speed, high-pressure switching that the high-pressure pipe is subject to creates pressure waves that are reflected between the delivery valve and injector nozzle seat. This pressure wave reflection causes local spiking of line pressure and may cause undesirable nozzle opening known as secondary injection or vacuum phases in the high-pressure pipe, causing cavitation.

Using a delivery valve with a restriction bore that is only a factor in the direction of return (spill) fuel flow can minimize pressure wave reflection. The return-flow restriction consists of a valve plate and a pressure spring arranged so that the restriction is only effective in the return direction when it dampens pressure spikes and vacuum phases.

CONSTANT-PRESSURE VALVE

Another means of dealing with the problems associated with pressure wave reflection (pressure spikes and vacuum phases) is to use a constant-pressure–type delivery valve. These are usually found on high-speed engines using direct injection (DI). Constant-pressure valves relieve the high-pressure pipe pressure by means of a single-acting, nonreturn valve set to a specific pressure. The specific pressure would define the residual line pressure. A diagram of a constant-pressure delivery valve is shown in **Figure 9-23.**

Governing

Bosch sleeve-metering, rotary distributor pumps are available with both variable speed and limiting-speed governing options. These are usually described by Bosch using the British terms:

- VS governor = all speed governor
- LS governor = min-max governor

A basic description of governor principles is provided a little earlier in this chapter, so it is not repeated here. Instead, we look at how the governor controls the positioning of the sleeve-metering collar under various operating conditions. In addressing the governors used on Bosch sleeve-metering distributor pumps, we reference a variable-speed governor. Remember, regardless of governor type, the position of the sleeve-metering collar defines the injection pump

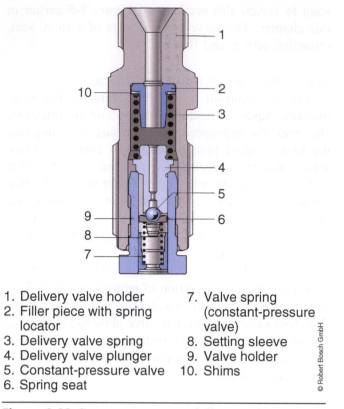

1. Delivery valve holder
2. Filler piece with spring locator
3. Delivery valve spring
4. Delivery valve plunger
5. Constant-pressure valve
6. Spring seat
7. Valve spring (constant-pressure valve)
8. Setting sleeve
9. Valve holder
10. Shims

Figure 9-23 Constant-pressure delivery valve.

effective stroke, so engine output is entirely dependent on this.

Governor Design. The governor assembly is attached to the governor drive shaft located in the governor housing. It consists of a flyweight carrier, **thrust collar**, and tensioning lever. When the flyweights rotate, they are forced outward due to centrifugal force. This radial outward movement is converted to axial movement of the thrust collar or what Bosch describes as a sliding sleeve. Thrust collar travel is allowed to act on the governor lever assembly, which is made up of a start lever, tensioning lever, and adjusting lever.

As in any mechanical governor, spring forces defined at the governor lever assembly act in opposition to the centrifugal force produced by the flyweights. Centrifugal force tends to reduce fueling/rpm, while spring force tends increase fueling/ rpm. Therefore, the interaction of spring forces and centrifugal force acting on the thrust collar (sliding sleeve) defines the positioning of the governor lever assembly. The governor lever assembly controls the position of the control sleeve or collar. As we learned earlier, the position of the control sleeve determines the plunger effective stroke that defines the quantity of fuel to be delivered. The governor assembly on a VE distributor pump is shown in **Figure 9-24.**

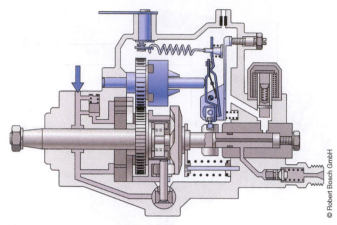

Figure 9-24 VE governor assembly.

Start-Up Fueling. When the engine is stationary, the flyweights and thrust collar are in their initial position. This results in the start lever being forced into the start position by the starting spring, moving the control sleeve on the distributor plunger to its start fuel position. When the engine is cranked, the distributor plunger travels through a complete working stroke before the cutoff cross drilling is exposed to end delivery. The result is a full-fuel delivery pulse.

Low-Idle Operation. Once the engine is running with the accelerator pedal released, the engine speed control lever shifts to the idle position. This positions the control sleeve to provide a short plunger effective pumping stroke and represents the lowest fuel delivery condition of a running engine. Idle speed can be adjusted independent of the accelerator pedal setting, and can be increased or decreased if temperature or load conditions require it.

Operation Under Load. During actual operation, the driver requests the required engine speed by accelerator pedal angle. If higher engine speeds are required, the driver pushes harder on the pedal, increasing the angle and therefore, the governor spring force. If a lower engine speed is required, she reduces the pedal angle. At any engine speed above idle, the start and idle springs are compressed completely, so they do not influence governing. Governed speed becomes the responsibility of the governor spring.

Under load, the driver sets the accelerator pedal at a specific position, increasing the pedal angle. If a higher engine speed is required, the pedal angle has to be further increased. As a result, the governor spring is tensioned, increasing the spring force available to counter the centrifugal force produced by the flyweights. This acts through the thrust collar governor levers to shift the control sleeve toward the full-fuel direction, increasing the injection pump effective stroke. As a result, injected fuel quantity increases and engine speed rises.

The control collar remains in the full-fuel position until equilibrium is established once again between the centrifugal force generated by the flyweights (now greater) and the governor spring forces that oppose it. Should engine speed continue to increase, the flyweights extend further, resulting in governor thrust collar movement that forces the control sleeve toward the no-fuel direction, trimming back fueling. The governor can reduce delivery fueling to no fuel, ensuring that engine speed limitation takes place.

During operation, assuming the engine is not overloaded, every position of the engine speed control lever relates to a specific engine speed: the governor manages that speed by having the ability to control the control sleeve in any position between full fuel and no fuel. In this way, the governor maintains desired speed. The speed at which the governor responds to a change in engine load in order to maintain desired engine speed is known as *droop*.

If during engine operation, engine load increases to the extent that when the control sleeve is in full-fuel position, engine speed continues to drop, the engine can be assumed to be overloaded and the driver has no option but to downshift.

Governor Break. During the downhill operation of a vehicle, the engine is driven by vehicle momentum, and engine speed tends to increase. This results in the governor flyweights moving outward, causing the governor thrust collar to press against the tensioning and start levers. Both levers react by changing position and pushing the control sleeve toward the no-fuel position until a reduced fuel equilibrium is established in the governor that corresponds to the new load/speed condition. At engine overspeed, the governor can always no-fuel the engine. With a variable-speed governor, any control sleeve position can be set by the governor in order to maintain desired speed.

Variable Timing Device

A hydraulically actuated timing device is located under the main pump housing at right angles to the pump longitudinal axis, as shown in **Figure 9-25.** The variable timing device is a speed-sensitive advance mechanism that defaults to the most retarded delivery position in the absence of hydraulic (fuel) pressure.

The timing device housing is closed with a cover on either side. A passage is located in one end of the

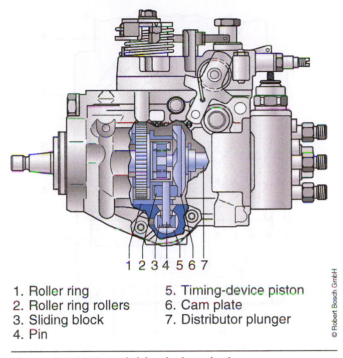

1. Roller ring
2. Roller ring rollers
3. Sliding block
4. Pin
5. Timing-device piston
6. Cam plate
7. Distributor plunger

© Robert Bosch GmbH

Figure 9-25 VE variable timing device.

timing device that allows fuel from the vane pump to enter. This fuel is allowed to act on the sectional area of the advance piston. On the opposite side of the piston, spring force opposes the hydraulic pressure of the fuel. The piston is connected to the roller ring by means of a sliding block and pin, enabling piston linear movement to be converted to rotational movement of the roller ring.

Timing Device Operation. The timing device is held in its initial or default position by the timing device spring as shown in **Figure 9-26A.** When the engine is started, the pressure control valve regulates fuel pressure so that it is exactly proportional to engine speed. As a result, this engine-speed–dependent fuel pressure is applied to the end of the timing device piston in opposition to the spring force acting on the other side of it.

At a predetermined vane pump fuel pressure, the advance piston overcomes the spring preload and shifts the sliding block and the pin that engages with the roller ring. The roller ring is rotated, moving its relative position with the cam plate, which results in the rollers lifting the rotating cam plate earlier. This action means that the actuation of the injection plunger is advanced. The maximum advance angle achievable in Bosch sleeve-metering distributor injection pumps is limited by timing piston linear movement and is usually 24 crank angle degrees. **Figure 9-26B** shows the maximum timing angle location of the advance piston.

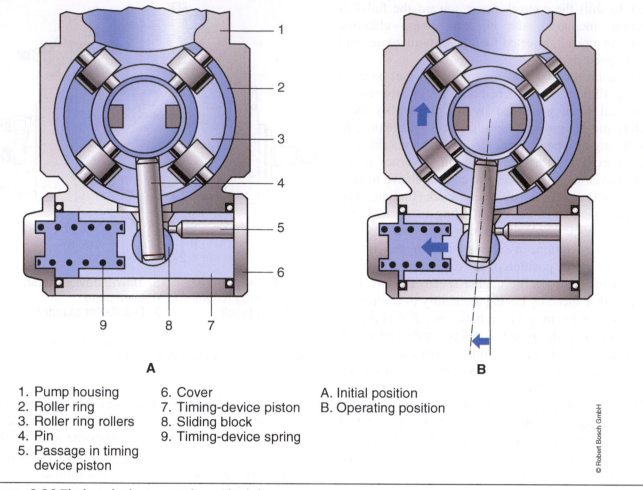

A

B

1. Pump housing	6. Cover
2. Roller ring	7. Timing-device piston
3. Roller ring rollers	8. Sliding block
4. Pin	9. Timing-device spring
5. Passage in timing device piston	

A. Initial position
B. Operating position

© Robert Bosch GmbH

Figure 9-26 Timing-device operating principle.

Add-On Modules and Shutdown Devices

Bosch sleeve-metering, distributor injection pumps are available with a variety of add-on modules and shutdown devices. Because of the modular construction of the pump, these supplementary devices can be added to optimize engine torque profile, power output, fuel economy, and exhaust gas composition. A brief description of add-on modules and how they affect engine operation follows. **Figure 9-27** is a schematic that shows how these add-on modules interact with the basic distributor pump.

Torque Control. Torque control relates to how fuel delivery is managed with respect to engine speed and the engine load requirement characteristic. Generally, engine fuel requirement increases when a request for higher engine speed is made, that is, the driver pushes her foot harder on the accelerator pedal. Fueling should then level off as actual engine speed approximates desired engine speed.

Note that the same position of the fuel control sleeve will result in slightly more fuel delivery at higher engine speeds than at lower rpms due to the throttling effect that occurs at the distributor plunger cutoff port. This means that if the fuel trim settings were set to produce maximum torque at low engine speeds, the engine would be overfueling at high engine speeds, resulting in smoking and possible engine overheating.

Conversely, if the fuel trim settings were set to produce optimum performance at rated speed, the engine would not be able to develop sufficient power at lower-than-rated speeds. Getting fueling quantities optimized throughout the engine operating range is known as torque control.

Positive Torque Control. Positive torque control would be required on a pump that delivered excess fuel at higher speeds. Positive torque control limits engine fueling in the upper portion of torque rise. This can be achieved by using a lower-tension spring in the delivery valve, altering the cutoff port geometry to

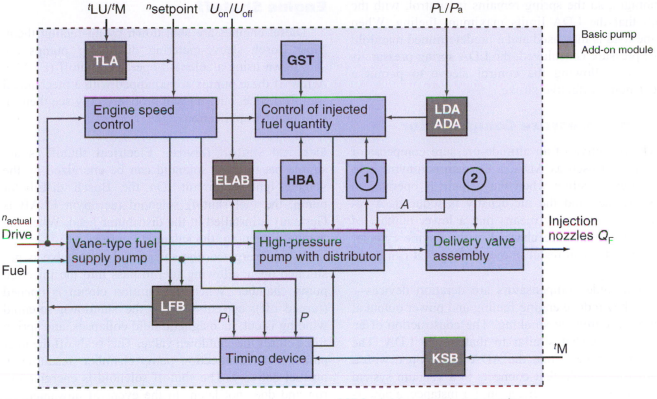

Figure 9-27 VE add-on modules.

LDA Manifold pressure compensator.
  Controls the delivery quantity as a function of the change-air pressure.

HBA Hydraulically controlled torque control.
  Controls the delivery quantity as a function of the engine speed (not for pressure-charged engines with LDA).

LFB Load-dependent start of delivery.
  Adaptation of pump delivery to load. For reduction of noise and exhaust gas emissions.

ADA Altitude pressure compensator.
  Controls the delivery quantity as a function of atmospheric pressure.

KSB Cold-start accelerator.
  Improves cold-start behavior by changing the start of delivery.

GST Graded (or variable) start quantity.
  Prevents excessive start quantity during warm start.

TLA Temperature-controlled idle speed increase.
  Improves engine warmup and smooth running when the engine is cold.

ELAB Electrical shutoff device.

A Cutoff port, n_{actual} Actual engine speed (controlled variable), $n_{setpoint}$ Desired engine speed (reference variable). Q_F Delivery quarterly, t_M Engine temperature, t_{LU} Ambient air temperature, P_L Change-air pressure, p_i Pump interior pressure.

①Full-load torque control with governor lever assembly.
②Hydraulic full-load torque control.

© Robert Bosch GmbH

produce a throttling effect, or using additional torque control springs in the governor.

Negative Torque Control. Negative torque control would be required in an engine that overfueled in the lower speed ranges but produced satisfactory performance at speeds closer to the rated speed of the engine. Negative torque control limits fueling in the lower portion of torque rise speed. It achieves this by governor spring pack trim or hydraulically, using pump housing fuel pressure.

Aneroid

The manifold pressure compensator (known by Bosch as an LDA, a German acronym) is a simple aneroid device. As such, it reacts to the manifold boost pressure generated by the turbocharger and essentially limits fueling until there is sufficient air in the engine cylinder to properly combust it.

Charge air is ported directly to the LDA assembly. The LDA is divided into two separate airtight chambers divided by a diaphragm. Manifold boost is applied to one side of the diaphragm while spring force is applied to the other. The diaphragm is connected to the LDA sliding pin, which has a taper in the form of a control cone. This is contacted by a guide pin through which levers act on the full-fuel stop setting for the control sleeve.

At lower engine loads, there is insufficient manifold boost developed to effect movement on the diaphragm, so the spring remains in control, with the result that the LDA limits maximum fueling. When engine load is increased and a predetermined manifold boost pressure is achieved, the LDA spring pressure is overcome, allowing the control sleeve to permit a longer pump effective stroke.

Altitude-Pressure Compensator

The objective of an altitude-pressure compensator (known by Bosch as ADA, a German acronym) is to limit engine fueling when the vehicle is operated at high altitudes and the air density is reduced. A reduction in air density means that a lower number of oxygen molecules are charged to the engine cylinder per cycle; this can result in overfueling that can cause smoking.

All altitude compensators are deration devices—that is, they reduce engine fueling and power output at altitude to eliminate smoking. The construction of an ADA is somewhat similar to that of the LDA. The major difference is that the ADA is equipped with a barometric capsule that connects to a vacuum system somewhere on the vehicle, as in, for instance, a power-assisted brake system circuit.

ADA Operation. Atmospheric pressure is applied to the upper side of the ADA diaphragm. A reference pressure from the barometric capsule is applied to the lower chamber in the ADA. The two chambers are separated by a diaphragm. Should a drop in atmospheric pressure occur such as would be experienced by driving a vehicle up a mountain, the barometric capsule (constant pressure) would be at higher pressure than that on the opposing side of the diaphragm. This pressure would cause the diaphragm to move the sliding bolt vertically away from the lower fuel stop and reduce engine fueling.

Engine Shutoff

Diesel engines are shut down by no-fueling them. Most Bosch sleeve-metering distributor pumps are shut down using a solenoid-operated shutoff (ELAB). A few of these pumps are equipped with a mechanical shutoff device, but you will probably only see them in off-highway applications.

Electrical Shutoff Device. Electrical shutoff is desirable because the solenoid can be energized by the vehicle ignition circuit. On the Bosch distributor pump, the fuel shutoff solenoid (acronym ELAB is German) is installed in the distributor head. When the engine is running, the solenoid is energized.

When energized, the solenoid shutoff valve is positioned open, allowing fuel to pass into the injection pump chamber. When the ignition circuit is opened (turned off), current flow to the shutdown solenoid winding is cut, its magnetic field collapses, and spring force closes the shutdown valve. This seals off the inlet passage to the injection pump chamber, resulting in no-fuel delivery. The shutoff solenoid is energized to run and does not latch. In the event of unwanted interruption of the electrical circuit, the engine will shut down.

Mechanical Shutoff Device. The mechanical shutoff device is located in the governor cover and has an outer and inner stop lever. The outer lever can be actuated by the driver from inside the vehicle. When the shutdown cable is actuated, both inner and outer levers swivel around a common pivot, causing the inner stop lever to push against the start lever of the governor-lever mechanism. This in turn, moves the control sleeve to the no-fuel position. Therefore, the distributor plunger's cutoff port remains open throughout the plunger stroke, and no fuel can be pressurized.

Summary

- High-pressure, liquid fuel injection to a diesel engine has been used since Robert Bosch designed the first system in 1927.

- Most inline, port-helix metering injection pumps are flange mounted to the engine block and gear driven at camshaft speed.

- The port-helix metering pump is driven through one full rotation (360 degrees) per full effective cycle of the engine (720 degrees in a four-stroke cycle).

- The pump camshaft is supported by main bearings and driven in the cambox, which also acts as a lubrication sump.

- Pump element actuating tappets are spring-loaded to ride the cam profiles, and cam geometry actuates the pump elements: one pump element dedicated to each engine cylinder.

- A pump element consists of a stationary barrel and a reciprocating plunger. The plunger is milled with a metering recess known as a helix or scroll.

- Plunger rotational position determines the point of register of the barrel spill port and the helix. The plungers are rotated in unison by a toothed rack meshed to slotted control sleeves, themselves lugged to the plungers.

- Plunger effective stroke begins at port closure and ends at port opening.

- Delivery valves separate the pump elements from each high-pressure pipe and act to retain dead volume fuel at pressure values approximating two-thirds NOP.

- Most later generation port-helix metering injection pumps have a variable timing mechanism that acts as an intermediary between the pump drive gear (on the engine) and the pump camshaft coupling.

- Hydromechanically managed injection pumps often incorporate an aneroid device and an altitude compensator to prevent more fuel from being injected into an engine cylinder than there is oxygen to burn it.

- Inline port-helix metering injection pumps must be accurately timed to the engine.

- There are two categories of rotary distributor pumps used today: the inlet-metering, opposed plunger-injection Roosa-type pump manufactured by Stanadyne or by Delphi Lucas CAV, and the sleeve-metering, rotary injection pump manufactured by Bosch.

- The main components of an *inlet-metering, distributor injection pump* are the vane pump, rotor and hydraulic head assembly, internal cam ring, opposed plungers, inlet metering valve, governor assembly, and timing advance.

- An opposed-plunger, inlet-metering injection pump has a single pump chamber actuated by a pair of opposed plungers: fuel quantity metered into the pump chamber determines the outward travel of the plungers, advancing fuel injection timing in respect of load.

- Fuel is moved through the fuel subsystem by a vane-type transfer pump, integral with the pump housing: this fuel is then routed through the fuel pump circuitry, directed to the metering valve and from there, to the pumping chamber. Pressure rise to injection pressures is developed in the pump chamber and unloaded to the rotor, from which it is distributed to the injectors.

- The main components of a Bosch *sleeve-metering, distributor injection pump* are a vane-type transfer pump, cam plate, single plunger-actuated pump chamber, distributor head, delivery valves, control sleeve, governor assembly, and advance mechanism.

- A sleeve-metering, distributor injection pump uses axial movement of a sleeve on the pumping plunger to alter the effective pumping stroke; the plunger both rotates and reciprocates to pressurize and distribute fuel to outlets in the distributor head.

- Movement of fuel through the fuel subsystem in a sleeve-metering, distributor injection pump is the responsibility of a vane-type transfer pump. Actual pump plunger stroke is defined by the actuating cam geometry, and the position of the metering or control sleeve determines the actual fuel pumped.

- The operating principles of the two types of distributor pumps presented in this chapter are distinct in the way in which they pump, meter, and distribute fuel, and it is important not to confuse the two systems during study.

Shop Exercises

1. Locate a diesel engine in your shop that uses a port-helix metering pump system and create a schematic on paper that maps the fuel system from the tank to the injectors.

2. Locate a diesel engine in your shop that uses a port-helix metering fuel system and check the port closure timing value. Use the manufacturer's service literature to verify the static timing value using one of the methods outlined in this chapter.

3. Locate a diesel engine in your shop that uses an inlet-metering, rotary distributor pump and create a schematic on paper that maps the fuel system from the tank to the injectors.

4. Locate a diesel engine in your shop that uses a sleeve-metering rotary distributor pump and create a schematic on paper that maps the fuel system from the tank to the injectors.

5. Fit a diesel engine timing light to any type of PLN fuel system equipped with an injection timing advance mechanism and run the engine from idle to high-idle rpms to verify the advance in crank angle degrees.

Internet Exercises

Use a search engine and research what comes up when you search the following:

1. Robert Bosch port-helix metering injection pumps.
2. Research the Web to learn what types of PLN pumps Dodge (Cummins B-series), Ford, and General Motors were using before they adopted CR fuel systems.
3. List three imported light duty diesels that used one type of rotary distributor pump.
4. Determine what classifications of engines can legally use PLN fuel systems post-2010 EPA emissions.
5. Use search engine navigation to identify two common tampering abuses that can result in increased power in mechanical PLN fuel systems.

Review Questions

Port-helix metering pumps

1. The beginning of the injection pump effective stroke in a port-helix metering pump is known as:
   A. Port closure
   B. Port opening
   C. Injection lag
   D. Afterburn

2. The mechanical device used to rotate the port-helix injection pump plungers in unison is known as a:
   A. Barrel
   B. Tappet
   C. Rack
   D. Control sleeve

3. Charging pressures in a port-helix metering injection pump are created by a(n):
   A. Supply pump
   B. Accumulator
   C. Vane pump
   D. Centrifugal pump

4. What component replaces the governor housing at the rear of a port-helix metering injection pump if it is managed by a computer?
   A. Electronic governor
   B. Rack actuator housing
   C. PCM
   D. Module

5. Static timing can also be referred to as:
   A. Port closure timing
   B. NOP timing
   C. Point of ignition
   D. Completion of injection

6. Which of the following components would limit engine fueling at high altitudes?
   A. Barometric capsule
   B. Aneroid
   C. Governor
   D. Variable timing

7. A port-helix metering injection pump managing a direct-injected diesel engine would typically be port closure timed to what position on the engine?
   A. 50 degrees BTDC
   B. 50 degrees ATDC
   C. 20 degrees BTDC
   D. 20 degrees ATDC

8. A common cause of diesel knock is:

    A. Prolonged injection lag   C. Prolonged ignition lag

    B. Afterburn   D. Preignition

9. What is the function of the retraction collar/piston on the delivery valve core?

    A. Guide the valve in the body   C. Limit pressure wave reflection

    B. Permit the valve to seal   D. Limit back leakage to the pump chamber
       before it seats

10. Which of the following does most to define the actual residual line pressure value in a running diesel engine
    using a port-helix metering PLN fuel system?

    A. The swept volume of the   C. The dead-volume fuel
       retraction collar/piston
                                        D. The length of the high-pressure pipe
    B. The mechanical force of the
       delivery valve spring

Rotary distributor pumps

11. If a four-stroke cycle diesel engine fueled by an inlet-metering, opposed-plunger fuel pump is rotated
    through a complete engine cycle (720 degrees), how many revolutions does the distributor pump turn?

    A. One-half revolution   C. Two revolutions

    B. One revolution   D. Four revolutions

12. What actuates the plungers of an inlet-metering, opposed-plunger fuel injection pump to create fuel injection
    pressures?

    A. Metered fuel quantity   C. External cam profile

    B. Charging fuel pressure   D. Internal cam ring

13. In an inlet-metering, opposed-plunger fuel pump, what causes the outward movement of the opposed
    plungers in the pump chamber during metering?

    A. Centrifugal force   C. Spring force

    B. Metered fuel quantity   D. Pump housing backpressure

14. How is an inlet-metering, opposed-plunger fuel pump lubricated?

    A. Dedicated supply of engine   C. Fuel from the vane pump
       lubricating oil
                                        D. Prelubed on assembly
    B. Pressurized engine lubrication
       oil

15. Technician A says that on an inlet-metering, opposed-plunger fuel pump, plunger actual stroke is constant
    regardless of fuel requirement. Technician B says that in the same pump, plunger effective stroke is
    determined by vane pump pressure. Who is correct?

    A. Technician A only   C. Both A and B

    B. Technician B only   D. Neither A nor B

16. How is the timing advance mechanism on a hydromechanical, inlet-metering, opposed-plunger fuel pump actuated?

    A. Hydraulically by the vane pump fuel pressure

    B. Hydraulically by engine oil pressure

    C. Pneumatically by manifold boost

    D. Manifold vacuum

17. When a Bosch sleeve-metering, rotary distributor pump is used on a six-cylinder, four-stroke cycle diesel engine, how many delivery valves are required?

    A. One

    B. Two

    C. Four

    D. Six

18. Technician A says that the pumping plunger used on a Bosch sleeve-metering, rotary distributor pump reciprocates. Technician B says that the pumping plunger used in this fuel system rotates. Who is correct?

    A. Technician A only

    B. Technician B only

    C. Both A and B

    D. Neither A nor B

19. What determines plunger actual stroke in a Bosch sleeve-metering, rotary distributor pump?

    A. Position of the sleeve-metering collar

    B. Cam geometry

    C. Rotational speed of the pump

    D. Metered fuel quantity

20. Technician A says that the cam geometry used on a Bosch sleeve-metering, rotary distributor pump is specific to the engine that it fuels. Technician B says, referring to the same pump, that the cam ring rotates at engine camshaft speed. Who is correct?

    A. Technician A only

    B. Technician B only

    C. Both A and B

    D. Neither A nor B

CHAPTER

10

Electronic Diesel Fuel Injection Systems

Prerequisites

Chapters 2 and **7.**
Note: This chapter should be studied in conjunction with **Chapter 12** to properly understand the systems discussed.

Learning Objectives

After studying this chapter, you should be able to:

- Describe the system layout and the primary components in the most common types of electronic fuel management systems.
- Identify the key features of electronic unit injector (EUI), hydraulically actuated electronic unit injector (HEUI), and common rail (CR) diesel fuel injection systems.
- Outline the roles played by subsystems in managing an EUI-fueled engine.
- Describe the operating principles of single-actuator EUIs.
- Outline some of the factors that govern the EUI fueling and engine output.
- Describe the HEUI system layout and the primary components.
- Outline the role played by the four HEUI subsystems.
- Describe the operating principles of a HEUI injection pressure regulator (IPR).
- Identify the differences between Caterpillar and Siemens HEUIs.
- Perform basic electronic troubleshooting on a HEUI system.
- Identify common rail (CR) diesel fuel systems.
- Identify some of the diesel engines that currently use common rail diesel fuel injection.
- Trace fuel flow routing from tank to injector on CR-fueled engines.
- Describe the operation of the inline and radial piston pumps used to achieve sufficient flow to produce rail and injection pressures in a typical CR system.
- Understand how rail pressures are managed in CR fuel systems.
- Outline the operation of an electrohydraulic injector (EHI).
- Describe how the PCM manages EUI duty cycle to control engine fueling.

Key Terms

accumulator

amplifier piston

calibration codes

calibration parameters

camshaft position sensor (CPS)

common rail (CR)

electrohydraulic injector (EHI)

electronic diesel controls (EDCs)

electronic unit injector (EUI)

engine family rating code (EFRC)

hard

hydraulically actuated, electronically controlled unit injector (HEUI)

injection pressure regulator (IPR)

injector driver module (IDM)

injector drivers

intensifier piston

multipulse injection

parameter

pilot injection

powertrain control module (PCM)

pre-injection metering (PRIME)

radial piston pump

rail

rail pressure control valve

rail pressure sensor

reference voltage (V-Ref)

single-actuator EUI

soft

swash plate pump

two-terminal EUIs

vehicle personality module (VPM)

INTRODUCTION

When diesel engines are managed by computer, there are two distinct levels of electronic controls:

- Partial authority systems: An existing hydro-mechanical system is adapted for computer controls while retaining all of the critical hardware components.
- Full authority systems: A fuel system that has been designed exclusively for computerized management.

When the distributor and inline hydromechanical fuel systems discussed in **Chapter 9** are adapted for electronic management, these are known as partial authority systems. They were much used during the 1990s in light duty diesels. Today however, partial authority systems are seldom seen on highway vehicles for the simple reason that they are unable to meet current emissions standards. All current on-highway light duty diesel engines are using full authority management systems. We use the term *full authority* because it describes a control system in which the engine management computer has full control over fueling. For this reason, the focus in this chapter will be on three types of full authority systems:

1. Electronic unit injector (EUI) fuel systems (used until 2010)
2. Hydraulically actuated, electronic unit injector (HEUI) systems (used until 2010)
3. Common rail (CR) fuel systems (almost universally used in today's LDDs)

While there are differences within each of these fuel systems, they can be studied at an elementary level with little referencing to individual manufacturers' systems.

Brief History of EUIs

Diesel **electronic unit injector (EUI)** systems have been used since the 1980s but primarily in heavy duty trucks. EUIs are still used in heavy duty diesels today, but they have evolved beyond the system described in this chapter. Until 2010, Volkswagon (VW) used Bosch EUIs in their popular turbocharged, direct-injected (TDI) engines. The EUIs described in this chapter are driven by a single actuator and should not be confused with the versions used in heavy duty diesel engines today. The EUIs described in this chapter use hydraulic injector nozzles that have hard opening and closing values that cannot be altered by computer controls.

Brief History of HEUIs

Hydraulically actuated, electronically controlled unit injectors (HEUIs) have been used in light duty diesel engines since the early 1990s. The first examples were used in Caterpillar-built engines found in some GM chassis. These were quickly followed by those used in Ford chassis, Power Stroke engines, built by Navistar International. Until 2004, all HEUI system components were manufactured by Caterpillar regardless of which manufacturer chassis they ended up in. However, from 2004, Navistar International engines (used in Ford and Navistar applications) began using Siemens-built HEUIs. In this chapter we will take a brief look at both Caterpillar and Siemens HEUIs.

Specifications	
Engine Code Letters	BEW
Design	4-Cylinder In-Line Engine
Displacement cu in (cm³)	116 (1896)
Output HP (kW)	100 (74) @ 4200 rpm
Torque lb ft (Nm)	177 (240) @1800 - 2400 rpm
Valves per Cylinder	2
Bore in (mm)	3.129 (79.5)
Stroke in (mm)	3.760 (95.5)
Compression Ratio	19.0 : 1
Firing Order	1-4-3-2
Engine Management	Bosch EDC 16
Fuel	Diesel
Exhaust Standard	Bin 10

© Volkswagen

Figure 10-1 Cutaway of a VW 1.9L, EUI-fueled diesel engine.

Brief History of CR Fueling

Electronically controlled, **common rail (CR)** diesel fuel injection systems were introduced on small-bore and automobile diesel engines in the late 1990s. A CR fuel system actually has a lot in common with the fuel injection system used on most current gasoline-fueled automobile engines, with the exceptions that:

- Fuel is direct-injected to the engine cylinder.
- Injection pressures are much higher, some exceeding 30,000 psi (2000 bar).
- Injection pressure is precisely controlled by the engine controller.

From 2010 onward, almost all light and heavy duty diesel engines meeting Environmental Protection Agency (EPA) standards use CR fueling. CR fuel management provides the powertrain control module (PCM) with much better control over injection cycles. More precise control over injection means better management of combustion in the engine cylinder. The result is an improved ability to meet:

- Diesel emissions standards
- Fuel economy expectations

In this chapter, we will take a close look at CR fuel systems following our study of EUI and HEUI fuel systems.

Hard and Soft Parameters

In addressing electronic management systems, the terms **hard** and **soft** will be commonly used. They are defined as follows:

- Hard: A set, unchangeable, value.
- Soft: A variable value usually controlled by computer.

The term **parameter** is also widely used; it simply means a *value* or *specification*.

EUI SYSTEM OVERVIEW

EUIs are mechanically actuated. The mechanical force required to produce injection pressures comes from a camshaft. The camshaft can either be cylinder block mounted or overhead. Mechanically actuated EUIs have an effective pumping stroke managed and switched by the **powertrain control module (PCM)**. Because the EUI plunger stroke is cam actuated, the PCM can only switch an effective pumping stroke while its plunger is being driven downward. This fact means that the PCM is limited to a hard limit window (represented by cam profile) within which it can select an effective stroke. EUI-fueled engines can be divided into the following subsystems for purposes of study:

- Fuel subsystem
- Electronic input circuit
- Management electronics (PCM)
- Output circuit

Figure 10-1 shows a cutaway view of a Volkswagen TDI engine with EUI fueling. Note the location of the camshaft in relation to the EUIs.

Fuel Subsystem

The fuel subsystem incorporates those components we studied in **Chapter 7.** In short, it is made up of those components that enable onboard fuel storage and the transfer of fuel from the fuel tank to the EUIs. In most systems, the fuel subsystem incorporates a return

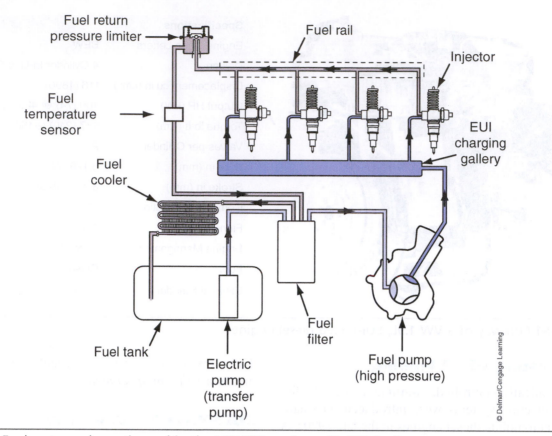

Figure 10-2 Fuel system schematic used in the VW TDI engine with EUI fueling.

circuit. This means that the fuel that is delivered to the EUIs, in addition to fueling the engine, is also used for:

- Cooling the EUI and surrounding area of the cylinder head.
- Lubricating the EUI internal components.

It is best introduced by looking at the fuel system schematic shown in **Figure 10-2.**

Fuel Subsystem Routing

The VW fuel subsystem shown uses two fuel pumps as follows:

- Low-pressure pump: Electrically driven vane pump located in fuel tank. The function of this pump is to charge the fuel filter and supply fuel to the charging pump.
- High-pressure pump: An engine-driven, fixed-clearance gear pump responsible for charging the injectors.

Auxiliary Fuel Subsystem Components. EUI fuel subsystems may be equipped with a range of auxiliary components that are designed to eliminate common maintenance problems. We will describe some of these here, but you should note that not all engines are equipped with them.

AUTOMATIC BLEEDING

An automatic bleeding valve purges any air in the fuel charge circuit on startup. During bleeding, any air in the charge circuit is routed to the return circuit.

SAFETY RELIEF VALVE

When used, a safety relief valve is usually located in the supply pump. It is designed to trip in the event of a blockage or partial restriction in the secondary circuit. When the pressure safety valve trips, it diverts fuel back into the primary circuit.

WATER DRAINAGE VALVE

A water drainage valve is located at the sump in the water separator. When water is detected by the water-in-fuel (WIF) sensor (WIF sensor operation is explained in **Chapter 12**), a dash warning light is illuminated. In most cases, an alert is posted on the digital dash display. When this occurs, the driver can activate a dash-located water drainage switch, providing that the following conditions are met:

- The ignition key is ON.
- The engine is not running.
- The parking brake is applied.

In older applications, it may be necessary to manually drain the water separator when a WIF alert is signaled.

In this case, the driver will obviously have to exit the cab to open the drain valve.

FUEL SUPPLY PRESSURE REGULATOR

Located downstream from the charge (high-pressure) pump. Maintains a consistent charging pressure to the EUIs.

FUEL RETURN PRESSURE LIMITING VALVE

Ensures the leak-off (return) circuit does not bleed dry.

FUEL TEMPERATURE SENSOR

The fuel temperature sensor is a thermistor (see **Chapter 12**) that signals fuel temperature in the leak-off circuit to the PCM. The negative temperature coefficient (NTC) thermistor is supplied with reference voltage (V-Ref) (5V DC) and returns a signal voltage that increases as thermistor resistance decreases with temperature rise.

FUEL COOLER

The fuel cooler is a plate-type heat exchanger located in the return circuit. At best, only half of the fuel routed to the EUIs is actually used to fuel the engine. Because the fuel plays a major role in lubing and cooling the EUIs, temperatures in the leak-off circuit can rise to 300°F (150°C). Fuel at these temperatures thins and loses lubricity and if not cooled, could cause system damage, especially when fuel tank levels are low.

INPUT CIRCUIT

Command and monitoring sensors and switches used by EUI-fueled engines do not vary much between the various manufacturers. An EUI-fueled engine requires one of two different types of throttle position sensors (TPSs):

- A standard potentiometer-type TPS supplied with V-Ref. This type of TPS uses a contact principle in which a wiper moves against a resistor. It returns an analog voltage signal representing a portion of the V-Ref that varies with accelerator pedal travel.
- A Hall effect TPS known as a noncontact TPS. These output a digital signal. The output signal returned to the PCM is pulse-width modulated (PWM) (see **Chapter 12**).

Management Electronics

The engine management controller used on EUI engines is usually known as a powertrain control module (PCM), but most Bosch systems use the term **electronic diesel controls (EDCs)** to describe their engine controller. All light duty highway vehicles make use of a control area network (CAN) powertrain data bus (explained in **Chapter 12**), so the EDC multiplexes with the chassis PCM via a proprietary bus line (also explained in **Chapter 12**).

PCM and EDC Responsibilities. The engine electronics are programmable with customer and proprietary data. The PCM is responsible for:

- outputting system reference voltages (V-Ref)
- receiving inputs from engine monitoring sensors
- receiving and broadcasting on the CAN data backbone
- logging proprietary data programming
- logging customer data programming
- performing all of the switching of actuators on the engine

The EDC incorporates the injector driver units, usually within the housing or in a separate module connected by a proprietary data bus. The driver units simply switch the EUIs on and off, usually by grounding out.

EUI Output Circuit

The output circuit of the PCM/EDC is used to convert the results of the controller processing cycle into action. In an EUI fuel system, the primary engine outputs are the EUIs and exhaust aftertreatment devices. Other outputs are V-Ref and any information broadcast to the data backbone, such as instrument cluster display data and powertrain management.

Electronic Unit Injectors

Electronic unit injectors (EUIs) have been used since the 1980s, and until 2007 they changed little in terms of their operating principles. Light duty diesel engines use **single-actuator EUIs**: they have a single solenoid actuator with two terminals. **Two-terminal EUIs** are mechanically actuated by an injector train in the same way cylinder valves are actuated. In an engine with an overhead camshaft, a rocker is used to actuate the EUI as shown in the VW TDI example shown in **Figure 10-3.** Effective stroke of the EUI is electronically controlled by the PCM. This is what is known as switching and is the responsibility of the injector drivers.

EUI Operation. The EUI is fitted to a cylindrical bore in the engine cylinder head. In terms of its functions, the EUI can be divided into three subsystems:

- Pumping circuit: mechanically actuated by cam profile

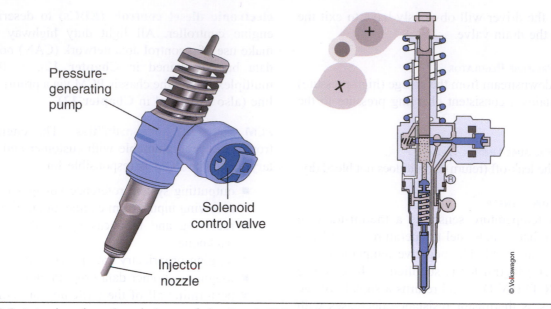

Pressure-
generating
pump

Solenoid
control valve

Injector
nozzle

© Volkswagon

Figure 10-3 Actual and sectional views of the Bosch EUI used on an earlier-generation VW TDI diesel engine.

■ Control circuit: EECU switched to manage effective stroke
■ Injector circuit: hydraulic multi-orifice nozzle

Charging pressure fuel from the fuel supply manifold (drilled passages in the cylinder head) enters the EUI through the fill port. This fuel circulates through the EUI and exits any time the engine is running. The EUI pump chamber is formed by the cam-actuated plunger and the stationary barrel. The EUI is actuated by cam profile. When the injector actuating cam is ramped off inner base circle (IBC) toward the cam nose, the EUI plunger is driven downward. This allows the plunger to act on the fuel in the EUI pump chamber. As the plunger is forced downward, it first closes off the fill passage and then starts to displace fuel in the EUI pump chamber. To begin with, this forces fuel out through the spill circuit. When the EUI solenoid is energized by the injector driver circuit in the PCM, the spill circuit is blocked, trapping fuel in the EUI. This begins an effective stroke.

An effective stroke can only occur when the EUI solenoid is energized. With the EUI solenoid energized and the plunger being driven downward, pressure rise takes place. Located below and connected by a passage to the EUI pump chamber is a hydraulic injector nozzle. When sufficient pressure rise has been created in the EUI pump chamber to open the nozzle, it opens to begin injection. Injection continues as long as the EUI solenoid is energized and the EUI plunger is being forced downward. The effective stroke ends when the PCM switches the EUI solenoid open. This opens up the spill circuit, initiating a rapid pressure collapse.

When there is insufficient pressure to hold the nozzle valve open, it closes, ending injection.

EUI Subcomponents. The following is a listing and brief description of the key components in a typical EUI.

1. Terminals: Connect to the injector drivers in the PCM.
2. Control cartridge: The actuator, a solenoid consisting of a coil and armature with an integral poppet control valve. A spring loads the armature open. Energizing the solenoid closes the armature/poppet control valve. These actions open and close the EUI spill circuit.
3. EUI tappet spring: Loads the EUI tappet upward. This enables the tappet/plunger actuation train to ride the cam profile and retract the tappet after a mechanical stroke.
4. Poppet control valve: A valve integral with the solenoid armature. When the poppet is closed, fuel is prevented from exiting the EUI through the spill port. This traps it in the EUI, enabling an effective stroke.
5. Plunger: The reciprocating member of the pump element. The plunger is lugged to the tappet, so it moves with it. The tappet is actuated by cam profile using the rocker as an intermediary as shown in **Figure 10-3.**
6. Barrel: The fixed member of the EUI pumping element. It is machined with the fill port, pump chamber, and a duct that connects the pump chamber with the injector nozzle.

7. Nozzle spring: Defines the nozzle opening pressure (NOP). The NOP is typically around 5,000 psi. The nozzle is loaded onto its seat by this spring, so its tension defines the *hard* value NOP.

8. Spacer or shims: These define nozzle spring tension and are used to set the NOP value. Thicker shims result in a higher NOP.

9. Upper nozzle assembly body: This is machined with the ducts that feed fuel to the pressure chamber of the nozzle valve.

10. Nozzle valve: The hydraulically actuated moving component of the nozzle. The nozzle assembly is a simple hydraulic switch that trips open and closed in accordance with the pressure it is subjected to. A full description of nozzle operation is provided in **Chapter 8.**

11. Tappet: The tappet is cam actuated: it is spring-loaded to a fully retracted position and moved through a mechanical stroke when actuated by cam profile.

12. Spill duct: The path through which fuel exits the EUI. Must be closed for an effective stroke to take place.

13. Spill control circuit: This is open when the control solenoid is de-energized, allowing fuel to spill from the EUI. When the PCM energizes the control solenoid, the spill circuit is blocked, trapping fuel in the EUI. Blocking off the spill circuit allows an effective stroke to take place.

14. Calibration port: Used only for factory bench setup. This should not be removed unless the EUI is connected to an EUI test bench.

15. Fuel duct: Connects the EUI pump chamber with the injector circuit.

16. Pressure chamber: The sectional area of the nozzle valve over which pressurized fuel acts to open the nozzle valve and begin injection.

EUI NOPs are typically around 5,000 psi. When the EUI actuating cam reaches peak lift, the injector train is unloaded. As the injector train ramps down the cam flank, the EUI spring lifts the plunger once again, exposing the fill port. Charging pressure fuel is then permitted to circulate throughout the EUI passages for purposes of cooling. In its lifted position, the plunger is ready for the next effective stroke.

EUI Summary

While EUI fuel systems were common in heavy trucks until recently, the only popular light duty application of this technology was in VW diesels. VW has recently abandoned EUI fueling for CR fuel systems. The reasons for doing this include the relative lack of control, increased complexity, and higher weight of EUIs when compared with equivalent CR fuel systems. An EUI fuel system represents a significant amount of extra weight tacked onto a 2L engine when compared to the weight of an equivalent CR fuel system.

HEUI

HEUI fuel systems were introduced in the early 1990s. **Hydraulically actuated, electronically controlled, unit injector (HEUI)** technology represented a new direction for the Caterpillar Engine Division because, for the first time, it launched into the business of supplying fuel system engineering and components to competitor engine manufacturers. While Caterpillar used HEUI in off-highway engines, it was also the fuel system used in the Cat-branded 3126 engine and the Navistar International 7.3L engine (444E), widely marketed as a Ford Power Stroke 7.3L engine.

Caterpillar's HEUI-fueled 3126 engines evolved into their C6 and C7 engines; the latter was sold with a HEUI fuel system up until 2007, when the engine was updated with a CR fuel system. The Navistar International version (i.e., Ford Power Stroke) of the 7.3L HEUI engine evolved to a smaller-displacement 6.0L engine boasting more horsepower and torque than its larger-displacement parent, and in 2004, the Cat HEUI injectors were replaced by slimmer, lighter units, manufactured by Siemens in Germany. Navistar International continued to supply engines to Ford until the 2010 model year, when Ford replaced these outsourced powerplants with a completely new Power Stroke 6.7, engineered and built in-house.

HEUI Principles of Operation

A disadvantage of a cam-actuated fuel injection system is that the fueling window available to the PCM is defined by the hard parameters of the injector train cam profile. Simply stated, if you have a cam-actuated pumping stroke, you can only inject fuel into the engine cylinder when the cam profile is moving the pumping plunger downward through its stroke. The HEUI system has no such limitation because it uses engine lube oil as hydraulic medium to actuate the fuel delivery pulse. The HEUI delivery stroke is actuated hydraulically, switched by the engine management PCM(s), and is not confined to any hard limits. HEUI technology showed that a camshaftless diesel engine was a possibility in the form of a number of prototypes, even though none were ever put into production.

HEUI Subsystems

For purposes of study, a HEUI management system can be loosely divided into four subsystems as follows:

1. Fuel supply system
2. Injection actuation circuit
3. HEUI assembly
4. Electronic management and switching

HEUI Overview

The HEUI system uses hydraulically actuated injectors. Oil from the engine lube circuit is ported to a high-pressure pump. The high-pressure pump boosts the engine oil pressure to much higher, but precisely controlled pressure values so that it can be used as the hydraulic medium to actuate injector plungers. Fuel is supplied to the HEUI injectors at a charging pressure, usually around 60 psi (4 bar). When the PCM switches the HEUI control cartridge, the high-pressure oil is admitted to, and acts on, the injector pump plunger, driving it through its pumping stroke. Fuel is thus pressurized in the HEUI pump chamber, and from there it is routed to a hydraulic injector nozzle to be atomized into the engine cylinder.

Fuel Supply System

The fuel supply system delivers fuel from the vehicle's tanks to the injector units. It consists of the following components:

- Fuel tank
- Fuel strainer
- Fuel filter
- Fuel transfer pump and hand primer
- Fuel supply gallery (feeds HEUIs)
- Charge pressure regulator

Fuel movement through the fuel supply system is the responsibility of a cam-driven plunger pump. The transfer pump pulls fuel from the chassis fuel tank(s) through a fuel strainer. It then charges fuel through a disposable cartridge-type fuel filter and feeds it to the fuel gallery of the fuel/oil supply manifold. A fuel pressure regulator at the fuel manifold outlet is responsible for maintaining a charging pressure of approximately 30 to 60 psi (2–4 bar). Fuel is cycled through the fuel supply system and the HEUIs are mounted in parallel from the fuel manifold. A hand priming pump is located on the filter pad in the event

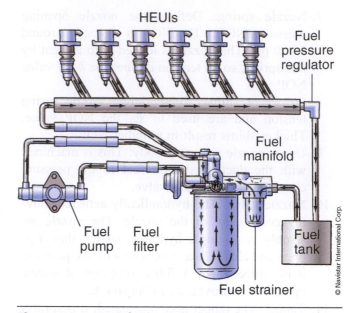

Figure 10-4 HEUI fuel supply system.

of loss of prime on most Caterpillar and International engines that use HEUI fueling. **Figure 10-4** shows a HEUI fuel subsystem.

Injection Actuation System

The HEUI system uses hydraulically actuated, electronically controlled injector assemblies to deliver fuel to the engine's cylinders. The hydraulic medium used to actuate the pumping action required of the injector is engine oil. The engine lubrication circuit provides a continuous supply of engine lube to the HEUI high-pressure pump. This is usually a gear-driven, swash plate hydraulic pump used to boost the lube oil pressure to values up to 4,000 psi (272 bar/ 27.5 MPa), depending on the system. The injection actuation circuit consists of the following components:

- Main engine lube circuit
- HEUI oil reservoir
- High-pressure (stepper) pump
- Injection pressure regulator (a PCM output)
- Oil manifold (feeds HEUI control cartridges)
- Actuation oil pressure sensor

Actuation Circuit Routing

Oil sourced from the main engine lube circuit is delivered to the high-pressure pump. Some manufacturer service literature refers to the high-pressure pump as a stepper pump. This pump receives oil from the engine lube circuit, then substantially boosts its pressure. Actual high-pressure oil values are managed by the injection pressure regulator (IPR). The IPR is controlled electronically (by the PCM) and actuated electrically (proportioning solenoid). It is a spool valve that manages the high-pressure oil value by either directing it to the actuation oil manifold (to increase pressure) or spilling it back to the engine oil sump (to reduce pressure).

High-Pressure Pumps. A **swash plate pump** is commonly used as the high-pressure pump. These use opposing cylinders and double-acting pistons actuated by a swash plate (circular plate, offset on a shaft to act as an actuating cam). Swash plate pump operating principles are similar to those of common automotive A/C compressors. The IPR manages the high-pressure oil pressure values at the pump outlet. Actuation pressures are managed between a low of 485 psi (33 bar) and a high of up to 4,000 psi (275 bar) by receiving all the oil pressurized by the high-pressure pump and spilling the excess to the oil reservoir.

Oil Manifold. From the high-pressure pump, oil is then piped to the oil gallery ducting within the fuel/oil supply manifold. From there, the oil is delivered to an exterior annulus in the upper portion of each HEUI. The HEUIs are mounted in parallel, fed by the high-pressure oil manifold. When the HEUI solenoid is energized, a poppet valve is opened by an electric solenoid within the HEUI; this permits the high-pressure lube to flow into a chamber and actuate a pumping stroke. At the completion of each duty cycle (injection), the HEUI is de-energized and it spills the actuation oil to the rocker housing, from which it drains back to the main engine oil sump. **Figure 10-5** shows how oil is routed through the HEUI actuation circuit.

Injection Pressure Regulator (IPR). The **injection pressure regulator (IPR)** is a spool valve, positionally controlled by a pulse-width modulated signal to achieve a range of injection control pressures (ICP). It operates at 400 Hertz. The PCM pressure regulating signal determines the magnetic field strength in the coil of the proportioning solenoid and therefore, the armature position. Integral with the armature is a poppet valve designed to control the extent to which high-pressure pump oil is optioned either to the oil manifold or to the drain port. It is a PCM output. **Figure 10-6** shows a Navistar International IPR in engine-off and engine-on positions.

HEUI Injector

The HEUI injector is an integral pumping, metering, and atomizing unit controlled by PCM drivers.

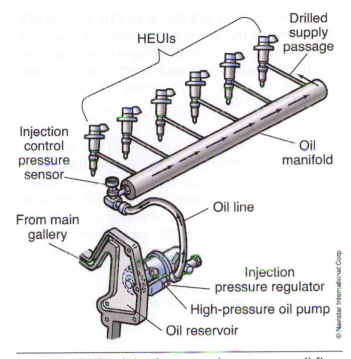

Figure 10-5 HEUI injection actuation pressure oil flow schematic.

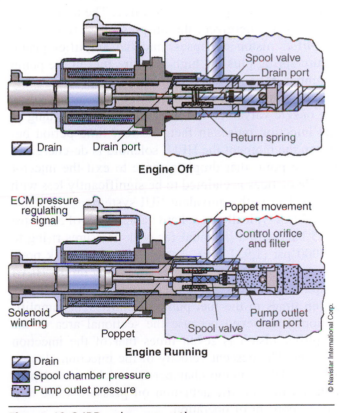

Figure 10-6 IPR valve operation.

The unit is best understood if you think of it as being an EUI that is actuated hydraulically rather than by mechanical cam profile. At the base of the HEUI is a hydraulically actuated, multi-orifice nozzle, almost identical to what you will find in an EUI. Control of a HEUI is by the PCM-switched control cartridge. This spends most of the cycle de-energized and therefore closed. The instant the control cartridge is energized, the HEUI actuation circuit opens, admitting high-pressure oil into the unit.

Amplifier or Intensifier Piston. The **amplifier** (International/Ford) or **intensifier** (Caterpillar) **piston** is responsible for creating injection pressure values. Because each manufacturer prefers a different term, for consistency we will use *amplifier piston* in this text. When the HEUI is energized, high-pressure actuation oil supplied by a stepper pump is admitted to the HEUI circuit, where it acts on the amplifier piston. This drives its integral plunger downward into the fuel in the pumping chamber, creating the fuel pumping stroke.

HEUI Effective Stroke. It will help if you connect the description that follows with the images in **Figure 10-7**. A duct connects the pump chamber with the pressure chamber of the injector nozzle valve. The moment the HEUI is de-energized, the oil pressure acting on the amplifier piston collapses, and the amplifier piston return spring plus the high-pressure fuel in the pump chamber retracts the amplifier piston, causing the almost immediate collapse of the pressure holding the nozzle valve open. This results in rapid ending of the injection pulse. In fact, the real time period between the moment the HEUI solenoid is de-energized and the point that droplets cease to exit the injector nozzle orifices is claimed to be significantly less with HEUI than with equivalent EUI systems.

HEUIs typically have hard value NOPs of 5,000 psi (345 bar) with a potential for peak pressures rising to 28,000 psi (1,931 bar). Actual NOPs and peak pressures vary by engine. Oil pressure acting on the HEUI amplifier piston is "amplified" (or "intensified") by seven times in the fuel pump chamber. This amplification is achieved because the sectional area of the amplifier piston is seven times that of the injection plunger. The descent velocity of the injection plunger into the HEUI pump chamber is variable and dependent on the specific actuation oil pressure value at a given moment of operation.

Because the PCM directly controls the actuation oil pressure value, it can therefore control injection pressure. The injection pressure determines the emitted droplet size. The higher the injection pressure, the smaller the droplets emitted from the HEUI. Control over droplet size provides what is known as the

rate-shaping ability of the HEUI. In short, rate-shaping provides the PCM with the ability to determine the extent of atomization to best suit the combustion conditions at any given moment of operation.

Collapse Phase. When the PCM de-energizes the HEUI control solenoid, the oil actuation pressure collapses as oil vents the HEUI to the rocker housing. Compared to an EUI with an older hydraulic injector nozzle, the fuel injection collapse phase in a HEUI is accelerated. This rapid pressure collapse reduces the injection of larger-sized droplets toward the end of an injection pulse.

Mechanical Pilot Injection. In the late 1990s, some versions of Caterpillar-built HEUIs were manufactured with plunger and barrel geometry that provided **pilot injection**. *Pilot injection* is a term used to describe an injection pulse that is broken into two separate phases. In a pilot injection fueling pulse, the initial phase injects a short-duration pulse of fuel into the engine cylinder, ceases until close to the moment of ignition, then resumes injection, pumping the remainder of the fuel pulse into the engine cylinder. Pilot injection is used as a cold-start and warmup strategy in most current fuel systems to avoid an excess of fuel in the engine cylinder at the point of ignition. This is important in diesel engines to minimize cold-start detonation. HEUI systems with the pilot injection feature are designed to produce a pilot pulse for each injection. The feature is mechanical and is achieved by machined center and cross drillings in the plunger to an annular recess. Caterpillar uses the term **PRIME** to describe its mechanical pilot injection feature. PRIME is a loose acronym for *pre-injection metering*. **Figure 10-7** shows a HEUI with the PRIME feature as it passes through the five phases of injection.

HEUI Oil and Fuel Manifold. Actuation oil and fuel at charging pressure are routed to the HEUI units by means of an oil/fuel manifold on the cylinder head(s). The HEUIs are inserted into cylindrical bores in the cylinder head. Each HEUI has a dedicated external annulus separated by O-rings to access the oil and fuel rifles.

HEUI Subcomponents

As mentioned previously, there are some differences between the Caterpillar and Siemens-built HEUIs. We will focus first on the Caterpillar HEUIs that were used in both Caterpillar and International

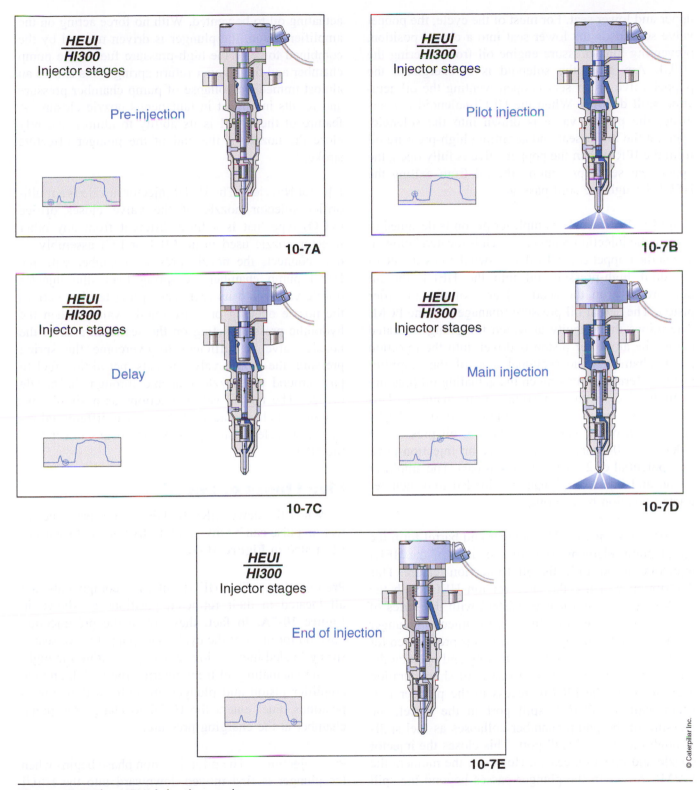

Figure 10-7 The HEUI injection cycle.

Navistar HEUI-fueled engines until 2004. We will look at the Siemens HEUIs in the next section.

Solenoid. The solenoid is switched by the PCM using a 100V-plus, coil-induced voltage. The HEUI electrical terminals connect the solenoid coil with

wiring to the PCM injector drivers. The injector drivers ground out the solenoid circuit to switch an effective stroke.

Poppet Valve. The HEUI poppet valve is integral with the solenoid armature. It is machined with an

© Caterpillar Inc.

upper and lower seat. For most of the cycle, the poppet valve seat loads the lower seat into a closed position, preventing high-pressure engine oil from entering the HEUI. Any time the solenoid is de-energized, the poppet valve upper seat is open, venting the oil actuation spill ducting. When the HEUI solenoid is energized, the poppet valve is drawn into the solenoid, opening the lower seat and admitting high-pressure oil from the IPR. When the poppet valve is fully open, the upper seat seals, preventing the oil from exiting the HEUI through the spill passage.

Amplifier Piston. The amplifier piston is designed to actuate the injection plunger, which is located below it. When the poppet control valve is switched by the PCM to admit high-pressure oil into the HEUI, the oil pressure acts on the sectional area of the amplifier piston. The actual oil pressure (managed by the PCM) determines the velocity at which the plunger located below the amplifier piston is driven into the injection pump chamber. The sectional area of the amplifier piston determines how much the actuating oil pressure is multiplied in the injection pump chamber. This value is specified as seven times in current HEUI systems. In other words, an actuating oil pressure of 3,000 psi (204 bar) would produce an injection pressure potential of 21,000 psi. (1,428 bar). The amplifier piston and injection plunger are loaded into their retracted position by a spring.

Plunger and Barrel. The plunger and barrel form the HEUI pump element. The first versions of the HEUI injectors did not offer the pilot injection feature. This description will use the later version HEUI with the PRIME feature, and it is consistent with the series of images shown in **Figure 10-7.** As the injection plunger is driven into the pump chamber, fuel is pressurized for a short portion of the stroke, actuating and opening the injector nozzle. The pressure rise is of short duration because when the PRIME recess in the plunger registers with the PRIME spill port in the barrel, the pressure in the pump chamber collapses as fuel spills through the PRIME spill port. This closes the injector nozzle and injection ceases. However, the moment the PRIME recess in the plunger passes beyond the spill port, fuel is once again trapped in the pump chamber and pressure rise resumes. This results in the injector nozzle being opened once again for the delivery of the main portion of the fuel pulse.

The fueling pulse continues until the PCM ends the effective stroke by de-energizing the HEUI solenoid. At this point, the poppet control valve is driven onto its lower seat, opening the upper seat and permitting the actuating oil to be vented. With no force acting on the amplifier piston, the plunger is driven upward by the combined force of the high-pressure fuel in the pump chamber and the plunger return spring. This causes an almost immediate collapse of pump chamber pressure and results in almost instantaneous nozzle closure. A feature of the HEUI is its ability to almost instantly close the nozzle at the end of the plunger effective stroke.

Injector Nozzle. The HEUI injector nozzle is a multi-orifice injector nozzle of the valve closes orifice (VCO) type that is a little different from any other injector nozzle used in an MUI or EUI assembly. A duct connects the nozzle pressure chamber with the HEUI pump chamber. A spring loads the injector nozzle valve onto its seat. The spring tension defines the nozzle opening pressure (NOP) value. When the hydraulic pressure acting on the sectional area of the nozzle valve is sufficient to overcome the spring pressure, the nozzle valve unseats, permitting fuel to pass around the nozzle seat and through the nozzle orifice. The nozzle valve functions as a simple hydraulic switch. Because of the nozzle differential ratio, the nozzle closure pressure is always lower than the NOP.

Five Stages of Injection

When the newer PRIME HEUIs are used, the injection pulse can be divided into five distinct stages as illustrated in **Figure 10-7.**

Pre-Injection. The HEUI internal components are all located in their retracted positions as shown in **Figure 10-7A.** In fact, they are in the pre-injection position for most of the cycle. The poppet valve seat is spring-loaded into the lower seat, preventing the high-pressure actuating oil from entering the HEUI, and the amplifier piston and plunger are both in their raised positions. Fuel enters the HEUI to charge the pump chamber at the charging pressure.

Pilot Injection. The pilot injection phase begins when the plunger is first moved downward into the HEUI pump chamber by actuation oil as it enters the HEUI circuit. The pressure rise thus created opens the injector nozzle to deliver a short pulse of fuel. The pilot injection phase ends when the PRIME recess in the HEUI plunger is driven downward enough to register with the PRIME spill port, causing the pump chamber pressure to collapse and the nozzle valve to close. **Figure 10-7B** shows the pilot injection stage.

Delay. The delay phase occurs between the ending of the pilot injection phase and restart of the fuel pulse. As the PRIME recess registers with the PRIME port in the pump barrel, pressure collapses. This pressure collapse results in nozzle closure. The objective is to cease fueling the engine cylinder while the prime pulse of fuel is vaporized and heated to its ignition point. It is important to note that the plunger is still being driven through its stroke during this phase because the HEUI poppet control valve is in the open position, and oil pressure continues to drive the amplifier piston downward. **Figure 10-7C** shows the delay stage of HEUI injection.

Main Injection. When the PRIME recess in the plunger passes beyond the PRIME spill port, fuel is once again trapped in the HEUI pump chamber because it can no longer exit through the spill port. The resulting pressure rise opens the injector nozzle a second time to deliver the main volume of fuel to be delivered in the cycle. In a HEUI injector with no PRIME feature, the plunger has no cross and center drillings and PRIME recess, so main injection begins when the plunger leading edge passes the spill port on its downward stroke, as shown in **Figure 10-7D**.

End of Injection. The end of injection begins with the de-energizing of the HEUI solenoid. The armature is released by the solenoid coil and a spring drives the poppet valve downward to seat on its lower seat. The instant the poppet valve starts to move downward, the upper seat is exposed, permitting the actuating oil inside the HEUI to spill. When the actuating oil pressure acting on the amplifier piston is relieved, the fuel pressure in the HEUI pump chamber combined with the plunger return spring collapses the fuel pressure almost instantly. Injection ends when there is insufficient pressure to hold the nozzle valve in its open position, and the three moving assemblies (poppet valve, amplifier/plunger, and nozzle valve) in the HEUI are all in their return positions outlined in the pre-injection phase. **Figure 10-7E** shows a PRIME HEUI injector during the end of the injection stage.

Tech Tip: The importance of the engine oil type and condition in a HEUI-fueled engine cannot be overemphasized. Because engine oil is the HEUI actuation medium, it should meet the manufacturer's temperature and operating condition specifications. Contaminated, high mileage, and degraded engine oil can cause HEUI injector malfunctions.

Siemens HEUI

As we said earlier, Navistar International began using the Siemens HEUI system for the 2004 model year. This included all Navistar-branded engines, including those they supplied to Ford as Power Stroke 6.0L engines. The operating principles of Caterpillar and Siemens HEUIs are identical, but the packaging of the Siemens model is a little different. The Siemens injectors use dual integral actuators built into the HEUI cylinder as shown in **Figure 10-8**. This provides for a sleeker design and a reduced actuation potential of 48 volts. The actuator coils are located on either side of a single armature with integral spool valve.

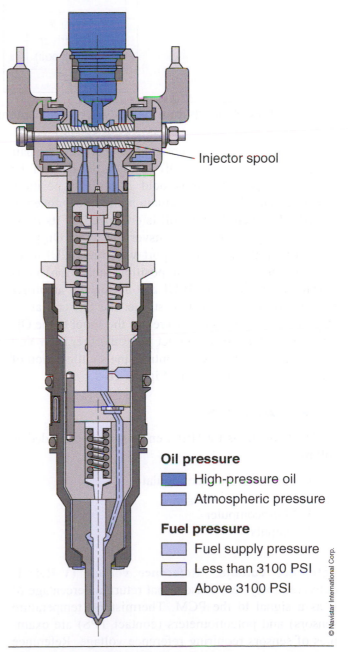

Injector spool

Oil pressure
- High-pressure oil
- Atmospheric pressure

Fuel pressure
- Fuel supply pressure
- Less than 3100 PSI
- Above 3100 PSI

© Navistar International Corp.

Figure 10-8 Siemens HEUI.

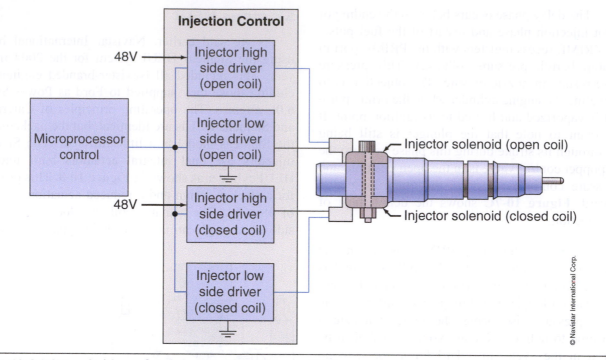

Figure 10-9 ECM switching of spool valve in a Siemens HEUI.

The spool valve shuttles horizontally between the pair of coils to either an On or Off position. For most of the cycle, the spool valve is held in its Off position, meaning that oil from the actuation circuit cannot enter the HEUI. When the On-coil is energized by its PCM driver, the spool shuttles transversely to its On position, which aligns an inlet port with the HEUI actuation oil inlet. In the On position, actuation oil is permitted to enter the HEUI and act on the amplifier piston. To end an effective stroke, the Off-coil at its opposite side is energized to return the spool to the Off position, taking the spool inlet port out of register. You can see the how the PCM controls the shuttle action of the Siemens spool valve in **Figure 10-9.**

PCM Functions

PCM functions for HEUI engines can be divided as follows:

1. Reference voltage regulator
2. Input conditioning
3. Microcomputer
4. Outputs

Reference Voltage. Reference voltage (V-Ref) is delivered to system sensors that return a percentage of it as a signal to the PCM. Thermistors (temperature sensors) and potentiometers (contact TPS) are examples of sensors requiring reference voltage. Reference voltage values used are at 5V DC pressure, and the

flow is limited by a current-limiting resistor to safeguard against a dead short to ground. Reference voltage is also used to power up the circuitry in Hall effect sensors used in the system, such as the **camshaft position sensor (CPS)**. There is a full explanation of the electrical principles of input circuit components in **Chapter 12.**

Input Conditioning. Signal conditioning consists of converting analog signals to digital signals, squaring-up sine wave signals, and amplifying low-intensity signals for processing. Part of this process includes cleaning up electronic noise induced in the circuit wiring.

Microcomputer. HEUI PCMs function similarly to other vehicle system management computers. They store operating instructions, control strategies, and tables of values, known as **calibration parameters.** They compare sensor monitoring and command inputs with the logged control strategies and calibration parameters and then compute the appropriate operating strategy for any given set of conditions. In multiple module units, a proprietary data bus connects the PCM with the personality and injector driver modules. For a full accounting of PCM processing, refer to **Chapter 12.**

Personality Module. Both Caterpillar and International Trucks describe their combined PROM and

EEPROM (see Chapter 12) as a **vehicle personality module (VPM)**. The VPM is customer and proprietary data programmable. The function of the VPM is to trim engine management to a specific chassis application and customer requirements. The **engine family rating code (EFRC)** is located in the VPM calibration list and can be read with an electronic service tool (EST); this identifies the engine power and emission calibration of the engine.

HEUI Injector Driver Module. The **injector driver module (IDM)** may be integral with the PCM or it may be a separate module. The IDM switches the HEUIs. It supplies:

- a constant 115-plus V DC supply, or
- a constant 48-plus V DC supply

to each HEUI. Which voltage value is used depends on whether a Caterpillar or Siemens HEUI injector is used in the system; we will refer to this as actuation voltage. The actuation voltage is created in the IDM by making and breaking a 12V DC source potential across a coil using the same principles employed by the ignition coil in a spark-ignited (SI) engine. The resultant voltage induced by the coil is stored in capacitors until discharged to the HEUIs.

The IDM controls the effective stroke of the HEUI by closing the circuit to ground. The direct control of the HEUI is managed by an output driver transistor in the IDM. When the fuel demand signal is delivered by the PCM processing cycle, the beginning of injection (timing) and fuel quantity is determined.

Injection Actuation Pressure

The PCM is responsible for maintaining the correct injection actuation pressure (IAP) during operation. This means monitoring and continually adjusting the high-pressure oil circuit responsible for actuating HEUI fueling. It does this by comparing *actual* injection pressure with *desired* injection pressure and using the injection actuation pressure solenoid valve to attempt to keep the two values close to each other.

- Actual IAP: Measured by IAP sensor and signaled to PCM.
- Desired IAP: Calculated by PCM.

Actual injection pressure is measured by a variable-capacitance pressure sensor and signaled to the PCM. Desired injection actuation pressure is based on the fueling algorithm (computed by PCM) effective at any given moment of operation. In the processing loop, the PCM will evaluate any differential between actual and desired actuation pressures and will modulate an

output signal to the injection actuation control solenoid valve to keep the values close.

HEUI Diagnostics

Diagnostic procedures for Caterpillar, International, and Ford versions of HEUI are distinct, as each has developed its own management software. We will refer to the troubleshooting software as diagnostic software and use the generic term service information system (SIS) to refer to online service information.

Diagnostic Software

You need the appropriate manufacturer software platform to read, program, and troubleshoot HEUI systems. The technician who attempts to troubleshoot any electronically managed engine without the diagnostic software will inevitably run into costly roadblocks. The diagnostic software will typically run the following test profiles:

- Test injector solenoid
- Test injection actuation pressure (IAP)
- Perform cylinder cutout tests
- Identify active faults
- Identify logged faults
- Identify logged events
- Display engine configuration data
- Rewrite customer programmable parameters
- Flash new software
- Print configuration and test results

Two typical tests distinct to HEUI systems are described here.

Injector Cutout Test. The diagnostic software can perform injector cutout tests on HEUI-managed engines. Cylinder cutout testing on HEUI-fueled engines is not unlike that used on EUI systems. Depending on the engine and model year, the software may choose a single-cylinder cutout or a multiple-cylinder cutout. Multiple-cylinder cutout tests produce a more comprehensive cylinder balance analysis.

The test sequence should be performed with the engine at operating temperature and with intermittent parasitic loads such as A/C disconnected. The test sequence for single-cylinder cutout on an eight-cylinder engine is described here:

1. Governor maintains programmed idle speed.
2. The diagnostic software turns off one HEUI injector at a time: this means the functioning seven HEUIs must increase their duty cycle if the specified engine rpm is to be maintained.

3. The diagnostic software measures the average duty cycle of the seven functioning HEUIs at each stage of the cutout test.
4. A test value is assigned for each HEUI tested.
5. The cutout test cycle is then repeated.

Tech Tip: When you make a warranty claim for a defective HEUI, a printout of the injector cutout test profile may be required for the warranty claim to be processed.

Injection Actuation Pressure Test. The injection actuation pressure (IAP) test checks the high-pressure oil pump performance and IAP valve operation. The test is performed at low idle. It functions by having the diagnostic software read desired IAP pressure versus actual IAP pressure using PCM data. One version of the IAP test sequence uses four desired pressure values:

1. 870 psi (60 bar)
2. 1,450 psi (100 bar)
3. 2,100 psi (145 bar)
4. 3,300 psi (228 bar)

and compares these with actual IAP values read by the IAP pressure sensor.

COMMON RAIL FUEL SYSTEMS

We are going to define a diesel common rail (CR) fuel system as one in which fuel injection pressure values are held in the **rail** that directly feeds the injectors. Fuel in the rail is pressurized to injection pressure values by a PCM-controlled pump. The pump is located upstream from the rail. The injectors used in a CR fuel system are electrohydraulic injectors (EHIs) such as those described in **Chapter 8.** At this time, there are two generations of CR diesel fuel systems:

- Simple CR systems in which the EHI does nothing more than open or close
- Amplified CR systems in which the EHI increases the rail pressure using an intensifier

Simple CR systems have been used in commercial diesel engines since the end of the 1990s, and continue to be far and away the type most used in current light duty diesel engines. For this reason, the coverage of CR fuel systems in this chapter will be confined to the simple common rail system used in current Ford, GM, Dodge (Cummins), VW, and Audi diesel engines. **Figure 10-10** shows a current Bosch CR system,

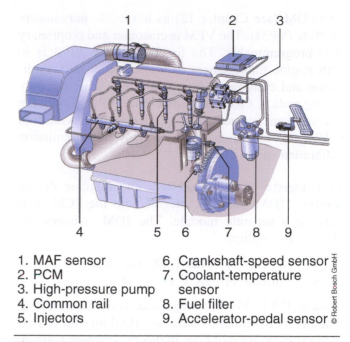

1. MAF sensor
2. PCM
3. High-pressure pump
4. Common rail
5. Injectors
6. Crankshaft-speed sensor
7. Coolant-temperature sensor
8. Fuel filter
9. Accelerator-pedal sensor

© Robert Bosch GmbH

Figure 10-10 Common rail injection system on a four-cylinder diesel engine.

widely used in automotive light duty diesels, to which you should refer during this introductory explanation.

Function of the Rail

In this text, the term *rail* will be used to describe the supply manifold or gallery that directly feeds all of the diesel fuel injectors. The basic principle of a diesel common rail fuel system can be compared to the way in which automotive gasoline fuel injection (GFI) systems operate. The major differences between GFI and diesel CR are that the diesel systems:

- operate at much higher pressures, some exceeding 30,000 psi (2068 bar).
- precisely manage rail pressure within a wide range of values.

In terms of managing and switching the injectors, there are many similarities between diesel and gasoline common rail injection. The similarities increase when the comparison is made between diesel CR and current direct injection (DI) gasoline fuel injection systems.

CR System Manufacturers

CR fuel system technology was launched by Bosch and Delphi Lucas, but these companies have been joined by others. In most cases, diesel engine manufacturers have opted to use one of the CR systems manufactured by a specialty diesel fuel system manufacturer rather than develop their own. The exception is Caterpillar, which manufactures its own CR system. Current systems are manufactured by:

- Bosch
- Caterpillar
- Delphi Lucas
- Denso
- Siemens

The Bosch CR system is used on most U.S.-built light duty diesels, and for this reason the Bosch system will be more commonly referenced.

Advantages of CR Diesel Fuel Systems

CR diesel fuel systems are simple. They use fewer components and consequently produce fewer failures. Not only is CR diesel fueling less complicated, but the system gives the engine control electronics better control of the power stroke. Better control over combustion results in:

- Lower emissions
- Improved fuel economy
- Lower engine noise levels
- Balanced engine cylinder pressures

CR Subsystems and Components

A typical diesel CR system consists of the following key components:

- Fuel subsystem: Stores and supplies fuel to the high-pressure rail pump.
- High-pressure pump: An engine-driven pump capable of sufficient flow to produce injection pressures up to, and exceeding, 30,000 psi (2,000 bar). Either a radial piston or multi-cylinder inline piston pump is used.
- PCM-controlled **rail pressure control valve**: A linear proportioning solenoid with an integral spool valve is typically used.
- A V-Ref–supplied **rail pressure sensor**: A variable capacitance electronic device that signals "actual" rail pressure to the PCM at any given moment of operation.
- Common rail: Stores fuel at injection pressures. Distribution lines connect the rail with EHIs.
- Electrohydraulic injectors (EHIs): These PCM-switched hydraulic valves inject fuel directly into the engine cylinders.

Figure 10-11 illustrates the Bosch CR fuel system on an eight-cylinder 2011 Ford Power Stroke engine, while **Figure 10-12** shows the system as it would appear on a typical four-cylinder engine.

CR Features

Some of the reasons CR diesel fueling systems have become common today are:

- They produce high injection pressures independent of engine speed.
- By controlling injection pressure, they control the size of the atomized droplets that exit the EHIs.
- EHIs can be switched on and off at high speeds. This allows up to seven injection "events" during a single power stroke.

CR Management Electronics

Common rail fuel systems feature full-authority electronic controls. These do not vary too much regardless of manufacturer. In all CR fuel systems, injected fuel quantity is defined by the PCM. Injected fuel quantity determines engine output. The actual output of a CR-managed engine depends on a range of input variables. These include:

- Accelerator pedal angle (fuel *request*)
- Operating temperatures
- Emission requirements

As in all full-authority engine management systems, the engine electronics can be divided into:

- Input circuit (sensors and switches)
- Processing circuit (the PCM)
- Output circuit (the actuators such as EHIs)

Light duty CR systems are networked to the powertrain data bus (see **Chapter 12**). This allows the PCM to network with other powertrain components and with other chassis data buses.

Input Circuit

The input circuit consists of sensors and switches. These signal data to the PCM. Sensors and switches can be divided into:

- Command devices. These would include the TPS, ignition control, and cruise switches.
- Monitoring devices. These would include the means used to signal pressure, temperature, engine speed, and low fluid levels to the PCM.

Both analog and digital signals are sent to the PCM.

Analog Inputs. Mass air flow, engine fluid and intake-air temperatures, engine pressure values, diesel particulate filter (DPF) pyrometer, and battery voltage are examples of analog signals. Analog signals have to be converted to digital values by an A/D converter in the PCM.

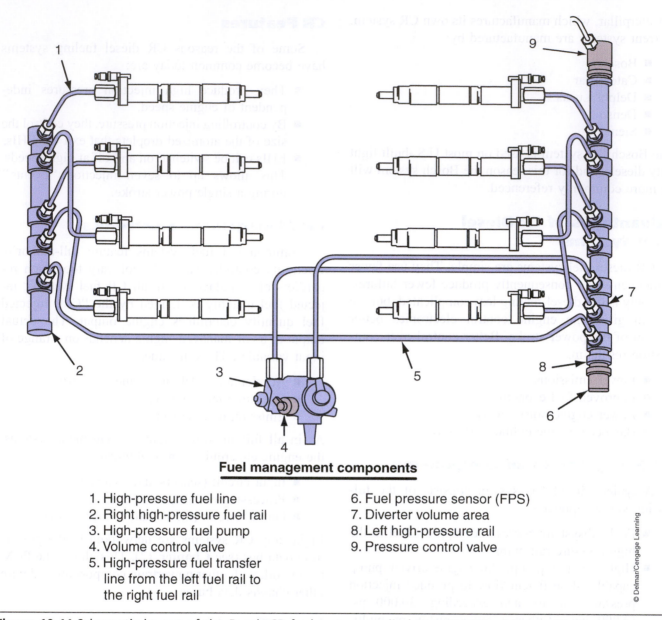

Fuel management components

1. High-pressure fuel line
2. Right high-pressure fuel rail
3. High-pressure fuel pump
4. Volume control valve
5. High-pressure fuel transfer line from the left fuel rail to the right fuel rail

6. Fuel pressure sensor (FPS)
7. Diverter volume area
8. Left high-pressure rail
9. Pressure control valve

© Delmar/Cengage Learning

Figure 10-11 Schematic layout of the Bosch CR fuel system applied to a 2011 Ford Power Stroke engine.

Digital Inputs. Digital input signals include On/Off signals and digital sensor signals such as Hall effect sensors. Digital signals can generally be responded to more quickly by the PCM because they bypass the analog/digital unit (ADU). Depending on the manufacturer, the TPS signal may either use an older-style analog signature or may produce a digital signal using one of two types of devices:

- Hall effect (noncontact) TPS broadcasting PWM signal
- Potentiometer (contact) with built-in I/C (integrated circuit) broadcasting PWM signal

Processing Circuit

The processing circuit is a function of the PCM. Input signals, stored memory, and the way a PCM has

been programmed all play a role in the processing cycle. In operation, the PCM maps an output strategy. This output strategy is put into effect by PCM-located drivers. These drivers switch actuators such as EHIs and the rail pressure control valve. All of today's PCMs have a write-to-self capability. This means that they can record fault codes and audit trails to internal memory or EEPROM. Because the PCM is also net-worked to the powertrain data bus, it can communicate with other chassis controllers.

Fuel Mapping

Managing injected fuel quantity in diesel engines is complex, but we can simplify this by saying that fueling uses a "closed" loop cycle. The cycle attempts to keep *desired* rail pressure and *actual* rail pressure as

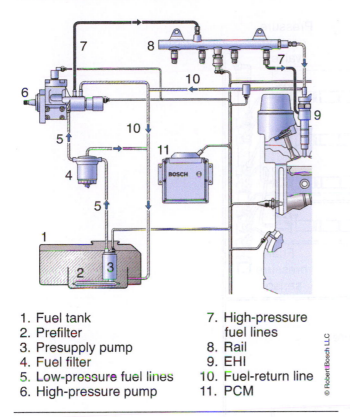

1. Fuel tank
2. Prefilter
3. Presupply pump
4. Fuel filter
5. Low-pressure fuel lines
6. High-pressure pump
7. High-pressure fuel lines
8. Rail
9. EHI
10. Fuel-return line
11. PCM

© Robert Bosch LLC

Figure 10-12 Schematic layout of a Bosch common rail diesel fuel system.

close to each other as possible. Desired and actual rail pressure can be defined as follows:

- Desired rail pressure: Calculated by the PCM. It is based on sensor inputs, emissions monitoring, and stored instructions in memory.
- Actual rail pressure: Measured by a pressure sensor and signaled to the PCM.

By continually comparing the desired and actual rail pressures, the PCM attempts to maintain desired rail pressure by controlling the rail pressure control valve.

Output Circuit

The output driver system, insofar as the fuel system management is concerned, is not that complex. The key components of the output circuit are the:

- Rail pressure management control valve
- EHIs

Rail Pressure Management Control. The rail pressure management control (RPMC) valve, as its name suggests, manages rail pressure. Each manufacturer has a different name for this valve. For instance, you can see in **Figure 10-13** that Ford chooses to call this a pressure control valve (PCV). However, we will stick with RPMC, the Bosch term. The RPMC valve is located downstream from the high-pressure pump and upstream

from the rail. RPMC valves are spool valves. They are controlled by pulse width modulation (PWM) to control their linear position. This has to be done with some precision. The valve is spring-loaded to default to no-fuel status. At no-fuel, the RPMC valve routes all fuel exiting the high-pressure pump directly to the return circuit. When energized by PCM output drivers, depending on the current flowed through its coil, an RPMC spool options fuel from the high-pressure pump to either the rail or the fuel return circuit. In order to precisely maintain desired rail pressure, it has to be capable of fast correction response times. RPMC valves function at high frequency (60 kHz). **Figure 10-13** is a schematic showing the Ford Power Stroke CR system with the PCV in high-pressure operating mode.

Injector Drivers. PCM-located **injector drivers** are used to switch the EHIs. They function similarly to drivers used to switch EUIs and HEUIs. The PCM outputs a PWM control current. This control pulse is usually spiked to around 100V DC. Current draw during actuation depends on whether solenoid or piezoelectric actuators are used in the EHI. EHIs with piezo actuators require less current and lower voltage.

CR Fuel Routing Circuit. Common rail fuel routing is modular. We can divide it up as follows:

- Fuel subsystem
- High-pressure pump
- Pressure accumulator or rail
- High-pressure distribution system
- Electrohydraulic injectors (EHIs)

Fuel Subsystem

Fuel subsystems vary by manufacturer, but the Bosch example shown in **Figure 10-12** is typical. All CR systems use a transfer pump. Bosch calls its transfer pump a presupply pump. In some small-bore diesels this can be located in the fuel tank.

Transfer Pump. The transfer or presupply pump may be located within the fuel tank or inline. At present, there are two possible versions:

- Electric roller-cell fuel pump located in the fuel tank. It functions similarly to the fuel supply pump in a gasoline-fueled automobile.
- External gear pump mechanically driven by the engine. The gear pump is often located behind, and driven by, the CR high-pressure pump.

In both cases, the fuel transfer pump is positive displacement and pushes fuel through the fuel subsystem to deliver it to the high-pressure stage.

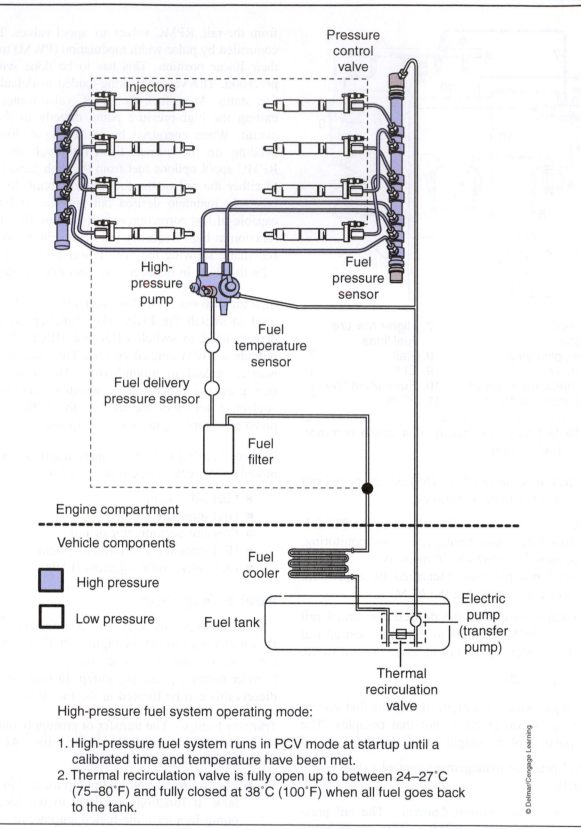

Pressure control valve

Injectors

High-pressure pump

Fuel pressure sensor

Fuel temperature sensor

Fuel delivery pressure sensor

Fuel filter

Engine compartment

Vehicle components

High pressure

Low pressure

Fuel cooler

Fuel tank

Electric pump (transfer pump)

Thermal recirculation valve

High-pressure fuel system operating mode:

1. High-pressure fuel system runs in PCV mode at startup until a calibrated time and temperature have been met.
2. Thermal recirculation valve is fully open up to between 24–27°C (75–80°F) and fully closed at 38°C (100°F) when all fuel goes back to the tank.

© Delmar/Cengage Learning

Figure 10-13 Schematic layout of the Power Stroke CR fuel system with the PCV set to maximize rail pressure.

Fuel Filter(s). The fuel filter is a manufacturer's responsibility, but it should meet the specifications of the CR fuel system manufacturer. In a Bosch CR application, the fuel filter should have a nominal entrapment capability of 8 microns, but manufacturers can specify higher entrapment values. All fuel filtration must occur upstream from the high-pressure pump. CR fuel systems should either use a fuel filter

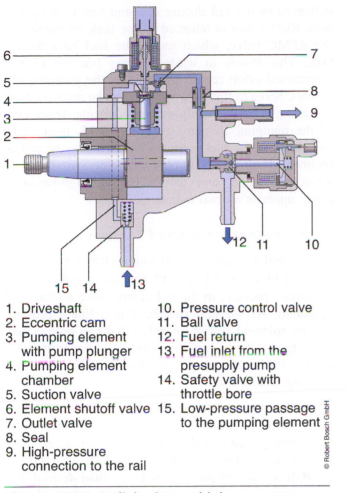

1. Driveshaft
2. Eccentric cam
3. Pumping element
   with pump plunger
4. Pumping element
   chamber
5. Suction valve
6. Element shutoff valve
7. Outlet valve
8. Seal
9. High-pressure
   connection to the rail
10. Pressure control valve
11. Ball valve
12. Fuel return
13. Fuel inlet from the
    presupply pump
14. Safety valve with
    throttle bore
15. Low-pressure passage
    to the pumping element

© Robert Bosch GmbH

Figure 10-14 Radial piston, high-pressure pump (schematic, longitudinal section).

with a water separator or integrate a separate water separator into the fuel subsystem. Collected water should be drained at regular intervals in systems equipped only with manual drains. Some water separators are equipped with a water-in-fuel (WIF) sensor that triggers a warning lamp indicating that water should be drained.

High-Pressure Pumps

There are some differences between the types of high-pressure pumps used by CR manufacturers, and while a Bosch CR system may be used on an engine, the high-pressure pump does not necessarily have to be Bosch manufactured. We can classify high-pressure pumps as:

- Radial piston pumps
- Inline piston pumps

The high-pressure pumps used in a CR diesel fuel system are simple in terms of the role they play. All they are required to do is create sufficiently high flow

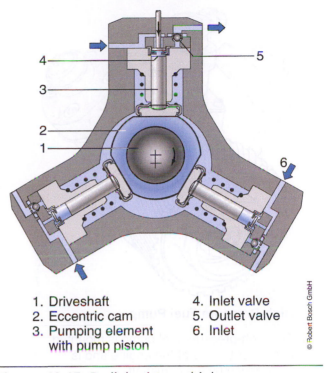

1. Driveshaft
2. Eccentric cam
3. Pumping element
   with pump piston
4. Inlet valve
5. Outlet valve
6. Inlet

© Robert Bosch GmbH

Figure 10-15 Radial piston, high-pressure pump (schematic, cross-section).

volume so that when subject to flow restriction, they achieve the required rail pressures. Pump output is unloaded to the rail. The high-pressure pump plays no role in metering and timing of the fuel delivery. **Figure 10-14** shows all the high-pressure stage components in a typical CR diesel fuel injection system, while **Figure 10-15** shows a cross-sectional view of the cylinders on the same pump. **Figure 10-16** shows an external view of the high-pressure pump used on a Ford Power Stroke with CR fueling.

Radial Piston Pumps. Bosch and Denso CR systems use a three-cylinder, **radial piston pump** (see **Figure 10-14, Figure 10-15,** and **Figure 10-16**) to produce the rail pressures required for system operation. Pressurized fuel from the high-pressure pump is unloaded by means of a high-pressure line into a tubular, high-pressure fuel accumulator that we call the common rail. The high-pressure pump continually generates the PCM-desired pressure at the rail, with the result that in contrast to conventional systems, fuel does not have to be specially compressed for each individual injection pulse. The advantage of using a radial piston pump is its compact design.

The high-pressure pump is installed on a diesel engine so that it can be driven by the engine timing gear train. It may be driven at either camshaft or crankshaft speed, depending on the system, but camshaft speed is more common. The high-pressure pump is

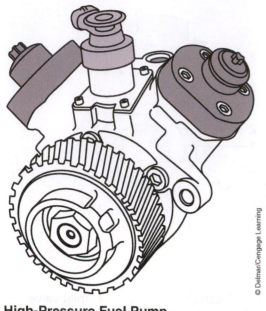

High-Pressure Fuel Pump

The high-pressure fuel pump is mounted in the front valley of the engine and is gear driven by the camshaft.

The high-pressure fuel pump is timed to the crankshaft and camshaft to optimize the effects of the high-pressure fuel pulses.

The high-pressure fuel pump is lubricated by diesel fuel.

Figure 10-16 External view of the two-cylinder, high-pressure pump used on the 2011 Ford PowerStroke. (*Text courtesy of Ford Motor Company*)

usually drive coupled to the timing gear train using a dedicated gear, but both chain and toothed-belt drives may be used in light duty applications. The pump is internally lubricated by the diesel fuel it pumps. **Figure 10-15** shows a schematic cross section of a radial piston, high-pressure pump; note how the pumping elements are actuated by the central camshaft.

Radial Piston Pump Components. Pressure rise is created in the high-pressure pump by three radially arranged pump elements, evenly offset at an angle of 120 degrees around the pump drive shaft. This means that each pump element is actuated once during a single pump rotation. This reduces stress on the pump drive mechanism. Each pump piston is actuated by cams machined into the drive shaft (**see Figure 10-15**). Fuel from the pump element is discharged through an outlet valve.

Fuel Delivery Rate. Because the high-pressure pump is designed to deliver the required volume of fuel for rated speed and load performance, excess fuel is

delivered to the rail during idle and low-load operation. Excess fuel is returned to the tank by means of the RPMC valve, which routes the fuel back to the tank. This results in wasted energy because of the mechanical effort required to actuate the pumping elements. Some of this parasitic loss can be recovered by switching off one of the pumping elements. When one of the pumping elements (**Figure 10-15,** item 3) is switched off, the fuel volume delivered into the rail is reduced. With one pumping element switched off when less engine power is required, the high-pressure pump operates on two cylinders.

Rail Pressure Control Valves

The rail pressure control valve (RPCV) is PWM actuated by the PCM. The PCM uses it to define the "desired" pressure in the rail at any given moment of operation. As we said earlier, it is a linear proportioning solenoid and spool valve that can option fuel unloaded by the high-pressure pump either to the rail or into the return circuit. Its operation can be summarized as follows:

- If *actual* rail pressure (signaled to the PCM by the rail pressure sensor) is higher than *desired* rail pressure, the RPCV opens to divert fuel from the rail, sending it to the return circuit.
- If *actual* rail pressure is lower than *desired* rail pressure, the pressure control valve closes, sealing the rail and permitting pressure to rise.

A sectional view of the PCM-controlled rail pressure control valve is shown in **Figure 10-17.**

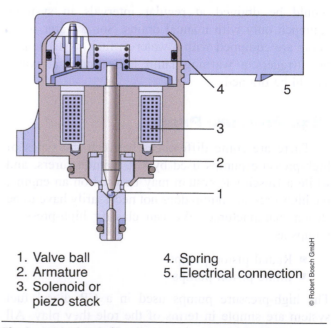

1. Valve ball	4. Spring
2. Armature	5. Electrical connection
3. Solenoid or piezo stack	

Figure 10-17 Bosch rail pressure control valve.

The Bosch rail pressure control valve is located on the high-pressure pump **(Figure 10-17)** and is flange mounted to it. Some systems use a pressure control valve located at the rail inlet. To seal the high-pressure rail from the return circuit, the control valve armature forces a ball into a seat to create a seal. Two forces act on the armature. Mechanical force is provided by a spring, and opposing this, electromagnetic force is created by the solenoid coil when energized. Two control loops are used to manage pressure control valve operation:

- A slow-response electrical control loop for setting (variable) mean pressure in the rail.
- A fast-response mechanical control loop to compensate for the high-frequency pressure fluctuations.

Pressure Control Valve Nonenergized. When desired rail pressure is higher than actual rail pressure, the rail pressure control valve must drop rail pressure. High pressure at the rail or at the high-pressure pump outlet acts on the pressure control valve via the high-pressure input. Because the nonenergized electromagnet in the control valve exerts no force, high-pressure fuel exceeds the spring force, opening the control valve and spilling rail fuel to the return circuit.

Pressure Control Valve Energized. When desired rail pressure is lower than actual rail pressure, the pressure control valve must allow rail pressure to rise. If the pressure in the high-pressure circuit is to be increased, the force of the electromagnet must combine with the mechanical force of the spring. The PCM energizes the pressure control valve, causing it to close and remain closed until equilibrium is established between desired and actual rail pressures. This means that a balance is reached between the high-pressure fuel forces on the one side and the combined forces of the spring and the electromagnet on the other. The valve then remains open and maintains rail pressure constant. Any change in the pump delivery quantity/engine load is compensated for by the valve assuming a different setting.

Volume Control Valve

The volume control valve (VCV) controls the volume of low-pressure fuel that enters the pumping chambers of the high-pressure pump via an inlet, one-way valve. The PCM regulates the VCV using a PWM signal. A high duty cycle indicates less volume is being commanded, while a low duty cycle indicates more fuel is being requested. **Figure 10-18** shows the

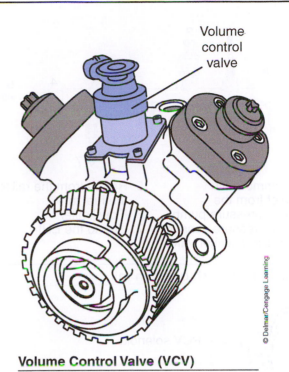

Volume control valve

Volume Control Valve (VCV)

The fuel VCV is mounted on the high-pressure fuel pump. The PCM controls the volume of low-pressure fuel that enters the inlet one-way check valve and two main pump pistons by activating the fuel VCV.

The PCM regulates fuel volume by controlling the on/off time of the fuel VCV solenoid. A high duty cycle indicates less volume is being commanded. A low duty cycle indicates a high fuel volume is being commanded.

Figure 10-18 Location of the pressure control valve on the current Ford PowerStroke high-pressure pump. (*Text courtesy of Ford Motor Company*)

location of the pressure control valve on the current Ford Power Stroke high-pressure pump.

Common Rail

The common rail or **accumulator** receives fuel from the high-pressure pump and, by means of dedicated lines, makes it available to the PCM-controlled EHIs. The accumulator feature of the rail means that even after an injector has discharged a pulse of fuel into the engine, the fuel pressure in the rail remains almost constant. The volume of fuel in the rail has a damping effect on the changes in rail pressure that occur as the injectors are actuated and the RPCV kicks in and out. The damping ability of the rail can be accounted for by its accumulator effect, which results from the compressibility factor of the fuel. **Figure 10-19**

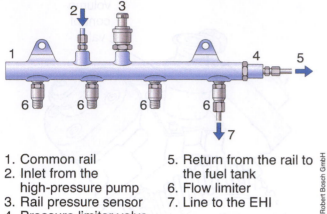

1. Common rail
2. Inlet from the high-pressure pump
3. Rail pressure sensor
4. Pressure limiter valve
5. Return from the rail to the fuel tank
6. Flow limiter
7. Line to the EHI

© Robert Bosch GmbH

Figure 10-19 Bosch common rail.

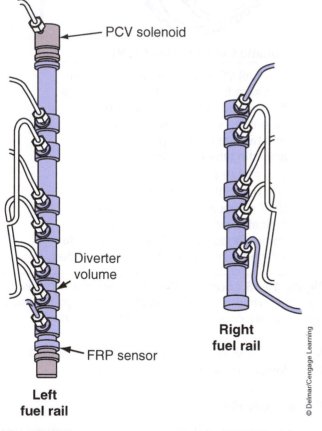

PCV solenoid

Diverter volume

FRP sensor

Left fuel rail

Right fuel rail

© Delmar/Cengage Learning

Fuel Rails

The fuel rails on the 6.7L Power Stroke® diesel engine are on the outboard side of the valve covers. The left fuel rail has the FRP sensor and the PCV solenoid. The right fuel rail does not have any sensors or solenoids.

The left fuel rail is longer in length due to a diverter volume section.

Figure 10-20 The pair of fuel rails used in the V8 Ford PowerStroke engine. (*Text courtesy of Ford Motor Company*)

is a line drawing of the Bosch common rail used on a four-cylinder engine, while **Figure 10-20** shows the pair of rails used in the current Ford Power Stroke V8 engine.

Rail Pressure Limits. Rail pressure is monitored by the rail pressure sensor and maintained at the desired value by the PCM pressure control valve. A pressure-limiter valve acts to limit the maximum fuel pressure in the rail to a specification that usually exceeds the intended peak rail pressure by a small margin. For instance, in the current Caterpillar CR C7 system, the highest specified rail pressure is 27,550 psi (1900 bar), so the pressure-limiter valve is designed to trip at 33,300 psi (2300 bar). Fuel in the rail is made available to the injectors by means of flow limiters, which prevent excess fuel from being injected. The rail and its critical components on a Bosch CR system are shown in **Figure 10-19.**

Rail Pressure Sensor. The rail pressure sensor consists of a sensor housing, integral printed circuit, and sensor element. Two types of rail pressure sensors are used. One type uses a piezo-resistive operating principle. This is more common. Variable capacitance–type sensors are also used. Both sensors are usually supplied with a 5 volt V-Ref. They are fully explained in **Chapter 12. Figure 10-21** shows the location of the fuel rail pressure sensor used on the Ford Power Stroke engine.

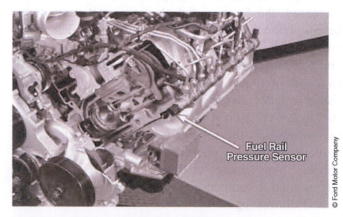

Fuel Rail Pressure Sensor

© Ford Motor Company

Fuel Rail Pressure (FRP) Sensor

The FRP is threaded into the front of the left fuel rail. The FRP sensor is a three-wire variable capacitance sensor. The PCM supplies a 5-volt reference signal, which the FRP sensor returns a portion of to indicate pressure. The FRP sensor continuously monitors fuel rail pressure to provide a feedback signal to the PCM.

Figure 10-21 The fuel rail pressure sensor used in the V8 Ford PowerStroke engine.

Fuel Rail Temperature (FRT)

The FRT is mounted in the fuel line that runs between the secondary fuel filter and the high-pressure fuel pump. It is green in color.

The PCM monitors the temperature of the fuel using the FRT before the fuel enters the high-pressure fuel pump.

Fuel temperature effects fuel viscosity. The PCM uses this information for more precise fueling no matter what the fuel temperature.

Figure 10-22 Location of the fuel rail temperature sensor on the Ford PowerStroke engine.

Flow Limiter. The flow limiter prevents continuous injection in the event that an EHI sticks in the open position. Once the fuel quantity leaving the rail exceeds a predetermined volume, the flow limiter shuts off the line to the problem injector. The flow limiter is a fail-safe device that is rarely engaged. When an EHI sticks in the open position, it is usually due to contaminated fuel.

Fuel Rail Temperature Sensor. The fuel rail temperature sensor is a thermistor located upstream from the rail. It is V-Ref powered and signals fuel temperature in the rail to the PCM. Temperature affects the viscosity of the fuel. **Figure 10-22** shows the location of the fuel temperature sensor on the current Ford PowerStroke engine.

High-Pressure Fuel Lines. The fuel lines used in CR systems must be capable of handling the maximum system pressures with a wide margin of safety. The injection lines are required to be of identical length and internal diameter. They are designed for one-time use. Because they are expensive, some thought should be given to the decision to remove them. They should never be shorted out when troubleshooting an engine miss. **Figure 10-23** shows the high-pressure lines used on the Ford Power Stroke engine.

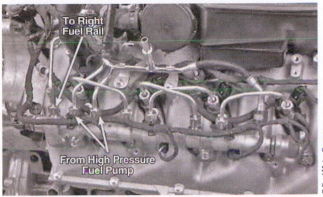

High-Pressure Fuel Lines

The high-pressure fuel lines run between the:
- high-pressure fuel pump and left fuel rail.
- right and left fuel rails.
- fuel rails and the fuel injectors on the outside of the valve covers.

Figure 10-23 High-pressure fuel lines used on a CR system.

WARNING *NEVER crack the high-pressure pipe nuts attempting to bleed them. It could prove to be dangerous, and manufacturers prefer CR high-pressure lines to be single-use devices (they yield to deform to shape on initial torque). CR fuel systems are designed to self-prime.*

Electrohydraulic Injectors

Electrohydraulic injectors (EHIs) were first introduced in **Chapter 8.** Two general types of EHIs are used in CR fuel systems. First-generation EHIs used a solenoid control valve, while more recent types use piezoelectric actuators. However, the general principles used by each type remain identical. We are going to examine the commonly used Bosch EHI (see **Figure 10-24**) for purposes of the description here. An EHI is subdivided as follows:

- Nozzle assembly
- Hydraulic servo system
- Actuator valve

Overview of Operation. In **Figure 10-24,** fuel at rail pressure is supplied to the high-pressure connection (4), to the nozzle through the fuel duct (10), and to the control chamber (8) through the feed orifice (7). The control chamber is connected to the fuel return (1) via a bleed orifice (6) that is opened by the actuator valve. With the bleed orifice is closed, hydraulic force acts on the valve control plunger (9) exceeding that at the

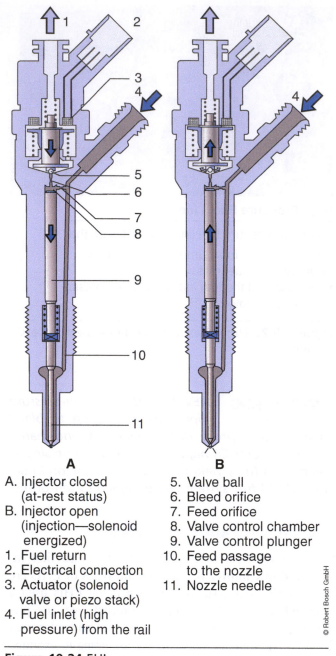

A

B

A. Injector closed
   (at-rest status)
B. Injector open
   (injection—solenoid
   energized)
1. Fuel return
2. Electrical connection
3. Actuator (solenoid
   valve or piezo stack)
4. Fuel inlet (high
   pressure) from the rail

5. Valve ball
6. Bleed orifice
7. Feed orifice
8. Valve control chamber
9. Valve control plunger
10. Feed passage
    to the nozzle
11. Nozzle needle

Figure 10-24 EHI.

nozzle-needle pressure chamber (located between the shank and the needle of the nozzle valve). Although the hydraulic pressure values acting on the top of the nozzle valve and in the pressure chamber are identical, the sectional area at the top of the nozzle valve is greater. As a result, the nozzle needle is loaded into its seated position, meaning that the injector is closed.

When a PCM signal triggers the injector control actuator valve, the bleed orifice opens. This immediately drops the control-chamber pressure and, as a result, the hydraulic pressure acting on the top of the nozzle valve (11) also drops. When hydraulic force

acting on top of the nozzle valve drops below the force on the nozzle-needle pressure shoulder, the nozzle valve retracts and allows fuel to pass around the seat to be injected through orifices into the combustion chamber. The hydraulic assist and amplification factor are required in this system because the forces necessary for rapid nozzle valve opening cannot be directly generated by an actuator valve alone. Fuel used as a hydraulic medium to open the nozzle valve is in addition to the injected fuel quantity, so this excess fuel is routed back to the tank. In addition to this fuel, some leak-by fuel losses occur at the nozzle valve-to-body clearance and at the valve plunger guide clearance. This leak-off fuel volume is also returned to the fuel tank via the fuel return circuit.

Injector Operating Phases. When the engine is stopped, all injector nozzles are closed, meaning that their nozzle valves are loaded onto their seats by spring pressure. In a running engine, injector operation takes place in three phases.

INJECTOR CLOSED

In the at-rest state, the actuator valve is not energized, and therefore the nozzle valve is loaded onto its seat by the injector spring combined with hydraulic pressure (from the rail) acting on the sectional area of the valve control plunger. With the bleed orifice closed, the actuator valve spring forces the armature ball check onto the bleed-orifice seat. Rail pressure builds in the injector control chamber, but identical pressure will be present in the nozzle pressure chamber. Given equal pressure acting on the larger sectional area of the nozzle control plunger (which is mechanically connected to the nozzle valve) and in the nozzle pressure chamber, this and the force of the nozzle spring combine to load the nozzle valve on its seat, holding the injector closed.

NOZZLE OPENING

The injector actuator valve is PWM energized by the PCM injector driver. The actuation voltage varies by manufacturer, but typically it spikes at around 100V DC. The current draw during injector energization depends on whether a solenoid or piezoelectric actuator is used; current draw is lower when piezoelectric actuators are used. Force exerted by the triggered actuator now exceeds that of the valve spring, and the control valve opens the bleed orifice. Almost instantly, the high-level pickup current to the actuator drops off to a lower hold-in current flow. As the bleed orifice opens, fuel flows from the valve-control chamber into the cavity above it and out to the return circuit.

This collapses the hydraulic pressure acting on the valve control plunger that was helping to hold the nozzle valve closed. Now, pressure in the valve-control chamber is much lower than that in the nozzle pressure chamber, which is maintained at the rail pressure. The result is a collapsing of the force that was holding the nozzle valve closed, and the result is that the nozzle valve opens, beginning the injection pulse.

The speed at which the nozzle needle opens is determined by the difference in the flow rate through the bleed and feed orifices. When the control plunger reaches its upper stop, it is cushioned by fuel generated by flow between the bleed and feed orifices. When the injector nozzle valve has fully opened, fuel is injected into the combustion chamber at a pressure very close to that in the fuel rail.

NOZZLE CLOSING

When the actuator valve is de-energized by the PCM, its spring forces the control valve downward, and the check ball closes the bleed orifice. The closing of the bleed orifice creates pressure buildup in the control chamber via the input from the feed orifice. This pressure should be the same as that in the rail, and now it exerts an increased force on the nozzle valve control plunger through its end face. This force, combined with that of the nozzle spring, exceeds the hydraulic force acting on the nozzle valve sectional area, and the nozzle valve closes, ending injection. Nozzle valve closing velocity is determined by the flow through the feed orifice. The injection pulse ceases the instant that the nozzle valve seats.

Multipulse Injection. EHIs lend themselves to **multipulse injection**. Multipulse injection is required in post-2007 diesel engines. Multipulse injection is multiple-shot injection pulses during a single power stroke.

The number of injection pulses that take place will depend on how the engine is being operated and the type of actuator used in the EHI. Piezoelectric actuators allow for much faster responses, so they are often used in CR injectors. Injection pressures produced during each injection pulse of multipulse injection events remain pretty stable during a single cycle, regardless of duration or volume. These pressures directly depend on the rail pressure that is being managed by the PCM at any given moment of operation. The action of opening and closing injectors during multipulse events causes little deviation in actual rail pressures.

Timing Pumps

High-pressure pumps used on CR systems may have to be timed to the engine they fuel. Timing the pump is required for reasons of mechanical balance. You should reference the manufacturer service literature when timing a pump to the engine.

Calibration Codes

Electronically managed injection systems fuel engines with precision. For this reason, most injector units are bench tested and assigned a **calibration code**. A calibration code is a precise evaluation of how an injector flows fuel when bench tested. The code may be alpha or numeric; it is usually scribed onto replacement injectors. When injectors are changed out for whatever reason, it is important that the PCM is reprogrammed with the new calibration code. Failure to do this can result in unbalancing the engine. Calibration codes are known by terms such as cal codes, E-trim, and other names, but they all amount to the same thing. In instances where a complete set of new injectors is installed in an engine, all of the injectors should be reprogrammed to the PCM by cylinder number.

Summary

- Most current light duty engines use either EUI or CR fueling. HEUI systems were phased out for on-highway use in 2007.

- EUI fuel systems are mechanically actuated and electronically controlled.

- The PCMs used to manage EUI-fueled engines are usually housed in a single module that contains both the computing and output driver hardware.

- The EUIs used in light duty applications use a single solenoid–type actuator. The single actuator controls the spill circuit.

- Single-actuator EUIs are often called two-terminal EUIs; they are switched by injector drivers using PWM.

- The HEUI engine management system was the first to use electronically controlled, hydraulically actuated injection pump units. Until the introduction of HEUI technology, pumping to diesel injection pressure values was always achieved mechanically.

- Effective injection pumping stroke in HEUI units is not confined by the hard limits of cam geometry as in EUI electronic management systems.

- Fuel charging pressure to the HEUIs is at values between 30 psi (2 bar) and 60 psi (4 bar).

- Engine-lubricating oil is used as the hydraulic medium for the HEUIs. The oil pressure is boosted to values of up to 4,000 psi (276 bar) by a swash plate–type hydraulic pump.

- HEUI-actuating oil pressure values are precisely controlled by the PCM, which switches the IPR to achieve oil pressure values of between 485 psi (33 bar) and 4,000 psi (276 bar).

- The IPR is a spool valve, positionally controlled by a pulse-width modulated PCM signal.

- The actuating oil pressure value determines the downward velocity of the HEUI pumping plunger. In turn, this determines the pumping pressure and therefore the injection rate and the emitted atomized droplet sizing.

- The HEUI solenoid controls a poppet valve that, when energized, admits high-pressure oil so that it acts on the amplifier piston.

- In Caterpillar-manufactured HEUIs, when the control solenoid is energized, actuation oil is admitted to the HEUI to create a pumping stroke. When the control solenoid is de-energized, the high-pressure oil is spilled to the rocker housing.

- In the more recent Siemens HEUI, a transverse spool actuated by a pair of coils controls the entry of oil to the HEUI internal circuit.

- HEUIs use multi-orifice, hydraulic injector nozzles with opening pressures around 5,000 psi (345 bar) and peak system pressures rising to between 18,000 and 28,000 psi (1,241 and 1,931 bar) depending on manufacturer.

- The HEUI PCM uses an IDM to switch the HEUI control solenoids.

- The IDM used to drive Caterpillar-manufactured HEUIs steps the actuation voltage to 115V DC while Siemens HEUIs require only 46V DC. In both cases, the voltage is stepped using induction coils.

- Common rail diesel fuel systems are currently used on a wide range of engines that extend from small- to large-bore highway diesels.

- Light duty common rail (CR) diesel fuel systems have full-authority engine management capability and are networked to the powertrain data bus. The powertrain bus can communicate with other chassis hard wire and optical chassis buses.

- The fuel subsystems that supply common rail fuel systems have few differences when compared with other current fuel supply circuits.

- The electronic controls on a typical CR system consist of an input circuit, processing hardware, and actuator circuit.

- The primary outputs that manage fueling on a CR system are the rail pressure control valve and the switching of the EHI actuators.

- Most Bosch CR systems use radial piston high-pressure pumps to supply the rail, but cam-actuated piston pumps are also used.

- Rail pressures on CR diesel fuel systems are managed by the PCM using a rail pressure control valve. The rail pressure control valve applies flow restriction to fuel discharged by the high-pressure pump and options it either to the rail or to the return circuit.

- The rail pressure control valve is a linear proportioning solenoid. It uses a spool valve to option fuel exiting the high-pressure pump outlet either to the rail or to the fuel return circuit.

- Actual rail pressures are signaled to the PCM by the rail pressure sensor.

- The PCM managing rail pressures in a typical CR system does so by calculating *desired* rail pressure. By monitoring actual rail pressure, the PCM attempts to keep desired and actual rail pressures as close to each other as possible.

- EHIs are not limited by hard value opening and closing pressures. Because they are opened and closed by PCM injector drivers, the actual pressure at nozzle opening is determined by the CR pressure.

- The actuators used in EHIs may be solenoid or piezo triggered. Piezo actuators are used in the latest CR systems because of super-fast response times to PCM driver signals.

- The actuators in EHIs are PWM switched by the PCM.

- Some CR high-pressure pumps must be timed to the engine. Failure to do this may unbalance the engine.

- The manufacturer's diagnostic software should be used to diagnose and program all full-authority electronic diesel engines.

Shop Exercises

1. Identify the electronically managed diesel engines in your shop and make a note of which type of fuel system each uses.

2. Locate a diesel engine that uses an EUI fuel system: consult the manufacturer service literature and identify the method required to install the EUIs: note whether calibration codes have to be programmed to the PCM when EUIs are changed out.

3. Locate a diesel engine that uses a HEUI fuel system. Diagram the main fuel system components from the fuel tank to the HEUIs and note the location of the high-pressure oil pump. Determine whether the engine uses Caterpillar or Siemens HEUIs.

4. Locate a diesel engine that uses a CR fuel system. Diagram the fuel system components from the fuel tank to the EHIs. See if you can identify any sensors in the fuel system that are visible without disassembly and map them into your diagram.

5. Use a specific CR-fueled engine and check the manufacturer recommended procedure for changing out injectors. Make a note of whether the high-pressure lines between the rail and the EHIs have to be replaced and whether the PCM has to have the EHI calibration codes reprogrammed.

Internet Exercises

Use a search engine and research what comes up when you search the following:

1. Volkswagen TDI fuel system

2. Ford Power Stroke 6.0 fuel system

3. Caterpillar HEUI fuel systems

4. Bosch diesel common rail fuel systems

5. Bosch animated common rail fuel system operation (YouTube and Bosch websites)

Review Questions

1. What force creates injection pressure values in an EUI injector?

   A. Pneumatic

   B. Mechanical

   C. Electrical

   D. Hydraulic

2. Which of the following does the solenoid control in an EUI fuel system?

   A. Fuel pressure

   B. Spill circuit

   C. Nozzle valve travel

   D. Tappet travel

3. What force creates injection pressure values in a HEUI injector?

   A. Pneumatic

   B. Mechanical

   C. Electrical

   D. Hydraulic

4. What type of pump is used to increase engine lube oil pressure to the values required by a HEUI?

   A. Gear

   B. Plunger

   C. Swash plate

   D. Diaphragm

5. Oil pressure values within the HEUI actuation circuit are controlled by:

   A. Pressure relief valve

   B. Accumulator

   C. IPR

   D. IDM

6. Which of the following best describes the type of injectors used with CR fuel systems?

   A. Pintle injectors

   B. Electrohydraulic injectors
   (EHIs)

   C. Poppet injectors

   D. Electronic unit injectors (EUIs)

7. Which of the following are advantages of the multipulse injection cycles enabled by CR fueling?

   A. Lower cold start emissions

   B. Lower noise levels

   C. Better fuel economy

   D. All of the above

8. Which of the following best describes the means by which the PCM drivers control the actuation of CR EHIs?

   A. V-Ref signal

   B. V-Bat signal

   C. Distributor spike

   D. Pulse width modulation (PWM)

9. Which of the following is the key component used to signal actual rail pressure to the PCM?

   A. Rail pressure sensor

   B. Rail pressure control valve

   C. Pressure limiter valve

   D. Flow limiter valve

10. What force is used to hold a CR EHI nozzle valve in its closed and seated position when the engine is running?

    A. Spring force only

    B. Electrical force only

    C. Combined hydraulic and electrical force

    D. Combined hydraulic and spring force

11 Charging and Starter Circuits

Learning Objectives

After studying this chapter, you should be able to:

- Identify the components in a vehicle cranking circuit.
- Identify charging circuit components.
- Explain the operating principles of magnetic switches, solenoids, and starter motors.
- Describe the construction of an alternator.
- Explain full-wave rectification.
- Full field an alternator.
- Measure AC leakage in the charging circuit.
- Verify the performance of an alternator.
- Navigate a charging circuit schematic.
- Test and troubleshoot a cranking circuit using voltage-drop testing.

Key Terms

A-circuit voltage regulator

armature

B-circuit voltage regulator

brushes

charging system

commutator

control circuit

delta

diode

external ground

field coils

full-wave rectification

ground circuit

half-wave rectification

insulated circuit

internal ground

load-dump

neutral safety switch

remote sensing

rotor

sensing voltage

slip rings

starter circuit

starter motor

starter relay

starter system

stator

thermostat

voltage-drop testing

voltage regulator

windings

wye

INTRODUCTION

A **charging system** consists of the batteries, alternator, **voltage regulator**, associated wiring, and the electrical loads of the engine and chassis. The purpose of the system is to recharge the batteries whenever necessary and to provide the current required to power the electrical components of the chassis. It does this by converting a portion of the mechanical energy of the engine into electricity.

A **starter system** is designed to turn an engine over until it can operate under its own power. A cranking system can be divided into two subcircuits known as the **control circuit** and the **starter circuit**. The objective of the cranking system is to energize a **starter motor** using energy from the vehicle batteries. The control circuit is activated either directly from the vehicle ignition key or by a dedicated starter button. The control circuit uses low current to switch the starter circuit. The switch used to bridge the control circuit and the starter circuit is a magnetic switch. The magnetic switch receives a low-current command signal that switches the high-current starter circuit. Once energized, the starter circuit uses full battery power to energize the starter motor. The arrangement of a typical vehicle electrical system is shown in **Figure 11-1.** Note how the battery connects with the starter motor, and identify the cranking system control circuit.

CHARGING CIRCUITS

In a charging system, the battery acts as a reservoir of electrical energy. Electrical energy is drawn from the battery to crank the engine and to power electrical systems. The electrical "pressure" in a battery is called voltage. You can also call this "potential." When fully charged, a wet cell lead acid 12-volt battery has 12.6V potential available across its terminals. As the battery is discharged, the voltage level drops. When battery potential reaches a preset voltage level, an electrical switch, called a *voltage regulator*, activates the alternator.

Electron Pump

The alternator acts as an electron pump. It "pushes" electrical current into the battery to restore the voltage level. On 12V systems, charging is typically at 14.0 to 14.5 volts to get the battery potential up to 12.6V. The alternator has to act against whatever voltage is in the batteries, so it must do so with higher potential. When the batteries are discharged, it does not take much alternator voltage to start current flowing into the battery. When the battery is close to full charge, it requires higher alternator potential to charge the battery. A regulator might turn the alternator on and off as many as 600 times per minute. When electrical demand is high, the alternator will stay on longer; when low, it will freewheel. This helps fuel economy since a 100-amp alternator can require 6 to 7 horsepower to turn it.

Alternator Construction

To generate electricity, the alternator uses this basic law of physics:

When magnetic lines of force move across a conductor (such as a wire or bundle of wires), an electrical current is produced in the conductor.

This principle is illustrated in **Figure 11-2.** Electricity can be induced in an electrical circuit by either moving the circuit wiring through the magnetic field of a magnet or by moving the magnetic field past the wiring. In both situations, the direction of electrical current flow is determined by the polarity of the magnetic field. The magnetic forces of the north pole of a magnet will force electrons to flow in one direction and the south pole of a magnet will force electrons to flow in the opposite direction. If the wiring is exposed alternately to both north and south poles, an alternating current will be produced.

Actual current flow induced depends on several factors: the strength of the magnetic field, the speed of the wire passing through the field, and the size and number of wires. In an alternator, the **rotor** provides the magnetic fields necessary to induce a current flow. The rotor spins inside a stationary coil of wires called a **stator**. The moving magnetic fields induce a flow of

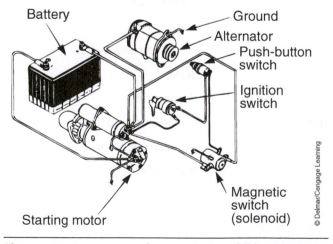

Battery
Ground
Alternator
Push-button switch
Ignition switch
Magnetic switch (solenoid)
Starting motor

© Delmar/Cengage Learning

Figure 11-1 Layout of a typical vehicle electrical system.

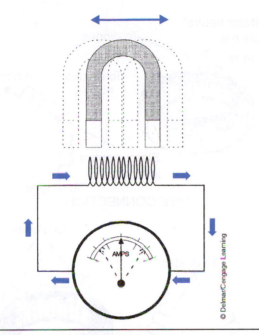

Figure 11-2 A magnetic field moving past a wire will induce current flow in the wire.

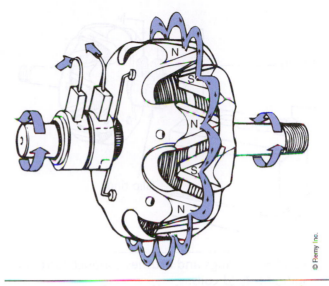

Figure 11-4 The interlacing fingers of the pole pieces create alternating north and south poles.

electrons through the stator wiring that is pumped to the batteries when they are in need of recharging. The following paragraphs explain how the rotor and stator and other alternator components work together to provide charging current to the battery. A cutaway view of a Delcotron alternator is shown in **Figure 11-3.**

Rotor

The rotor is the only moving component within the alternator. It is responsible for producing the rotating magnetic field. The rotor consists of a coil, two pole pieces, and a shaft. The magnetic field is produced when current flows through the coil; this coil is simply a series of **windings** wrapped around an iron core. Increasing or decreasing the current flow through the

coil varies the strength of the magnetic field, which in turn defines alternator output. The current passing through the coil is called the *field current.* It is usually 3 amperes or less.

The coil is located between the interlocking pole pieces. As current flows through the coil, the core essentially becomes a magnet. The pole pieces assume the magnetic polarity of the end of the core that they touch. Thus, one pole piece has a north polarity and the other has a south polarity. The extensions on the pole pieces, known as the *fingers,* form interlacing magnetic poles. A typical rotor has 14 poles, 7 north and 7 south, with the magnetic field between the pole pieces moving from the north poles to the adjacent south poles **(Figure 11-4).** The more poles a rotor has, the higher the alternator output will be.

Slip Rings and Brushes

The wiring of the rotor coil is connected to **slip rings**. The slip rings and **brushes** conduct current to the rotor. Most alternators have two slip rings mounted directly on the rotor shaft; they are insulated from the shaft and from each other. A spring-loaded carbon brush is located on each slip ring to carry the current to and from the rotor windings **(Figure 11-5).** Because the brushes carry only 1.5 to 3.0 amperes of current, they do not require frequent maintenance. This is in direct contrast to generator brushes, which conduct all of the generator's output current and consequently wear rapidly.

Stator

The stator is made up of many conductors, or wires, into which the spinning rotor induces voltage. The

Figure 11-3 Cutaway view of a Delcotron alternator.

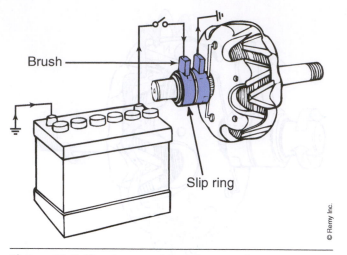

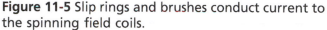

Figure 11-5 Slip rings and brushes conduct current to the spinning field coils.

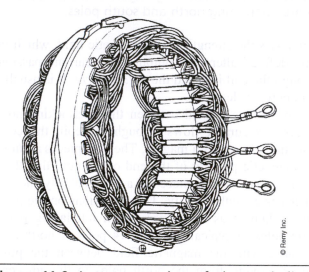

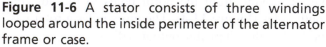

Figure 11-6 A stator consists of three windings looped around the inside perimeter of the alternator frame or case.

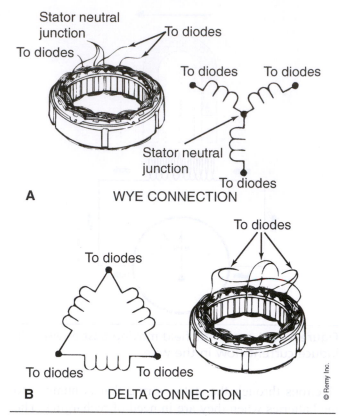

Figure 11-7 The stator windings are wired together in one of two arrangements: (A) A wye configuration; or (B) a delta configuration.

wires are wound into slots in the alternator frame, with each wire forming several coils spaced evenly around the frame **(Figure 11-6)**. There are as many coils in each wire as there are pairs of north and south rotor poles.

Windings. The wires are grouped into three separate bundles, or windings. The coils of the three windings are staggered in the alternator frame so that the electrical pulses created in each coil will also be staggered. This produces an even flow of current out of the alternator. In an alternator rated at 90 amps output, each winding or phase will pump out 30 amps. The ends of each winding are attached to separate pairs of diodes, one positive and the other negative. They are also wired together in one of two configurations: a **wye** shape or a **delta** shape **(Figure 11-7)**.

End Frame Assembly

The end frame assembly, or housing, is made of two pieces of cast aluminum and contains the bearings for the end of the rotor shaft. The drive pulley and fan are mounted to the rotor shaft outside the drive end frames. Each end frame also has built-in air ducts so the air from the rotor shaft fan can pass through the alternator. A heat sink—called a *rectifier bridge*, or **diode** holder—containing the diodes is attached to the rear end frame; heat can pass easily from these diodes to the moving air **(Figure 11-8)**. Because the end frames are bolted together and then bolted directly to the engine, the end frame assembly provides the electrical ground path for many alternators. This means that anything connected to the housing that is not insulated from the housing is grounded.

ALTERNATOR OPERATION

As the rotor, driven by a belt and pulley arrangement, rotates inside the alternator, the north and south poles (fingers) of the rotor alternately pass by the coiled windings in the stator. Only a very small air gap separates the rotor from the stator, so the windings are

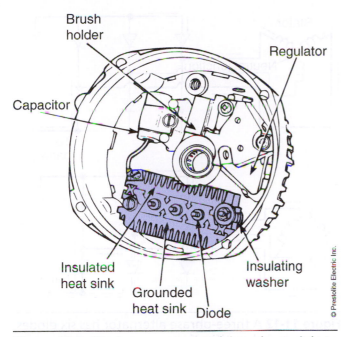

Figure 11-8 Diodes are mounted in a heat sink to keep them cool.

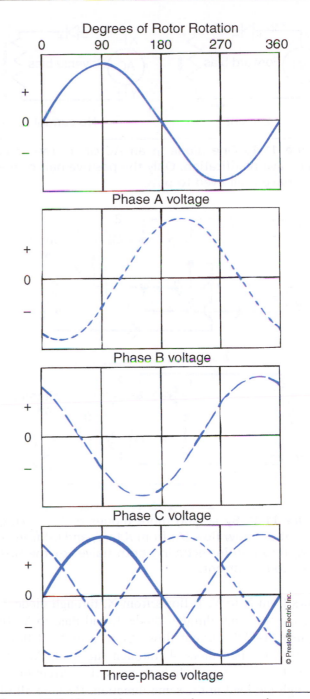

Figure 11-9 Alternating current is seen as a sine wave on an oscilloscope.

subjected to a maximum amount of the magnetic force fields. As each pole alternately passes by the coils, magnetic lines of force cause electrons to flow in the wires. As a north pole passes by a coil in the winding, electrons first flow in one direction. As the next pole, having a south polarity, passes by the coil, the flow of electrons changes direction. In this way, an alternating current (AC) is produced. When viewed on an oscilloscope, the alternating pulses of current are seen as a sine wave **(Figure 11-9).** The positive side of the wave is produced by a north pole and the negative side of the wave is produced by a south pole. Because an alternator has three windings staggered around the rotor, a three-phase sine wave is produced.

AC to DC

A battery can only be charged using direct current (DC). Most engine or chassis accessories also are powered by DC. In order for the alternator to provide the electricity required, the AC it produces must be converted, or rectified, to DC. While generators accomplish this using a mechanical commutator and brushes, alternators achieve this electronically with diodes. A *diode* can be thought of as an electrical check valve. It permits current to flow in only one direction—positive or negative. When a diode is placed in a simple AC circuit, one-half of the current is blocked. In other words, the current can flow from X to Y as shown in **Figure 11-10,** but it cannot flow from Y to X. When the voltage reverses at the start of the

next rotor revolution, the current again can pass from X to Y, but not back. This type of output is not efficient since current would be available only half of the time. Because only 50 percent of the AC voltage produced by the alternator is being converted to DC, this is known as **half-wave rectification.**

Full-Wave Rectification

By adding more diodes to the circuit, the full wave can be rectified to DC. In **Figure 11-11A,** current

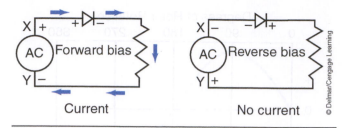

Figure 11-10 One diode in an AC circuit results in half-wave rectification. Only the positive half of the sine wave is allowed to pass.

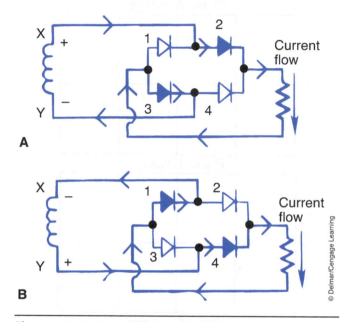

Figure 11-11 By adding more diodes to the circuit, current can now flow (A) from X to Y, and (B) from Y to X, resulting in rectifying both halves of the sine wave to DC current.

flows from X to Y. It flows from X, through diode 2, through the load, through diode 3, and then to Y. In **Figure 11-11B**, current flows from Y to X. It flows from Y, through diode 4, through the load, through diode 1, and back to X. In both cases, the current flows through the load in the same direction. Because all of the AC is now rectified, this is known as **full-wave rectification**. However, there remain brief moments when the current flow is zero; therefore, most alternators use three windings and six diodes to produce overlapping current pulses, thereby ensuring that the output is never zero.

Three-Phase Alternator

In a typical three-phase alternator, there are three coils and six diodes (**Figure 11-12**). At any time, two of the windings will be in series and the third winding will be neutral, doing nothing. Depending on the combination of coils and the direction of current flow,

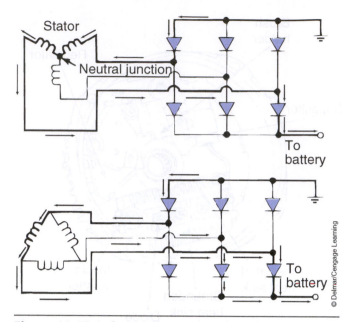

Figure 11-12 A three-phrase alternator has six diodes to provide full rectification.

one positive diode will rectify current flowing in one direction and a negative diode will rectify the current flowing in the opposite direction. This is true for both wye and delta configurations. The windings in a delta stator are arranged in parallel rather than series circuits. The parallel paths permit more current to flow through the diodes, thereby increasing the output of the alternator.

Brushless Alternators

Many alternators do not use brushes to deliver voltage to the field windings of the rotor. In this type of alternator, the field windings do not rotate with the rotor. They are held stationary while the rotor turns around them. The rotor itself retains sufficient residual magnetism to energize the stator when the vehicle engine is first started. Part of the voltage induced in the stator windings is then diverted to the field windings, which energizes the electromagnetic field. As the strength of the stationary field windings increases, the magnetic field of the rotor also increases, and stator output reaches its specified potential.

Many current alternators use a brushless design because they produce better longevity. The output ratings in current vehicles are necessarily higher because of the more complex electrical and electronic circuits.

Voltage Regulators

For more than 50 years prior to the introduction of solid-state voltage regulators in the early 1970s, mechanical regulators were used to control alternator (or generator) output. Like so many other mechanical components that have been replaced with electronic parts, the mechanical voltage regulator is slow and imprecise when compared to today's transistorized regulators. A wiring schematic of a solid-state voltage regulator is shown in **Figure 11-13**.

A voltage regulator defines the amount of current produced by the alternator and the voltage level in the charging circuit. It does this by turning the field circuit off and on. Without a voltage regulator, the batteries would be overcharged, and the voltage level in the electrical system would rise to a point where lights would burn out and fuses and fusible links would blow. Controlling the voltage level is particularly important as electronic components are added to vehicle electrical systems. Microprocessors and electronic sensors and switches are easily damaged by voltage spikes and high-voltage levels.

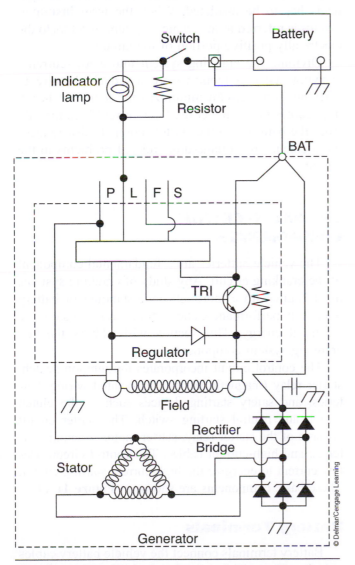

Figure 11-13 Schematic of AC Delcotron CS voltage regulator operation.

Sensing Voltage. The voltage regulator receives battery voltage as an input. This is called the **sensing voltage**; it allows the regulator to sense and monitor the battery voltage level. When the battery voltage rises to a specified level (approximately 13.5 volts), the regulator will turn the field current off. When the battery voltage drops below a set level, the regulator will again turn the field circuit back on. This on/off cycling of the field circuit takes place hundreds of times each second.

Remote Sensing. Due to resistance in the battery cables, the voltage output by the alternator drops by the time it arrives at the battery. This voltage drop can be as much as 0.5V. When an alternator is equipped with **remote sensing**, a second sensing wire reads actual voltage at the battery. This second wire signals the alternator to compensate for any voltage drop. The result is that battery charge times are reduced by half.

Temperature Factors. The ability of a battery to accept a charge varies with temperature. A cold battery needs a greater charging current to bring it up to capacity. In warm or hot weather, less current is needed. So the voltage regulator uses a thermistor to vary the voltage level at which the regulator will switch the field circuit on and off.

Types of Field Circuits. The field circuit, which is controlled by the voltage regulator, might be one of two types. If it is an **A-circuit voltage regulator**, it is positioned on the ground side of the rotor. Battery voltage is picked up by the field circuit inside the alternator. The regulator turns the field circuit off and on by controlling a ground. A **B-circuit voltage regulator** is positioned on the feed side to the alternator. Battery voltage is fed through the regulator to the field circuit, which is then grounded in the alternator. The regulator turns the field circuit off and on by controlling the current flow from the battery to the field circuit. Most voltage regulators are mounted either on or in the alternator. Some charging systems might have a regulator mounted separately from the alternator. A typical electronic regulator is an arrangement of transistors, diodes, zener diodes, capacitors, resistors, and thermistors.

Most alternators with internal regulators use current generated by the alternator stator to energize the field circuit. The regulator controls a ground to turn the field circuit off or on. Because the stator generates AC current, a set of three diodes is used to rectify the current to DC. These three diodes are often referred to as a *diode trio*. A separate sensing circuit delivers battery voltage to the regulator to control the on/off cycle of the alternator. **Figure 11-13** is a schematic that shows how current flow is managed by a typical voltage regulator. Battery current is delivered to the regulator through the sensing circuit.

Overvoltage Protection.

The best overvoltage protection device on a vehicle electrical system is the battery, which helps condition system voltage and smooth out high voltage spikes. Emergency operation without the battery or with deficiencies on the insulated side of the battery circuit can produce extreme voltage transients (short-duration spikes) that can damage both solid-state and electrical components. Some electrical circuits are susceptible to **load-dump**, a condition that occurs when current flow to a major electrical consumer is opened, creating a voltage spike.

Zener diodes can be installed in the rectifier bridge to limit high-energy voltage spikes. These can be used elsewhere in the electrical system to protect voltage-sensitive components. Other solid-state overvoltage protection devices can be used to protect the alternator ground circuit from voltage spikes; these are designed to short the alternator to ground at the excitation winding.

> **CAUTION** *Many alternators are not provided with reverse-polarity protection. Reverse-polarity connections such as those caused by jump-starting can destroy alternator diodes and numerous other chassis solid-state devices.*

STARTING CIRCUIT COMPONENTS

As stated at the beginning of this chapter, there are five basic components and two distinct electrical circuits **(Figure 11-1)** in a typical cranking system. The components are:

- Battery
- Key switch (or starter button)
- Battery cables
- Magnetic switch
- Starter motor

A starter motor draws high current from the batteries, typically upto 400 amperes. This current flows through the heavy-gauge cables that directly connect the battery to the starter. The driver switches the current flow using the ignition key. However, if the high-current cables were to be routed from the battery to the ignition key and then on to the starter motor, the voltage drop caused by resistance in the cables would be too great.

The starter motor is energized directly by the vehicle battery. The function of the starter motor is to convert electrical energy into sufficient mechanical torque to turn a high-compression diesel engine. The output torque from the starter motor pinion is imparted to ring gear teeth on the engine flywheel. A cranking circuit must be capable of starting an engine even under the most demanding conditions, such as cold weather extremes.

Ground and Insulated Circuits

Almost all current highway vehicles use negative chassis ground electrical circuits. In such a system, the entire chassis is used as the ground or negative path for current flow. This is known as the **ground circuit**. An advantage of chassis negative ground systems is that only the positive side of the current path to components has to be insulated. When the term **insulated circuit** is referred to in a vehicle system, it refers to the electrically positive portion of the circuit.

Dividing the cranking system into a low-current–carrying switching (control) circuit and a high-current–carrying starter circuit minimizes the length of heavy duty cables that conduct the high-current load directly from the batteries to the starter motor. It also reduces the incidence of voltage-drop–related problems in the circuit.

CRANKING CIRCUIT COMPONENTS

The vehicle batteries are a fundamental component of the cranking circuit. Any study of cranking systems must begin with a thorough understanding of battery operation, performance, and testing. Batteries store the electrical energy. This section will address the key cranking system components.

The control circuit incorporates the ignition switch, starter relay (magnetic switch), associated wiring, batteries, and safety starting devices such as the clutch switch and neutral starting switch. The starter circuit consists of the starter relay, starter motor solenoid, batteries, and high-current cables. The control circuit uses low current to energize the high-current starting circuit. These basic components are shown in **Figure 11-14.**

Battery Terminals

Battery terminals connect the vehicle battery cables to battery posts. They are the source of more cranking-circuit malfunctions than all other cranking circuit

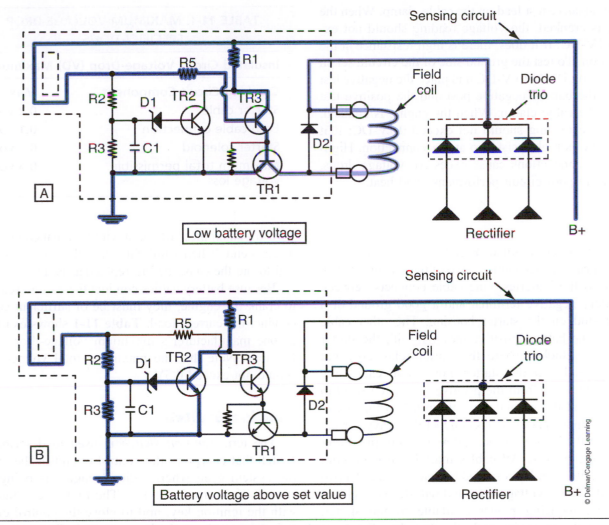

Figure 11-14 (A) A basic cranking circuit and (B) a cranking circuit with a thermostat.

problems combined, yet experienced technicians somehow find ways of skipping this key step in cranking-circuit troubleshooting. Battery terminals can either be lead collets that are bolted to threaded or recessed battery posts, or lead clamps that are bolted to lead battery posts. In most cases, the copper battery cable is soldered into a socket integral with the lead terminal. Terminals that are crimped or clamped to the socket tend to have higher failure rates. Battery terminals are susceptible to oxidation and corrosion, both of which are insulators that can create high resistance. High resistance anywhere in the cranking circuit will result in cranking-circuit problems.

Cleaning Terminals.   If any evidence of corrosion is observed on visual inspection, a battery terminal should be removed from the post and cleaned. Cleaning a battery post requires that acidic corrosion be neutralized and washed away using a solution of water and baking soda. Next, both the terminal clamp inside

bore and the battery post can be prepped using battery wire brushes. The objective of any battery terminal is to maximize the surface contact area between the connections—that is, provide the largest possible path for electron flow. When reinstalling a battery clamp, the bolt should be sufficiently torqued so that the clamp is moderately deformed to achieve this. When installing threaded terminals, the fastener should be torqued to the manufacturer specification.

Testing Terminals.   The only way to determine whether a battery terminal is doing its job is to test it by measuring voltage drop. **Voltage-drop testing** will be explained many times over in this book because it is a dynamic test. Voltage-drop testing is performed with a circuit energized, so when you are testing battery terminals it should be done while attempting to crank the engine. To test the insulated side of the circuit, set an auto-ranging digital multimeter (DMM) to V-DC and place the positive test lead on the battery positive post and the negative test lead on the cable clamp. When the engine is cranked, the voltage reading should not exceed 0.1V-DC; if it does, there is high resistance at the connection. To test the ground side of the circuit, set an auto-ranging DMM to V-DC and place the negative test lead on the battery negative post and the positive test lead on the cable clamp. When the engine is cranked, the voltage reading should not exceed 0.1V-DC; if it does, there is high resistance at the connection. High-resistance connections cause excessive voltage drops, resulting in poor circuit performance and heat.

Cables

The cranking circuit in **Figure 11-14** requires two or more heavy-gauge cables. Two of these cables attach directly to the batteries. One cable connects between the battery negative terminal and a good ground or a ground stud on the starter housing. The other cable connects the battery positive terminal with the starter relay. On vehicles where the starter relay does not mount directly on the starter motor, two cables are needed. One runs from the positive battery terminal to the relay and the second from the relay to the starter motor terminal. These cables conduct the heavy current load from the battery to the starter and from the starter back to the battery. All cables must be in good condition. Cables can be corroded by battery acid. Corrosion will cause a voltage drop and will decrease circuit amperage, reducing power available to the starter. Contact with the engine and other metal surfaces can fray the cable insulation. Deteriorated insulation can result in a dead short that can damage electrical components. A short to ground is the cause of many dead batteries and can result in fire. Cables must be heavy enough to carry the required current load.

When checking cables and wiring, check any fusible links in the wiring. Some vehicles are equipped with fusible links to protect wiring from overloads. Fusible links are different in construction from a fuse, but they operate in much the same way. The most common type consists of a wire with a special non-flammable insulation. Wire used to make fusible links is ordinarily two sizes smaller than the wire in the circuit it is designed to protect. When a fusible link is subjected to a current overload, the insulation becomes charred.

The largest fusible link is usually located at the starter solenoid battery terminal. From this terminal, current is distributed to the remainder of the vehicle electrical system. A second fusible link joins this battery terminal to the main body harness and protects the vehicle wiring. This link may take several forms, from a

TABLE 11-1: MAXIMUM VOLTAGE-DROP SPECIFICATIONS

Insulated Circuit Voltage-Drop (VD) Maximums

Cranking Circuit Component	Max VD
Starter cable	0.2 volt
Each cable connection	0.1 volt
Starter solenoid	0.3 volt
Maximum total permissible voltage loss	0.5 volt

wire to a small piece of metal with terminal connections on each end. When a link fails, troubleshoot the system and locate the cause before replacing the link.

Because battery cables conduct the current required to crank the engine, they must be of sufficient size to conduct the current load. **Table 11-1** shows examples of one manufacturer's maximum voltage-drop specifications. Note that the maximum total voltage drop should not exceed 0.5V.

Ignition Switch

The term *ignition switch* is used to describe the switch that energizes the control circuit in the cranking system even when a diesel engine is being discussed. There are two types. The first type is integral with the ignition key, and to close the control circuit to initiate cranking, the key is turned. The second type is a push-button and requires that the ignition key circuit be closed before the push-button starter is powered. In both cases, the switch is spring-loaded to the open position: This means that it must be either turned (by key) or pushed (by thumb) against the spring pressure to close the starter control circuit and initiate cranking.

When the ignition switch is closed, the coil in the **starter relay** (magnetic switch) is energized, enabling current flow in the high-current starter circuit. In some newer chassis, ignition switch status is broadcast on the chassis data bus. Always observe the manufacturer service procedures when troubleshooting and replacing ignition switches.

Starter Relay

The starter relay is a magnetic switch that enables the control circuit to open and close the high-current cranking circuit. It is either mounted at a remote location from the starter or built into the front end of the starter solenoid assembly. Remote-mounted starter

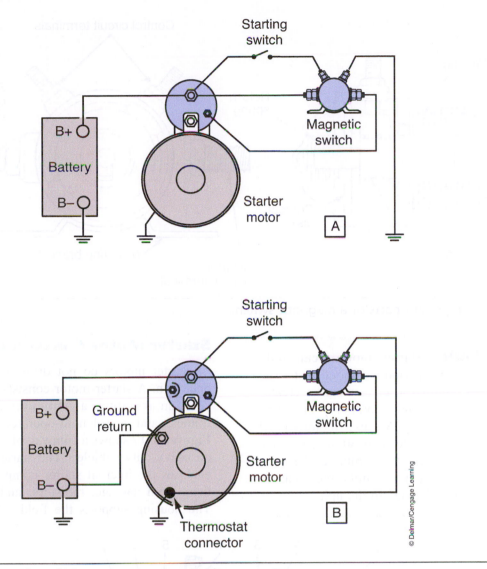

Figure 11-15 A magnetic switch, or starter relay circuit.

relays are located close to the battery or starter to keep the cables as short as possible. **Figure 11-15** shows a starter relay mounted on the vehicle firewall near the windshield wiper motor on the driver's side. On some vehicles, the starter relay can be mounted on a frame crossmember.

The starter relay consists of an electromagnet plunger, contact disc, and two springs **(Figure 11-16)**. The electromagnet has a hollow core in which the plunger moves.

Starter Relay Operation. When the operator turns the ignition switch to the crank position, battery current flows through the starting switch to a control circuit terminal on the starter relay. Control circuit current flows through the windings in the electromagnet, creating a magnetic field that pulls the plunger into the hollow core. Spring pressure then forces the

contact disc against the starter circuit terminals, closing the circuit. High-amperage current flows through contacts to the starter motor.

When the engine starts, the ignition switch is released, and the control circuit is opened. This deactivates the electromagnet in the starter relay. The return spring forces the plunger out of the hollow core, which moves the contact disc away from the starter circuit terminals, interrupting current flow to the starter motor.

Starter Motors

The starter motor converts electrical energy from the battery into mechanical energy for cranking the engine. The starter is an electric motor designed to operate under great electrical loads and produce high torque.

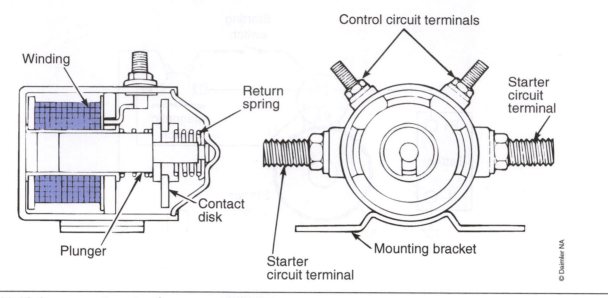

Figure 11-16 Component parts of a magnetic switch.

Shop Talk: Starter motors can only operate for short periods without rest. The high current required to drive a starter creates considerable heat, and continuous operation will cause overheating. A starter motor should never operate for more than 30 seconds at a time and should rest for 2 minutes between cranking cycles. This permits the heat to dissipate without damage to the unit.

Starter Motor Construction

Starter motors do not differ much in design and operation. A starter motor consists of a housing, **field coils**, an **armature**, a **commutator**, and brushes, end frames, and a solenoid-operated shift mechanism. **Figure 11-17** shows a cutaway of a typical starter used on heavy duty vehicles. The starter housing or frame encloses the internal starter components and protects them from damage, moisture, and foreign materials. The housing supports the field coils **(Figure 11-18)**

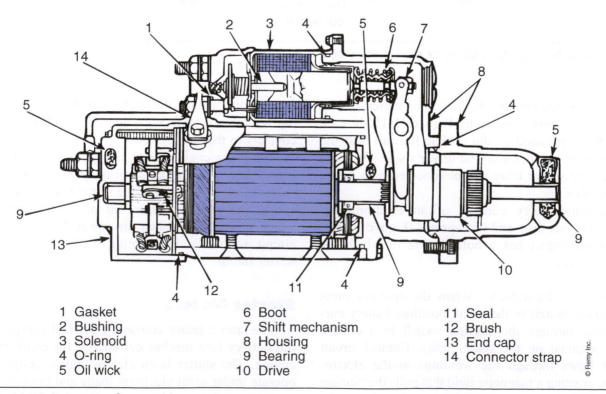

1 Gasket	6 Boot	11 Seal
2 Bushing	7 Shift mechanism	12 Brush
3 Solenoid	8 Housing	13 End cap
4 O-ring	9 Bearing	14 Connector strap
5 Oil wick	10 Drive	

Figure 11-17 Cutaway of a cranking motor.

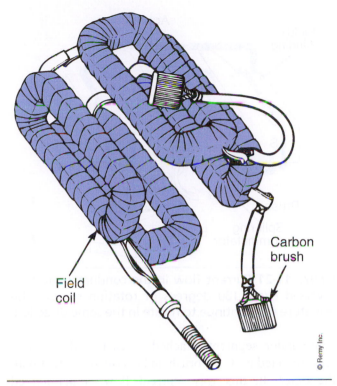

Figure 11-18 Four field coils used in a starter motor.

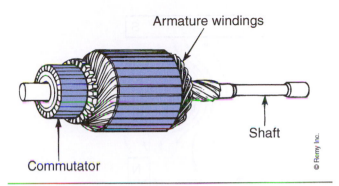

Figure 11-19 Starter motor armature and commutator.

and forms a conducting path for the magnetism produced by the current passing through the coils.

Field Coils. The field coils and their pole shoes are securely attached to the inside of the iron housing. They are insulated from the housing but are connected to a terminal that protrudes through the outer surface of the housing. The field coils and pole shoes are designed to produce strong stationary electromagnetic fields within the starter body as current is passed through the starter. These magnetic fields are concentrated at the pole shoes. Field coils will have a N or S magnetic polarity depending on the direction the current flows. The coils are wound around respective pole shoes in opposite directions to generate opposing magnetic fields. The field coils connect in series with the armature winding through the starter brushes. This design permits all current passing through the field coil circuit to also pass through the armature windings.

Armature. The armature **(Figure 11-19)** is the rotating component of the starter. It is located between the drive and commutator end frames and the field windings. When the starter motor operates, the current passing through the armature produces a magnetic field in each of its conductors. The reaction between the armature's magnetic field and that of the field coils causes the armature to rotate. This creates the torque required to crank the engine.

Armature Windings. The armature has two main components: the armature windings and the commutator. Both mount to the armature shaft. The armature windings are not made of wire. Instead, heavy flat copper strips that can handle heavy current flow are used. The windings are constructed of several coils of a single loop each. The sides of these loops fit into slots in the armature core or shaft, but are insulated from it. Each slot contains the side of two of the coils. The coils connect to each other and to the commutator so that current from the field coils flows through all of the armature windings at the same time. This action generates a magnetic field around each armature winding, resulting in a repulsion force all around the conductor. It is this repulsion force that causes the armature to turn.

Commutator. The commutator assembly presses onto the armature shaft. It is made up of heavy copper segments separated from each other and the armature shaft by insulation. The commutator segments connect to the ends of the armature windings. Starter motors have four to twelve brushes that ride on the commutator segments and carry the heavy current flow from the stationary field coils to the rotating armature windings via the commutator segments. The brushes are held in position by a brush holder.

Starter Operation

The starter motor converts electric current into torque or twisting force through the interaction of magnetic fields. It has a stationary magnetic field (created by passing current through the field coils) and a current-carrying conductor (the armature windings). When the armature windings are placed in this stationary magnetic field and current is passed through the windings, a second magnetic field is generated with its lines of force wrapping around the wire **(Figure 11-20).** Since the lines of force in the stationary magnetic field flow in one direction across the winding, they combine on one side of the wire,

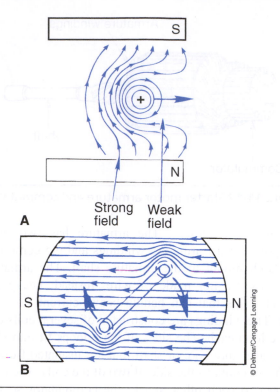

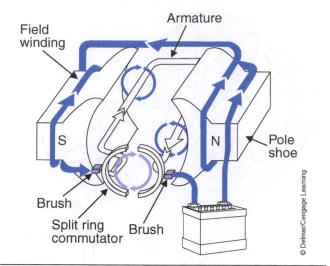

Figure 11-21 Current flow in the conductor must be reversed every 180 degrees of rotation so that the armature will continue to rotate in the same direction.

Figure 11-20 (A) When a current-carrying conductor passes through a magnetic force field, the field is deflected to one side of the conductor. This creates pressure on one side of the wire and forces the conductor to move away from the strong force field. (B) When a loop of wire is placed between two magnets, the intersection between the force fields causes the loop to rotate around its axis.

increasing the field strength, but are opposed on the other side, weakening the field strength. This creates an unbalanced magnetic force, pushing the wire in the direction of the weaker field.

Current Flow. Because the armature windings are formed in loops or coils, current flows outward in one direction and returns in the opposite direction. This means the magnetic lines of force are oriented in opposite directions in each of the two segments of the loop. When placed in the stationary magnetic field of the field coils, one part of the armature coil is pushed in one direction while the other part is pushed in the opposite direction. This causes the coil, and the shaft to which it is mounted, to rotate. Each end of the armature windings is connected to one segment of the commutator. Two carbon brushes are connected to one terminal of the power supply. The brushes contact the commutator segments conducting current to and from the armature coils **(Figure 11-21).**

As the armature coil turns through a half revolution, the contact of the brushes on the commutator causes the current flow to reverse in the coil. The

commutator segment attached to each coil end will have traveled past one brush and is now in contact with the other. In this way, current flow is maintained constantly in one direction, while allowing the segment of the rotating armature coils to reverse polarity as they rotate. This ensures that the armature will turn in one direction.

Armature Segments. In a starter motor, many armature segments must be used. As one segment rotates past the stationary magnetic field pole, another segment immediately takes its place. The turning motion is made uniform and the torque needed to turn the flywheel is constant rather than fluctuating, as it would be if only a few armature coils were used. Starter motors are series motors. Current is first flowed through the field coils, then routed to the armature. This produces maximum torque at initial start. All series-wound motors produce maximum torque at low rpm close to stall.

Starter Solenoids

The solenoid-operated shift mechanism **(Figure 11-22)** is mounted in a solenoid housing that is sealed to keep out oil and road splash. The case is flange mounted to the starter motor housing. It contains an electromagnet with a hollow core. A plunger is installed in the hollow core, much like the starter relay described earlier. The solenoid performs two functions.

- When energized, plunger movement acts on a shift lever that shifts the starter motor drive pinion into mesh with the engine flywheel teeth so that the engine can be cranked.

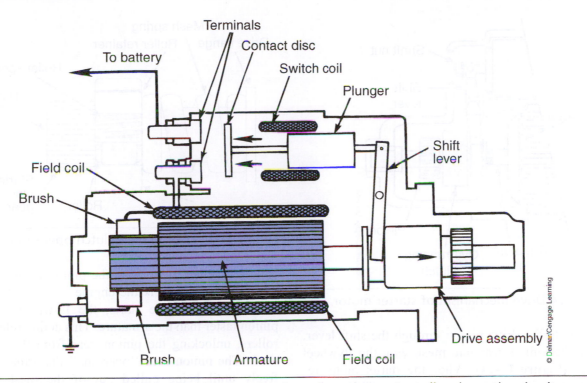

Figure 11-22 The solenoid engages the drive mechanism when the contact disc closes the circuit across the terminals.

- The solenoid also closes a set of contacts that allows battery current to flow to the starter motor; this ensures that the pinion and flywheel are engaged before the starter armature begins to rotate.

Solenoid Windings. The solenoid assembly has two separate windings: a pull-in winding and a hold-in winding **(Figure 11-23)**. The two windings have approximately the same number of turns but are wound from different-gauge wire. Together these windings produce the electromagnetic force needed to pull the

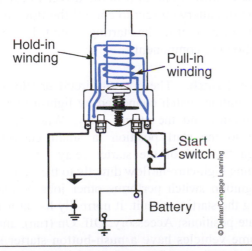

Figure 11-23 The starter solenoid has two windings: a pull-in winding and a hold-in winding.

plunger into the solenoid coil. The heavier-gauge pull-in windings draw the plunger into the solenoid, while the lighter-gauge hold-in windings produce enough magnetic force to hold the plunger in this position.

Current Routing. Both windings are energized when the starting switch is turned to the start position. If the system has a remote starter relay, battery current passes through the relay to the "F" terminal on the solenoid. When the plunger contact disc touches the solenoid terminals, the pull-in winding is deactivated. At the same time, the plunger contact disc makes the motor feed connection between the battery and the starting motor, directing full battery current to the field coils and starter motor armature for cranking power.

Shop Talk: In almost all diesel engine starters, the solenoid performs the function of a relay. The control circuit is wired to the windings of the solenoid. Battery current is routed directly to the starter motor through the drive solenoid.

Initiating Cranking. As this electrical connection is being made at the terminal end of the solenoid, the mechanical motion of the solenoid plunger is being

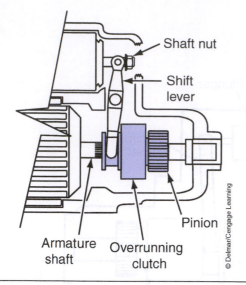

Figure 11-24 Drive mechanism of starter motor.

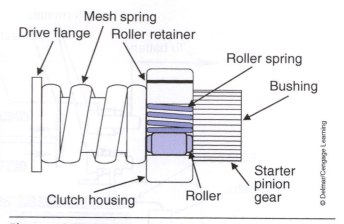

Figure 11-25 Overrunning clutch operation.

transferred to the drive pinion through the shift lever, bringing the pinion gear into mesh with the flywheel ring gear **(Figure 11-24).** When the starter motor receives current, its armature starts to turn. This motion is transferred through an overrunning clutch and pinion gear to the engine flywheel, and the engine is cranked.

With this type of solenoid-actuated direct-drive starting system, teeth on the pinion gear might not immediately mesh with the flywheel ring gear. If this occurs, a spring located behind the pinion compresses so that the solenoid plunger can complete its stroke. When the starter motor armature begins to turn, the pinion teeth will mesh with the flywheel teeth under spring pressure.

Overrunning Clutches. The overrunning clutch performs an important job in protecting the starter motor. When the engine starts and runs, its speed increases. If the starter motor remained connected to the engine through the flywheel, the starter motor would spin at very high speeds, destroying the armature windings. To prevent this, the starter must be disengaged from the engine as soon as the engine turns more rapidly than the starter. But with a solenoid-actuated drive mechanism, the pinion remains engaged with the flywheel until current stops flowing to the starter. In these cases, an overrunning clutch is used to disengage the starter. A typical overrunning clutch is shown in **Figure 11-25.**

The clutch housing is internally splined to the starting motor armature shaft. The drive pinion turns freely on the armature shaft within the clutch housing. When the clutch housing is driven by the armature, the spring-loaded rollers are forced into the small ends of their tapered slots and wedged tightly against the pinion barrel. This locks the pinion and clutch housing

solidly together, permitting the pinion to turn the flywheel and, thus, crank the engine.

When the engine starts, the flywheel spins the pinion faster than the armature. This action releases the rollers, unlocking the pinion gear from the armature shaft. The pinion then "overruns" the armature shaft freely until being pulled out of the mesh without stressing the starter motor. Note that the overrunning clutch is moved in and out of mesh with the flywheel by a linkage operated by the solenoid.

Thermostats. Some starter motors are equipped with a thermostat (refer back to **Figure 11-14B**). The **thermostat** monitors the temperature of the motor. If prolonged cranking causes the motor temperature to exceed a safe threshold, the thermostat opens and the starter current is interrupted. The starter motor, then, will not operate until the motor cools and the thermostat closes.

Control Circuit

The control circuit allows the driver to use a small amount of battery current to control the flow of a large amount of current in the starting circuit. It consists of the following components.

Ignition Switch. The control circuit usually consists of an ignition switch connected by light-gauge wire to the batteries and the starter relay. When the key is turned to the start position, a small current flows through the coil of the starter relay, closing it and switching high-current flow directly to the starter motor. The ignition switch performs other jobs besides controlling the starting circuit. It normally has at least four separate positions: Accessory, Off, On (run), and Start.

Some vehicles have a push-button starter switch. Battery voltage is available to the switch when the ignition switch is in the On position. When the push-button is depressed, current flows through the control circuit to the starter relay coil. Electronic push-button starter switches may require that data bus wake-up is complete, and either the clutch or brake to be fully depressed before the control circuit closes to begin cranking.

Neutral Safety Switch. The **neutral safety switch** prevents vehicles with automatic transmissions from being started in gear. Neutral safety switches are located in either of two places in the control circuit. One position is between the ignition switch and the starter relay. Placing the transmission in Park or Neutral will close the switch so current can flow to the starter relay. The neutral safety switch can also be connected between the starter relay and ground so that the switch must close before current can flow from the starter relay to ground.

Starter Relays. The starter relay covered earlier in the chapter is the point in the cranking circuit where the control circuit and starter circuit come together. Low current in the control circuit passes through the starting switch and neutral safety switch to energize the starter relay and activate the starter circuit.

Reduction-Gearing Starter Motors

A recent innovation in engine starter technology has been the introduction of reduction-gearing starters. This enables a significant size and weight reduction of the starter. In a reduction-gearing starter motor, a planetary gear set is used to achieve a reduction ratio of between 3:1 and 4:1. In this type of starter, the sun gear is the input (the armature stub), the planetary carrier is the output (drive pinion), and the ring gear that surrounds the four planetary gears is held stationary. A planetary gear set arrangement such as this provides for maximum rpm reduction and maximum output torque.

ELECTRICAL TROUBLESHOOTING

This section covers some of the basics of troubleshooting cranking and charging circuits. When diagnosing electrical problems on multiplexed vehicles (J1850), it becomes critically important to adhere to the manufacturer's troubleshooting procedure. Failure to do this can result in costly repairs.

Charging System Failures and Testing

A malfunction in the charging system results in either an overcharged battery or an undercharged battery.

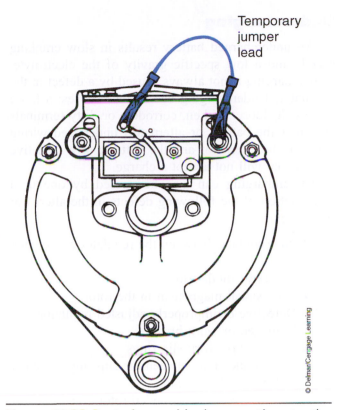

Temporary jumper lead

© Delmar/Cengage Learning

Figure 11-26 Restoring residual magnetism to the rotor.

Shop Talk: Alternators that energize the field circuit with current produced by the stator windings rely on residual magnetism in the rotor to initially energize the stator when the engine is being started. During handling or repair, this residual magnetism can be lost. It must be restored before testing the system. This is done by connecting a jumper wire between the diode trio terminal and the alternator output terminal as shown in **Figure 11-26**.

Overcharging

An overcharged battery will produce water loss, eventually resulting in hardened plates and the inability to accept a charge. Overcharging can be caused by one or a combination of the following:

- Defective battery
- Defective or improperly adjusted regulator
- Poor sensing lead contact to the regulator or rectifier assembly.

A high percentage of overcharging problems are caused by either a defective or improperly adjusted voltage regulator.

Undercharging

An undercharged battery results in slow cranking speeds and a low specific gravity of the electrolyte. Undercharging is not always caused by a defect in the alternator. Undercharging has various causes: a loose drive belt; loose, broken, corroded, or dirty terminals on either the battery or alternator; undersized wiring between the alternator and the battery; or a defective battery that will not accept a charge.

Undercharging can also be caused by one or a combination of the following defects in the alternator field circuit:

- Poor contacts between the regulator and carbon brushes
- Defective diode trio
- No residual magnetism in the rotor
- Defective or improperly adjusted regulator
- Damaged or worn brushes
- Damaged or worn slip rings
- Poor connection between the slip ring assembly and the field coil leads
- Shorted, open, or grounded rotor coil
- Open remote sensing wire
- Problems with alternator drive pulleys or belts

Undercharging can also be the result of a malfunction in the generating circuits. One or more of the stator windings (phases) can be shorted, open, or grounded. The rectifier assembly might be grounded, or one or more of the diodes might be shorted or open.

Testing the Charging System

Determining battery's state of charge is the first step in testing the alternator. The battery must be at least 75 percent charged before the alternator will perform to specifications. (Some manufacturers specify that the battery must be 95 to 100 percent charged before testing the alternator.) Recharge the battery if the charge is low; if it will not accept a charge, a fully charged test battery should be installed in its place before testing the charging system.

Sourcing Problems. After verifying that the battery is fully charged and is functioning correctly, the charging system can be performance tested. There are basically three tests, although some manufacturers might specify more. First, the output of the alternator is tested. If the output is below specifications, the voltage regulator is bypassed and battery current is wired directly to the field circuit of the rotor. This is called *full fielding* the rotor. If this corrects the problem, the fault is in the regulator; if the output remains low with the regulator

bypassed by full fielding, the alternator might be defective. However, before condemning the alternator or regulator, voltage-drop tests should be performed on the system wiring to determine if high resistance could be the cause of the charging problem.

Alternator Output Testing

To test the maximum output of an alternator, follow these steps:

1. Disconnect the ground battery cable from the alternator.
2. Install an ammeter in series with the alternator output terminal and the ground battery cable or use an inductive pickup. Also install a voltmeter and a carbon pile across the terminals of the battery **(Figure 11-27)**.
3. Start the engine and run it fast enough to obtain maximum output from the alternator (typically above 1,500 rpm).
4. Turn on all accessories and increase the load on the carbon pile until the voltage in the system drops to 12.7V. This will cause the regulator to send full-fielded voltage to the rotor, maximizing the alternator output.

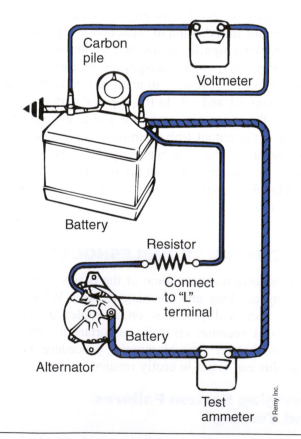

Figure 11-27 Test connections for an ammeter, voltmeter, and carbon pile.

5. Compare the amperage reading with the manufacturer's specifications. If the output is within 10 percent of rated amperage, the alternator is good. If the output is more than 10 percent below specs, full-field test the alternator.

Shop Talk: An inductive (amp) clamp ammeter is often used instead of a series ammeter. It is clamped onto the ground battery cable instead of connected in series with it. This method is preferred because the cable is not required to be disconnected.

Full-Field Testing the Alternator

By applying full battery voltage directly to the field windings in the rotor, it can be determined whether or not the regulator is the cause of an undercharging condition. There are two variations of this procedure that apply to alternators with external regulators. If the field circuit is grounded through the regulator (an A circuit), the regulator is disconnected from the field terminal on the alternator and a jumper is connected between the terminal and a ground. If the alternator receives battery voltage through the regulator (a B circuit), the regulator is disconnected from the field (F) terminal and a jumper is connected to the terminal and to the insulated battery terminal. In either case, the regulator circuit is bypassed completely, and full battery voltage is available to the rotor.

Now, repeat the procedure outlined for testing the alternator output. With the load applied, observe whether or not the alternator output rises to rated amperage. If it does, the alternator is functioning correctly, and the regulator must be replaced. If it does not, the alternator must be further tested to determine the cause of undercharging.

CAUTION *When testing the output of a full-fielded alternator, carefully observe the rise in system voltage. Because the current output is not regulated, battery voltage can quickly rise to an excessive level, high enough to overheat the batteries, causing electrolyte to spew from the vent holes and possibly damage sensitive electronic components. Do not allow system voltage to rise above 15V.*

Shop Talk: Some alternators with remote-mounted electronic regulators are connected to the regulators by a wiring harness and a multipin connector. Full fielding the field circuit is accomplished by removing the connector from the regulator and connecting a jumper wire between two pins (terminals) in the harness connector (consult manufacturer service literature to correctly identify the pin assignments). Doing so bypasses the regulator, sending battery current directly to the field circuit.

Full Fielding Internally Regulated Alternators

Full-field testing is not possible on some alternators with internal voltage regulators. To isolate and test the regulator on these types of alternators, the alternator must be removed from the vehicle and disassembled. Other internally regulated alternators have a hole through which the field circuit can be tested. The test will require the use of a short jumper with insulated clips and a stiff paper clip wire or a $\frac{1}{32}$-inch drill bit (Figure 11-28).

1. With the engine off and all electrical accessories turned off, measure the voltage across the battery terminals. Make a note of the reading.
2. Start the engine and run it at the speed necessary to generate full alternator output.
3. Connect a short jumper to the alternator negative output terminal and to the shank of the straightened clip wire or drill bit.

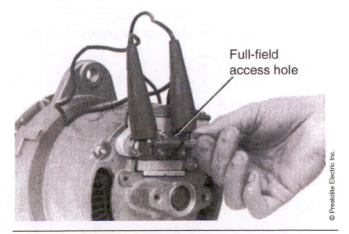

Full-field access hole

© Prestolite Electric Inc.

Figure 11-28 Full fielding an alternator with an integral voltage regulator.

4. Insert the wire or bit into the full-field access hole as far as it will go and make a note of the voltage reading. If a fault in the regulator or diode trio is causing the undercharged condition, the alternator output should climb to within 10 percent of its rated output.

> **CAUTION** *Never full field an alternator without applying an electrical load to the batteries. Whenever the voltage regulator is bypassed, there is nothing to control peak alternator output; this can cause voltage spikes that can damage both electrical and electronic components.*

Stator Winding Testing

If full fielding the alternator does not solve the undercharging condition, the regulator is probably okay and the problem is within the alternator. On some alternators, such as the one shown in **Figure 11-29,** the individual stator windings can be tested without removing the alternator from the vehicle. The windings are tested by connecting an AC voltmeter across the alternator AC terminals 1 and 2, 1 and 3, and 2 and 3. These readings should be approximately the same if the stator is okay. If the readings vary, the stator probably has an open or short. If the readings are approximately the same, the stator is okay and some other fault exists in the alternator, causing the undercharging condition. The alternator will have to be removed from the vehicle for further testing if the problem is not found in the circuit wiring.

Diode Trio Test. If an alternator with an internal regulator is undercharging, the problem could be a defective diode trio. If one or more of the diodes are defective, full-field voltage will not be delivered to the rotor, resulting in low alternator output. Some alternators have a diode trio that can be removed without

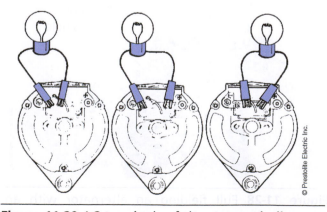

Figure 11-29 AC terminals of the stator windings.

removing the alternator from the vehicle. This permits a known good diode trio to be installed in its place. If this corrects the undercharging problem, the fault has been located.

Worn Brushes. Undercharging might also be caused by worn or corroded brushes or contact pads in the regulator. On some alternators, the regulator can be removed to provide access to the brushes and the contact pads. If the brushes appear burned, cracked, or broken, or if they are worn to a length of $^3/_{16}$ inch or less, they should be replaced. Brushes should also be replaced if the shunt lead inside the brush spring is broken. If the brushes are okay, the contact caps on the brushes and the contact pads on the regulator should be cleaned, using 600-grit (or finer) grade sandpaper or crocus cloth.

Regulator Circuit Testing

In addition to providing current to the field windings of the rotor, the voltage regulator must also keep the system voltage within a predetermined range, typically 13.5 to 14.5 on a 12V system. The regulator must reduce the output of the alternator when necessary to keep system voltage below the set level. If it does not, the battery will be overcharged. To test system voltage regulation, follow this procedure:

1. Connect a volt-amp tester to the batteries as described earlier under "Alternator Output Testing."
2. Start the engine and run it at the speed necessary to achieve full output from the alternator.
3. Observe the voltage readings and current output. As voltage approaches 13.8 to 14.5 volts, alternator output should slowly decrease. When the batteries are fully charged, the output of the alternator should be low. If alternator output remains high after system voltage reaches its specified peak, the voltage in the system will rise above 15V. Batteries will overheat, begin to gas, and might even start spewing electrolyte through vent holes. The problem might be a defective regulator, a short in the field circuit, or high resistance in the wiring. For example, unwanted resistance in the sensing wire will cause the voltage regulator to read system voltage at a lower value than it actually is. This would result in the alternator generating current to fully charged batteries.

Test Alternator Ground Circuit. To test the ground of the regulator, install a voltmeter between the case of the regulator (if externally mounted) and the battery ground. Start the engine and measure the voltage difference between the two points. With the engine running, there should not be any difference between the two ground points. The voltmeter should read zero.

Test Sensing Circuit. If the ground side tests okay, measure the voltage supplied to the regulator by the sensing circuit. It should be equal to battery voltage. If resistance is high on the ground side or if voltage is low on the sensing (supply) side of the regulator, the problem might be loose or corroded connections or a partial open in the circuit. If resistance and supply voltage meet specifications, the overcharging condition is the fault of either the alternator or the regulator.

Check for High Resistance in Charging Circuit. If there is excessive resistance in the charging circuit, the alternator might not fully charge the battery during peak load periods. The design of the alternator limits current output to a maximum level. High resistance in the circuit will prevent full current from reaching the battery. During peak load demand, the batteries will not be fully charged and, over time, can be damaged.

To test voltage drop in the charging circuit, connect the volt-amp tester as explained earlier and load the battery to full alternator output. Then, connect a voltmeter from the output terminal of the alternator and the insulated terminal of the battery. The voltage drop between the two points should be low (0.2V on alternators rated at 14 and 0.5 volts on alternators rated at 28V is a typical maximum). If the voltage drop is higher than the manufacturer's specifications, there might be loose or corroded connections in the circuit or a fusible link might be degenerating.

The ground side of the circuit should also be tested. Move the voltmeter leads to the alternator casing and the battery ground pole. The voltage drop in this case should be zero. If not, the alternator mounting might be loose or corrosion might be built up between the casing and mounting bracket.

AC Leakage Test. An important test of the diode rectifier bridge is the AC leakage test. Over time, the diode bridge can start to "leak" AC current to ground. Typically, AC leakage should not exceed 0.3V-AC, but check the manufacturer maximum specification. Use a DMM and switch to AC voltage. Place one test lead on the alternator-insulated terminal and connect the other lead to chassis ground. Record the voltage specification produced when the engine is run at above 1,500 rpm. A within-specification-but-high reading can be an indication of an alternator that will fail sooner rather than later.

© Delmar/Cengage Learning

Figure 11-30 Poly-V belt-driven Delcotron 24SI alternator on a Cummins ISB engine.

> *Tech Tip:* A commonly overlooked cause of an undercharging alternator is worn or loose belts. Checking the belt drive should be one of the first checks when diagnosing an undercharge condition. **Figure 11-30** shows a poly-V belt-driven alternator on a Cummins ISB engine.

Hand-Held Diagnostic Service Tools

Recently, a variety of hand-held diagnostic tools have been introduced to simplify and increase the accuracy of charging circuit diagnostics. Most function on similar principles, and manufacturers have come to regard them as a benchmark for assessing component warranties. It should be noted that these are not 100 percent reliable, especially in cases where circuit wiring may be corroded. One of the more common of these instruments is the Midtronics inTELLECT EXP. The operation of the inTELLECT EXP is outlined in **Figure 11-31.** When using this class of hand-held diagnostic instruments, the technician is required to do no more than make the appropriate connections and read the test results.

Cranking Circuit Diagnostics

The cranking circuit requires testing when the engine will not crank, when the engine cranks slowly, or when the starter motor will not turn.

Preliminary Checks. Cranking output obtained from the motor is affected by the condition and charge of the battery, the wiring circuit, and the engine cranking requirements. The battery should be checked and charged as needed before testing. Ensure that the batteries are rated to meet or exceed the vehicle manufacturer's recommendations. The voltage rating

The **Internal Batteries Status Indicator**, which appears in the screen's top left corner, lets you know the status and charge level of the analyzer's 6 1.5 V batteries. The X shown in the figure shows that the EXP is powered by the battery you are testing to conserve the internal batteries.

Press the two **Soft Keys** linked to the bottom of the screen to perform the functions displayed above them. The functions change depending on the menu or test process. So it may be helpful to think of the words appearing above them as part of the keys. Some of the more common soft-key functions are SELECT, BACK, and END.

When you first connect the EXP to a battery it functions as a voltmeter. The voltage reading appears above the left soft key until you move to other menus or functions.

In some cases, you can use the **Alphanumeric Keypad** to enter numerical test parameters instead of scrolling to them with the **ARROW** keys.

You will also use the Alphanumeric Keys to create and edit customer coupons. The keypad includes characters for punctuation. To add a space, press the **RIGHT** and **LEFT ARROW** keys simultaneously.

The **Title Bar** shows you the name of the current menu, test tool, utility, or function.

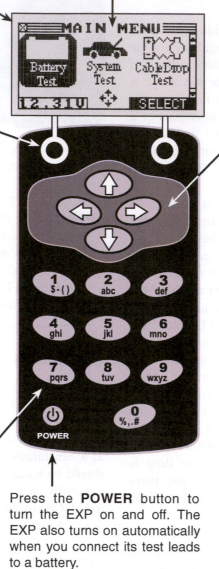

Press the **POWER** button to turn the EXP on and off. The EXP also turns on automatically when you connect its test leads to a battery.

Whichever way you turn on the EXP, it always highlights the icon and setting you last used for your convenience.

The **Selection Area** below the **Title Bar** contains items you select or into which you enter information. The area also displays instructions and warnings.

The **Directional Arrows** on the display show you which **Arrow Keys** to press to move to other icons or screens. The Up and Down Directional Arrows, for example, let you know to press the **UP** and **DOWN ARROW** keys to display the screens that are above and below the current screen.

The Left and Right Directional Arrows let you know to use the **LEFT** or **RIGHT ARROW** keys to highlight an icon for selection.

Another navigational aid is the **Scroll Bar** along the right side of the screen. The position of its scroll box tells you which menu screen you are viewing.

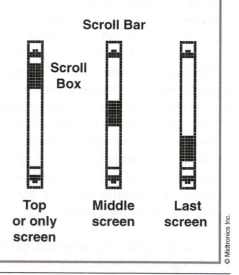

Figure 11-31 Main menu and keypad orientation of the Midtronics inTELLECT EXP.

of the batteries must also match the voltage rating of the starter motor.

Shop Talk: The starter should not be operated if the voltage at the battery tests below 9.6V. Some leasing companies now use

a voltage-sensing module to prevent starter operation if voltage is below 9.6V, to prevent drivers from destroying starters.

Check the wiring for clean, tight connections. Loose or dirty connections will cause excessive voltage drop.

Clean and tighten all connections as necessary. The cranking system cannot operate properly with excessive resistance in the circuit.

The engine crankcase should be filled with the proper weight oil as recommended by the engine manufacturer. Heavier-than-specified oil combined with low temperature lowers cranking speed to the point where the engine will not start.

Check the starting switch for loose mounting, damaged wiring, sticking contacts, and loose connections. Check the wiring and mounting of the safety switch, if so equipped, and make certain the switch is properly adjusted. Check the mounting, wiring, and connections of the starter relay and starter motor.

In starters equipped with a thermostat, you must check the thermostat circuit if the starter does not crank. Check the resistance between the two thermostat terminals on the motor. The ohmmeter should read close to zero. If it does not, the thermostat is open-circuited. Do not check the thermostat when the starter motor is hot: The thermostat operates by opening the circuit when a specified temperature is exceeded; this makes cranking impossible until the temperature has dropped below the threshold.

Troubleshooting the Cranking Circuit

Systematic troubleshooting is essential when servicing the starting system. A high percentage of "defective" starters returned on warranty claims function to specification when tested. This results from poor or incomplete diagnosis of the starting circuit. **Table 11-2** itemizes a systematic approach to starting circuit diagnosis. Testing the starting system can be divided into area tests, which check voltage and current in the entire system, and more detailed pinpoint tests, which target one particular component or segment of the wiring circuit.

Starter Relay Testing.
The starter relay bypass test is a simple method of determining if the relay is operational. This test should be performed when the starter motor does not activate when the ignition key is in the Start position (or when the starter button is depressed).

Ensure that the transmission is in neutral. Connect a jumper cable around the starter relay to bypass the relay. If the engine cranks with the jumper installed, the starter relay or the control circuit may be defective. A starter relay also can be checked by connecting a voltmeter across the winding, from the push-button key start connection to one of the mounting bolts that attaches the switch to the bulkhead. Have someone hold the starting switch closed. If the voltmeter reading

is 0 volt, check for an open circuit. If the voltmeter reading is less than 11 volts, check for corroded or loose connections. Repair or replace any damaged wires. Then, check the voltage to the starter relay again. If the voltmeter reading is now 11 volts or more, the relay and control circuit are okay. If less than 11 volts, perform a voltage-drop test across the power terminals.

Cranking Current Testing.
This on-engine cranking current test measures the amperage that the starter circuit draws to crank the engine. This amperage reading can be useful in isolating the source of certain types of starter problems. Before beginning, verify that the batteries are fully charged.

1. Connect the leads of a volt-amp tester.
2. Set the carbon pile to its maximum resistance (open). Install an inductive pickup clamp if equipped.
3. Crank the engine and observe the voltmeter reading.
4. Stop cranking. Adjust the carbon pile until the voltmeter reading matches the reading taken in step 3.
5. Note the ammeter reading.

Shop Talk: If the analyzer uses an inductive pickup (amp clamp), ensure that the arrow on the inductive pickup is pointing in the right direction as specified on the ammeter. Then, crank the engine for 15 seconds and observe the ammeter reading.

Compare the reading obtained during testing to the manufacturer's specifications. **Table 11-3** summarizes the most probable causes of high or low current draw. If the problem appears to be caused by high resistance in the circuit, test as shown in the next section.

Control Circuit Testing.
The control circuit test verifies the performance of the wiring and components used to activate the starter relay. The control circuit can be checked by connecting a voltmeter across the coil terminals of the solenoid or starter relay. Remove the positive battery cable from the solenoid/starter when performing the test. This allows both windings of the solenoid to energize while preventing the cranking motor from turning. If there are other leads connected to the main positive terminal of the solenoid, you will have to remove these and temporarily

TABLE 11-2: TROUBLESHOOTING A CRANKING CIRCUIT

Problem	Possible Cause	Tests and Checks	Remedy
Engine cranks slowly or unevenly.	1. Weak battery.	1. Perform battery open circuit voltage and load voltage tests. Perform battery load tests (capacity). Check capacity and voltage ratings against engine requirements.	1. Service, recharge, or replace defective battery.
	2. Undersized or damaged cables.	2. Perform visual inspection.	2. Replace as needed.
	3. Poor starter circuit connections.	3. Perform visual inspection for corrosion and damage.	3. Clean and tighten. Replace worn parts.
	4. Defective starter motor caused by high internal resistance.	4. Perform cranking current test and no-load test.	4. If cranking current is under specs, proceed with no-load bench testing.
	5. Engine oil too heavy for application.	5. Check oil grade.	5. Change oil to proper specs.
	6. Seized pistons or bearings.	6. Check compression and cranking torque.	6. Repair as needed.
	7. Overheated solenoid or starter motor.	7. Check for missing or damaged heat shields.	7. Replace shield. Service as needed.
	8. High resistance in starter circuit.	8. Use cranking current test, insulated circuit test, and ground circuit tests to pinpoint area of high resistance.	8. Replace defective components.
	9. Poor starter drive/ flywheel engagement.	9. Perform visual inspection of drive and flywheel components.	9. Replace damaged components.
	10. Loose starter mounting.	10. Perform visual inspection.	10. Tighten as needed.
Engine does not crank.	1. Discharged battery.	1. As listed above.	1. As listed above.
	2. Poor or broken cable connections.	2. As listed above.	2. As listed above.
	3. Seized engine components.	3. As listed above.	3. As listed above.
	4. Loose starter mounting.	4. As listed above.	4. As listed above.
	5. Open in control circuit.	5. Perform control circuit test to locate "open" or high resistance.	5. Repair or replace components as needed.
	6. Defective starter relay.	6. Perform starter relay bypass test.	6. Replace starter relay if engine cranks when bypassed.

(Continued)

	TABLE 11-2: (CONTINUED)		
Problem	Possible Cause	Tests and Checks	Remedy
	7. Defective starter motor caused by internal motor malfunction.	7. Perform starter relay bypass test.	7. Replace starter motor if engine will not crank when starter relay is bypassed.
Starter motor spins but does not crank engine.	1. Defective starter drive.	1. Perform starter drive test.	1. Replace starter drive.
	2. Worn or damaged pinion gear.	2. Perform visual inspection of components.	2. Replace starter drive.
	3. Worn or damaged flywheel gears.	3. Perform visual inspection of flywheel.	3. Replace as needed.
Starter does not operate or movable pole shoe starter chatters or disengages before engine has started.	1. Battery discharged.	1. Perform battery load test.	1. Recharge or replace.
	2. High resistance in starting circuit.	2. Perform cranking current test, insulated circuit test, and ground circuit tests to pinpoint area of high resistance.	2. Replace defective components.
	3. Open in solenoid or movable pole shoe hold-in winding.		3. Replace solenoid or movable pole shoe starter.
	4. Worn solenoid unable to overcome return spring pressure.		4. Replace solenoid. Install lighter return spring.
	5. Defective starter motor.	5. Perform visual inspection.	5. Replace starter motor.
Noisy starter cranking.	Loose mounting.		Tighten mounts. Correct alignment.

TABLE 11-3: RESULT OF CRANKING CURRENT TESTING	
Problem	Possible Cause
Low current	Undercharged or defective battery
	Excessive resistance in circuit due to faulty components or connections
High current draw	Short in starter motor
	Mechanical resistance due to binding engine or starter system component failure or misalignment

WARNING *When performing this test, do not operate the solenoid for extended periods of time because severe overheating will result.*

If the voltage available at the relay terminals is lower than specifications, check the control circuit wiring and components for high resistance.

High resistance in the solenoid switch circuit will reduce current flow through the solenoid windings, which can cause improper functioning of the solenoid. In some cases of high resistance, it might not function at all. Improper functioning of the solenoid switch will generally result in the burning of the solenoid switch contacts, causing high resistance in the starter motor circuit.

Check the vehicle wiring diagram, if possible, to identify all control circuit components. These normally include the starting switch, safety switch, starter drive solenoid winding, or a separate relay drive.

reconnect them to the battery cable. Operate the starting circuit and read the voltage across the solenoid coil terminals. This should be at least 10 volts for a 12V system and 20V for a 24V system.

While someone holds the starting switch in the start position, connect the voltmeter leads across each wire or component. A voltage drop exceeding 0.1 volt across any one wire or switch indicates high resistance. If a high reading is obtained across the neutral safety switch used on automatic transmissions, check the adjustment of the switch according to the manufacturer service literature.

Voltage-Drop Tests. If the control circuit is operating properly, voltage-drop tests can be made on the starter circuit. Voltage-drop testing locates any source of excessive resistance in the starter circuit.

To perform voltage-drop tests, an accurate low-range voltmeter is required. The meter range should be 2 or 3 volts full scale, and it should be equipped with leads that are long enough to reach the various points being checked. One of the leads should be equipped with a sharp probe so that when battery cables are being checked, it can be jabbed into the battery post. This will allow the drop across the clamp and post to be measured along with the cable drop. When connecting the voltmeter to the switch or motor, connect it to the terminal stud rather than to the terminal so that the drop across the connection will also be measured. Also, connect the positive voltmeter lead to the part of the circuit that is more positive and the negative lead to the more negative point.

GROUND RESISTANCE TEST

The first test to be performed on starters is the ground circuit resistance check. Determine whether the starter uses an **external ground** circuit (grounded through starter frame to engine) or **internal ground** circuit (dedicated negative cable). When working with an internally grounded starter motor, one lead of the voltmeter is connected to the ground terminal of the battery and the other lead is connected to the ground post of the starter **(Figure 11-32)**. Read the voltmeter while the engine is cranked. If the voltage reading exceeds 0.2V, excessive resistance exists in the ground circuit. Further voltage-drop checks must be made between the battery ground terminal and the starter ground to pinpoint the source of the unwanted resistance. The same problem can be isolated in externally grounded starters by measuring the voltage drop across the following connections:

- Ground cable connection at the battery.
- Ground cable connection at the engine.
- Starter bolt connection to the engine.
- Contact between the starter main frame and the end frames.

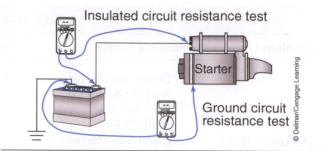

Insulated circuit resistance test

Ground circuit resistance test

Figure 11-32 Starter circuit testing.

- Any connection between the ground terminal and the starter base. Dirt, acid corrosion, loose connections, or any other contaminant can cause excessive resistance.

If the ground circuit resistance check does not expose any problems with resistance, the insulated circuit resistance check can be performed. The positive lead of the voltmeter is connected to the positive terminal of the battery. Then the engine is cranked. While the engine is being cranked, the other voltmeter lead is brought into contact with the starter input terminal.

A voltmeter reading higher than 0.5V indicates high resistance in one of the following components or connections of the insulated circuit:

- Cable connection at the battery
- Cable connection at the solenoid
- Cable
- Starter solenoid

Isolate the cause of excessive resistance by performing additional voltage checks across these possible sources. Repair or replace any damaged wiring or faulty connections.

Shop Talk: When testing starter circuits, use the manufacturer-recommended method of preventing the engine from starting. When performing voltage-drop tests, make sure you record the results to two decimal places. Record the results by writing them down so you can easily check them to spec afterward.

No-Load Tests

When testing indicates that a starter malfunction is the cause of the no-start or hard-start condition, the starter must be removed from the vehicle for additional testing, the first of which should be a no-load test. The no-load test is used to identify specific defects in the starter that can be verified with tests when

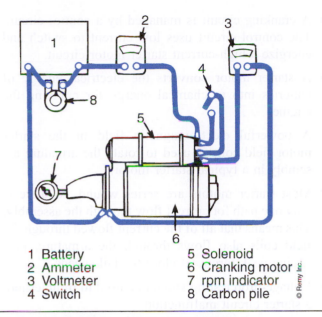

1 Battery
2 Ammeter
3 Voltmeter
4 Switch
5 Solenoid
6 Cranking motor
7 rpm indicator
8 Carbon pile

© Remy Inc.

Figure 11-33 No-load test connections.

disassembled. Also, the no-load test can identify open or shorted fields, which are difficult to check when the starter is disassembled. The no-load test also can be used to indicate normal operation of a repaired motor before installation.

To perform a no-load test, first clamp the starter motor in a bench vise. Then, connect the test equipment as shown in **Figure 11-33.** Connect a voltmeter from the motor terminal to the motor frame, and use an rpm indicator to measure armature speed. Connect the motor and an ammeter in series with a fully charged battery of the specified voltage, and a switch in the open position from the solenoid battery terminal to the solenoid switch terminal. Close the switch and compare the rpm, current, and voltage reading with specifications. It is not necessary to obtain the exact voltage specified because an accurate interpretation can be made by recognizing that if the voltage is slightly higher, the rpm will be proportionately higher, with the current remaining essentially unchanged. However, if the exact voltage is desired, a carbon pile connected across the battery can be used to reduce the voltage to the specified value. If more than one 12V battery is used, connect the carbon pile to only one of

the 12V batteries. If the specified current draw does not include the solenoid, deduct from the ammeter reading the specified current draw of the solenoid hold-in winding. Disconnect only with the switch open. Interpret the test results as follows:

1. Rated current draw and no-load speed indicates normal condition of the cranking motor.
2. Low free speed and high current draw indicate the following:
   - Too much friction. Tight, dirty, worn bearings, bent armature shaft, or loose pole shoes, allowing armature to drag.
   - Shorted armature. This can be checked further after disassembly.
   - Grounded armature or fields. Check further after disassembly.
3. Failure to operate with high current draw indicates:
   - A direct ground in the terminal or fields.
   - Seized bearings. This can be determined by turning the armature by hand.
4. Failure to operate with no current draw indicates:
   - Open field circuit. This can be checked after disassembly by inspecting terminal connections and tracing the circuit with a test lamp.
   - Open armature coils. Inspect the commutator for burned bars after disassembly.
   - Broken brush springs, worn brushes, high insulation between the commutator bars, or other causes that would prevent good contact between the brushes and commutator.
5. Low no-load speed and low current draw indicate high internal resistance due to poor connections, defective leads, dirty commutator, and the causes listed under step 4.
6. High free speed and high current draw indicate a shorted field. If shorted fields are suspected, replace the field coil assembly and check for improved performance.

Summary

- A vehicle charging system consists of batteries, alternator, voltage regulator, associated wiring, and the electrical loads of the chassis.

- The purpose of the charging system is to recharge the batteries whenever necessary and to provide the current required to power the electrical components.

- When the charging system fails, the result is either undercharged or overcharged batteries.

- The subcomponents of an alternator consist of stator, rotor, slip rings, brushes, and rectifier.

- A magnetic field is established in the rotor windings, and this is used to induce current flow in a stationary stator.

- Slip rings conduct current to the rotor to establish a magnetic field.

- Because the brushes in slip rings conduct very low current to the rotor windings, they outlast the brushes used in now obsolete generators.

- The alternator rectifier is located in the end frame assembly.

- Most current alternators use solid-state electronic voltage regulators.

- When the voltage regulator is shorted out of the circuit, the alternator is full fielded and will produce maximum output.

- The *sensing voltage* of the charging system should be battery voltage at any given moment of operation.

- AC leakage testing checks the performance of the diodes in the rectifier and helps predict alternator failures.

- A starter circuit cranks an engine until it can operate under its own power.

- A cranking circuit is managed by a control circuit. The control circuit uses low current to switch and energize a high-current starter motor circuit.

- A starter motor converts the electrical energy of batteries into mechanical energy for cranking the engine.

- A powerful electromagnetic field in the starter motor field coils is used to rotate the armature assembly in a typical starter motor.

- Most starter motors are series wound, so there is only one path for current flow through the assembly. This means that all of the current flowed through the field coils also flows through the armature, producing peak torque at close to stall speeds.

- Voltage-drop testing should be used to troubleshoot a starter circuit malfunction.

- A starter motor found to be defective should be removed from the engine and either rebuilt or replaced.

- Adhere to the manufacturer's service literature when diagnosing charging and cranking circuit problems. Failure to do this can result in component damage and costly repairs.

- Many basic charging and cranking circuit malfunctions can be identified using hand-held diagnostic instruments such as the Midtronics inTELLECT ESP.

Review Questions

1. What part of an alternator produces the magnetic field?

   A. Stator   C. Brushes

   B. Rotor   D. Poles

2. What is used in an alternator to rectify AC to DC?

   A. A stator   C. Diodes

   B. A regulator   D. A rotor

3. What is the function of slip rings in an alternator?

   A. To conduct current to the rotor   C. To conduct current to the armature

   B. To conduct current to the stator   D. To act as bearings to support the rotor

4. Why do the brushes in an alternator last much longer than brushes in a DC generator?

   A. They conduct much less   C. They rotate at slower speeds.
      current.
                                      D. They rotate at faster speeds.
   B. They conduct much more
      current.

5. When a diode is in reverse bias, what is happening to current flow through it?

   A. Current is blocked.   C. Current passes through it.

   B. Current is converted to AC.   D. Only the positive sine wave passes through it.

6. What component is used to switch field current on and off in an alternator?

   A. A diode bridge   C. An armature

   B. A stator   D. A voltage regulator

7. When the term *sensing voltage* is referred to in alternator operation, which of the following best describes it?

   A. 5 volts DC   C. Battery voltage

   B. 5 volts AC   D. Diode leakage voltage

8. When an alternator is full fielded, which of the following components is bypassed?

   A. The stator   C. The slip rings

   B. The rotor   D. The voltage regulator

9. When a carbon pile is used to load the electrical system to test the charging circuit, how much load should be applied?

   A. Enough to drop system   C. The rated amperage specification
      voltage to 12.7 volts
                                      D. Enough to drop system voltage to 9.6 volts
   B. One-half the CCA rating of
      the batteries

10. What is the function of the control circuit in a vehicle cranking circuit?

   A. Regulate current flow   C. Switch the starter motor to crank the engine
      through the charging circuit
                                      D. Manage the cranking motor voltage
   B. Limit overcharging of the
      batteries

11. What component prevents a starter motor from being turned by the engine at speeds that could damage the motor?

   A. Rotor   C. Pinion gear

   B. Override clutch   D. Solenoid gear

12. What is the usual maximum permitted *total* voltage drop through the insulated side of the starter circuit?

   A. 0.05 volt   C. 0.5 volt

   B. 0.1 volt   D. 1.0 volt

13. What is the usual maximum permitted *total* voltage drop through the ground side of the starter circuit?

   A. 0.05 volt   C. 0.5 volt

   B. 0.1 volt   D. 1.0 volt

14. Technician A says that a starter relay is a magnetic switch. Technician B says that the control circuit in a starter relay conducts the starter motor current load. Who is correct?

   A. Technician A only   C. Both A and B

   B. Technician B only   D. Neither A nor B

15. Technician A says that series-wound starter motors are commonly used on diesel engines. Technician B says that heavy duty starters produce peak torque when driven at the highest rpms. Who is correct?

   A. Technician A only   C. Both A and B

   B. Technician B only   D. Neither A nor B

CHAPTER

12 Engine Electronics

Prerequisites

Good understanding of basic electricity.

Learning Objectives

After studying this chapter, you should be able to:

- Describe the circuit layout of an electronically managed diesel engine.
- Explain what is meant by the input, processing, and output circuits in engine management electronics.
- Identify and explain the operation of critical input circuit devices.
- Outline the stages of a computer processing cycle.
- Describe how memory is managed in a diesel engine PCM.
- Identify the different types of memory used in vehicle electronics.
- Define the role played by output circuit devices in a typical diesel engine management system.
- Explain the differences between customer and proprietary data programming.
- Define *CAN bus multiplexing* and how it is used to network the engine electronics with other chassis systems.
- Identify the different types of CAN buses used on vehicles powered by light duty diesel engines.
- Describe the procedure required to connect an electronic service tool (EST) to the chassis data bus.

Key Terms

actuator

assembly line data link (ALDL)

body control module (BCM)

CAN 2.0

CAN-A

CAN-B

CAN-C

central processing unit (CPU)

chopper wheel

Class II bus

command sensors

communications adapter (CA)

controller area network (CAN)

critical flow venturi (CFV)

customer data programming

data communication link (DCL)

diagnostic trouble codes (DTCs)

digital display unit (DDU)

download

electronic service tool (EST)

electronically erasable, programmable read-only memory (EEPROM) engine/electronic control module (ECM)

engine/electronic control unit (ECU)

handshake

injector driver

local interconnect network (LIN)

mass air flow (MAF) sensors

Mastertech vehicle communications interface (MVCI)

microprocessor

monitoring sensors

multiple diagnostic interface (MDI)

multiplexing

NO$_x$ sensor

packet

platinum resistance thermometers (PRTs)

potentiometer

powertrain control module (PCM)

programmable ROM (PROM)

proprietary data programming

pulse wheel

pyrometer

random-access memory (RAM)

read-only memory (ROM)

reference voltage (V-Ref)

reprogram

resistance temperature detectors (RTDs)

SAE J1850

SAE J1939

SAE J1962 connector

state of health (SOH)

strategy

telematics

thermistor

thermocouples

throttle position sensor

tone wheel

universal asynchronous transmit and receive (UART)

variable capacitance

vehicle communications interface (VCI)

water-in-fuel (WIF) sensor

INTRODUCTION

Almost every diesel engine today is controlled by computer. Almost every engine certified for on-highway use since 1998, has been computer controlled. Onboard vehicle computers are referred to using the following terms:

- **Powertrain control module (PCM)**
- **Engine/electronic control module (ECM)**
- **Electronic/engine control unit (ECU)**

The term generally favored by automotive manufacturers is PCM, so this acronym will be used except when referring to a specific manufacturer's product in which the original equipment manufacturer (OEM) prefers a different term. Depending on the manufacturer, a PCM may drive powertrain components other than the engine, but this will be explained later in this chapter. Generally, but not always, the **body control module (BCM)** will act as a gateway for external (bus-to-bus) and internal (intrabus) networking. A generation ago, chassis external communications were limited to picking up a radio signal. Today, they are complex two-way transactions that include WiFi, Bluetooth, GPS (one-way), satellite (two-way), and vehicle tracking technologies. **Telematics** is the term that describes two-way mobile vehicle communications that use cellular, WiFi, Bluetooth, and satellite technology. A well known example of telematic communications is GM's popular OnStar.

Engine Controller

The PCM is the engine controller. Like most vehicle system controllers, it is packaged into a small box known as a module. The PCM contains a **microprocessor**, a means of storing data or "memory," and in most cases, output or switching apparatus. It is usually but not always mounted somewhere on the engine because it makes sense to have it close to the systems it monitors and devices it controls. Pretty much everything we say about the PCM that manages an engine can also be applied to home computer systems.

In the early days of vehicle computers, the only system that was electronically managed was the engine. This is not true today; most chassis systems are computer controlled. They are also networked to each other by means of data buses. Every controller with an address on the data bus can communicate with other modules on the bus. We use the term **multiplexing** to refer to networking between modules on a vehicle data bus. In addition, today's vehicles can be equipped with up to five separate data buses that use both hard-wire and optical technologies.

Data Processing

Vehicle computers do exactly what your home computer does: They manage information. A vehicle computer may be very simple, such as that required to manage a pulse wiper circuit. On the other hand, it may be more complicated if it is required to manage the many systems required to run a diesel engine. In a typical computer, information is managed in three distinct stages:

1. Data input: The circuits that get sensor, data bus, and switch status data fed into the PCM processing cycle.
2. Data processing: Takes place within the PCM; input data is correlated with program instructions contained within memory banks to generate output **strategy**.
3. Outputs: When data processing has produced an output strategy, this has to be converted into action using drivers.

Most vehicle PCMs today are housed within a single module, but they do not have to be; there are examples in which one module houses a memory bank, another contains the microprocessor, and a third houses output drivers. Most PCMs manage all the engine and emissions control systems on the vehicle. They may also manage other powertrain components, or alternatively, they may be networked to the chassis data bus.

INPUT CIRCUIT

PCM inputs can be divided into:

- Sensor telemetry
- Data bus messages
- Switch status inputs

The good news about the way in which engine input circuits operate is that they vary little between manufacturers. However, each manufacturer tends to prefer its own way of describing its components. That said, the approach in this textbook will be to attempt to make generic descriptions wherever possible. Inputs to the PCM can be divided into two categories:

- Monitoring: Feeding status data back to the PCM processing cycle.
- Command: Providing instructions to the PCM processing cycle.

Monitoring sensors are responsible for watching over circuit conditions such as:

- Temperature
- Pressure
- Speed

Command sensors and switches are used by the operator to provide the management electronics with instructions. The oil pressure sensor (OPS) signal sent to the PCM is an example of a monitoring input. The throttle position sensor (TPS) is an example of a command input.

Sensors

Sensors are the hardware used to signal input data to a computer's processing cycle. Sensors are usually automatic feedback devices: they signal the status of conditions that change, such as temperature, pressure, speed, and position. In many cases, they are devices that are supplied with a reference voltage (V-Ref), and they return a portion of the V-Ref back to the PCM. Or they may be devices that generate their own electrical signal. To begin, we should have an understanding of what is meant by V-Ref.

Reference Voltage. The term **reference voltage** is almost always referred to as **V-Ref**. Actually, V-Ref is a PCM output. Regardless of manufacturer, V-Ref is specified as 5V-DC. More correctly, it is a voltage as close to 5V-DC as the PCM can maintain through conditions that vary, such as temperature and battery voltage (V-Bat). If you were to measure V-Ref with an accurate digital multimeter (DMM) you might measure a value such as 5.014V-DC when the engine is cold, or 4.987V-DC when at operating temperature. Because the PCM knows exactly what V-Ref is at any given moment of operation, it can therefore accurately interpret the signals that are returned to it.

Sensors Using V-Ref

We will begin by taking a look at some sensors that receive V-Ref. There are four types:

Thermistors. **Thermistors** precisely measure temperature. To do this they use a variable resistor. The resistance varies according to temperature. The voltage supplied to the thermistor is V-Ref. They are two-wire, two-terminal devices, but bear in mind that what appears as a single sensor may actually incorporate two or more sensing devices:

- V-Ref
- Signal

There are two types of thermistors. Each is classified by whether resistance increases or decreases when temperature rise occurs:

- NTC (negative temperature coefficient): Temperature goes up, resistance goes down.
- PTC (positive temperature coefficient): Temperature goes up, resistance goes up.

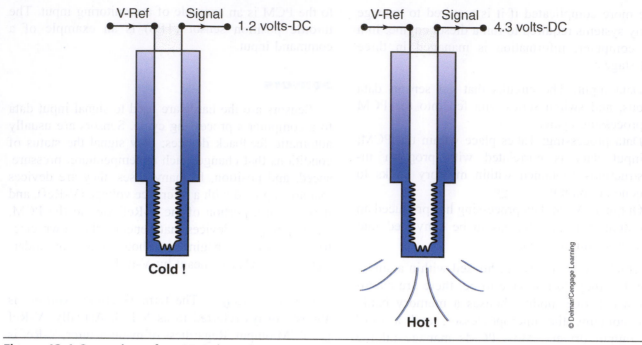

© Delmar/Cengage Learning

Figure 12-1 Operation of an NTC thermistor.

Although each manufacturer has its own preferences in describing sensors, some common examples of thermistors used on diesel engines are:

- Air cleaner temperature (ACT)
- Coolant temperature (CT)
- Ambient temperature (AT)
- Oil temperature (OT)
- Boost air temperature (BAT)
- Manifold charge (MC)
- Throttle charge temperature (TCT)

Mostly, NTC-type temperature sensors are used on diesel engine systems. This means that as temperature increases, resistance decreases, and therefore the signal voltage returned to the PCM increases. In summary, we can say that in an NTC thermistor, as temperature rises, the output voltage signal rises proportionally.

Figure 12-1 shows how a typical NTC-type thermistor operates to signal temperatures on a diesel engine.

Pressure Sensors. The two types of pressure sensors used in engine management electronics are:

- Variable capacitance
- Piezo-resistive

Variable Capacitance

Variable capacitance pressure sensors are three-wire sensors that are supplied with V-Ref. In addition to V-Ref, they have ground and signal terminals.

Typically, the device consists of a ceramic disc behind which is a flat conductive spring and steel disc. Pressure acts on the ceramic disc and moves it either closer to or farther away from the steel disc. This varies the capacitance of the device and as a consequence varies that portion of V-Ref that is returned as a signal to the PCM.

Variable-capacitance pressure sensors can be used as:

- Oil pressure sensors (OPS)
- Turbo-boost pressure (TBP) sensors
- Barometric pressure (BP) sensors
- Fuel pressure (FP) sensors
- Rail pressure (RP) sensors
- Manifold actual pressure (MAP) sensors
- Mass air flow (MAF) sensors (in pairs)

Figure 12-2 shows how a variable-capacitance–type pressure sensor responds to low and high pressure by outputting different return signals.

Piezo-Resistive

Piezo-resistive sensors may also be used to signal pressure values. This type of pressure sensor is often used to measure turbo-boost and cylinder pressure. They provide greater accuracy than an equivalent variable-capacitance sensor. Piezo-resistive pressure sensors are sometimes referred to as Wheatstone bridge sensors, a reference to the digital chip that outputs the signal. In a piezo-resistive sensor, V-Ref is used to supply a constant voltage across a bridge as

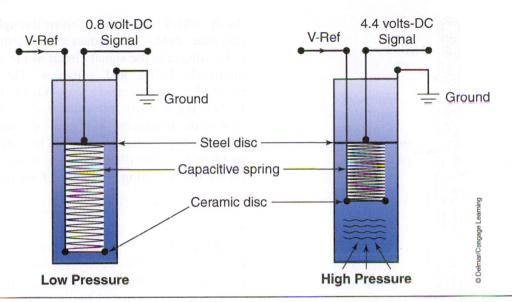

Figure 12-2 Operation of a variable-capacitance pressure sensor. Note how the signal voltage changes in response to a rise in pressure.

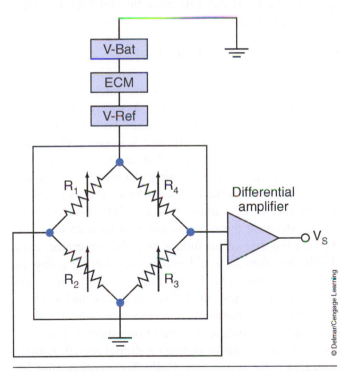

Figure 12-3 Piezo-resistive pressure sensor: Wheatstone bridge circuit. The output signal is represented by V_s.

shown in **Figure 12-3**. A silicon diaphragm, on which the pressure to be measured acts, is stretched to connect at four equidistant points of the Wheatstone bridge. When no pressure acts on the diaphragm, the sensing resistance will be equal and the bridge can be said to be balanced. Pressure causes the diaphragm to deflect increasing resistance across the sensing resistors, unbalancing the bridge, which produces a net

voltage differential. This net voltage differential is signaled to the PCM. **Figure 12-4** shows the piezo-resistive cylinder pressure sensor integrated into a glow plug used on Volkswagen diesel engines.

Position Sensors. A **potentiometer** is one means of signaling position to the PCM. A potentiometer is a three-wire device:

- V-Ref
- Ground
- Signal

The output signal of the potentiometer is proportional to the movement of a mechanical device. Potentiometers are voltage dividers. The moving mechanical device (say, an accelerator pedal) moves a contact wiper over a variable resistor. As the wiper is moved over the variable resistor, the resistance path changes. This means that V-Ref is divided between the signal (sent to the PCM) and ground.

Potentiometers are used in many **throttle position sensors (TPS)**. A disadvantage of a potentiometer-type TPS is the mechanical contact between the wiper (moved with accelerator pedal) and the variable resistor. This mechanical contact results in wear that may eventually cause failures. **Figure 12-5** shows a TPS that uses a potentiometer principle of operation.

Induction Pulse Generator. An induction pulse generator does not require a V-Ref input to function. Instead it generates its own signal using the same principles as those of an electric motor. Induction pulse generators are used to signal the speed and

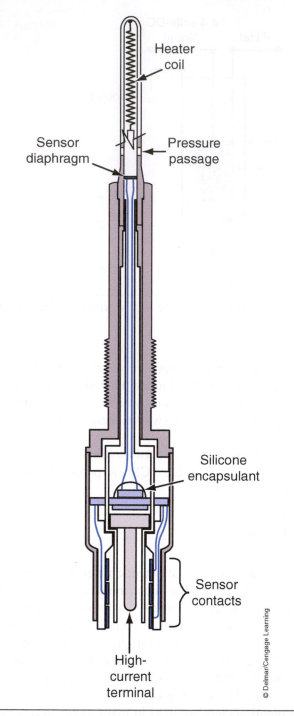

As the wheel rotates, it is driven through a stationary magnetic field. This induces an alternating current (AC) voltage in the signal circuit as the magnetic field alternately builds and collapses. The AC voltage pulses that are generated are signaled to the PCM. The PCM is not so much interested in the voltage value as it is in the frequency. The faster the pulse wheel rotates, the higher the frequency. The frequency is used to identify a specific rotational speed. **Figure 12-6** shows the operating principle of an induction pulse generator.

SIGNALING POSITION

Inductive pulse generators can also be used to signal position. Simply by adding or subtracting a tooth in one position on the pulse wheel, a specific position can be signaled by the change in frequency. An example would be on a camshaft position sensor. By adding a tooth to the pulse at the number 1 TDC position, the PCM can reference this for fuel injection timing purposes.

Hall Effect Sensors. Hall effect sensors generate a digital signal and may be used to signal both speed and position. They may be used to accurately signal both linear and rotational positions. A rotary Hall effect sensor has a rotating disc machined with timing windows or vanes. The rotating disc is positioned to pass through a magnetic field, alternately blocking and opening the field. The resulting changes in reluctance produce an on/off effect. In a rotary Hall effect sensor, the rotating disc is known as a *pulse wheel* or **tone wheel**. Because these terms are also used to describe the rotating disc of an induction pulse generator, care should be taken to avoid confusion. We are going to use the term *tone wheel* here. The frequency and width of the signal provides the PCM with speed and position data. To signal position data, the tone wheel uses a single narrow window or vane at one specific position. The Hall effect sensor outputs a digital signal. Unlike inductive pulse generators, Hall effect sensors must be powered up. Power-up is usually accomplished by feeding V-Ref to the device.

Figure 12-4 VW piezo-resistive cylinder pressure sensor integrated into the glow plug.

sometimes position of anything that rotates. This includes engine shafts such as camshafts and crankshafts. These types of sensors are also used in many non-engine systems for functions such as wheel speed signaling.

Their operation is simple enough. A toothed disc known as a **pulse wheel**, tone wheel, or **chopper wheel** is connected to the rotating component. The pulse wheel is manufactured with evenly spaced teeth.

Noncontact TPS. Hall effect sensors can also be used as throttle position sensors. Their advantage over a potentiometer TPS is that there is no mechanical contact between the moving components. The result is that they tend to be more reliable and less prone to wear. In a Hall effect TPS, a noncontact sliding shutter alternately blocks and exposes the Hall effect magnetic field to the signaling semiconductor sensor. This is known as a linear position Hall effect sensor. The use

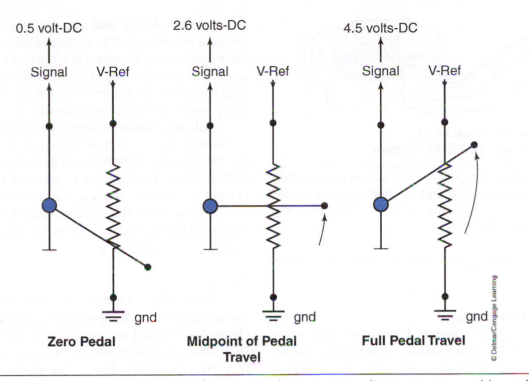

Zero Pedal | **Midpoint of Pedal Travel** | **Full Pedal Travel**

Figure 12-5 An accelerator position sensor of the potentiometer type shown at zero-, mid-, and full-travel positions.

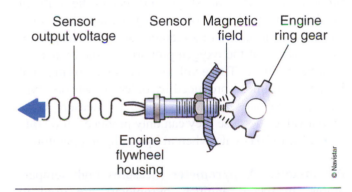

Figure 12-6 Induction pulse generator shaft speed sensor.

of linear position Hall effect sensors is not limited to noncontact TPS: they may also be used for such things as signaling crankcase oil level and transmission shift position.

MAF Sensors. Mass air flow (MAF) sensors are required on all current diesel engines. They have been used for decades in gasoline-fueled engine applications. Their objective is to provide the PCM with accurate measurement of the mass (weight) of air delivered to the engine cylinders. MAF sensors may be classified as:

- "Hot wire"
- Critical flow venturi
- Vortex flow

CRITICAL FLOW VENTURI

A **critical flow venturi (CFV)** MAF sensor is also known as a pressure differential flow sensor. It is the most common type of MAF sensor used on diesel engines, and it functions based on the relationship between inlet pressure and flow rate through a venturi. A venturi is a throat-like device located in a flow passage. It is designed to accelerate gas flow and decrease pressure. It uses the same flow principle that allows a carburetor to operate. When the pressure is known at the entry point to the venturi and at its exit point, the differential pressures can be used by the PCM to calculate the weight (mass) of the air flow. Some manufacturers refer to these as delta pressure sensors. **Figure 12-7** illustrates the operating principle

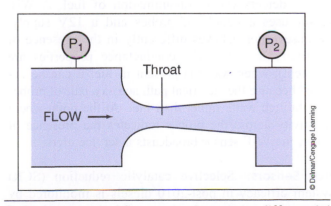

Figure 12-7 Critical flow venturi, pressure differential MAF sensor.

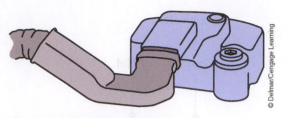

Mass Air Flow/Intake Air Temperature (MAF/IAT) Sensor

© Delmar/Cengage Learning

Figure 12-8 The MAF and IAT sensors are often integrated into a single unit located in the air inlet tube downstream from the air filter. (*Text courtesy of Ford Motor Company*)

of a critical flow venturi, pressure differential MAF sensor.

SUMMARY OF MAF OPERATION

MAF sensors are designed to work with other temperature and pressure sensors in the intake circuit. This allows the PCM to compensate for changing conditions such as:

- Outside air temperature
- Humidity
- Barometric pressure (altitude)
- Turbo-boost

For instance, cool air is denser than warm air, so the signal from an NTC-type ambient temperature sensor is used by the PCM to accurately compute actual air flow data. Humidity always affects air density because humid air is denser than dry air. However, because humid air has an increased cooling effect on the sensing element, no other compensation is required. **Figure 12-8** shows the location of the combination MAF and intake air temperature (IAT) sensor on a current Power Stroke diesel engine.

Water-in-Fuel Sensors. A **water-in-fuel (WIF) sensor** detects water contamination of fuel. A WIF sensor uses a couple of probes and a 12V supply. Because water behaves differently in the presence of electricity than fuel (its conductance properties are different), when water is present in fuel it can be detected because the electrical path across a pair of probes acts through water rather than fuel. At the point where resistivity across the probes indicates the presence of water, the WIF sensor broadcasts a service alert.

NO_x Sensors. Selective catalytic reduction (SCR) system efficacy in post-2010 diesels is monitored by **NO_x sensors**. At present, the most common method of calculating NO_x dump into the exhaust gas makes use

of a galvanic device that actually senses the oxygen levels in the exhaust. The operating principle is that of older zirconium dioxide (ZO_2) sensors used as O_2 sensors in gasoline-fueled automobiles until recently. Manufacturers are currently in the process of developing more accurate, resistance-driven NO_x sensors.

A galvanic sensor produces an electric current by chemical action. A pair of zirconium dioxide sensors are located, one upstream one downstream, from the SCR canister. Each sensor is exposed to the engine exhaust gas on one side and ambient air on the other; the differential in the readings signaled by the upstream and downstream sensors can then be used to calculate SCR efficiency.

CONSTRUCTION

A typical ZO_2 sensor is constructed of a zirconium dioxide ceramic material with gas-permeable platinum electrodes exposed on one side to the exhaust gas (measuring electrode) and on the other to ambient air (reference electrode). The zirconium dioxide begins to conduct oxygen ions at a temperature value of around 300°C (550°F). At this temperature, as the oxygen proportion in the exhaust gas and that in the ambient air differ, a small DC voltage is generated due to the electrolytic properties of zirconium dioxide. The greater the *difference* of the oxygen proportions, the higher the voltage produced. These voltage values may or may not be similar to those produced by older O_2 sensors; typically, technicians are not required to measure these output values because they can only produce valid **state of health (SOH)** data when at operating temperatures.

Pyrometers. A **pyrometer** measures high temperatures. In recent engines, pyrometers are used to monitor the extreme temperatures that result from regenerating diesel particulate filters (DPFs) so they may located upstream, within, and downstream from the DPF canister.

THERMOCOUPLE PRINCIPLE

A pyrometer uses a **thermocouple** principle. This requires that two dissimilar insulated wires (often pure iron and constantin [55% copper, 45% nickel]) connected at each end to form a continuous circuit. The two junctions are known as the hot end and the reference end. The two dissimilar wires are wound together at the hot end where temperature sensing is required. At the other end, the wires connect to a millivoltmeter. Whenever the junctions are at different temperatures, a small current flow takes place, the voltage increasing with difference in temperature. At the reference end where the millivoltmeter is located,

**Millivoltmeter
with Gauge Display**

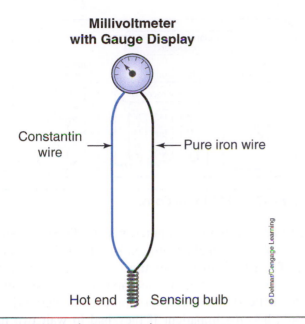

Constantin wire → ← Pure iron wire

Hot end   Sensing bulb

© Delmar/Cengage Learning

Figure 12-9 A thermocouple pyrometer.

the display gauge reads the temperature that the voltage correlates to. **Figure 12-9** shows the construction of a typical pyrometer.

Resistance Thermometers.

Resistance temperature detectors (RTDs), or resistive thermal devices, measure the change in the electrical resistance of gases that occurs with temperature change. They are made of platinum, so these devices are commonly known as **platinum resistance thermometers (PRTs)**. RTDs can replace thermocouple-type pyrometers where temperatures do not exceed 600°C (1100°F). This is on the threshold of peak DPF temperatures, so some manufacturers have used these in place of pyrometers. An image of a typical RTD appears in **Chapter 13** of this book.

Switches

Switches that signal a request for change of status to the PCM can usually be classified as command inputs. Switches may be:

- Electromechanical: Switching alters the electrical status of output.
- Smart: Sends messages via a bus line rather than switching analog voltage values to signal a change in status.

Switches can be classified into the following categories.

Switches Grounding V-Ref.

An example of a switch that grounds a V-Ref input would be that of a coolant level sensor. This switch is designed to warn an operator that the engine coolant level is becoming dangerously low. It receives a V-Ref signal from the PCM that grounds through the coolant in the upper radiator tank. Should the coolant level drop below the sensor level, the reference signal loses its ground. After a preprogrammed time period, for example, eight seconds, an alert is displayed. The time lag is required to prevent a temporary loss of ground caused by braking to open the circuit and triggering a PCM response. The PCM responds with whatever action it is programmed with—electronic malfunction alert, engine derate, or engine shutdown.

Manual Electromechanical Switches.

Manual electromechanical switches control electrical circuit activity by opening and closing circuits. There are many examples of electromechanical switches on the dash of a vehicle. They are used by the operator to control vehicle functions. Some examples would be the ignition key, engine retarder mode switches, and the cruise control switches. Switches that are controlled by the driver are sometimes called command switches.

Smart Switches.

Smart switches use digital signals to indicate a change in status. The signal produced by a smart switch may be automatically generated by a change in status condition or be generated by a mechanical action such as an operator toggling a switch. The advantage of a smart switch is that they are responsible for sending a message rather than changing electrical status. This means that several switches can share a single wire to send their messages. The messages are sent to the PCM, which then computes how those messages will be used. The advantage of smart switches is that they can significantly reduce the number of wires required by an electronic management system. Some vehicle manufacturers primarily rely on smart switches while others make almost no use of them. This is expected to change.

State of Health.

Late-model PCMs constantly sample and monitor sensor inputs by performing self-tests and checks that ensure that signal inputs fall within acceptable parameters. Processing logic can usually produce state of health (SOH) reckoning on sensor performance that helps determine whether a problem is due to a sensor malfunction or an abnormal operating condition. For this reason, it is important to use the manufacturer's diagnostic software rather than rely on scan tool-produced **diagnostic trouble codes (DTCs)** alone.

Powertrain Control Module (PCM)

The PCM receives battery power from the PCM power relay through the chassis connector. Ground is provided through the chassis connector and also includes a case ground.

Figure 12-10 Location of the bulkhead-mounted PCM on a 2011 Power Stroke. (*Text courtesy of Ford Motor Company*)

THE PCM

The powertrain control module (PCM) has the following functions:

- Filter and prepare input signals for processing.
- Retain the memory data required by the system.
- House the processing hardware.
- Convert the results of processing into action using drivers.

In most cases, the PCM is located on the engine or close to it. This is not always the case, but it does makes sense to keep it close to the systems it is monitoring and controlling. **Figure 12-10** shows the location of the PCM on a Ford PowerStroke; this is mounted on the chassis bulkhead.

Central Processing Unit

The "brain" of the processing cycle is the **central processing unit (CPU)**. It manages the processing cycle and performs all the program instructions and high-speed calculations required by the system. We are going to use the word *processing* to describe the various functions of a CPU. The instructions the CPU relies on are retained in memory banks in the same way that your home computer retains memory on chips and a hard drive. Memory in a PCM can be classified as:

- Nonvolatile: Permanent and semipermanent memory
- Volatile: Electronic memory that requires an electrically active circuit

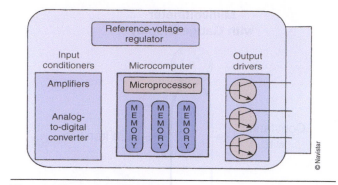

Figure 12-11 ECM functions.

Just like your home computer, when a PCM is booted by a wake-up signal or ignition key, the CPU works to transfer data from nonvolatile to electronic memory. You can bet that this has to happen a lot faster on a vehicle PCM than on your home computer; drivers would become impatient if they had to wait two minutes for Windows to launch! The primary functions of the CPU are:

- Manage the processing cycle.
- Organize incoming input data in main memory.
- Fetch and carry data from memory.
- Process data to produce outcomes.
- Generate switching commands to output drivers.

Figure 12-11 is a schematic showing some of the functions of a typical diesel engine PCM. Note that the output drivers are shown as transistors: these enable a very small signal current from the processor to control the high-current circuits required to manage the PCM output circuit.

PCM Memory

Data simply means information. Data is retained electronically, magnetically, and optically in diesel engine PCMs. We will take a basic look at data by type and function in this section. We categorize data in vehicle PCMs as follows:

Random Access Memory. **Random access memory (RAM)** is often called *main memory*. This is because the CPU can only manipulate data when it is retained electronically. The amount of RAM data that a computer system can retain plays a major role in determining total computing power. At startup, RAM is electronically loaded with the vehicle management operating system instructions and all necessary running data retained in other data categories (ROM/PROM/EEPROM). The CPU can access RAM at high speed, so all the key program instructions have to be logged into this memory segment.

VOLATILITY

Because RAM data is electronically retained, it is *volatile*. Another way of saying this is that RAM data storage is temporary. It requires power-up. If the circuit that supplies the RAM chip(s) is opened, the data in RAM is lost. Most current vehicle PCMs use only fully volatile RAM; in other words, when the ignition circuit is opened, all RAM data is dumped.

Sensor data such as coolant temperature, oil pressure, and boost pressure are also logged into RAM while the engine is running. Provided that the values input by all these sensors fall within a normal range, they are simply monitored by comparing them with system normal parameters. If they were to fall out of a normal range (as defined by program instructions), then other action would be required. This other action might include changing the power output of an engine, alerting the driver to a possible failure condition, or logging a code.

NV-RAM

A second category of RAM is used in vehicle systems. This is known as nonvolatile RAM (NV-RAM) (or KAM—keep-alive memory). NV-RAM, like RAM, is electronically retained. This means that NV-RAM requires power-up to retain its memory. The power-up circuit usually functions outside the ignition circuit, most often using a direct feed from the battery. For instance, when you disconnect a battery in some vehicles, the preset radio station selections have been recorded in NV-RAM, so these have to be reestablished. Codes and failure strategy are the types of data written to NV-RAM in systems that use it. Many older engines wrote OBD correction and wear algorithms to NV-RAM: if the battery were to be disconnected in one of these engines, these profiles would have to be reestablished before the engine would resume fully normal operation.

Read-Only Memory. Read-only memory (ROM)

data is magnetically or optically retained. It is designed not to be overwritten—that is, it can be regarded as permanent. A majority of the total data retained in the PCM is held in ROM. In an engine PCM, the master program for running the engine is loaded into ROM. ROM chips can be mass produced and designed to manage a whole family of engines built by one manufacturer. However, engines within that family cannot be run by ROM data alone. To actually make an engine run in a specific chassis application, the ROM data requires further qualification from data loaded into PROM and EEPROM.

Programmable ROM. Programmable ROM (PROM)

is magnetically or optically retained data. It can be a chip, a set of chips, or a card, pin-socketed into the PCM motherboard. PROM can sometimes be removed and replaced, especially in older systems for which it is often the only method of changing programming. PROM qualifies ROM to a specific chassis application. In the earliest diesel engine management systems, programming options, such as idle shutdown time, could only be altered by replacing the PROM chip. This is not the case today.

Electronically Erasable PROM. The **electronically erasable programmable read-only memory (EEPROM)** data category holds all the data that might have to be changed on a routine basis. It is usually magnetically retained in current systems, but this will probably change in the near future. EEPROM provides the PCM with a write-to-self capability. This allows it to log fault codes, audit trails, tattletales, and failure strategies in a way that is more permanent than NV-RAM. Other examples of the type of data that may be programmed to EEPROM are tire rolling radius, governor options, cruise control limits, road speed limit, and many others. EEPROM data can be rewritten on an as-needed basis, and some fields may be password protected. Usually, only the owner password is required to access the EEPROM and make any required changes.

PCM Output Drivers

The results of processing operations have to be converted to action. This is accomplished by switching units usually located within the PCM and **actuators** located on the components to be switched. In a PCM managing a diesel engine, **injector drivers** are required to switch injectors on and off. This means that PCM processing commands have to be converted to electrical signals. These electrical signals are used to control injector output. PCM logic processing commands are low potential, and to drive output devices, higher voltages are required. For this reason, transistors are used as amplification switches.

OUTPUT CIRCUIT

Because most domestic light duty diesels are using almost identical fuel systems mostly made by one manufacturer, the output hardware does not vary significantly. The PCM uses a combination of signals and electrical impulses to put the results of processing into action. The output circuit always begins in the PCM or in a module bus—connected to it. To make devices such as injectors and pressure valves operate under PCM control, actuators are required. Some examples of actuators are:

- Solenoids: Consist of a coil and an armature. The coil is usually stationary and the armature moves within it. A simple solenoid has two status conditions, Off or On—usually spring-loaded to default to the Off position when no current is flowed through the coil.

- Proportioning solenoids: Coil and armature devices capable of precise linear or rotary positioning. To ensure the accuracy of the position, most proportioning solenoids are equipped with a means of signaling *actual* position while the PCM driver attempts to control current flow to the proportioning solenoid to maintain desired position. An example of a linear proportioning solenoid would be the rail pressure control valve used in the high-pressure pumps on common rail (CR) fuel systems.

- Stepper motors: Brushless electric motors capable of precision positioning of a shaft (and whatever is connected to it). Gas flow gates and doors such as those used in EGR systems are controlled by stepper motors.

- Piezoelectric actuators: Much used in recent years in diesel fuel injection systems. Can be used to replace any solenoid, providing faster response times and requiring less current, but at this moment in time, they tend to be a little bulkier.

- Field effect transistors (FETs): These act as relays—when signaled with a very low-current signal they can switch high-current circuits.

MULTIPLEXING

Networking is a term we use to describe communications in a general sense. Communications may be between:

- Two persons
- Groups of persons
- Person(s) and machine
- Machine and machine

Earlier in this chapter, we defined *multiplexing* as the networking of multiple controllers on a data bus. The communications may be one- or two-way. Multiplex communications consist of messages sent between different electronic control modules (ECMs) connected to a data bus. A passenger or light duty commercial vehicle data bus consists of one or two wires. The wires carry digitally coded messages. The wire or pair of wires can only carry one message at a time, although it can do so at a very high speed. For this reason, we describe vehicle data buses as serial buses.

Serial Bus

A serial bus is used on vehicles for the simple reason that it minimizes message conflicts that could "confuse" the processing cycle of the various controllers involved in a message transaction. Every message pumped down the data bus is preceded by a "priority" tag. The priority tag on a message packet is a statement of how important it is. This hierarchy of messages is crucial to avoiding conflicts in processing outcomes. The chassis buses that concern powertrain operation with their respective transaction speeds in kilobits per second (kb/ps) are as follows;

- CAN-A single wire: Transfer speeds up to 33.3 kb/ps. Nonhigh performance such as GM-LAN. Slow by today's standards.
- CAN-B single and two wire: Transfer speeds up to 125 kb/ps. Rapidly becoming obsolete.
- CAN-C two wire: Transfer speeds up to 1 Mb/ps.

CAN-C and J1939. The powertrain and emissions control data bus used in most automotive chassis today is based on **controller area network (CAN)** generation II, or **CAN 2.0**. Automobile manufacturers call this data bus CAN-C. CAN architecture was jointly developed by Bosch and Intel, and it is universal. There are other buses on vehicles today, some hard wire, some optical. Up to five communication buses can be used on a loaded, top-of-the-line vehicle. However, only two communication buses concern the operation of a current light duty diesel engine:

- CAN-C: The passenger vehicle high-speed bus using CAN 2.0 architecture
- J1939: The heavy duty high-speed bus used on commercial vehicle chassis using CAN 2.0 architecture

All CAN-C buses use the same communication language. CAN-C is used as the powertrain bus by the majority of North American, Asian, and European light vehicles sold in our market after 2010. However, because light duty diesels are sometimes engineered for use in commercial vehicles, they may use the CAN-C equivalent that is designed for commercial trucks. This is known as J1939. The good news is that CAN-C and J1939 both use nearly identical communications protocols. The bad news is that the automotive manufacturers tend to make it more difficult than their heavy duty counterparts to access their data buses. So, if you are more familiar with navigating J1939, you are likely to experience a little more difficulty with proprietary CAN-C buses. An example of a vehicle equipped with both CAN-C and J1939 is a

Dodge pickup truck powered by a Cummins ISB engine, which will be equipped with:

- CAN-C to manage Dodge powertrain functions
- J1939 to manage the engine and all the data that is streamed into it

A Cummins ECM is used on the Dodge 2011 pickup chassis. This ECM has an address on both J1939 and CAN-C. The Cummins ECM acts as the gateway to either bus.

Why Communicate? Take the example of a pickup truck equipped with a separate bus to run engine functions. Ask it to perform an everyday operation, such as remote start, and the following is the sequence of events:

1. The driver makes the request for remote start using the vehicle key fob. The key fob sends a microwave message to the remote control lock receiver (RCDLR).
2. The RCDLR transmits a request to the body control module (BCM), the gateway to the CAN-C bus. This request wakes or boots the buses on the chassis.
3. The BCM runs a state of health check on its input circuit devices and networked modules that include the ECM (J1939) and all its input sensors, the transmission control module (TCM on CAN-C), theft deterrent module (TDM on CAN-C), driver door module (DDM on CAN-C), and liftgate control module (LGM on CAN-C).
4. If the state of health and security info check OK, the BCM (CAN-C) messages the ECM (J1939) to execute the start-engine strategy. The BCM also messages the heating, ventilating, and air-conditioning module (HVAC-M) to prepare it to function at its preprogrammed settings. If the vehicle is equipped with comfort personalization logic, the BCM will also command seat, mirror, and foot pedal position actuators.
5. Once the engine is started, the BCM will wake any other buses that the chassis might be equipped with, such as infotainment, telematics, and lighting management.

Types of Buses. Communications on a data bus may be generally classified as:

- Free flow (data exchange between modules)
- One way (master to slave)

Any free flow module can both broadcast and receive messages off the bus it has an address on. A slave module cannot broadcast messages directly onto the bus but instead must communicate with whatever bus it is mastered by. For instance, if a passenger door ajar condition is detected by the passenger door module, it cannot broadcast this data directly to the body control module (BCM); instead, it messages the driver door module (DDM). The DDM then messages the BCM, which can then message the instrument panel display unit (IPDU), which displays the information to the driver.

Bus Categories. A wide range of communications buses have been used in vehicles, mostly beginning in 1996 when OBD II was introduced. Although it is not a function of this textbook to review every type of bus used in modern vehicles, diesel engines are often specified in vehicles that are supposed to last longer than their gasoline-fueled counterparts. For this reason, some of the common buses will be reviewed.

UART

Universal asynchronous transmit and receive (always known as **UART**) is a serial one-wire bus using a reference voltage (V-Ref) to bias it at 5V-DC. Message architecture is achieved by modules with addresses on it pulling the V-Ref value low, thereby creating a recessive bit. This results in relatively slow (by today's standards) messaging with maximum speeds of around 8 kb/s.

Class II

Class II buses became common after the advent of OBD II when it was adopted as the compliant scan tool data line. Class II is a single-wire bus line with a default zero volt status. Modules with an address on a Class II bus line pull the voltage to 7V-DC to create the high logic bit change. Communication speeds top out at 10.2 kb/s.

LIN

Local interconnect networks (LINs) are used mainly to connect a master module to a slave module. They are one-wire buses, so speed and message complexity are limited. Many can be used on the chassis, but LIN usage on post-2007 diesel engines is limited. Operational voltages are based on a 12V-DC high pulled to a 5V-DC low.

Bus Architecture. Bus architecture refers to how modules are arranged and communicate within the bus rather than how communications take place between separate chassis buses. The following configurations are used for vehicle communication buses:

- Star: Self-explanatory in that a group of modules network through a central "star" through

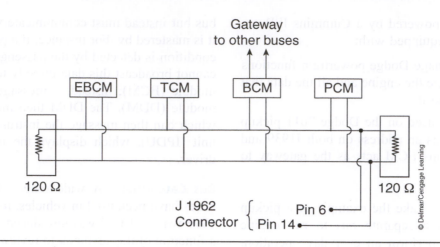

Figure 12-12 Bus topography: example of a typical serial data bus.

which all intramodule transactions must take place.

- Loop: Messages pass through each module in the loop, ending up at the module that originally broadcast the message. Removing or adding modules to the serial circuit does not affect operation so long as the loop is maintained.
- Hybrid: The features of a star and loop network are combined.

CAN-C Architecture.

CAN-C is a two-wire serial bus. It is based on CAN 2.0 protocols and architecture. Because it is a serial bus, only one message can be pumped down the bus at one time. However, the fact that messages travel down the bus at nearly the speed of light means that a great number of them can be broadcast on the bus within a time frame of one second. The advantages of a two-wire bus are:

- Lower operating voltages resulting in less self-induction.
- A twisted wire pair keeps the wires close to each other, diminishing the chances of radiation interference (radio frequency and electromagnetic).

CAN-C was originally used for powertrain and anti-lock braking system (ABS) management in light duty vehicles, but in current applications more modules are added. A current CAN-C network can handle up to 15 modules and can communicate with other chassis buses through a gateway. **Figure 12-12** is a schematic showing a typical serial data bus to which a number of controllers are connected. Each controller connected to the bus is said to have an address on that bus.

Packets

The messages broadcast on the bus are known as **packets**. The digital structure of a packet is standardized

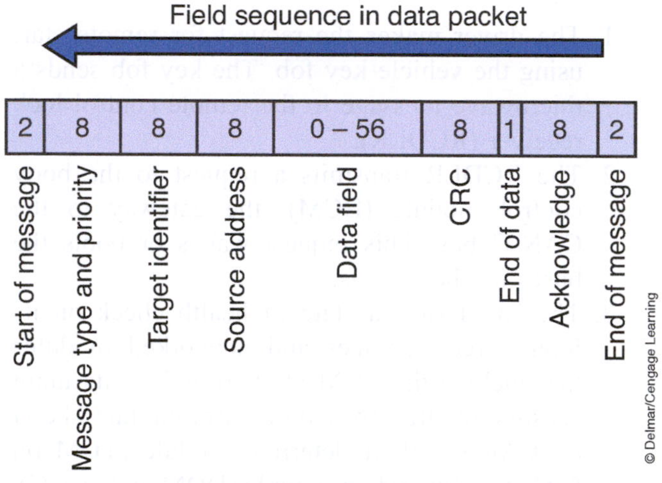

Figure 12-13 Message bit encoding of a data packet.

so that electronic systems made by different manufacturers can communicate with each other. One of the first fields built into a packet is a declaration of how important it is. This is known as its arbitration field. It is key to making a data bus function. Many of the messages broadcast on the bus are not that important. But when a chassis event such as directional yaw is detected, in order to put into effect evasive strategies such as directional stability control (DSC) and engine derate, messages from the DSC module and brake control electronics get some very high priority on the data bus. Because the bus is a serial bus, prioritizing the importance of messages minimizes the chances of message conflicts. To use a simplistic term, it prevents logjams. **Figure 12-13** shows the digital construction of a CAN bus packet.

Standardization

Data bus and electronic protocols are standardized in highway heavy equipment markets. This standardization has been managed by the SAE and automobile manufacturers, but it is a cooperation that has been necessary due to the fact that manufacturing today is global, meaning that parts are sourced worldwide. More important, they are outsourced. So is most of the engineering. This was not so in generations past. There are few differences between the sensor circuits of VW, Chrysler, GM, and Ford chassis on their light duty diesels. SAE J-standards cover almost every system found on a modern vehicle.

J-Standards

J-standards cover communications and compatibility of almost every component and system on a modern vehicle. They are established by industry consensus. SAE J-standards (surface vehicle recommended practice) dictate the hardware and software rules that apply to multiplexing data exchanges between modules. The rules are covered by the following J-standards:

- **CAN-A.** A slow first-generation multiplexing data bus. Can be considered obsolete.
- **CAN-B.** A first-generation, multiplexing data bus. Depending on manufacturer, used up until 2010. A one-wire bus run at speeds of 40 kb/ps or less.
- **SAE J1850.** A Class II bus that may use one or two wires, though two is usual. Bus speeds run at baud rates between 10 and 125 kb/ps.
- **CAN-C.** Based on a high-speed CAN 2.0 data bus. Covers the set of rules that incorporate both *software rules* and *hardware* standards. The light duty/automotive equivalent of current J1939 buses: both are based on CAN 2.0 vehicle communication rules. Runs at speeds between 125 kb/ps and 1 Mb/ps with 500 kb/ps being typical.
- **SAE J1939.** Based on a high-speed CAN 2.0 data bus. This covers the set of standards that incorporate both *software rules* and *hardware* standards for heavy duty highway equipment. The heavy duty equivalent of CAN 2.0 vehicle communication rules. Current versions of J1939 run at speeds between 125 kb/s and 1 Mb/ps with 500 kb/ps being typical.

In addition, manufacturers may use proprietary data buses. A proprietary data bus is usually designed to communicate with just one other controller and usually can only be accessed through that controller. The terms *master* and *slave* controllers may used to describe them. For instance, a fuel-injection controller may be networked to an engine controller on a high-speed proprietary bus: all troubleshooting of the fuel-injection circuits must take place via the engine controller. In order to add a module to the bus, it must be connected to a vacant stub on the bus.

J1962 Connector

To work with any controllers with an address on the data bus you must first connect to the bus. The data communication link (DCL) is used to access the data bus. Various terms are used to describe DCLs, and the specific term used depends on the manufacturer and the generation of the connector. The current OBD II data link is an assembly line data link (ALDL) 16-pin connector also known as an SAE J1962 connector, and in this text we will generally address this specific connector. A J1962 data connector and its cavity pin assignments are shown in **Table 12-1;** the CAN-C high and low pins are bolded.

After connecting to the DCL, the technician must verify that a handshake connection exists with the target module. The term *handshake* is used because it confirms that the connection is capable of two-way communication. Certain types of communication on the CAN-C bus using some types of EST (including a labscope) requires identifying these pins by function, and we will take a closer look at this in **Chapter 15.**

TABLE 12-1: J1962 CONNECTOR CAVITY PIN ASSIGNMENTS							
1	2	3	4	5	**6**	7	8
9	10	11	12	13	**14**	15	16

1. Manufacturer discretion
2. Bus positive + J1850 (CAN-A and CAN-B)
3. Ford DCL +
4. Chassis ground
5. Signal ground
6. **CAN-C high (J2284 and ISO 15765-4)**
7. K-line of ISO 9141 and ISO 14230
8. blank (manufacturer discretion)
9. blank (manufacturer discretion)
10. Bus negative – for J1850 PWM
11. Ford DCL –
12. blank (manufacturer discretion)
13. blank (manufacturer discretion)
14. **CAN-C low (J2284 and ISO 15765-4)**
15. L line of ISO 9141 and ISO 14230-4
16. V-Bat (battery positive)

Connecting to the Chassis Data Bus

Depending on the vehicle manufacturer, the technician is often required to access the chassis data bus even if it is to do something as simple as performing a routine service. Although there is more than one way of doing this, connecting to a chassis data bus usually requires the following procedure on a current vehicle:

- A Windows environment PC loaded with manufacturer-specific software.
- A **communications adapter (CA)** to act as a serial link between the PC and the DCL.
- USB to CA cable.
- CA to J1962 data-link connector.

Depending on what you are trying to achieve while connected to the data bus, it may also be necessary to be connected to the manufacturer's data hub via the Internet. Former generations of light vehicle electronics required more primitive tools (such as scan tools) to connect to the data bus: however, these are disadvantaged by low speed, little memory, high cost, and upgrading limitations.

Electronic Service Tools

An **electronic service tool (EST)** is simply a way of referring to the wide range of data communication tools used by technicians. EST can reference any of the following shop tools:

- PC loaded with proprietary software and proprietary communications adapter (CA), such as GM's **multiple diagnostic interface (MDI)**.
- PC loaded with generic software (such as the Bosch **vehicle communications interface [VCI]**) and a generic CA such as Bosch's **Mastertech vehicle communications interface (MVCI)**.

- A proprietary scan tool used to access one manufacturer's electronic systems, such as the GM Tech II.
- A generic scan tool used to access multiple manufacturers' electronic systems, such as the Bosch Mastertech MTS 3100 scan tool.
- An in-dash **digital display unit (DDU)**. Never overlook this: it is a great first-level diagnostic option when troubleshooting.

ESTs will be more closely examined in **Chapter 15**.

PCM PROGRAMMING

Just as your home computer can be programmed and **reprogrammed**, so can vehicle PCMs. Vehicle PCM programming can be divided into two categories:

- Customer data programming
- Proprietary data programming

In a general sense, the programming category indicates who owns the right to make the change. **Customer data programming** is *owned* by the owner of the vehicle; tire revolutions per mile is an example of a programmable customer data field. **Proprietary data programming** is owned by the manufacturer of the vehicle; the fuel map file is an example of proprietary data programming. It should be noted that all automotive manufacturers prefer that technicians be trained on their specific diagnostic and service support software before they try to perform any type of programming. However, at the most basic level, customer data programming can be performed by the driver. This is what happens when one requests a set road speed using cruise control. At a more advanced level, proprietary data programming has to be executed by **downloading** files from the manufacturer data hub and transferring these files to the module onboard memory.

Summary

- All current vehicles use computers to manage their engines and most other chassis systems.

- We use acronyms such as PCM, ECM, and ECU to describe the computers used on vehicles.

- A vehicle with multiple electronically managed systems networks them to a chassis data bus. This allows different system controllers to talk with each other using multiplexing technology.

- Computer processing takes place in three stages: data input, data processing, and outputs.

- RAM or main memory is electronically retained. It is therefore volatile.

- The master program for engine management is usually written to ROM.

- PROM data is used to qualify the ROM data to a specific chassis application.

- EEPROM is used for write-to-self capability, retaining customer and proprietary programming, holding failure strategy, logging codes, and the recording of audit trails.

- Multiplexing technology uses standard communication protocols regardless of manufacturer.

- The current multiplex data bus used in light vehicle applications is known as CAN-C. This is based on the international CAN 2.0 and boasts transaction speeds up to 1 Mb/ps; currently, 500 kb/ps bus speeds are typical.

- Multiplexing reduces hardware and makes vehicle operation more efficient.

- Input data is divided into command data and system monitoring data.

- Thermistors precisely measure temperature.

- Variable-capacitance–type sensors are used to measure pressure.

- Throttle position sensors use either a potentiometer or the Hall effect operating principle.

- Hall effect sensors generate a digital signal and are used to signal either rotating or linear position and speed data.

- Induction pulse generators are used to signal rotational speed data.

- The PCM is usually responsible for regulating reference voltage, conditioning input data, processing, and driving outputs.

- Engine management PCMs can be mounted on the engine itself or in a remote location, such as on a bulkhead or under the dash.

- PCM outputs require drivers to actuate a range of output devices. Output devices are known as actuators. Some examples of actuators are solenoids, proportioning solenoids, stepper motors, FETs, and piezo-actuators.

- Connecting to the chassis data bus requires using a scan tool or on more recent vehicles, a PC loaded with manufacturer software, a CA, and J1962 connector.

Internet Exercises

Check out some of the terms used in this chapter on Wikipedia and www.howstuffworks.com. Then use a search engine to see what you can discover with the following key words:

1. Piezoelectricity

2. CAN 2.0 data bus architecture

3. Multiplexing

4. MasterTech and Tech II diagnostic platforms

5. Bosch MCVI diagnostic software

Shop Tasks

1. Identify an EST, CA, and J1962 connector. Make a note of the location of the J1962 connector on three different manufacturer vehicles.

2. Establish an EST connection to a diesel-powered vehicle chassis data bus. Make a note of all the chassis buses that can be accessed with the EST you are using.

3. Identify all the controllers networked to the powertrain data bus.

4. Using the PCM address on the powertrain bus, disconnect any easy-to-access sensors on the engine; make a note of how the management system responds. Should the engine shut down, record this as an outcome, reconnect the sensor, and restart the engine.

5. With the EST-connected engine running, surf whatever fields you can access. Make a note of anything unusual. Shut the engine off. With key-on, explore whatever fields you can access and note whether you can effect any programming changes: DO NOT reprogram any fields without first consulting with your instructor.

Review Questions

1. Which of the following internal PCM components manages the processing cycle?
   A. CRT
   B. RAM
   C. CPU
   D. ROM

2. Which of the following memory types is volatile?
   A. RAM
   B. ROM
   C. PROM
   D. EEPROM

3. In which memory component would you find the master program for engine management written to in a typical PCM?
   A. RAM
   B. ROM
   C. PROM
   D. EEPROM

4. What is the maximum transaction speed of current CAN-C communications?
   A. 16.7 kb/ps
   B. 100 kb/ps
   C. 500 kb/ps
   D. 1 Mb/ps

5. Which of the following components conditions V-Ref?
   A. PCM
   B. Voltage regulator
   C. EUI
   D. Personality module

6. What type of data does a thermistor produce?
   A. Pressure data
   B. Temperature data
   C. Rotational speed data
   D. Rotational position data

7. Which of the following would a Hall effect sensor most likely be used for?
   A. Pressure data
   B. Temperature data
   C. Fluid flow data
   D. Rotational or linear position data

8. What type of information is produced by induction pulse generators?
   A. Pressure data
   B. Temperature data
   C. Rotational speed data
   D. Altitude data

9. Which of the following components could be used to signal accelerator pedal travel?
   A. Rotary pulse generator
   B. Potentiometer
   C. Thermistor
   D. Variable-capacitance sensor

10. Which of the following words best explains what *multiplexing* is on a vehicle chassis?
    A. WiFi
    B. Handshaking
    C. Networking
    D. Broadband communication

CHAPTER

13 Emission Controls

Learning Objectives

After studying this chapter, you should be able to:

- Define the origin of the word *smog*.

- Define *photochemical smog* and describe the conditions required to create it.

- Identify some common tailpipe emissions.

- Outline the operating principles of EGR, oxidation catalytic converters, reduction catalytic converters, diesel particulate filters, and selective catalytic reduction (SCR).

- Analyze diesel engine smoke by appearance.

Key Terms

active regeneration

AdBlue

aqueous urea

California Air Resources Board (CARB)

carbon dioxide (CO_2)

carbon monoxide (CO)

catalyst

catalytic converter

closed crankcase ventilation (CCV)

diesel exhaust fluid (DEF)

diesel particulate filter (DPF)

dosing

driver display unit (DDU)

Environmental Protection Agency (EPA)

hydrocarbons (HC)

manual regeneration

nitrogen dioxide (NO_2)

NO_x adsorber catalyst (NAC)

oxides of nitrogen (NO_x)

ozone

palladium

particulate matter (PM)

passive regeneration

photochemical smog

platinum resistance thermometer (PRT)

positive crankcase ventilation (PCV)

pyrometer

regeneration cycle

resistance temperature detector (RTD)

rhodium

selective catalytic reduction (SCR)

self-regeneration

smog

sulfur dioxide (SO_2)

thermocouple

urea

volatile organic compounds (VOCs)

INTRODUCTION

The standards that apply to diesel engine tailpipe emissions fall under the jurisdiction of the **Environmental Protection Agency (EPA)**. The state of California continues to lead the way for the remainder of the jurisdictions in North America when it comes to defining vehicle emissions standards and enforcing them. The California agency responsible for emissions legislation and enforcement is the **California Air Resources Board**, or **CARB**. Any new engine introduced to the marketplace requires testing to EPA and CARB standards prior to certification.

Importance of CARB

For many years CARB (www.arb.ca.gov/) was important to engine manufacturers because they obviously wanted to sell engines in the state with the largest population and, despite recent problems, the most powerful economy. Recently, the influence of CARB has increased with other states adopting CARB standards. This list of "green" states has risen to over 20 states and represents well over half of the gross domestic product of the U.S. economy. This makes it unlikely that engine manufacturers can afford to make engines that do not meet CARB standards, and the bottom line is that the dirty diesel engines of a generation ago have been squeezed into the *almost* clean diesel engines of our world today!

Photochemical Smog

Photochemical smog (also known as photosynthetic and photoelectric smog) is produced mainly by vehicle tailpipe gases. After it has been formed, photochemical smog can be seen as yellowish/light brown haze. The effects of photochemical smog include crop damage, eye irritation, and breathing problems in animals and humans.

Formation. Photochemical smog is produced in two stages. First, **hydrocarbons (HC)** and nitrogen oxides from vehicle tailpipes react with sunlight. This produces ozone. Ozone then begins to react with hydrocarbon gas to produce smog. Ozone by itself is highly toxic in low concentrations, and the smog that results from exposure to light is a major problem in America today.

WHAT IS SMOG?

The word **smog** comes from the words fog and smoke. Throughout the world there are two main types of smog. The first type is **sulfur dioxide (SO_2)** smog produced by the burning of sulfurous fuels such as coal and heavy oils. Sulfurous smog conditions are aggravated by dampness, so foggy conditions can make it worse. Today, it is produced mostly by industry, especially in areas burning industrial coals and heavy oils. Forty years ago, diesel fuel contained enough sulfur to make diesel-powered vehicles a significant contributor (mostly from heavy commercial trucks). However, over the years the sulfur in diesel fuel has been gradually reduced to almost nothing so that today, diesel-powered highway vehicles using ultra-low-sulfur diesel (ULSD) fuels contribute little to sulfur dioxide smog formation.

Global Warming

Carbon dioxide (CO_2) is classified as a greenhouse gas. It is *not* by definition a harmful emission—after all, we expel it from our lungs every time we exhale. However, carbon dioxide emissions are associated with global warming, and many legislators want to regulate them. This is a problem. Reducing carbon dioxide emissions will mean taking a serious look at the chemical composition of the fuels we burn. At present, most of the fuels we use are carbon based. Because both gasoline and diesel fuel are composed of approximately 85 percent carbon, when legislators talk about reducing (or taxing) CO_2 emissions, the only current methods involve increasing fuel efficiency.

Ingredients of Smog

Combustion is a chemical reaction. When a diesel is burned, it uses whatever oxygen is available in ground-level air. This oxygen is known as the reactant. When oxygen reacts with diesel fuel (during combustion) it is oxidized. This oxidation reaction forms:

- H_2O (water)
- CO_2 (carbon dioxide)

Neither the water nor carbon dioxide are harmful results of the combustion process. For this reason, if these two gases were the only products of burning fuel in an engine cylinder, the term *perfect* combustion would be used to describe what had happened. If you take a look at the *ingredients* in any engine cylinder at the point of ignition, there is more nitrogen by weight than either oxygen or fuel: the hope is that the fuel can be oxidized without involving nitrogen in the process. Harmful emissions are produced when the combustion of an HC fuel is not "perfect."

Imperfect Combustion. When a hydrocarbon fuel is not completely oxidized, gaseous HC, particulate HC, and **carbon monoxide (CO)** result. These emissions tend to be associated with rich air-fuel ratios (AFRs) and colder engine temperatures. When emitted from the tailpipe of a vehicle these chemicals are all classified as noxious emissions.

In addition, nitrogen can become involved in the combustion process under certain conditions. Within the normal combustion temperature range of typical diesel engines, this tends to be when AFR is leaner and temperatures are higher. When nitrogen is oxidized, it forms a number of different compounds known collectively as NO_x, but **nitrogen dioxide (NO_2)** presents the most problems. Nitrogen dioxide is a key to the formation of ozone and acid rain. All NO_x emissions are classified as noxious emissions by the EPA.

Hydrocarbons. The HC emitted in the combustion process of an internal combustion engine burning a liquid or gaseous petroleum product may be gaseous, liquid, or solid (particulate) state. Hydrocarbons are only observable in liquid or solid states when they may be seen as white smoke (liquid emission) or black smoke (particulate emission). A more detailed analysis of diesel engine smoke emission appears toward the end of this chapter. In the days before rigid emissions standards, a diesel engine commonly emitted HC in liquid state (that is, condensing in the exhaust gas) at cold startup and in particulate state under conditions of overfueling or air starvation.

Particulate Matter. Anything in the exhaust gas that is in the solid state can be classified as **particulate matter (PM)**. Carbon soot and dust form PM. Incomplete combustion is the main culprit in the formation of PM. These minute particulate compounds can cause respiratory problems in humans and animals. PM emission is always visible from a diesel engine. The objective of diesel particulate filters (DPFs) examined later in this chapter is to eliminate PM.

Volatile Organic Compounds. **Volatile organic compounds** or **VOCs** are hydrocarbons in the gaseous state that boil off fuels during production, distribution, and pumping. Diesel-fueled vehicles contribute VOCs in a small way: fuel tanks exposed to the high heat of a summer's day vaporize the more volatile fuel fractions to the extent that the CN can degrade. Automobiles and commercial trucks account for approximately 30 percent of the VOCs found around population centers, but most of this is gasoline sourced. VOCs are more likely to be an atmospheric problem in hot weather conditions when the most volatile fractions of fuels (especially gasoline) more readily boil off. VOCs react with sunlight to produce ground-level ozone.

Carbon Monoxide. Carbon monoxide is a colorless, odorless, tasteless, and highly poisonous gas. It is a result of the incomplete combustion of a hydrocarbon fuel. Carbon monoxide can be readily combusted to form harmless CO_2. Approximately two-thirds of the atmospheric levels of CO in North America can be attributed to vehicle emissions, mostly from gasoline engines. Exposure to CO can impair brain function, cause fatigue, and may be fatal in high concentrations.

Ozone. Atmospheric oxygen is diatomic, that is, it combines to form molecules consisting of two oxygen atoms (O_2). **Ozone** is triatomic, that is, three oxygen atoms covalently bond to form a molecule of ozone (O_3). It occurs naturally in small quantities in the Earth's stratosphere, where it absorbs solar ultraviolet radiation. Ozone can be produced by passing a high-voltage electrical discharge (arc) through air: you can sometimes smell it during a thunderstorm. It is manufactured commercially using an electric arc process. As a pollutant, ozone results from the photochemical reaction between NO_x and various hydrocarbons (especially those categorized as VOCs). It is known to irritate the eyes and mucous membranes. Exposure to ozone at low-level concentrations can be a health threat.

Oxides of Nitrogen. Nitrogen makes up nearly 80 percent of what we call *air* at ground level. Combustion in an engine cylinder requires only the oxygen contained in the mixture we call air. Ideally, we would like the nitrogen in the mixture to exit the tailpipe exactly as it came in. However, given the wrong conditions, nitrogen may become involved in the combustion process, and when it does, it oxidizes.

When nitrogen reacts during the combustion process, just like the HC fuel, it oxidizes. It forms nitrous oxide (N_2O), nitric oxide (NO), and nitrogen dioxide (NO_2) when oxidized. All **oxides of nitrogen** produced in the cylinders of engines are known as NO_x. On- and off-highway vehicle engines are the source of about 60 percent of the NO_x in our atmosphere. NO_x directly affects persons with respiratory problems. This is why the EPA has so aggressively attacked NO_x emissions from diesel engines in its certification requirements. **Figure 13-1** shows the EPA diesel emissions story over a 40-year period extending from 1970 to 2010.

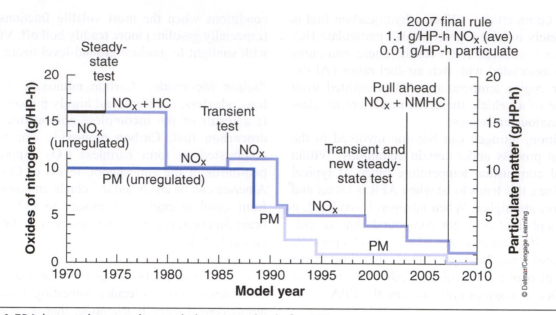

Figure 13-1 EPA heavy duty engine emissions standards for NOₓ.

Smog Summary

We can summarize the consequences of vehicle emissions as follows:

$$NO_x + VOC_s + Sunlight = Ozone$$
$$Ozone + PM = Smog$$

DIESEL ENGINE EMISSION CONTROLS

All engines today rely primarily on computers to manage emissions at acceptable levels. In addition, current diesel engines rely on a full range of internal and external emission control devices that include:

- Oxidation catalysts
- Reduction catalysts
- Particulate filters
- Urea injection systems

Factors Influencing Emissions

The reason that the engine management computer plays such a major role in controlling emissions in diesel engines has to do with injection timing and the ignition timing it controls. The study of combustion in engine cylinders can become complicated, but we can say the following:

- Retarded timing tends to reduce tailpipe NOₓ. Severely retarded timing can greatly increase HC emissions.
- Advanced timing tends to reduce HC. Severely advanced timing can greatly increase NOₓ emissions.

Timing that is out of spec by as little as 1 degree of crank angle can increase NOₓ or HC emissions by as much as 10 percent, depending on whether it is advanced or retarding. In addition, dumping all of the fuel required for one power stroke in a continuous pulse tends to create combustion inefficiencies. All current engines meeting current EPA highway diesel engine standards *dose* the engine during combustion using multipulse injection; because the injection process must occur within 3 milliseconds or less, computer controls are required to manage multipulsing. Temperature also has a major effect on exhaust emissions, and once again, managing an engine with respect to temperature variables becomes a job which can only be managed by electronic controls.

External Emission Controls

While the engine management computer manages the combustion process in a diesel engine, a wide range of external emission control devices are also used that are similar to those on a gasoline-fueled automobile engine. If you take a look at any highway diesel engine series manufactured in the year 2001, and compare it with its series equivalent today, you would have difficulty identifying them as being in the same series. Today's diesel engines use many different external emission controls. In this section we will take a look at some of these devices.

Cooled-EGR

Until 2004, diesel engine manufacturers had been able to meet NOₓ emissions requirements without the

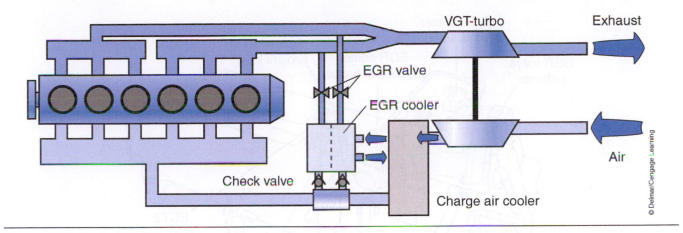

Figure 13-2 Typical C-EGR system schematic.

use of exhaust gas recirculation (EGR) systems primarily because the standards were not that tough. Since 2004, almost all highway diesels have used some form of EGR, usually cooled-EGR (C-EGR). In post-2010 diesels, EGR may be used in conjunction with other NO_x control devices.

EGR dilutes the intake charge with "dead" gas. The dead gas is routed into the engine cylinders from the exhaust system. This dead gas creates a dead zone inside the engine cylinder. The idea is to reduce NO_x emissions by lowering combustion heat. It makes the cylinder volume available for combustion smaller. When rerouting (hot) exhaust gas back into the engine cylinders, it makes sense to cool it as much as possible. For this reason, most engine manufacturers use cooled-EGR, usually known by its acronym of C-EGR. **Figure 13-2** shows a schematic of a typical C-EGR system.

C-EGR Components. A typical C-EGR system on a diesel engine has to be PCM controlled because the mixture percentage varies according to how the engine is being operated. This ranges between 0 and depending on the manufacturer, may be as high as 50 percent. Typically a C-EGR system consists of:

- A heat exchanger (engine coolant is used)
- PCM-controlled mixing chamber (mixes boosted air with exhaust gas)
- Mass air flow sensor
- Plumbing to route engine coolant through the heat exchanger
- Piping to route exhaust gas to the mixing chamber

When C-EGR is used on diesel engines, it must be managed by the PCM because the mixing ratios depend on how the engine is being operated at any given moment. Because coolant temperature also has

a range of variability, some C-EGR systems have a bypass circuit to block coolant flow from the C-EGR heat exchanger. **Figure 13-3** shows the Ford Power Stroke C-EGR unit, its features, and how it is integrated into the secondary cooling circuit. **Figure 13-4** shows how EGR gas is routed into the intake manifold through the throttle body on the same engine.

CATALYTIC CONVERTERS

Catalytic converters were not common on diesel engines sold outside of California before 2002, but today they are standard. By definition, a **catalyst** is a substance that enables a chemical reaction without itself undergoing any change. For many years, automobile catalytic converters have been two-stage, three-way converters, but it is only since 2007 that diesels have incorporated two-stage devices. The two stages are:

- Oxidation stage: Attempts to oxidize HC and CO not oxidized in the engine cylinder.
- Reduction stage: Attempts to reduce NO_x back to elemental nitrogen and oxygen when NO_x compounds have been formed during combustion.

Oxidation Stage

When cylinder end gas contains HC and CO, the oxidizing stage of the catalytic converter attempts to oxidize (combust/burn) these to harmless H_2O and CO_2. **Platinum** and **palladium** are both oxidation catalysts. Oxidizing catalysts enable what is called catalytic afterburning. This sometimes requires the addition of fresh air into the converter assembly. Peak temperatures in oxidation converters can be as high as $1,000°C$. It is not unusual for an oxidation converter to

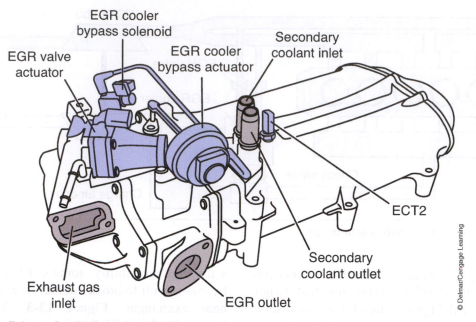

Exhaust Gas Recirculation (EGR)

The EGR system allows cooled (inert) exhaust gases to re-enter the combustion chamber, which lowers combustion temperatures and oxides of nitrogen (NO_x) emissions.

EGR system control is based off an air system model to estimate the percentage of exhaust gas in the cylinder. The PCM looks at engine temperature, intake pressure, exhaust pressure (EP), RPM, and engine load to determine the EGR flow rate. The ratio of MAP and EP is used by the PCM to estimate a desired EGR valve position. The desired position is compared to the actual and the duty cycle is adjusted to meet that desired position for the required EGR flow rate. If the rate is not achieved with EGR valve position, the intake throttle body closes to a desired position, reducing intake manifold pressure. Reducing the intake manifold pressure increases the pressure ratio allowing more exhaust to fill the intake manifold at a given EGR valve position. As more exhaust gas is introduced into the intake manifold the amount of air measured by the mass air flow (MAF) sensor is decreased.

The 6.7L has a hot side EGR valve due to it being before the EGR cooler. Once past the EGR valve, the exhaust gas is either directed through the EGR cooler or bypasses the EGR cooler. This is done by the PCM controlling the EGR cooler bypass solenoid which turns vacuum on or off to the actuator on the bypass door. The EGR outlet temperature (EGRT) sensor measures the temperature of the exhaust gas leaving the system for cooler effectiveness and bypass control.

Figure 13-3 Ford Power Stroke C-EGR unit and features. (*Text courtesy of Ford Motor Company*)

glow red during nighttime operation, especially when an engine is under load.

A diesel oxidation catalytic converter functions identically to its gasoline-fueled version. The actual amount of heat generated within the converter is important because the unit is located upstream from two other key emission control devices, both of which require a minimum heat level to function properly. **Figure 13-5** shows the oxidation catalyst used on a Ford pickup powered by a diesel engine.

Reduction Stage

Whether a reduction stage converter is used on an engine depends on the generation of diesel: most post-2010 diesels use a different means to control NO_x emissions, as we discuss later in this chapter. Where NO_x is present in the exhaust gas, the reduction stage in a catalytic converter attempts to reduce this back to nitrogen and oxygen. **Rhodium** is the usual reduction catalyst and has been used by automobile manufacturers since the 1970s. A disadvantage of rhodium is that it can only function as a reduction catalyst when the cylinder burn is close to a stoichiometric ratio, because no excess oxygen should be present. All diesel engines operate with excess air, meaning that unreacted oxygen is discharged into the exhaust. Manufacturers that use rhodium-based catalytic converters have to ''enrich'' the mixture locally in the reduction converter. They do this by **dosing**. Dosing requires excess fuel to be injected into the exhaust gas and combusted. This has

Intake Throttle Body

The intake throttle body is mounted on the lower intake manifold.

The intake throttle body promotes flow of EGR gases to the intake manifold by creating a differential between exhaust pressure and intake pressure.

Figure 13-4 Intake throttle body used on the Ford Power Stroke intake system showing how EGR gases are routed into the intake.

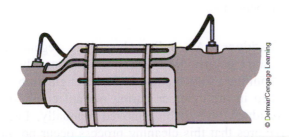

Figure 13-5 Diesel oxidation catalyst.

to happen upstream from the reduction converter. Reduction catalytic converters are not common on diesel engines because technology has progressed to superior NO_x reduction equipment.

NO_x Adsorber Catalysts. These were used up until 2010 by some diesel engine manufacturers. A **NO_x adsorber catalyst (NAC)** also relies on dosing. *Adsorption* is a chemical term that is best defined as *adhesion*. When a substance is adsorbed it does not undergo any chemical change. NO_x adsorber catalysts (NACs) use base metal oxides to initially collect or "store" NO_x. They then use a rhodium reduction catalyst to reduce NO_x back to nitrogen and oxygen during a powertrain control module (PCM)-managed reduction phase. In other words, a NAC functions in two stages:

- NO_x storage: NO_x is adsorbed to the base metal oxide substrate during the normal lean burn operation of the diesel engine.

- NO_x reduction: The engine PCM temporarily operates the engine operation in a rich AFR mode or injects (doses) fuel into the exhaust system to simulate a rich AFR allowing a rhodium catalyst to reduce the NO_x to elemental nitrogen (N_2), elemental oxygen (O_2), and water vapor (H_2O).

NAC Regeneration Cycle. The key to enabling an NAC to function in its regeneration cycle is to temporarily eliminate excess oxygen from the exhaust. NACs require the use of the ULSD fuel. Sulfur at any levels can damage NACs. Some NACs require occasional desulfation even when only the appropriate ULSD fuel has been used. Desulfation temporarily requires raising the exhaust temperatures using fuel injected directly into the exhaust system. The good news is that these devices were only used from 2007 to 2010, and then only on certain engines.

DIESEL PARTICULATE FILTERS (DPFS)

Diesel particulate filters (DPFs) have been mandatory on every highway diesel engine meeting 2007 emissions and beyond. A DPF is an aftertreatment device designed to eliminate soot produced by the engine cylinder combustion process. First, it is necessary to define *soot* and *ash*:

- Soot: Incompletely combusted, hydrocarbon solids. Combustible.
- Ash: Completely combusted, hydrocarbon solids. Not combustible.

The basic function of a DPF is to collect (by entrapment) soot, then combust it into ash. Combustion of collected soot is known as regeneration. Because a small amount of ash results from combusting soot, all DPFs require service intervention to remove these solids from time to time. **Figure 13-6** shows the DPF used on a Ford Power Stroke engine.

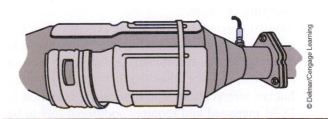

Figure 13-6 Diesel particulate filter (DPF) used on a Power Stroke diesel engine oxidation catalyst.

Types of DPFs

There are two general types of DPFs used on light duty diesels:

- Catalyzed. This is the more common type of DPF. A catalyzed DPF is usually built into the aftertreatment canister that also contains multistage catalytic converters within a housing that used to be known as the muffler. Catalyzed DPFs will be found on most highway diesel-powered vehicles and are capable of both passive and active regeneration cycles.
- Low-temperature particulate filters. These are usually aftermarket devices fitted to pre-2007 highway diesels. Many are fitted to school buses. While capable of significantly reducing PM, they usually do not achieve this to EPA 2007 standards.

DPF Operating Principles

The operational idea behind a DPF is that engine-emitted soot first collects on the walls of the device. Engine manufacturers design DPFs to function primarily in self-regeneration mode. This means that when soot collection reaches a threshold level, it is burned off in what is known as a **regeneration cycle**. There are three general categories of regeneration cycle:

- Passive regeneration. **Passive regeneration** or **self-regeneration** (the terminology varies by engine manufacturer) cycles occur when exhaust temperatures are sufficiently high during normal operation that regeneration can take place unassisted by additional fuel or air injection to the exhaust gas.
- Active regeneration. When the operating environment is not conducive to a self-regeneration cycle, regeneration can also occur assisted by the injection of some fuel (diesel). Sometimes this additional fuel is ignited by a spark plug or by induced high exhaust temperatures. **Active regeneration** uses fuel sourced from the fuel subsystem and delivered at the specified charging pressure.
- Manual regeneration. **Manual regeneration** cycles have to be switched either by the vehicle driver or by the technician. In fact, they are not really manual other than the fact that they require human intervention. The manual regeneration is PCM managed and requires additional fuel to be discharged into the exhaust gas upstream from the DPF.

DPF Management

Ideally, all DPF regeneration should be passive. A vehicle running down the highway at normal speeds should be capable of producing sufficient heat to generate a passive regeneration, especially when optioned with a short period of rich air-fuel ratio (AFR) operation. With this type of passive regeneration, the driver is usually alerted to the regeneration event by a message on the **driver display unit (DDU)** and DPF dash warning light. Operating temperatures of the combined oxidation converter and DPF can exceed 1,100°F (600°C), so most of these devices have plenty of heat shielding.

Frequency of Cycles. Regeneration cycles are designed to occur at set intervals. These set intervals may be as often as once every hour, or as infrequent as once every eight hours of operation, depending on the engine and its power ratings and how the engine is being operated. Drivers should be informed about the expected intervals of the regeneration cycles of the equipment they operate. During both passive and active regeneration, there will be an increase in exhaust gas temperatures.

Ash Residues. As indicated earlier, regeneration leaves some ash residues. These ashes primarily originate from the additive package in the engine lube that is burned in the cylinder during normal combustion. Ash residues have to be removed manually. Federal law requires that this cleaning process occur no more frequently than once a year or 150,000 miles (whichever comes first). Having said that, some manufacturers are aiming to have DPF off-chassis service intervals that exceed 300,000 highway miles.

DPF Temperature Monitoring

Temperature monitoring in DPFs is by **thermocouples (pyrometers)** or by **resistance temperature detectors (RTDs)**. Which of the two methods is used depends on the expected peak temperatures, because RTDs have a threshold operating temperature that is close to peak DPF regeneration temperature.

Pyrometers. A thermocouple pyrometer has a hot end and a sensing end. Two dissimilar wires are arranged to form a circuit: they are wound together at the hot end. When the hot end is exposed to heat, a small voltage is created. This small voltage is read by a millivoltmeter at the sensing end and displayed as temperature values. A thermocouple device must be replaced as a complete unit when diagnosed as failed.

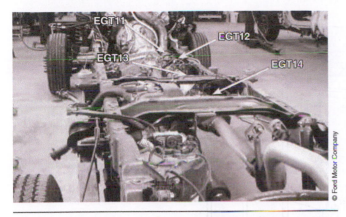

Figure 13-7 RTDs used to monitor exhaust system temperatures on a stripped-down pickup truck chassis.

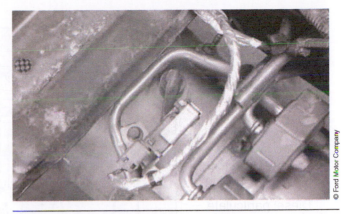

Figure 13-8 Location of the differential pressure sensor used upstream from the DPF: used as a primary reference to determine when a regeneration event is required.

The DPF is a complex assembly within which the regeneration cycles and temperatures have to be precisely managed.

Resistance Thermometers. **Resistance temperature detectors (RTDs)** or resistive thermal devices measure the change in the electrical resistance of gases that occurs with temperature change. They are made of platinum, so these devices are commonly known as **platinum resistance thermometers (PRTs)**. RTDs can replace thermocouple-type pyrometers where temperatures do not exceed 600°C (1,100°F). This is on the threshold of peak DPF temperatures, so some manufacturers have used these in place of pyrometers. They function at greater accuracy than pyrometers. **Figure 13-7** shows the locations of RTDs used to monitor exhaust system temperature on a stripped-down pickup truck chassis.

DPF Pressure Sensor

The DPF pressure sensor is a differential-type pressure sensor that references the pressure upstream from the DPF to atmospheric pressure. It is an input to the PCM. It is used as a primary reference to determine when the DPF is sufficiently restricted (clogged) to indicate that a regeneration event is required. **Figure 13-8** shows the upstream location of a DPF pressure sensor before the DPF canister.

DPF Cleaning Stations

Most of the diesel engine manufacturers state that their DPFs will require routine in-shop cleaning using special equipment. This is required mainly to remove the ash that accumulates from repeated regeneration cycles. The DPF filter assembly is placed in the cleaning station. This high-flow device back pressurizes

the system for about 20 minutes. The total DPF cleaning procedure usually takes less than two hours. Skill levels required to perform the cleaning are not high, but safety precautions must be observed. **Figure 13-9** shows a Donaldson generic DPF cleaning cart designed to clean some units. Most manufacturers seem to favor an exchange system rather than to perform the cleaning within general service facilities.

CAUTION *An appropriate respirator should be worn when servicing DPF components. Submicron ash particulate is known to cause respiratory problems.*

SELECTIVE CATALYTIC REDUCTION (SCR)

Like DPFs, **selective catalytic reduction (SCR)** is also an exhaust gas aftertreatment process. While the DPF addresses particulate HC, an SCR system attempts to "reduce" oxides of nitrogen (NO_x) back into nitrogen and oxygen. Although SCR has been used in Europe for a decade, the EPA did not approve SCR use until 2007. The reason for the reluctance on the part of the EPA in approving SCR systems is that to function, this type of system depends on using a "consumable," specifically urea. The urea, which is in aqueous form, has to be routinely refilled, just like fuel. This means that the EPA approval came with some conditions. However, most light duty diesel engines meeting EPA 2010 on highway emission standards are using SCR: an exception is some versions of the Cummins ISB engine found in Dodge pickups. **Figure 13-10** shows a schematic of SCR components and operation.

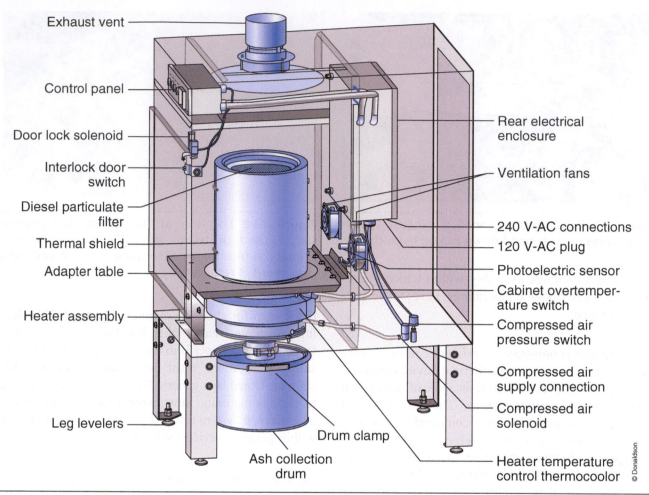

Exhaust vent

Control panel

Door lock solenoid

Interlock door switch

Diesel particulate filter

Thermal shield

Adapter table

Heater assembly

Leg levelers

Ash collection drum

Drum clamp

Rear electrical enclosure

Ventilation fans

240 V-AC connections

120 V-AC plug

Photoelectric sensor

Cabinet overtemperature switch

Compressed air pressure switch

Compressed air supply connection

Compressed air solenoid

Heater temperature control thermocoolor

© Donaldson

Figure 13-9 A Donaldson DPF cleaning system.

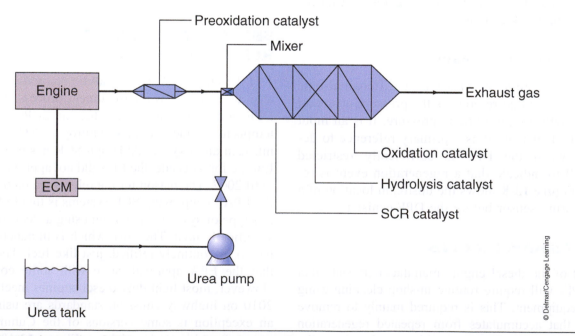

Preoxidation catalyst

Mixer

Engine

ECM

Exhaust gas

Oxidation catalyst

Hydrolysis catalyst

SCR catalyst

Urea pump

Urea tank

© Delmar/Cengage Learning

Figure 13-10 Schematic of SCR components and operation.

Aqueous Urea

SCRs use consumable **urea** to achieve what the rhodium catalyst does in the gasoline-fueled engine. Urea is composed of crystallized nitrogen compounds sourced from natural gas. The urea is in a solution with water, known as **aqueous urea**. The term used to describe this solution in North America is **diesel exhaust fluid (DEF)**; for this reason, we will use this term in this text. The term used to describe aqueous urea in Europe is **AdBlue**. Some manufacturers with European connections are opting to use this term in North America, but there is no difference between the substance used in Europe and that used here. **Figure 13-11** shows an overhead view of a typical small vehicle DEF tank: the cap on this tank is always blue.

The DEF is injected into the exhaust gas stream by a computer-controlled injection system. After injection, the urea-based DEF reduces to ammonia, which itself reacts with NO_x compounds, *reducing* them back to their original oxygen and nitrogen. For this reason, DEF is sometimes referred to as *reductant*.

DEF Storage. In light commercial and passenger vehicle applications, the DEF is stored in tanks large enough not to require replenishment between oil changes when the vehicle is used in typical operating conditions. An aqueous urea solution freezes at 12°F (−8°C), so it must be freeze protected in winter operation. **Figure 13-12** shows a cross-sectional view of a Ford DEF storage tank and the heating circuit that freeze protects it. The DEF is consumed at a rate that varies between approximately 1.5 percent and 10 percent of the fuel used, the variability depending on how the engine is being operated, that is, how much NO_x has to be reduced. The DEF pump is located within the tank (see **Figure 13-13**); at

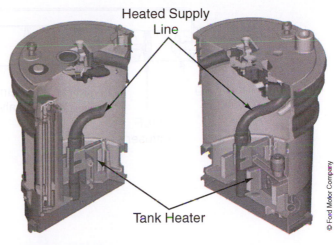

Reductant Heaters

Below a specified temperature the PCM commands the glow plug control module (GPCM) to activate the heaters in the reductant system. The reductant system has heaters in the tank, pump, and lines. The heaters in the tank thaw the DEF if it is frozen and allow it to flow to the pump without freezing. The heaters in the pump and lines allow the DEF to flow to the injector without freezing.

Figure 13-12 Heating system used in a DEF storage tank.

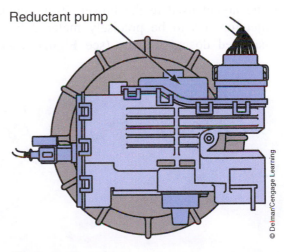

Reductant pump

The reductant pump supplies urea to the dosing module. One unique function of the pump is that when the ignition is turned off, the pump pulls all of the reductant out of the lines. This prevents damage to the lines if the reductant was to freeze.

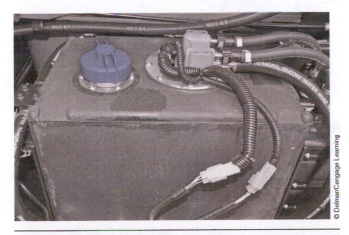

Figure 13-11 Overhead view of a typical small highway vehicle DEF tank.

Figure 13-13 Location of the DEF pump. (*Text courtesy of Ford Motor Company*)

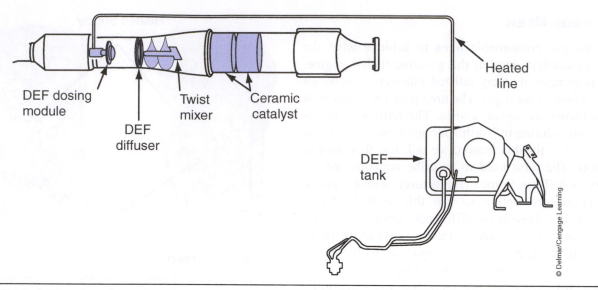

Figure 13-14 SCR system layout and components.

ignition key-off, the pump operates in reverse and purges the DEF solution out of the lines to minimize the chances of freezing.

SCR Management

SCR is managed by the engine PCM. The DEF is pumped from the storage tank and injected upstream from the converter/DPF/muffler assembly. **Figure 13-14** shows the layout used in the Power Stroke engine. DEF injection has to be precisely metered by the PCM-managed dosing module (see **Figure 13-15**).

Too much urea can result in ammonia discharge through the exhaust system. Too little results in NO_x emission.

DEF Injection and Mixing. The DEF module communicates with the PCM on a proprietary CAN bus. It manages the dosing of DEF based on PCM-derived logic commands. Injected DEF is atomized into the exhaust gas stream, then passed through a spiraled mixing chamber as shown in **Figure 13-16.**

Reductant Dosing Module

The reductant dosing module is controlled by the PCM. The reductant dosing module injects reductant into the exhaust system to reduce NO_x coming out of the tailpipe. The injector is made to resist the corrosive properties of the reductant.

Figure 13-15 DEF dosing module.

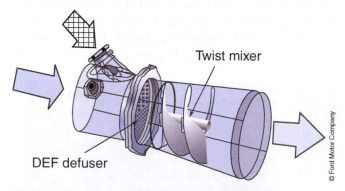

Reductant Exhaust Mixer

There is an exhaust mixing system in the exhaust stream to mix the reductant with the exhaust gas. The mixer is made up of an atomizer and a twist mixer. The atomizer breaks up and vaporizes the reductant droplets. The twist mixer evenly distributes the reductant in the exhaust gases for maximum efficiency.

Figure 13-16 DEF mixing chamber.

NO$_x$ sensor

The NO$_x$ sensor is used primarily to sense O$_2$ and NO$_x$ concentrations in diesel exhaust gas. The sensor is mounted in a vehicle's exhaust pipe, perpendicular to exhaust gas flow. The sensor is mounted downstream of the SCR and DPF. The sensor interfaces with the NO$_x$ sensor module that controls the sensor and heater circuits.

Figure 13-17 NO$_x$ sensor location downstream from the aftertreatment canisters.

NO$_x$ Sensor Module

The NO$_x$ sensor module is mounted to the vehicle frame under the body. It controls the NO$_x$ sensor mounted in the diesel aftertreatment exhaust system downstream of the SCR and DPF. It communicates to the PCM via CAN2 to report NO$_x$ and O$_2$ concentrations as well as sensor and controller errors.

Figure 13-18 NO$_x$ sensor module located on chassis frame.

Monitoring SCR Effectiveness. SCR system effectiveness in reducing NO$_x$ is fully monitored by the engine management electronics. Most light duty diesel engines use at least one NO$_x$ sensor (see **Figure 13-17**). The so-called NO$_x$ sensor is actually an oxygen (O$_2$) sensor: it detects unreacted oxygen in the exhaust gas and sends an analog voltage signal back to either the PCM or more commonly, a NO$_x$ sensor module (see **Figure 13-18**). When a separate NO$_x$ sensor module is used, it is networked to the engine PCM by means of a CAN bus line through which data can be exchanged. The NO$_x$ sensor signals the oxygen detected in the exhaust gas stream to the NO$_x$ sensor module, which then uses the data it has on the mass of air entering the engine to calculate the NO$_x$ ultimately discharged. The operation of oxygen and NO$_x$ sensors is studied in **Chapter 12**.

Monitoring Emissions Circuit Effectiveness. OBD II requires that the emissions control effectiveness be monitored and codes logged should problems be detected. In most cases, the PCM manages the emissions control circuit, but this is not required. **Figure 13-19** shows the physical location of the sensors on a post-2007 Duramax and outlines the electrical circuits that connect those sensors to the engine control module.

Maintaining DEF Levels

A requirement for the EPA approval of SCR was that manufacturers establish a set of fail-safe measures

to ensure that operators did not willingly run the system without DEF. These measures include engine de-rate and no-go status depending on how low the DEF level is and how long the engine has been operated without it. This subject is covered fully in **Chapter 14**, which addresses servicing and maintenance.

Positive Crankcase Ventilation

Highway diesel engines that meet emissions standards for 2007 onward are equipped with a **positive crankcase ventilation (PCV)** system. Because light duty diesel engines are in some cases engineered by the manufacturers of heavy duty diesel engines, the term **closed crankcase ventilation (CCV)** is sometimes used, but both terms refer to the same system. The objective of a PCV system on a diesel engine is the same as that on a gasoline-fueled engine, that is, to prevent venting to atmosphere of crankcase gases. Crankcase gases consist of:

- Blow-by gas from engine cylinders
- Boil-off gases from lubricant

PCV Operation. Diesel PCV systems are usually plumbed upstream from the turbocharger impeller housing so that some pull is exerted on the crankcase. The PCV piping is routed through a filter assembly to prevent crankcase impurities from being drawn into the intake circuit. Manufacturer recommendations for PCV service intervals should be observed.

Sensor Locations

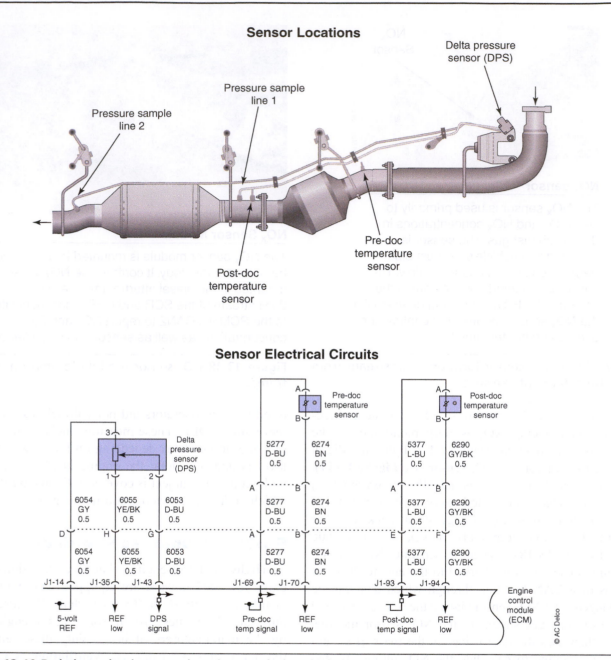

Figure 13-19 Emissions circuit sensor locations and electrical circuits on a 2007 Duramax 6600.

MISCELLANEOUS FACTORS THAT LIMIT NOXIOUS EMISSIONS

In addition to the post-combustion devices examined to this point in the chapter, other engine components can significantly influence the output of noxious emissions. The technician is required to have some understanding of how key components affect emissions in cases where the engine fails an emissions test.

Charge Air Cooling

Effective cooling of intake air lowers combustion temperatures, making it less likely that the nitrogen in the air mixture is oxidized to form NO_x. Air-to-air charge air coolers cool air more effectively than those that rely on engine coolant. Note that anything that compromises the charge air cooler's ability to cool will result in higher NO_x emissions. This is why codes are logged when charge air heat exchangers become plugged.

Variable-Geometry Turbochargers

Variable-geometry turbochargers (VGTs) that perform effectively over a much wider load and rpm range can make a significant difference to both HC and NO_x emissions, especially when PCM controlled, by providing the ability to manage boost on the basis of the fueling and emissions algorithms. VGTs are used rather than constant-geometry turbos in almost every 2010-compliant highway diesel engine—most manufacturers have been using them for at least a decade. When a constant-geometry turbocharger is used today, it is usually as one of the pair used in series turbocharged engines.

Low-Headland Pistons

Low-headland-volume pistons raise the upper compression ring close to the leading edge of the piston crown. This keeps the headland gas volume close to minimum. Headland gas volume tends to be unclean and can increase HC emissions. The use of low-headland-volume pistons by engine designers has resulted in some radical design changes in diesel engine pistons within a short time. Most diesel engine manufacturers favor trunk-type steel pistons today, such as the Mahle Monotherm, examined in **Chapter 3,** which feature low-headland volumes.

SMOKE ANALYSIS

At this point we should say something about how tailpipe smoke looks to the observer's eye. The way we see smoke depends on the *state* of the emission. All matter has three states: vapor or gaseous, liquid,

and solid. It is possible to emit matter in all three states from a diesel engine tailpipe.

- **Gaseous emission.** To the observer's eye, this is "clean" exhaust smoke. And it may be clean—or it may be contaminated with any amount of noxious pollutant in a vapor state. Because light is not affected by passing through a gas, we see nothing when we look at exhaust in a 100 percent gaseous state exiting the tailpipe. For instance, harmful NO_x emission cannot be observed as it discharges from the exhaust.
- **Liquid emission.** To the observer's eye, liquid condensing in the exhaust gas stream appears white. When light attempts to pass through condensing liquid droplets in the exhaust it either *reflects* or *refracts*, making it appear white to the observer. White smoke may be completely unburned fuel, or it may be water or antifreeze. Because water is a product of combusting a hydrocarbon fuel, if the weather is cold enough, this can be condensed to steam at the tailpipe exit.
- **Solid emission.** To the observer's eye, solids (we call these *particulates*) contained in the exhaust gas stream appear black. Light cannot pass through particulate suspended in the exhaust gas, making it appear black to the observer. Black smoke is caused by toasted fuel or oil exiting the exhaust.

Summary

- Emissions standards that engine manufacturers must meet are set nationwide by the EPA. The influence of CARB has greatly increased in recent years because its standards have been adopted by more than 20 states, with others scheduled to follow.

- The word smog is derived from the words *smoke* and *fog*.

- Photochemical smog is formed by gaseous and particulate hydrocarbons (HC) combining with ozone and oxides of nitrogen (NO_x) followed by a period of exposure to sunlight.

- Because creating photochemical smog requires a period of exposure to sunlight, relatively still air is

required for its formation. This means it is more prevalent during static summer heat waves.

- Vehicle emissions are the largest single contributor to the photochemical smog in most geographic areas of North America.

- VOCs are hydrocarbon fractions boiled off fuels during their production, transportation, and storage. VOCs are a more severe problem in high-temperature conditions.

- Ozone is known to be toxic in small concentrations.

- Diesel engines today rely on exhaust gas recirculation (EGR), NO_x adsorber catalysts, oxidation and reduction catalytic converters, diesel particulate filters (DPFs), and selective catalytic reduction (SCR), along with computer control of combustion, to comply with emissions standards.

- The DPF entraps soot particles. The entrapped soot is then combusted in what is known as a regeneration process.

- DPF regeneration can be passive (occurs during normal vehicle operation), active (requires dosing fuel and sometimes spark ignition), or manual (technician or operator managed).

- SCR is used on most highway light duty diesel engines that meet 2010 EPA emissions standards.

- SCR reduces NO_x and uses an aqueous urea solution known as DEF to achieve this.

- DEF is a consumable, and most manufacturers ensure that enough of it can be retained in an onboard storage tank so that it only has to be replenished at each oil change.

- DEF is injected into the exhaust stream in precisely metered quantities, managed either directly by the PCM or by a module networked to the PCM by a CAN bus line.

- The effectiveness of SCR is monitored by a NO_x sensor.

- Observation of diesel engine exhaust can be used to determine the state of the emission.

- Clear tailpipe emission indicates that the exhaust is entirely in a vaporized state. White smoke indicates liquid emission in the exhaust gas, while black smoke indicates solid emission or PM in the exhaust gas.

Internet Tasks

1. Download Al Gore's "An Inconvenient Truth" and make some notes.

2. Google some opposing points of view to that of Al Gores'.

3. Formulate your own opinions on items 1 and 2, making your points in bullet form.

4. Use a search engine to locate information about the Air Quality Index (AQI) for the geographic area you reside in. How does your area compare with other regions of North America? With Europe? With Asia?

5. Log onto Wikipedia and see what you can learn about SCR.

6. Log onto howstuffworks and check out DPFs.

Shop Tasks

1. Select a diesel-powered chassis, note the model year, and make a list of the external emissions hardware used on the engine.

2. Using a vehicle equipped with a DPF (model year after 2007), connect an EST to the data bus and perform a manual, active regeneration using the manufacturer's recommended procedure.

3. Disconnect the DEF tank level sensor in an SCR-equipped vehicle. Make a note of exactly what outcomes are displayed on the DDU and on an EST.

Review Questions

1. Which of the following compounds is *not* currently classified by the EPA as a harmful tailpipe emission?

   A. NO_x C. PM

   B. Hydrocarbons   D. Carbon dioxide

2. Which of the following is present in the largest quantity by weight in the diesel engine cylinder during combustion?

   A. Oxygen

   B. Nitrogen

   C. Fuel

   D. Carbon monoxide

3. Which tailpipe emission does an SCR system attempt to reduce?

   A. Particulate matter (PM)

   B. Oxides of nitrogen

   C. Hydrocarbons

   D. Ozone

4. What does an NO_x sensor actually measure at its probe?

   A. NO_x

   B. PM

   C. H_2O

   D. O_2

5. Which of the following is the most common form of oxygen at ground level?

   A. O

   B. O_2

   C. O_3

   D. O_4

6. Which of the following results from "perfect" combustion of a HC fuel when it is burned with oxygen?

   A. Carbon monoxide and carbon dioxide

   B. Carbon dioxide and nitrogen dioxide

   C. Carbon dioxide and water

   D. Nitrogen dioxide and water

7. When a diesel engine is operated at lower-than-normal temperatures, which of the following tailpipe emissions is likely to increase?

   A. Oxides of nitrogen

   B. Carbon dioxide

   C. Ozone

   D. Hydrocarbons

8. When a diesel engine is operated at higher-than-normal temperatures, which of the following tailpipe emissions is likely to increase?

   A. Oxides of nitrogen

   B. Carbon dioxide

   C. Ozone

   D. Hydrocarbons

9. Which of the following compounds is classified as a greenhouse gas responsible for contributing to global warming?

   A. Oxides of nitrogen

   B. Hydrocarbons

   C. Carbon dioxide

   D. Sulfur dioxide

10. Which of the following describes what happens in an *oxidation*-type catalytic converter?

    A. Nitrogen in NO_x is burned

    B. CO and HCs are burned

    C. HCs are filtered out

    D. PM is filtered out

CHAPTER

14 Servicing and Maintenance

Prerequisites

Chapters 3 through **7.**

Learning Objectives

After studying this chapter, you should be able to:

- Explain why it is sometimes important to connect to a chassis data bus even when performing routine service work.
- Outline the procedure required to break in a light duty diesel engine.
- Respond appropriately when a diesel fuel tank gets filled with gasoline.
- Service the air intake system in a light duty diesel engine.
- Identify the appropriate oil to use in a light duty diesel engine.
- Identify the steps required to perform an engine wet service.
- Perform an engine oil and filter change.
- Mix and test coolant and service an engine cooling system.
- Identify the appropriate fuel to use in a diesel engine.
- Refuel and prime a diesel engine.
- Respond to a water-in-fuel alert and service a water separator.
- Service and replenish an SCR system with DEF.
- Identify when it is appropriate to service a DPF.

Key Terms

biodiesel

Bosch MasterTech

Bosch VCI-CA

dash display unit (DDU)

diagnostic trouble code (DTC)

diesel exhaust fluid (DEF)

diesel particulate filter (DPF)

ethylene glycol (EG)

multiple diagnostic interface (MDI)

predelivery inspection (PDI)

propylene glycol (PG)

restriction gauge

selective catalytic reduction (SCR)

service information system (SIS)

service literature

Tech II

ultra low sulfur diesel (ULSD)

Volkswagen 507.00 specification

water-in-fuel (WIF) sensor

waterless engine coolant (WEC)

INTRODUCTION

In many of today's light duty trucks and diesel-powered vans, manufacturers prefer that a data bus connection be made with the chassis data bus as a first step to routine servicing. This connects the vehicle with the manufacturer's data hub, and a report card on the vehicle is established. This chapter will take a brief generic look at servicing and maintenance, but there is no substitute for using the vehicle manufacturer's service literature when undertaking maintenance procedures on the shop floor. When undertaking any kind of service work on today's vehicles, it makes sense to check if there are any active **diagnostic trouble codes (DTCs)** before servicing the vehicle. Some current vehicles also carry a data log with an address on the chassis data bus. The data log can be used to track maintenance, service, and warranty information, so it should be referenced each time service work is undertaken.

The objective of this chapter is to briefly outline some typical maintenance practices as they apply to diesels equipped with light duty diesel engines. It begins with engine break-in. If you are familiar with the break-in procedure used on heavy duty (HD) commercial diesel engines, the process used with light duty diesels differs significantly due to the different construction materials used. While break-in of HD diesels usually requires a short period of maximizing cylinder pressures using a dynamometer, light duty diesels should be nursed somewhat for the first 500 to 1,000 highway miles (10 to 20 engine hours).

Service Literature and Diagnostic Software

The days of hardcopy service manuals and scan tools are all but over. Most (but not quite all... yet) manufacturers today have, or are in the process of, eliminating scan tools (such as the GM **Tech II** or generic **Bosch MasterTech**) and replacing them with specialty personal computer (PC) software loaded onto a laptop. This arrangement requires the use of a communications adaptor (CA) such as the generic **Bosch VCI-CA** or its proprietary equivalent such as the GM **multiple diagnostic interface (MDI)**. In this text, the term *electronic service tool (EST)* will be used to refer to any data bus access tool, whether it is a scan tool or PC based.

When PC-based diagnostic software is used in conjunction with an online link to the manufacturer's data hub, the technician always has access to the latest information pertinent to a service or repair procedure,

along with a log specific to the vehicle being worked on, which provides the opportunity to receive alerts to recalls. Full coverage of diagnostic software and online **service information systems (SISs)** is provided in **Chapter 12**. In this chapter, the term **service literature** will be used to represent service information, whether that information is available in electronic or hardcopy format.

STARTUP AND ENGINE BREAK-IN

Although most light duty diesels continue to use cast iron piston rings, an extensive break-in is usually not required. Manufacturers suggest that the vehicle not be driven continuously at the same road speed for the first 1,000 miles (1,600 km) or so of operation. In addition, a new vehicle should be driven at least 500 miles (800 km) before towing a trailer or hauling maximum rated loads. In addition, it is important to use the specified engine oil during break-in. The oil should not be contaminated by any kind of additive during break-in. Friction modifier compounds or aftermarket break-in oils should never be added to diesel engine oils: they may interfere with piston ring seating and result in an engine that burns oil.

Post–Break-In Checkup

However thoroughly a **predelivery inspection (PDI)** has been performed, there is always a chance something has been missed or components have shifted and changed during initial operation. For this reason, a post–break-in checkup after having put a thousand or so miles (1,600 km) on a new engine is recommended. The only instrument required to undertake this checkup is a set of eyeballs, but based on the inspection, some minor adjustments may have to be made even on the healthiest engine. Check for the following:

- Oil leaks
- Coolant leaks (hot and cold)
- Hose clamp integrity
- Belt tightness
- Air delivery pipes and hoses
- Radiator and heat exchanger leaks
- Wiring harness chaffing

Startup

Using the correct procedure to start a diesel engine is important, especially when temperatures drop below 32°F (0°C). The engine must have the correct grade of oil for the temperatures it must operate in. All computer-controlled engines use cold start strategy to limit smoke emission and engine loading when cold. Cold start strategy usually disables the accelerator pedal, allowing the software startup map to be fully in command of the start strategy. Accelerator pedal input is usually disabled for a set period, such as 30 seconds, or until a preset engine oil temperature has been achieved. Note also that some electronic automatic transmissions may inhibit shifting to first and reverse when temperatures are below 0°F (–17°C) until the automatic transmission fluid (ATF) warms.

When starting the engine in temperatures below –15°F (–26°C), the engine should be idled for at least five minutes before attempting to drive it. Before pushing the start button or turning the key, the gearshift lever should be in park or neutral, the parking brake set, and the brake pedal depressed. It is never required that the accelerator pedal be depressed during startup in any chassis built since 1997.

Extreme Cold Weather Starting. Most manufacturers recommend that an engine block heater be used any time temperatures drop below –10°F (–23°C). Some manufacturers recommend the use of cetane improvers in extreme cold weather operation. These should be nonalcohol based. Generally, an engine should not be cranked for longer than 10 to 15 seconds because starter damage may occur. Leave a full 30 seconds between each cranking phase. Avoid the use of starting fluids such as ether: these are designed for heavy duty diesels and may blow out the rings on light duty, automotive diesels.

WARNING *Never add gasoline, gasohol, kerosene, or alcohol to diesel fuel. This can cause fire hazards, engine performance problems, and costly fuel system damage.*

Start Sequence

1. Turn the key to On but do not crank the engine. This usually activates the glow plug circuit if the engine is so equipped. Wait until the glow plug preheat indicator turns off, or wait until you are prompted by the **dash display unit (DDU)**, before cranking.
2. It is common for the glow plugs to remain on for a short period of time after engine start. If the engine is not started before the glow plug activation time ends, the glow plugs will likely have to be reset by turning the key to Off.
3. After the engine starts, allow it to idle for around 15 seconds and at least 1 minute in

winter conditions. Check the oil pressure. With some types of automatic transmissions in sub-zero conditions, shifting may be inhibited until the transmission fluid has been warmed to a threshold temperature.

Jump-Starting. The procedure for jump-starting light duty diesels differs little from that used to jump-start a gasoline engine. However, depending on the manufacturer, there may be more than one battery. In these cases, one of the batteries is usually identified as the primary battery, and this should be used as the external connection point. The recommended procedure is to connect the positive jump clamp first, followed by the negative jump clamp. Most manufacturers recommend that the ground connection be made directly to battery ground and not to chassis ground; this minimizes voltage surges that can be caused when a high-resistance ground is made. High-voltage surges can damage chassis computer equipment.

CAUTION *Never allow the vehicles to physically contact each other during a jump-start procedure. The only contact points between the two vehicles should be the jumper cable clamps.*

Water-in-Fuel Alert. It is always possible for water-contaminated diesel fuel to be pumped into fuel tank(s) during refueling, especially when it is purchased from refuelers with lower fuel volume turnover. Most modern automotive diesels are equipped with **water-in-fuel (WIF) sensors** and a water separator to remove water from the fuel within the fuel subsystem. The WIF warning light will illuminate when the water separator has exceeded a threshold of water accumulation. Should the WIF light illuminate when the engine is running, the engine should be shut down and the water separator drained. Operating the vehicle with the WIF warning light illuminated can result in costly fuel-injection system failures. Draining the water separator with the engine running may allow air to be pulled into the fuel system.

WARNING *Do not attempt to drain a water separator while an engine is running. Fuel may ignite if the separator is drained while the engine is running.*

DEF Alert. When the ignition is put into the key-on position, the **diesel exhaust fluid (DEF)** warning light will briefly illuminate and extinguish. When the aqueous urea fluid is either contaminated and/or low, the DEF light will remain illuminated and other intervention strategies will progressively kick in. If the vehicle runs out of DEF, it will enter a trim back mode that limits road speed and engine rpm. Depending on the manufacturer, engine output may be limited to idle only. The DEF must be replenished: it is a legal and operational requirement. Full coverage of DEF maintenance appears later in this chapter.

AIR INTAKE SYSTEM MAINTENANCE

For every gallon of diesel fuel by volume combusted by a diesel engine, between 10,000 and 15,000 volumetric gallons of air pass through the air intake system. The variance depends on how the engine is being operated. For this reason, a diesel engine air intake system must be closely scrutinized during operation; forgetting about the air intake system between service intervals is an invitation to costly problems. Because many workhorse pickup trucks are operated in agricultural or construction environments, special attention should be paid to air filter **restriction gauges**. Grain chaff is a special problem and can plug a brand-new filter in a single eight-hour workshift. Engines operated in these types of conditions should be fitted with an aftermarket precleaner.

Air Filter Restriction Gauge

The air filter restriction gauge is usually located on the air filter canister; the display readout may be located on the gauge itself or may be set on the dash so the driver can immediately be alerted if there is a problem. A restriction gauge measures pressure values below atmospheric (vacuum) and therefore identifies the restriction that has to be exerted on air to pull it into the turbocharger compressor housing. The more the air filter is restricted (clogged), the higher the restriction value; the restriction value is displayed in inches of water (in./H_2O).

Most air filter restriction gauges are located under the hood on light duty diesel vehicles, so the hood must be raised to check the readings. Manufacturers recommend that an air filter restriction gauge be checked every 5,000 miles (8,000 km) to 7,500 miles (12,000 km), but the frequency should be increased when a vehicle is operated in hostile environments (to engines) such as ag or construction: daily inspections and resets are advised under these conditions. The restriction gauge may be calibrated in in./H_2O or in

dumbed-down gradients that culminate in "change filter."

> **Tech Tip:** An engine air filter functions at its best efficiency just before it fails. Do not change filters unnecessarily when the gauge indicates the unit is close to failure.

Engine performance and fuel economy are compromised when the maximum restriction is reached. Using compressed air to blow out air filter elements should be avoided. In most cases, this practice damages the filter element. You should also note that the physical appearance of a filter element is no guide as to whether it is functional. Learn to rely on the restriction gauge: if in doubt, you can double-check by fitting a reliable external gauge or water-filled manometer.

Resets. Most inlet restriction gauges are equipped with a reset button. When a gauge indicates that it is getting close to its maximum spec, depress the reset button. Check shortly afterward. However, note that a valid restriction reading will not occur until the engine is run at full load because it has to be pulling the maximum amount of air through the filter. Note that running a diesel engine at no-load high idle represents only a small percentage of peak loading. After installation of a new filter element, the restriction gauge should be reset.

Snow and Rain. Depending on the location of the air filter and whether a precleaner is fitted to the air intake system, vehicle operation in heavy snowfall or heavy rain can channel excessive quantities of snow/water into the filter element. This can temporarily plug and saturate the filter element, but the condition is not terminal, although engine performance may be severely compromised. Take the following action:

- Snow: At the earliest opportunity, open the hood and clear the snow and ice from the air filter housing inlet. Do not attempt to remove the filter element at this point because you will likely damage it. Reset the restriction gauge and run the engine.
- Water saturation. After heavy rain the filter element may saturate, resulting in reduced performance, but the air filter will usually dry within 15 to 30 minutes at highway speeds. While saturated, the restriction gauge will produce high readings, so at the earliest opportunity, pop the hood and reset the air filter restriction gauge.

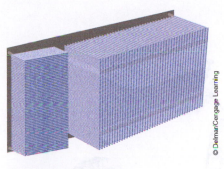

Air Inlet Components

Air Filter

The air filter is located on the passenger side of the engine compartment in front of the battery.

The air filter housing includes a non-electrical filter minder to measure inlet restriction. When the filter element becomes contaminated beyond useful limits, the filter minder visually indicates the need for replacement.

Figure 14-1 Typical air filter assembly. (*Text courtesy of Ford Motor Comapny*)

Air Filter Replacement

It should be noted that failure to use the specified air filter element may result in severe engine damage, as can attempting to run an engine without an air filter. **Figure 14-1** shows the air filter used on a 2011 Ford Power Stroke engine. To replace an air filter element use the following general procedure, but note that it will vary from engine to engine:

1. Locate the mass air flow (MAF) sensor electrical connector on the air inlet pipe and unplug it. Unlock the clip on the connector, then squeeze and pull it off the air inlet pipe.

2. Release the four clamps that secure the cover to the air filter housing. Push the air filter cover forward and up to loosen it.

3. Pull the air filter element from the air filter housing.

4. Sometimes a foam pre-element is used. In some cases, this can be washed rather than replaced, but go by the manufacturer's recommendation. If the foam filter is not to be replaced, ensure that it is clean and in place.

5. Install the new air filter element. Handle it with care and ensure that the groove seal on the pleated paper filter seals at both ends of the element to housing surfaces.

6. Replace the air filter housing cover and secure the clamps. Make sure you do not crimp the filter element edges between the air filter housing and cover. The tabs on the edge must be properly aligned into the slots.

7. Reconnect the mass air flow sensor electrical connector to the inlet pipe. Make sure the locking tab on the connector is in the locked position.

ENGINE LUBE SERVICE

The engine lubrication system must be routinely serviced. **Figure 14-2** shows a typical lubrication circuit used in a current light duty diesel. Using the specified engine lube oil is important, especially for diesel engines equipped with **diesel particulate filters (DPFs)**. All DPF-equipped engines must use American Petroleum Institute (API) service categories CJ-4 or CJ-4/SM. Failure to use a CJ-4 category lube can either plug or destroy a DPF. In addition to the API category, it is important to use oil of the correct viscosity for the temperature and conditions to which the engine will be subjected.

Proprietary Lube Specifications

Manufacturers may specify an oil formulation that in their opinion exceeds general industry standards. An example would be the **Volkswagen 507.00 specification** oil, an extended-service lubricant recommended for VW and Audi diesel engines. The 507.00 lube is close to our CJ-4 lube standard, but it has slightly lower ash and a lower general level of wear-enhancing additives that suit heavy truck but not light automotive applications. Online forums may debate the advantages of either lubricant, but generally we can conclude that API-CJ-4 tends to be thicker and conducive to longer engine life while VW 507.00 might provide fractionally better fuel economy. Typically, the following are recommended options for post-2007 (DPF-equipped) diesel engines:

- Severe duty, year-round usage: SAE 5W-40 API CJ-4
- Extended service: VW 507.00 SAE 5W-40
- Moderate duty, nonwinter usage: SAE 15W-40 API CJ-4

Most manufacturers recommend that an engine block heater be used when temperatures drop below −10°F (−23°C) when SAE 5W-40 is used. SAE 15W-40 begins to thicken at 30°F (−1°C) and turns to grease as temperatures descend: it is not recommended for use in typical North American winter conditions, even when vehicles are garaged. Observing this precaution is especially important when considering an oil for use

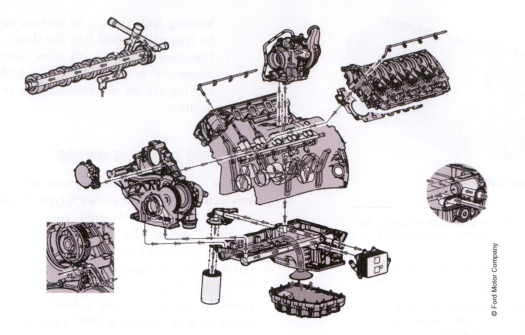

© Ford Motor Company

Lubrication System Oil Flow

- Oil is drawn from the oil pan through the pickup tube. The oil is then routed through a passage cast into the upper oil pan then to the oil pump inlet.
- From the oil pump, oil is directed to the oil cooler and then to the oil filter.
- The main oil passage in the rear of the engine block feeds the right, left and the camshaft galleries.
- Right oil gallery feeds the:
  - rocker arm oiling manifold for the right cylinder head.
  - cam followers and hydraulic lifters on the right side.
  - piston cooling jets on the right side.
  - crankshaft main bearings,
    - a separate oil passage for each main bearing.
    - also used to lubricate the connecting rod bearings.
  - turbocharger.
- Left oil gallery feeds the:
  - rocker arm oiling manifold for the left cylinder head.
    - An oil passage connected to the gallery going up to the left cylinder head also provides engine oil to the:
      - vacuum pump.
      - meshed gears of the crankshaft, camshaft, and high-pressure fuel pump.
      - cam followers and hydraulic lifters on the left side.
      - piston cooling jets on the left side.
- Camshaft oil gallery feeds the camshaft bearings.

Figure 14-2 PowerStroke 6.7L lubrication circuit.

in the high-revving diesels used in light passenger vehicles.

Changing the Engine Oil and Oil Filter

Most modern vehicles are equipped with a smart oil condition monitor such as the Ford Intelligent Oil Life Monitor™ that prompts the operator when an oil change is required. When a DDU message appears indicating OIL CHANGE REQUIRED, the engine oil, oil filter, and fuel filters should be changed.

Check Engine Oil Level. Most manufacturers recommend that the engine lube level be checked every time the vehicle is refueled, but the reality is that this seldom occurs. However, it should be done while the engine is being broken in because there is a tendency for engines to burn oil while in the process of seating rings. Manufacturers will inform operators that it is normal to add some oil between oil changes, so engine oil levels should be checked at least once a week.

The procedure to accurately check the engine oil level is as follows:

1. The engine should be at normal operating temperature, indicated by the engine coolant temperature gauge.
2. Park the vehicle on a level surface, then turn off the engine and open the hood.
3. Wait at least 20 minutes after engine shutdown to ensure that the oil contained in the upper sections of the engine has drained back to the oil pan.
4. Withdraw the dipstick, wipe it clean with a lint-free wiper, and reinsert it fully.
5. Withdraw the dipstick again and read oil level on both sides of dipstick to determine the actual engine oil level.
6. The oil level should be maintained within the crosshatch area of the dipstick by adding oil as required. The lower end of the crosshatch is usually the minimum oil level and the upper end of the crosshatch indicates the maximum oil level. Never overfill. The consequences of excessively high oil levels can be as severe as excessively low levels. On most engines, the distance from the top to the bottom of the crosshatch area on the dipstick represents 1.0 quart (.95L).

Procedure to Change Engine Oil

1. Place a catch sump under the engine large enough to contain the engine oil with a suitable margin of error. Unscrew the oil filter (see **Figure 14-3**) and oil pan drain plug and drop the oil. Wait for the oil to properly drain.

WARNING *Take care not to sustain burns by handling hot oil or filters with bare hands.*

2. Replace the filter. Smear oil over the filter sealing gasket and tighten exactly according to the instructions. Make sure it is not over-tightened: this can damage the filter. Note: some manufacturers want the oil filter primed (filled with oil) on their engines with turbochargers.

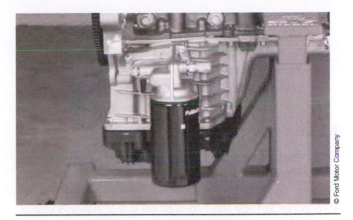

Figure 14-3 Oil filter assembly bolted to side of oil pan.

3. Reinstall the oil pan drain plug, replacing the sealing washer or O-ring if used.

Note: Some newer oil pan drain plugs only require one-quarter turn to either remove or install. Usually just a ⅜-inch socket drive is required to remove or install the plug. Ensure that the plug is not overtightened during installation.

---

4. Refill the engine with new oil of the appropriate specification. For the specified quantity, consult the manufacturer's service literature.
5. Reset the smart oil service life indicator, such as the Intelligent Oil Life Monitor™ on a Power Stroke.

WARNING *Used engine oil is a known carcinogen. Protect your skin by wearing gloves and washing with soap and water after performing a wet service on an engine.*

Defining Severe Service Operation. When a diesel engine is subjected to severe service operation, the oil should be changed more frequently even when on-board smart lube oil monitoring is used. Some of the factors that determine severe service are:

- Frequent or extended idling (over 10 minutes per hour of normal driving)
- Low-speed operation/stationary use
- Vehicle is operated in sustained ambient temperatures below −10°F (−23°F) or above 100°F (38°F)
- Frequent low-speed operation, in consistent heavy traffic at speeds less than 25 mph (40 km/h)

- Operating in severe dust or agricultural chaff conditions
- Operating the vehicle off road
- Hauling a trailer over 1,000 miles (1,600 km)
- Sustained, high-speed driving at full loads

Oil Change Intervals

It is not the function of any textbook to identify oil change intervals because there is so much variance between the manufacturers and there are so many differences among the types of oil used. Always check the manufacturer recommendations and specifications. With engines using the HEUI fuel system, oil is used as hydraulic media (to actuate injector pumping) and because of this, the consequences of extending an oil change interval or using an inappropriate oil (high-detergent oils aerate more easily) can produce immediate driveability problems. Some oils are much cheaper than the manufacturer-recommended oil (likely to break down earlier), while others are more costly, such as synthetics (suited for severe duty and high performance). The one thing that can be said is that extending oil change intervals can result in shortened engine longevity.

COOLING SYSTEM SERVICE

The cooling system in a modern vehicle can be the source of many problems if it is not properly maintained. The basics would be to ensure that the proper coolant was used and that its chemistry was routinely checked using the manufacturer's required procedure. Coolants break down. They break down faster in today's diesel engines than they might have a generation ago because of the extensive use of aluminum with the cooling system of the vehicle. During servicing, engine coolant should be checked for:

- Level
- Concentration (freeze point protection)
- Additive strength (corrosion inhibitor)
- Acidity

Coolant

As we said in **Chapter 6**, coolants may be purchased either as concentrate or premixed. That said, most antifreeze sold over the counter today continues to be as concentrate. Where **ethylene glycol (EG)** antifreeze is used, it should be maintained at a concentration of 50:50 antifreeze and water, which should provide freeze protection down to $-36°C$ ($-34°F$).

It also provides antiboil protection up to $130°C$ ($265°F$). Temperatures below $-36°C$ ($-34°F$) are classified as subarctic, and for operation in the central northern states, central northern Canada and Alaska, using **propylene glycol (PG)** should be considered. PG tends to be used more often in commercial vehicles operating in subzero and arctic conditions. The antifreeze protection of PG increases with its percentage in a solution, unlike EG; increasing the percentage of EG over water beyond 60 percent actually raises the freeze point temperature.

The type and condition of coolant run in an engine have a big influence on how long that engine is going to last. With this in mind, it is important to say something about tap water. Tap water varies from town to town and district to district. In other words, it is a substance of variable and unknown chemistry. Avoid using it except in an emergency, and after the emergency, flush the engine cooling circuit. Assuming that the engine will be using an EG-based antifreeze, the options are:

- Premix extended-life coolant (ELC). Just make sure it is never contaminated with other antifreezes or water. Requires little testing during its service life. Best choice. The problem is that anyone can use the term ELC to describe coolant, so it is a question of buyer beware when shopping for this product. The original premix ELC was manufactured by Texaco.
- Long-life EG concentrate mixed with distilled water. Distilled water costs little and pays for itself many times over by increasing the life of the engine. Must be routinely tested and chemically adjusted; prone to galvanic effect, which can destroy an engine if left. Examples of long-life EG are Dexcool and Pentosin G-12 (VW and Audi). Next best choice to ELC.
- **Waterless Engine Coolant (WEC).** This premixed EG base coolant was introduced a few years ago and is gaining some popularity. Despite higher upfront cost, WEC is cost effective over the life of an engine because it never has to be changed or tested—providing it is never contaminated with water. WEC is currently manufactured by Evans.
- EG mixed with tap water. Must be routinely tested because it is subject to degradation, which can be rapid when the water used has a high mineral content. A poor choice and over the long-term, the most costly. The sad thing is that this malpractice is featured in many manufacturer dealerships.

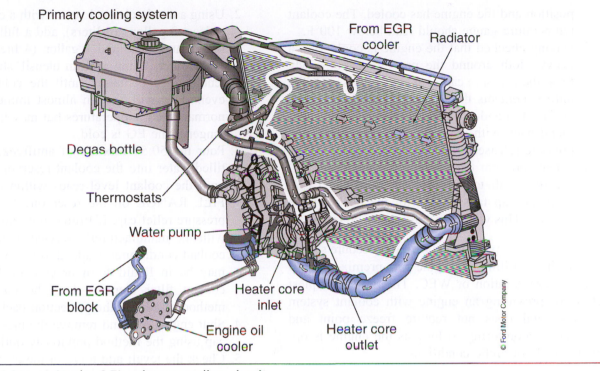

Primary cooling system
From EGR cooler
Radiator
Degas bottle
Thermostats
Water pump
From EGR block
Heater core inlet
Engine oil cooler
Heater core outlet
© Ford Motor Company

Figure 14-4 Power Stroke 6.7L primary cooling circuit.

As outlined in **Chapter 6**, a coolant mixture does much more than provide antifreeze and antiboil characteristics. For this reason, ensure that only coolants approved by the manufacturer are used in the engine cooling system. **Figure 14-4** shows the primary cooling circuit of the dual cooling systems used on the Power Stroke 6.7L engine.

Coolant Color. The color of antifreeze means nothing. Despite attempts over the years to standardize antifreeze color, this has not been achieved. Antifreeze may be dyed yellow, blue, green, red, or any other color. The only color of significance in engine coolant is muddy brown: this usually indicates that the coolant has degraded and should be exchanged before it begins to claim internal engine components.

Coolant Concentration Testing. Coolant concentration can be tested with a hydrometer, a specialized antifreeze tester (such as a Rotunda Antifreeze Tester), or a refractometer. Refractometers **(see Chapter 6)** produce greater accuracy and do not require a temperature-adjustment calculation.

Coolant Level Testing

When the engine is cold, check the level of coolant in the reservoir(s). Some vehicles use a primary and secondary cooling system, and it is important to regard each system separately. When cooling system

maintenance has been overlooked, the main engine (primary) or secondary coolant reservoir may become low or empty. Should either reservoir become low or empty, add the specified coolant to the appropriate level.

Coolant Precautions. General guidelines:

- Never use alcohol, methanol, brine, or any engine coolants with an alcohol or methanol base.
- Avoid using recycled coolant unless it is from a known reliable source.
- Ensure that the engine has cooled (coolant temperature gauge reading below 100°F (38°C) before attempting to unscrew the coolant pressure relief cap.
- When replenishing or replacing a coolant mixture, make sure the mixing of EG and distilled water takes place before pouring the solution into the cooling system.

Adding Coolant

Treat engine coolant with a high level of respect and it will pay dividends in increased engine longevity.

Remove Pressure Cap. To remove a cooling system pressure cap, use the following procedure:

1. Before attempting to remove the pressure cap, make sure the ignition circuit is in the key-off

position and the engine has cooled. The coolant temperature gauge should read below 100°F.

2. Having checked that the engine is cool, wrap a heavy cloth around the pressure cap. Slowly twist the pressure cap counterclockwise (CCW) until it contacts the first set of stops; this will allow any residual pressure to release.

3. Stand back with your face protected while the pressure releases.

4. When sure that the pressure has been relieved, use the cloth to twist the pressure cap CCW until the cap tabs align with the neck release notches. This will permit the cap to be removed.

Add Premix Coolant. The following procedure should be used when refilling the engine with premixed, long-life antifreeze solution or WEC. This is the simplest method of providing an engine with coolant system protection and does not require freeze point and maintenance monitoring so long as the engine is operated outside of arctic conditions.

1. Remove the pressure relief cap from the engine or secondary coolant reservoir using the procedure previously outlined.

2. Unseal a jug of long-life diesel engine coolant or premix WEC. Pour into the coolant reservoir until the coolant level reads within the COLD FILL RANGE. Do not overfill.

3. Reinstall the pressure relief cap.

4. Start and run the engine at 2,000 rpm for a minimum of two minutes.

5. Shut engine off, and remove the pressure relief cap using the method previously outlined.

6. Check the level: add more premix to ensure that the coolant level is within the COLD FILL RANGE as identified on the reservoir. Do not overfill.

Tech Tip: When using premix long-life coolants and WECs, make sure the container is sealed (new) or at most, has been opened no more than a week. The additives used in premix coolant boil off at ambient temperatures and can react to atmospheric oxygen.

Mix/Add a 50:50 Coolant Solution. Use the following procedure to refill or add a 50:50 EG-to-distilled water mixture to an engine or secondary cooling system after either has been drained or become low.

1. Remove the pressure relief cap from the engine or secondary coolant reservoir using the method previously outlined.

2. Using a newly cleaned vessel with a capacity of at least 2 gallons (8 liters), add a full container of EG antifreeze and 1 gallon (4 liters) of distilled water. Using a clean utensil, stir the contents of the container until the color appears even. This will happen almost immediately at normal room temperatures but may take a little longer if the EG is cold.

3. Pour the 50:50 mixture of antifreeze and distilled water into the coolant reservoir, filling it until the coolant level reads within the COLD FILL RANGE on the reservoir. Reinstall the pressure relief cap. If filling from empty, check with the manufacturer's service literature how coolant conditioner is added to the system—this may be in liquid form or contained within a coolant filter element. Use the manufacturer method of adding the protection package.

4. Shut engine off, and remove the pressure relief cap using the method previously outlined.

5. Check the level: add more of the 50:50 coolant solution to ensure that the coolant level is within the COLD FILL RANGE as identified on the reservoir. Do not overfill.

FUEL SYSTEM MAINTENANCE

Diesel fuel systems are damaged as often as they are helped by products known as diesel fuel conditioners. While these products are not exactly snake oil, they seldom do much good and may create performance problems. Diesel fuels meeting either the ASTM D975 (diesel) or the ASTM D7467 (biodiesel) industry specifications should be used: such fuels do not require additional additives unless they have been allowed to break down. Outside of North America, use fuels meeting EN590. Most diesel engine manufacturers stipulate that they are not responsible for any repairs required to correct the effects of using diesel fuel conditioner. The only exception is much older hydromechanical systems, designed for an era of diesel fuels with much higher lubricity than currently approved fuels. Alcohol-based conditioners should be especially avoided and only used in an absolute emergency to address water freeze-up problems in the tank.

CAUTION *Never put known contaminated fuel into a vehicle fuel tank and expect additives to take care of the problems. Have degraded/contaminated fuel disposed of according to your state guidelines for waste disposal.*

Ultra-Low-Sulfur Diesel Fuel

The only fuel available at the pumps today for on-highway usage is **ultra low sulfur diesel (ULSD)**; sometimes the acronym ULS is used in place of ULSD. ULSD has a maximum sulfur content of 15 ppm. This can also be expressed as 0.015 percent maximum sulfur.

Fueling the Vehicle. Diesel fuel should be purchased from a reputable station that preferably sells large volumes of diesel fuel. Diesel fuel is seasonally adjusted by refiners, and running summer fuel under winter conditions is inviting problems. Most automotive highway vehicles powered by diesel engines are equipped with fuel fill pipes that can receive fuel up to 20 gallons per minute from a 1⅛-inch nozzle. Attempting to pump fuel at higher flow rates may result in premature nozzle shutoff or spitback.

Truck stops have high-flow pumps and nozzles designed for heavy duty, commercial vehicles; these may not be suitable for refueling smaller diesel-powered vehicles. If the refueling nozzle shuts off repeatedly while attempting to refuel, wait 5 to 10 seconds, then use a slower rate of flow by not depressing the nozzle trigger to full travel.

CAUTION *Diesel fuel should never be stored in a galvanized container. Diesel fuel reacts with the zinc, forming zinc compounds that can be pumped through the fuel system and can destroy the injectors.*

To fuel the vehicle:

1. Turn the engine off.
2. Turn the filler cap counterclockwise until it spins off.

WARNING *The fuel system may be under pressure. If the fuel filler cap is venting vapor or if you hear a hissing sound, wait until it ceases before completely removing the fuel filler cap. Failure to do this can result in pressurized fuel spraying out of the tank.*

3. Pull on the cap to remove it from the fuel filler pipe.
4. To reinstall the cap, align the tabs on the cap with the notches on the filler pipe and insert.
5. Turn the filler cap clockwise one-quarter of a turn until its ratchet clicks at least once.

Fuel Filler Caps. If a fuel filler cap has to be replaced, it must be replaced with one specified for the fuel system on the vehicle. The vehicle warranty may be voided if the correct fuel filler cap is not used because excessive pressure or vacuum may build up in the fuel tank.

Priming the Fuel System. Depending on the type of fuel system, when air enters the fuel subsystem it may be necessary to prime the fuel system. When air enters the fuel subsystem, it is usually due to:

- Running out of fuel
- Failure to prime a fuel filter at servicing

Self-Prime

It should be noted that most more recent, light duty diesel engines are designed to be self-purging. Self-purge–capable engines use an electric priming pump. To run self-prime, turn the key on and do not crank. Repeat this six times in succession. Then crank the engine until it starts. When the engine fires, it may run rough and produce white smoke while air remains in the system.

Priming the Fuel Subsystem

When priming an engine with no self-prime feature, use the following procedure:

1. Remove all the filter(s) upstream from the transfer pump and fill clean, filtered diesel fuel.
2. Loosen the exit line downstream from the transfer pump and crank the engine until air-free fuel exits the line. If this does not occur after two cranking sequences of 15 seconds, a hand or electrical priming pump will have to be fitted in series with the circuit close to the fuel tank. After prime has been established in the primary circuit, tighten the exit connection at the transfer pump.
3. Loosen the exit line from the secondary filter. Crank the engine over until air-free fuel exits the line: this should occur within two cranking sequences of 15 seconds. Retighten the exit line from the secondary filter.
4. Loosen the secondary circuit feed line to the fuel injection pumping gallery or low-pressure rail. Crank the engine over until air-free fuel exits the line: this should happen almost immediately. Retorque the line.
5. Crank the engine until it fires making sure a cranking cycle does not exceed 15 seconds. If the engine fails to start, the problem is likely not caused by loss of fuel system prime.

WARNING *During the entire priming procedure, you are working with the fuel subsystem. Do not confuse the fuel subsystem with high-pressure injection circuits, such as those found on CR fuel systems. Separating high-pressure injection lines may be dangerous and result in the need for costly replacement parts.*

Tech Tip: Do not crack high-pressure lines on PLN systems to prime diesel engines; using the preceding priming procedure should be enough because the injection pressure circuits on most fuel systems will self-prime after fuel subsystem prime is established.

Putting Gasoline into a Diesel Fuel Tank. This occurs daily, so something has to be said here about it. The mistake is usually identified at the completion of the fill-up. Take one of the following courses of action:

- If the diesel fuel tank is empty at the time of the fill-up, do not try to start the engine. Get the vehicle removed on a hook to the nearest garage. Flush the fuel system thoroughly using diesel fuel. Change all the fuel filters.
- If the diesel fuel tank is empty at the time of the fill-up and the engine starts but splutters to shutdown: ensure that the engine is key-off. Get the vehicle towed on a hook to the nearest garage. Flush the fuel system thoroughly using diesel fuel. Change all the fuel filters. To be sure that the injectors are not damaged, they should be pulled and performance checked in a fuel injection shop.
- If the diesel fuel tank is more than 25 percent full at the time of the error, allow the fuel to settle for 15 minutes. Gasoline is lighter than diesel fuel and does not readily form a solution with it; it should settle above it. Because fuel is pulled from the bottom of the tank, driving a short distance without incurring damage is possible. Drive directly to the nearest diesel shop (do not drive for more than 20 minutes) and flush the fuel system thoroughly using diesel fuel. Change all the fuel filters.

Biodiesel

Most light duty diesel engines manufactured after 2007 can be operated on **biodiesel** cuts of up to

20 percent: this is identified as B20. In addition to B20, B5 (5 percent biodiesel) and B10 (10 percent biodiesel) may also be available depending on geographic region, and all are approved for most post-2007 chassis.

Biodiesel is more likely to be available in agricultural communities, and sometimes B100 (100 percent) is available. At this time, none of the automotive manufacturers approve B100, and using it can void your warranty. Biodiesel, in any concentration, is less stable than petroleum-based diesel fuel, so there are some advisories that should be followed when using it:

- Never store biodiesel fuel in the vehicle fuel tank for longer than one month: biodiesels are more susceptible to biological attack, which can rapidly degrade the fuel quality.
- If cold temperature fuel gelling issues are observed, change fuel to petroleum-based diesel.
- Never use unrefined oils, waste fat, or cooking greases as fuel.
- If the engine/vehicle is to be parked or stored for longer than one month, drain the fuel tank(s), fill with a pure petroleum-based diesel fuel, and then run for at least 30 minutes.
- Never put home heating oil, agricultural fuel, kerosene, or jet fuel into a diesel fuel tank.
- Volatile fuels such as gasoline, gasohol, or alcohol may damage engines and fuel systems and represent a serious fire hazard.

Water Separator Service

A fuel filter and water separator assembly are often combined into a single unit, such as the diesel fuel conditioner module (DFCM) discussed in **Chapter 7.** Locating the unit for servicing may not always be straightforward because they are often tucked away under a frame rail to provide a measure of protection. Consult the owner's service literature to locate the unit.

Water should be drained from the module assembly whenever the driver alert warning light is illuminated by the WIF sensor (see **Figure 14-5**); this is usually accompanied by a message displayed on the driver display unit (DDU). In the case of the Ford 2011 Power Stroke water separator, the alert is signaled when the separator drain bowl has accumulated approximately 0.32 pints (150 ml) of water. If water is allowed to exceed this level, it may be pumped through the fuel system and result in serious fuel injection equipment damage. The typical

Figure 14-5 Location of WIF sensor in water separator sump.

procedure required to drain a water separator is as follows:

1. Stop the vehicle and shut off the engine.

> **WARNING** *The vehicle must be stopped with the engine off when draining a water separator because fuel may ignite if the engine is running. In addition, air can be pulled into the fuel system if the water separator is drained while the engine is running, resulting in shutdown.*

2. Locate the water separator and place an appropriate container under the drain port.
3. Rotate the drain tap counterclockwise until the O-ring is visible. Allow the separator to drain (see **Figure 14-6** for location of drain tap) for approximately 25 seconds or until clean fuel is observed. Rotate the drain clockwise to close off the flow.
4. When the drain valve is fully tightened, remove the container from under the vehicle.
5. Restart the engine. The WIF alert message and/or warning light should no longer be displayed on the DDU.

Fuel Filter Service

Most current light duty diesel engines are equipped with two fuel filters. It is usual to replace both filters at the same time, usually with a lube change. Routine fuel filter changes are an important part of engine maintenance, and failing to adhere to scheduled maintenance can produce engine performance issues and fuel-injection system problems. In this section we will outline the procedure required to service the two

Water Drain Valve

The water drain valve is located on the bottom of the DFCM.

To drain water that has accumulated in the DFCM, turn the drain valve to the open position.

Figure 14-6 Location of drain valve in the water separator.

filters on a Ford Power Stroke engine. The filters are identified as:

- The diesel fuel conditioner module (DFCM) filter
- Engine-mounted fuel filter

Servicing the DFCM Filter. The DFCM filter is the primary fuel filter (see **Figure 14-7**), and it is located upstream from the transfer pump. It is located in the lower portion of the DFCM housing.

Removal:

1. Drain the DFCM using the procedure outlined in the section that precedes this: drain the fuel into a container and dispose of in an environmentally approved waste oil container.
2. Remove the lower portion of the DFCM housing (filter bowl) by rotating it counterclockwise

Figure 14-7 Primary fuel filter.

(CCW) using a 32-mm socket. Depending on the amount of seal swelling, removal of the filter bowl may be noisy and may require some effort. Replace the seal before reinstalling the filter/bowl assembly.

3. Remove and discard the old fuel filter element.
4. Carefully clean the mating surfaces using a lint-free rag.

Installation:

1. Install the new filter element into the filter bowl tabs. Replace the seal on the DFCM header (top portion of DFCM).
2. Reinstall the lower portion of the housing by slowly turning it clockwise (CW) onto the DFCM housing, allowing fuel to soak into the fuel filter element. Tighten the lower housing until it contacts the mechanical stop. (Note: The engine will not run properly if the DFCM fuel filter is not installed in the housing.)
3. The system will have to be purged of air after the replacement of the filter; use the procedure outlined earlier in this chapter titled "Priming the Fuel System."

Servicing the Engine-Mounted Fuel Filter. On the Power Stroke, the engine-mounted fuel filter is the secondary filter, located downstream from the transfer pump. It is a plastic disposable cartridge.

Removal:

1. Disconnect both fuel lines by squeezing the connector tabs and pulling the lines straight off. Although the fuel system is not pressurized when the engine is not running, some residual pressure may exist in the fuel lines because it takes time for the pressure to bleed down. For this reason, a shop rag should be wrapped around the filter connectors as they are loosened.
2. Loosen the bracket bolt.
3. Rotate the filter CCW until it unlocks from the bracket.
4. Pull the filter straight out from the bracket and discard the filter cartridge.

Installation:

1. Install the new filter cartridge into the filter bracket. The filter has two locking tabs: one is located on the bottom and the other on the side approximately 180 degrees from the bracket bolt. Line this tab up with the slot and the bottom should follow. Turn the filter CW to lock it in place.

2. Tighten the bracket bolt until the filter cartridge sits snug in the bracket.
3. Reconnect both fuel lines.
4. The system will have to be purged of air after the replacement of the filter. Use the procedure outlined earlier in this chapter on "Priming the Fuel System."

Using a fuel which has higher-than-average impurities may require more frequent replacement of the fuel filters. After servicing the filters, a no-start or rough-running engine may indicate that air is entering the system through the filter bowl seal or drain. Ensure that the drain is tight.

Low Fuel Pressure. Most current engines using CR fuel systems are equipped with a low fuel pressure detection system. If the DDU displays LOW FUEL PRESSURE, the following service routine should be run:

Cold start or cold operation below 32°F (0°C): If this message appears during a cold start or during cold operation up to 10 minutes after the initial start, monitor the DDU. If it disappears and does not reappear after the engine has fully warmed up, the low fuel pressure message is most likely caused by waxed or gelled fuel. To repair a fuel gelling problem, consult the manufacturer's service literature; generally, this will advise against using alcohol-based additives. Most manufacturers approve of at least one type of anti-gel additive, providing they are not used as a general preventive.

If a low fuel pressure message persistently appears after refueling during cold start and winter operation, then disappears when the engine has warmed up, the cause is probably summer fuel or a nonappropriate fuel (such as kerosene).

Low Fuel Level. If a low fuel level message appears on the DDU once the engine is warm, the tank level is at or very near empty, so refueling is required. Should the message return when the engine is at running temperature after refueling, the fuel subsystem is likely restricted, and the filters should be replaced regardless of the maintenance schedule interval. If replacement of the fuel filter(s) does not correct the low fuel pressure alert, the problem is likely fuel subsystem restriction at the pickup tubes or transfer plumbing.

Combination Filter/Separators. Combination fuel filter and water separators are common in original equipment, and because of their effectiveness, as aftermarket add-ons. These units will in addition contain

© Delmar/Cengage Learning

Figure 14-8 The combination secondary fuel filter and water separator used on a Cummins ISB 6.7L engine.

fuel heaters, water separators, and pressure sensors. **Figure 14-8** shows the combination unit used on some versions of the Cummins 2010 ISB 6.7L engine.

SELECTIVE CATALYTIC REDUCTION

All post-2010 light duty highway diesel engines are equipped with a **selective catalytic reduction (SCR)** system to minimize NO_x emissions. SCR uses aqueous urea (see **Chapter 13**) as a reduction agent; aqueous urea is known as diesel exhaust fluid (DEF) in North America and as AdBlue in Europe. The SCR system automatically injects DEF into the exhaust system, but it is essential to maintain the DEF level.

DEF Level

For the SCR system to operate properly, the DEF level must be maintained. Generally, the DEF tank should be filled during each oil change service interval in light duty diesel operation, but there are circumstances that require this to be more frequent. This would include certain driving styles, hauling a trailer, off-highway operation, and a heavy foot.

The PCM monitors the fluid level available in the DEF tank. Running a system check in the driver digital display electronics will indicate to the operator whether the DEF level is okay or if it is less than half full. A message will automatically appear in the display window when the DEF level is low. When this message appears, the DEF tank must be refilled. If the vehicle is operated when the DEF level is either low or contaminated, it defaults to a trimback mode in which road speed is limited. At the first trimback, vehicle

speed is limited to 55 mph (89 km/h), progressing to a further trimback to 50 mph (80 km/h). Under these conditions, the vehicle should be driven with caution and the DEF replenished immediately. Should the DEF tank become completely empty (or contaminated), the vehicle speed will be limited to idle speed once stopped.

The driver display unit is designed to display a sequence of messages regarding the amount of DEF available in the tank. A systems check displays messages indicating the DEF available in the following sequence:

- Okay.
- Under one-half full.
- Warning message displaying the mileage (kilometers) remaining as the fluid in the DEF tank nears empty with 800 miles (1,287 km). Vehicle drives normally.
- Warning message and audio alert as DEF projected driving distance drops below 300 miles (483 km) along with road speed trimback accompanied by a message posted to the DDU.
- Continued operation without refilling the DEF tank will result in the engine being limited to idle-only operation after refueling. At this point, it is required that a minimum of 0.5 gallons (1.9L) of DEF be added to the tank to exit the idle-only mode, but the vehicle will still be in speed-limiting mode until the tank is refilled.
- To exit the vehicle speed-limiting or idle-only condition, normal vehicle operation will only resume after the DEF tank is refilled.

Tech Tip: When refilling the DEF tank from empty, there may be a short delay before the PCM detects the increased level of DEF. This has to occur before the trimback mode is exited and full power is returned.

Filling the DEF Tank from a Container. The vehicle is equipped with a DEF tank with a blue-capped filler port (see **Figure 14-9**) usually located next to the diesel fuel fill inlet. The tank can be filled using a nozzle at a DEF filling station (similar to fuel fill) or using a portable DEF container with a spout.

WARNING *Do not put DEF in the fuel tank. This can result in damage and void the warranty.*

© Delmar/Cengage Learning

Figure 14-9 The DEF fill cap and tank used on a class 4 truck: the fill cap is always blue.

Tech Tip: Immediately wipe away any DEF that has spilled on painted surfaces with water and a damp cloth to prevent damage to the paint.

DEF can be purchased at most manufacturer dealerships, many service stations, and all highway truck stops. Only DEF certified by API meeting ISO 22241 should be used in the DEF tank, so look for the API certification trademark on dispensers and containers.

WARNING *Refill DEF in a well-ventilated area. When opening the cap on the DEF container, ammonia vapors may escape. Ammonia can irritate skin, eyes, and mucous membranes. Inhaling ammonia vapors can cause burning to the eyes, throat, and nose and cause coughing and watery eyes.*

Use the following step-by-step procedure to refill the DEF tank:

1. Remove the cap from the DEF container. Remove the spout from the bottle and insert the straw end into the bottle. Ensure that the arrow above the nut is aligned with the bottle handle and the small tube end extends into the far corner of the bottle. Twist the spout nut on the container until it is tight.
2. Open the DEF filler port on the vehicle by twisting open the blue cap counterclockwise.
3. Lift and hold the DEF container, without tipping, and insert the spout into the DEF filler port until the small black seal on the spout is fully seated into the DEF filler port.

4. While pouring the DEF into the tank, the fluid level on the container will be seen to drop.
5. When the vehicle DEF tank is full, the fluid level in the bottle will cease to drop. This indicates that the fluid is no longer flowing.
6. When the level in the DEF container has ceased to drop, return the container to a vertical position slightly below the DEF filler port and allow any DEF to drain from the spout. Do not attempt to continue to add DEF to the tank by shaking or repositioning the container to induce flow because this can result in spillage and overfilling.
7. Once the spout has drained, remove it from the DEF filler port and install the blue cap on the DEF filler port. Remove the spout from the DEF container and install the cap back on the bottle.
8. If the container has been emptied, recycle it and the spout. If some DEF remains in the container, retain it and the spout for later use. Ensure that the spout stays clean.
9. Use water and a damp cloth to wipe away any DEF that has spilled on painted surfaces.

DEF Refill from a Fuel Station Nozzle. Filling the DEF tank using a nozzle is similar to a normal fuel fill. The nozzle is designed to shut off automatically when the tank is full. Do not continue to fill the tank as this may cause spilling and may overfill the tank, which can cause damage.

Tech Tip: Some filling station nozzles may hinder the filling of a DEF tank due to a magnetic mechanism in the nozzle. This is not a problem with most North American–built 2010 vehicles. Where it is a problem, another refueling station or portable container will have to be used to fill the tank.

Filling DEF in Winter Conditions. DEF freezes at 12°F (−11°C), but that said, most vehicles designed for operation in North America are equipped with an automatic preheating system. This allows the DEF system to function below 12°F (−11°C). When a DEF-equipped vehicle is not in operation for an extended period of time with temperatures at or below 12°F (−11°C), the DEF tank may freeze. If the tank is overfilled and it freezes, it could be damaged. DO NOT OVERFILL.

DEF Storage. DEF should be stored away from direct sunlight and in temperatures between 23°F (−5°C) and

68°F (20°C). Always ensure that the DEF is protected from exposure to temperatures below 12°F (−11°C). DEF containers should not be stored in the vehicle because in the event of a leak, DEF can damage interior components and release an ammonia odor inside the vehicle. DEF should never be diluted with water or any other liquid. It is normal to smell an ammonia odor when removing the container or tank caps.

Tech Tip: DEF is a nonflammable, nontoxic, colorless, and water-soluble liquid. However, refilling DEF should always be undertaken in a well-ventilated area.

DIESEL PARTICULATE FILTER SERVICE

All highway diesels (and many off-highway diesels) manufactured since 2007 are equipped with diesel particulate filters (DPFs). As described in **Chapter 13**, DPFs entrap combustion soot and then burn this off in regeneration cycles. What remains of the burn-off are residual ashes, which collect over time. Each manufacturer has slightly different guidelines for DPF service, so the following is general. A DPF may need to be removed for ash cleaning at approximately 120,000 miles (192,000 km), depending on how the vehicle is operated and how ash levels are monitored. In addition, the DPF may need to be replaced at approximately 250,000 miles (400,000 km), and again this depends on engine/vehicle operating conditions. In most cases, the engine management system will set an alert to advise the operator that the DPF requires service attention.

DPF Regeneration

The frequency with which a DPF requires regeneration depends on whether it is original equipment or aftermarket, and the manufacturer. In most cases (but not all), DPF regeneration cycles are driven by the engine software. In some cases, the regeneration is designed to occur while the vehicle is on the highway operating under conditions of normal speed and load, but in others, the regeneration cycle must be initiated and managed by the technician using an electronic service tool (EST). When a DPF alert is displayed on the dash display unit (DDU) or by dash warning lights, you must consult the manufacturer's service literature and follow the recommended procedures precisely.

Summary

- Some manufacturers prefer that the first step in any routine service procedure is to connect an EST to the chassis data bus and establish a handshake between the vehicle and the manufacturer's online data hub.

- The procedure required to break in a light duty diesel engine varies, but generally this requires a period of 1,000 miles (1,600 km) in which the engine is not overworked or operated at a consistent highway speed.

- Make sure you understand exactly how to respond when diesel fuel tanks get filled with gasoline.

- Servicing of air filters in light duty diesel engines should be determined by the status of an inlet restriction gauge which monitors vacuum pull.

- Selecting the appropriate oil for use in today's diesel engines is critical; most DPF-equipped engines require the use of an SAE CJ-4 engine lubricant.

- A light duty diesel engine wet service, at minimum, requires testing air filter restriction, changing the fuel and coolant filters, and performing an engine oil and filter change.

- Where engines use a coolant mixture of EG and distilled water, the mixing should be performed before pouring the mixture into the cooling system.

- The only fuel available for on-highway use today is that classified as ultra low sulfur (ULS) with a maximum sulfur content of 15 ppm. Most manufacturers approve of the use of ASTM-rated biodiesel cuts up to B20.

- When a WIF alert is broadcast to the DDU, the water separator sump should be drained to avoid the costly damage that may occur if water is pumped through the fuel system.

- The SCR system is engine controller–managed, and when the DEF levels run low, a series of alerts are broadcast on the DDU. If the alerts are ignored and the DEF is not replenished, the engine may be first derated, then inhibited to idle-only operation.

- A DPF requires servicing when it becomes loaded with the soot that results from regeneration cycles.

Always use the manufacturer's required procedure to service the DPF. This may require the replacement of the DPF.

Internet Tasks

1. Check out the Ford, GM, and Chrysler websites and locate whatever information you can on each company's service information systems.

2. Using the same web pages and their internal search engines, research the ESTs and the software they want used on their products.

3. Check out the Volkswagen, Audi, BMW, and Nissan websites and make a list of each company's preferred diagnostic ESTs, software, and service information systems.

4. Investigate the testimonial evidence (online chat) of the consequences of using a non CJ-4 engine oil in an engine equipped with a DPF.

5. Investigate the testimonial evidence (online chat) of the consequences of putting substances other than DEF into an SCR tank.

Review Questions

1. Which of the following engine lubricants must be used if an engine is equipped with a DPF?
   - A. Any synthetic lubricant
   - B. Any biodiesel lubricant
   - C. API CI-4
   - D. API CJ-4

2. Which of the following would best describe the color of DEF?
   - A. Blue
   - B. Clear
   - C. Amber
   - D. Red

3. What quantity of oil does the distance from the top to the bottom of the dipstick crosshatch area on a typical light duty, automotive diesel engine represent?
   - A. 1.0 quart (.95L)
   - B. 2.0 quarts (1.9L)
   - C. 4.0 quarts (3.8L)
   - D. 8.0 quarts (7.5L)

4. Technician A says that the fastest way to break in a light duty diesel engine is to add a charge of aftermarket break-in solution to the engine oil. Technician B says that manufacturers recommend adding fuel conditioners to diesel fuel tanks to improve the winter driveability of diesel engines. Who is correct?
   - A. Technician A only
   - B. Technician B only
   - C. Both A and B
   - D. Neither A nor B

5. A diesel fuel tank has been filled with gasoline and the engine run until it quits. Which of the following should be done?
   - A. Flush fuel system with diesel fuel.
   - B. Replace the fuel filter(s).
   - C. Remove and service the fuel injectors.
   - D. All of the above.

6. Which of the following solutions would provide the highest level of antifreeze protection?

   A. Pure distilled water

   B. A 50:50 solution of EG and distilled water

   C. A 100 percent solution of PG

   D. A 100 percent solution of EG

7. Technician A says that EG is always colored green. Technician B says that PG is always colored red. Who is correct?

   A. Technician A only

   B. Technician B only

   C. Both A and B

   D. Neither A nor B

8. Technician A says that DEF is available at most diesel refueling stations. Technician B says that DEF can be obtained from most manufacturer dealerships. Who is correct?

   A. Technician A only

   B. Technician B only

   C. Both A and B

   D. Neither A nor B

9. Which of the following is the correct course of action when a WIF alert is posted to the DDU?

   A. Drain the water separator sump.

   B. Replace the fuel filters.

   C. Replace the water separator.

   D. All of the above.

10. When should a diesel engine air filter be changed?

    A. At every engine wet service

    B. Every time the oil filter is changed

    C. Every six months

    D. When the restriction gauge indicates it is beginning to plug

CHAPTER

15 Diagnostics and Testing

Prerequisites

Chapters 10 and 12.

Learning Objectives

After studying this chapter, you should be able to:

- Define the acronyms *EST* and *SIS*.

- Identify the different types of EST in current usage and the levels of access and programming capability of each.

- Use a DMM to test some everyday input circuit components such as pulse generators, thermistors, and potentiometers.

- Connect a scan tool or PC and CA to a chassis data bus and establish a handshake connection.

- Identify how to access GM, Ford, Dodge, VW, BMW, and other manufacturers' online SIS and diagnostic software.

- Interpret the pin assignments on a J1962 DLC.

- Analyze exhaust gas smoke emission by color, and relate some typical engine performance malfunctions to smoke color.

- Identify common operator and technician abuses of engine and fuel systems.

- Develop a checklist for tackling lack-of-power complaints specific to an engine or chassis system.

- Profile some of the most common engine malfunction symptoms and their typical causes.

- Diagnose common diesel engine mechanical malfunctions.

- Troubleshoot a no-communication problem with chassis bus modules on both one-line and CAN buses.

- Connect a labscope CA to a CAN bus to determine normal and abnormal voltage spreads.

- Learn how to identify and repair some common bus line problems using a DMM.

Key Terms

active codes

assembly line data link (ALDL)

black smoke

blink codes

blue smoke

breakout box

breakout T

CAN-C

communications adapter (CA)

cylinder contribution test (CCT)

cylinder leakage test

data communication link (DCL)

data link connector (DLC)

diagnostic routines

diagnostic trouble code (DTC)

digital display unit (DDU)

digital multimeter (DMM)

dynamometer

electronic cylinder balance test

electronic service tool (EST)

flash codes

global diagnostic system (GDS)

ground strap

handshake

historic codes

inactive codes

J1962 connector

master gauge

master pyrometer

open circuit

platinum resistance thermometers (PRTs)

resistance temperature detectors (RTDs)

scan tool

scopemeter

sequential troubleshooting chart

service information systems (SIS)

thermistor

troubleshooting

white smoke

INTRODUCTION

The term **troubleshooting** is generally used to describe noninvasive (meaning that as little of the engine is disassembled as possible) methods of determining the cause of an engine problem. These methods can vary from educated guesswork to highly structured procedures, such as those required when diagnosing electronic engine management systems.

Today's engines are computer controlled. Yes, the engine is a mechanical device, but it is entirely controlled and monitored by electronics. For that reason, in this chapter we will take a look at electronic approaches to troubleshooting diesel engine problems *first*. In a large majority of cases, electronically driven **diagnostic routines** will lead the technician to the source of the problem.

- *Step one* is to make sure that you have the manufacturer's recommended electronic tooling, software, and service information in your possession. It is a fool's game and possibly costly to proceed without them.
- *Step two* is to actually use this electronic tooling, software, and service information in the manner outlined by the manufacturer.

Technicians today do not consult their mechanical instincts and experience as a first step; these qualities only come into play when the electronic diagnostic routines fail to produce a conclusive result. So connect the chassis data bus by means of the **data communication link (DCL)** and check for **diagnostic trouble codes (DTCs)**. Follow the manufacturer's routines to source the problem. Never skip steps in the routine. Do not rely on previous experience of a similar problem to short circuit a troubleshooting path.

Tooling and Training

There will be occasions when strict adherence to a manufacturer's diagnostic routines fails to produce a conclusive result. At this point, the power of the human brain comes into play and can exercise some of its special abilities as yet beyond the capabilities of computers and their programmers. This is why the power of the human mind is required to manage diagnostic processes.

Electronic diagnostic routines require two things to function effectively:

1. The manufacturer's recommended **electronic service tool (EST)**, access to the manufacturer's (usually online) **service information system (SIS)**, and diagnostic software.
2. A technician who has been trained to use the manufacturer's recommended tooling, service information, and diagnostic software.

Sounds simple. But a failure at item 1 is a common failure of the service repair industry today, and it ends up costing vehicle owners millions of dollars a year. A failure at item 2 is also a problem. Manufacturers' electronic tooling and software are not intuitive because unlike Apple, auto manufacturers have not spent millions of dollars making it do so. Instead, they have engineered it to be effective in the hands of technically minded persons who fully understand vehicle systems. The best technician in the world cannot diagnose the complex computer and networking problems that modern vehicles present without having been properly trained. And the tooling and training are likely to be different from one manufacturer to the next.

This chapter will begin by taking a look at ESTs and then go on to look at some generic diagnostic routines. Diagnostic routines fall into two categories. The first is step-by-step, structured by software. The second is less structured and only comes into play when the limits of electronic troubleshooting have been exceeded; it depends more on having a good understanding of failure profiles. As a technician today, you will be required to possess all the skills that the technician of your parents' generation had, plus all those computer skills just emphasized.

ELECTRONIC SERVICE TOOLS

The acronym EST (electronic service tool) can be generally used in the automotive industry to describe a range of electronic service instruments ranging from onboard diagnostic/malfunction lights to sophisticated computer-based communications equipment. ESTs may be generic and capable of working with multiple manufacturers' equipment, or they may be proprietary, such as the hand-held GM Tech II scan tool, designed to work exclusively with one manufacturer's chassis electronics. Generic scan tools such as the Bosch MasterTech are widely used in the industry but tend to have limited functionality with late-model chassis electronics, and they may not be enabled for programming changes.

The automotive repair industry has relied on scan tools for the past couple of decades, but this is rapidly changing. The bottom line is that scan tools have too little computing power and speed to effectively diagnose and reprogram modern automobile data bus–driven systems, along with limited or no Web access. In addition, these instruments are costly and cumbersome to update/upgrade compared with laptop computers. Low-cost, powerful PCs whose hardware and software can be cheaply upgraded are the way of the present and the future of automobile diagnostics.

Onboard Diagnostic Lights

Blink or **flash codes** (manufacturers use both terms) are an onboard means of troubleshooting using a dash or PCM-mounted electronic malfunction light or check engine light (CEL). Usually, only **active codes** (ones that indicate a malfunction at the time of reading) can be read, but some also display **historic (inactive) codes**. This is the simplest method of accessing the data bus in a vehicle that has no dash **digital display unit (DDU)**.

Data Connector Standards

The first step in communicating with the chassis data bus is to connect with it using an appropriate **data**

TABLE 15-1: J1962 CONNECTOR CAVITY PIN ASSIGNMENTS							
1	2	3	4	5	**6**	7	8
9	10	11	12	13	**14**	15	16

1. Manufacturer discretion
2. Bus positive + J1850 (CAN-A and CAN-B)
3. Ford DCL +
4. Chassis ground
5. Signal ground
6. **CAN-C high (J2284 and ISO 15765-4)**
7. K-line of ISO 9141 and ISO 14230
8. Blank (manufacturer discretion)
9. Blank (manufacturer discretion)
10. Bus negative – for J1850 PWM
11. Ford DCL –
12. Blank (manufacturer discretion)
13. Blank (manufacturer discretion)
14. **CAN-C low (J2284 and ISO 15765-4)**
15. L line of ISO 9141 and ISO 14230-4
16. V-Bat (battery positive)

link connector (DLC). DLCs have passed through several generations, but the connection on a current vehicle is usually made by means of a 16-pin, **J1962 connector**. A data connector also goes by the name **assembly line data link (ALDL)**.

A connection to the chassis data bus is usually made by a compatible mating connector, but in cases where it is required to couple a lab scope or DMM with the connector, the cavity pin assignments should be known. **Table 15-1** identifies the cavity pin assignments on a standard J1962 connector; the CAN-C high and Can-C low addresses on the DLC have been bolded.

Connecting to the Chassis Data Bus

Depending on the vehicle manufacturer, the technician is often required to access the chassis data bus even if it is to do something as simple as performing a routine service. Although there is more than one way of doing this, connecting to a chassis data bus usually requires the following procedure:

- A Windows PC with Web access, loaded with manufacturer-specific software
- A communications adapter (CA) to act as a serial link

- USB to CA cable
- CA to the J1962 DLC

Depending on what you are trying to achieve while connected to the data bus, it may also be necessary to connect to the manufacturer's data hub via the Internet.

Communications Adapters

A **communications adapter (CA)** acts as a serial link and transducer to connect chassis electronics with a PC loaded with the appropriate software required to work with data bus–gleaned information. CAs used to access CAN should be J2534 compliant (on any light duty vehicle), meaning that theoretically they should all speak the same language. However, although it is mandatory that generic ESTs access OBD II data (relating to emissions only), manufacturers often use lockout security to ensure that any nonemissions–related data requires a specific CA to be connected to the bus. **Figure 15-1** shows how a CA interfaces between the chassis data bus on a light-duty vehicle and a PC loaded with proprietary software.

Scopemeters

Scopemeters and graphing meters are designed to display waveforms produced by a variety of electrical

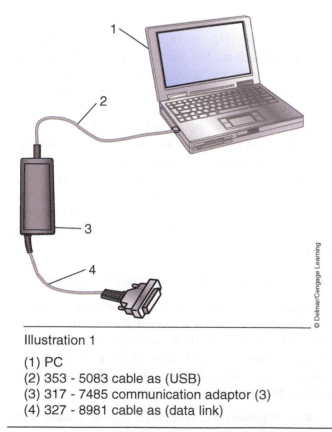

Illustration 1

(1) PC
(2) 353 - 5083 cable as (USB)
(3) 317 - 7485 communication adaptor (3)
(4) 327 - 8981 cable as (data link)

Figure 15-1 The CA interface between the chassis data bus on a light truck and a PC loaded with proprietary software.

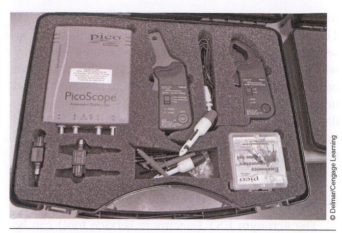

© Delmar/Cengage Learning

Figure 15-2 Pico Scopemeter kit that consists of a CA and PC-based software to display waveforms. The PicoTech software that interacts with the CA is downloaded from the Web.

signals. They do not get used that often on the shop floor today because manufacturer diagnostic software can usually do the same job more rapidly when it is important to interpret a waveform anomaly. However, they may be used to determine whether activity on a communication bus line falls within its specified range or whether injector driver actuation pulses spread a normal or abnormal profile. In addition, scopemeters are useful instruments in a learning environment because they help students to "see" electrical signals.

As a learning tool, a scopemeter provides a visual image to the waveform produced by a PWM driver command for an injector: because most of us are visual learners, this can help retention. **Figure 15-2** shows a Pico Scopemeter that consists of a CA and PC-based software to display waveforms. Using a scopemeter to verify CAN bus communications is covered toward the end of this chapter.

Scan Tools

Generic and proprietary **scan tools** are microcomputer-based ESTs designed to read and reprogram controllers on a data bus, providing they are loaded with the appropriate software cartridge. These are usually tough, portable, and have become widely accepted by technicians working in the automotive industry. However, they are severely handicapped when it comes to working with the complex networks found on modern vehicles. For example, a General Motors (GM) Tech II scan tool functions at a maximum speed of 18 kb/s with a maximum main memory (RAM) of 32 megabytes (10 megabytes is standard). When such an instrument has to interact with a data bus that transacts at speeds over 50 times faster, it obviously has limitations.

Personal Computers (PCs). All the automobile manufacturers are either making, or are in the process of making, the generic PC and Windows environment operating systems their diagnostic and programming tool of choice. PCs are inexpensive, easily upgradable, and have vast computing power and speed when compared with scan tools. Hardware upgrades are simple, and software upgrades can be downloaded online from the manufacturer data hub.

PCs are connected to a vehicle data bus through the chassis data connector; a communications adapter (CA) or serial link is also required, depending on the system and the production year. Some are capable of wireless connectivity, but observe the cautions that appear a little later in this chapter. Chassis manufacturers use a MS Windows environment to run their software. All of the manufacturers offer comprehensive courses on their own management systems; although navigating manufacturer software has become progressively more user-friendly, technicians are advised to take manufacturer courses so they can get the best out of the system. Most of these courses can be taken online with manufacturer data hub access.

Web-Based SIS

As we have said earlier in this book, most manufacturers prefer that the first step in any troubleshooting procedure be to log onto their Web-based service information system (SIS), whether you are working on an electronic engine or one of an earlier vintage. Subscription fees are charged; some manufacturers offer a daily rate, and typical maximum annual rates are often around $3,000.

Diagnostic Software

Each engine manufacturer produces its own diagnostic software, and if you are to do anything more than just scan a data bus, you must use this software along with the appropriate EST. Many manufacturers combine their diagnostic software with their SIS, so both are accessed from the same portal using the same subscription. Each manufacturer uses its own system-specific software; an example is the **global diagnostic system (GDS)** used by GM. Although GM is careful to point out that GDS does not replace the Tech II, they add that it is required for post-2010 GM vehicles.

SIS and Diagnostic Web Software

Most manufacturers bundle their Web SIS and diagnostic software into a single interactive package. For good reason, in the best software packages, each complements the other. The packages are made available to the service repair industry by subscription. The costs of subscribing can be relatively high, but some manufacturers sell 24-hour access for as little as $25. Annual subscriptions are more likely to cost up to $3,000, but there is some considerable variability between manufacturers. In addition to the manufacturer Web-based subscription, companies such as Mitchells and AllData offer no-frills options at lower cost, but often with some serious limitations. Some examples of engine manufacturer subscription SIS and diagnostic software packages are:

- BMW Tech info www.bmwtechinfo.com (subscription)
- Caterpillar Service Information System www.sis.com (subscription)
- Chrysler Dodge www.installers-mopar.com (free)
- Chrysler Dodge www.techauthority.com (subscription)
- Cummins QuickServe Online www.quickserve.cummins.com (subscription)
- Ford www.motorcraftservice.com (subscription)
- General Motors www.acdelcotds.com (subscription)
- Navistar ISIS and MD www.evalue.internationaldelivers.com (subscription)
- Volkswagen www.erwin.vw.com (free)
- Volkswagen www.vwparts.com (subscription)

Most manufacturer SISs are updated daily. They tend to be easy to navigate. In most cases, information searches are by engine serial number and may include searches of related parts data. It is important to note that most online SISs support engines manufactured before the website service existed, so try to make a habit of using them. Needless to say, some are better than others. You should note that some versions of the preceding software may have limited functionality; software for dealerships, body builders, and rental applications may all be different, depending on the level of programming required.

EST Wireless Connectivity. Manufacturers are increasingly making use of ESTs that can make wireless connections to the chassis data bus. The current generation of wireless ESTs tend to produce some communications glitches, so while some troubleshooting operations are made easier and pose little risk, when performing programming it is best to ensure that you have a hard wire backup in the event of a communications failure. Dropping an Internet connection midway through a data hub–driven programming sequence can disable a PCM/ECM, resulting in a downed vehicle with a dysfunctional control module(s).

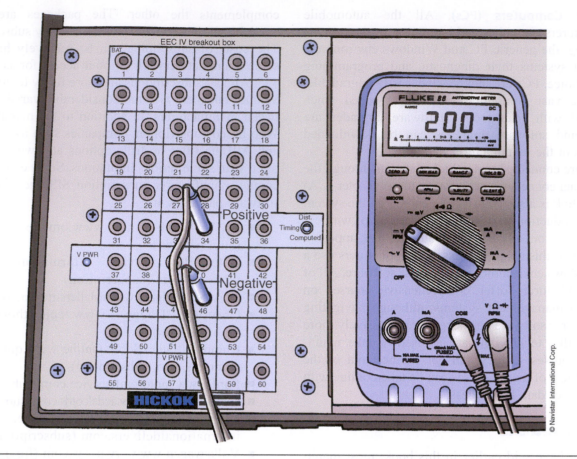

Figure 15-3 Navistar breakout box, which is used in conjunction with a DMM.

Troubleshooting Toolkit

In order to effectively use electronic troubleshooting methods, the technician should be capable of using:

- The manufacturer online service information system (SIS)
- The manufacturer diagnostic software
- The appropriate EST and CA
- A DMM (and should possess a basic understanding of electrical and electronic circuitry)
- Electrical schematics
- The correct tools, including terminal spoons, breakout Ts, and breakout boxes

Making use of the correct tools is often overlooked. When the use of a breakout box is mandated, make sure that it is used; the consequence of not using it is costly damage to wiring and connectors. Also, it should be remembered that electrostatic discharge can damage solid-state components, so it is good practice to wear a **ground strap** when opening up any housing that contains a microprocessor, especially if you are inside the vehicle when doing so. When you are testing separated weatherproofed connectors with a DMM, ensure that the sockets have been correctly identified

before inserting test leads, terminal spoons, or breakout Ts. Use socket adapters/spoons where necessary to avoid damaging terminal cavities. Never spike wires when testing circuits.

DIGITAL MULTIMETERS

A **digital multimeter (DMM)** (see **Figure 15-3**) is simply a tool for making electrical measurements. DMMs may have any number of special features, but essentially they measure electrical pressure (volts), electrical current quantitative flow (amps), and electrical resistance (ohms). A good-quality DMM with minimal features may be purchased for as little as $100; as the features, resolution, and display quality increase in sophistication, the price increases proportionally. A DMM is one of the most important tools in the technician's toolbox. Surprisingly, too many technicians do not understand how powerful a diagnostic instrument a good-quality DMM can be.

Most electronic circuit testing *requires* the use of a DMM. Avoid using analog multimeters when attempting to diagnose circuit malfunctions; you may find analog multimeters in some older shop VATs (volts-amps testers), and these can damage solid-state components. Reliability, accuracy, and ease of use are all factors that should be considered when selecting a DMM for purchase. Some options the technician may wish to consider are a protective rubber holster (will greatly extend the life of the instrument!), analog bar graphs, and enhanced resolution. This section deals with the practice of using DMMs; a knowledge of basic electricity and the meter itself is assumed.

Some Typical DMM Tests

Always perform tests in accordance with the vehicle manufacturer's specifications; never jump sequence or skip steps in **sequential troubleshooting charts**. Most DMM tests on vehicle electronic systems will be used in conjunction with a scan tool or PC-driven diagnostic routines.

The following tests assume the use of a Fluke 88 DMM, but there are many other good automotive DMMs, including the faster and less costly Fluke 87.

Engine Position, Cam, and Crank Position Sensors.
This procedure tests Hall effect sensors, so make sure you correctly identify the sensor (it could be an analog inductive pulse generator):

a. Cycle the ignition key, then off.
b. Switch meter to measure V-DC/rpm.
c. Identify the ground and signal terminals at the Hall sensor. Connect the positive (+) test lead to the signal terminal and the negative (−) test lead to the ground terminal. Crank the engine. At cranking speeds, the analog bar graph should pulse; at idle speeds or above, the pulses are too fast for bar graph readout.
d. Press the duty cycle button once. Duty cycle can indicate square wave quality, with poor-quality signals having a low duty cycle. Functioning Halleffect sensors should have a duty cycle of around 50 percent, depending on the sensor. Check to specifications.

Potentiometer-Type TPS.
Resistance test:

a. Key off.
b. Disconnect the TPS.
c. Select Ω on the DMM. Connect the test probes to the signal and ground terminals. Next, move the accelerator through its stroke while observing the DMM display.
d. The analog bar should move smoothly without jumps or steps. If it steps, there may be a bad spot in the sensor.

Voltage test:

a. Key on, engine off.
b. Set the meter to read V-DC. Connect the negative lead to ground.
c. With the positive lead, check the reference voltage value and compare to specifications. Next, check the signal voltage (to the PCM) value through the accelerator pedal stroke. Check values to specification. Also observe the analog pointer; as with the resistance test, this should move smoothly through the accelerator stroke.

Note: This test will not work on those systems that digitize the signal to produce a PWM input to the PCM, nor will it work on "noncontact" TPSs that use a Hall effect principle.

Inductive Pulse Generator Sensors.
Inductive pulse generator–type sensors are often used as shaft speed and position sensors. They may also be known as magnetic or variable reluctance sensors. They function similarly to a magneto. A toothed wheel cuts through a magnetic field, which induces an AC signal voltage in the signal (output) terminal: the PCM reads the frequency of the AC signal to determine a shaft speed. The AC voltage rises proportionally with rotational speed increase and ranges from 0.1V up to 5.0V AC, the peak voltage depending on what shaft rpm the sensor is measuring. Vehicle speed sensors (VSS), engine speed sensors (ESS), and anti-lock braking system (ABS) wheel speed sensors all use this method of determining rotational speed. Test using the V-AC switch setting, locating test leads across the appropriate terminals, then spin up the device.

Min/Max Average Test for NO$_x$ and O$_2$ Sensors.
The foregoing test can be used to test the operation of some types of NO$_x$ sensors used on the DPFs on diesel engines produced later than the 2007 model year, but the manufacturer specifications and test instructions must be observed.

a. Key on, engine running, DMM set at V-DC. Select the correct voltage range.
b. Connect the negative test lead to a chassis ground and the positive test lead to the signal wire from the lambda sensor. Press the DMM Min/Max button.

c. Make sure that the engine is at operating temperature (thermostat open but not close to the high end of normal operating temperatures). Run at no load, varying the rpm between low and peak torque rpm to enable the DMM to sample a scatter of readings.

d. Press the Min/Max button slowly three times while watching the DMM display. Record the voltage output values. You will have to correlate the voltage readings to manufacturer specifications, and these may difficult to locate. However, you can conclude that higher levels of oxygen dump result in lower voltage readings, and vice versa.

e. Next, hold the engine rpm to no load, high idle to induce a lean burn condition. Repeat steps c and d to read the average voltage. Average voltage readings should drop lower, indicating a lean condition.

f. Oxygen/NO$_x$ sensor tests may be performed while road testing the vehicle, but when testing NO$_x$ sensors, you should consult the manufacturer literature.

Thermistors. Most **thermistors** used in computerized engine systems are supplied with V-Ref (5V) and have a negative temperature coefficient (NTC), meaning that as the sensor temperature increases, its resistance decreases. They should be checked to specifications using the DMM ohmmeter function and an accurate temperature measurement instrument.

Applying DMM Tests. Manufacturers seldom suggest random testing of suspect components. The preceding tests are typical procedures. Circuit testing in today's computerized engine management systems is highly structured and part of sequential troubleshooting procedure. It is important to perform each step in the sequence precisely; skipping a step can invalidate every step that follows.

Breakout Boxes and Breakout Ts

The DMM is often used in conjunction with a **breakout box** or **breakout T**. Breakout devices are designed to be teed into an electrical circuit to enable circuit measurements to be made on both closed (active) and de-energized circuits. The objective is to access a circuit with a test instrument without interrupting the circuit. A breakout T is a simpler diagnostic device that is inserted into a simple two- or three-wire circuit such as that used to connect an individual sensor, while a breakout box accesses multiple-wire circuits for diagnostic analyses of circuit conditions.

Most of the electronic engine management system manufacturers use a breakout box that is often inserted into the interface connection between the engine electronics and chassis electronics harnesses. The face of the breakout box displays a number of coded sockets into which the probes of a DMM can be safely inserted to read circuit conditions. Electronic troubleshooting sequencing is often structured based on the data read by a DMM accessing a circuit. A primary advantage of breakout diagnostic devices is the fact that they permit the reading of an active electronic circuit, for instance, while an engine is running. **Figure 15-3** shows the breakout box and DMM setup Navistar recommend for troubleshooting their HEUI engine (Ford engines up to 2010) electronics.

WARNING *When a troubleshooting sequence calls for the use of breakout devices, always use the recommended tool. Never puncture wiring or electrical harnesses to enable readings in active or open electronic circuits. The corrosion damage that results from damaging wiring insulation will create problems later on, and the electrical damage potential can be extensive.*

Connector Dummies and Spoons. Diagnostic connector dummies are used to read a set of circuit conditions in a circuit that has been opened by separating a pair of connectors. The dummies are manufactured by the electrical/electronic connector manufacturer as a means of accessing the circuitry with a DMM without damaging the connector sockets and pins. Diagnostic spoons are curved and are designed to be inserted behind a connector without damaging the insulation boot. When the correct dummies and spoons are used, energized circuits can be tested without inflicting damage, provided some care is exercised.

Tech Tip: When performing a multiple-step electronic troubleshooting sequence on a large multiterminal connector, photocopy the coded face of the connector(s) from the service manual and use the copy as a template. The alphanumeric codes used on many connectors can be difficult to read, and using a template is a good method of orienting the test procedure.

DMM Circuit Testing

Many auto technicians learn to use DMMs on-the-fly, that is, they acquire their understanding on an as-needed basis, with the consequence that most do not properly understand the potential of a high-end DMM. It does not make sense to spend $500 on a complex DMM that will never be used to perform tasks that a DMM one-fifth of the cost will handle. The next section introduces some simple tasks that can be performed by low-end DMMs using components readily available in introductory electricity/electronics labs.

Testing Resistors. Most carbon resistors used in electronic circuits are color-coded. Although individual resistors would seldom be tested in the real world of the auto shop, testing them in a teaching/learning environment can be a great way of getting to know the ohmmeter function of your DMM.

You can grab a pile of resistors and check the color-coded resistance value against the actual measurement. First, you have to understand the color codes:

Color	Value
Black	0
Brown	1
Red	2
Orange	3
Yellow	4
Green	5
Blue	6
Violet	7
Gray	8
White	9

In the preceding list, note that the darkest colors like black and brown are used for the lowest numeric values (0 and 1) moving up to white, which is given the highest numeric value. These codes are standardized by the Electronics Industries Association (EIA) and are also used for capacitors. In addition, the colors gold and silver are used to rate tolerance or the percentage amount that similarly rated resistors can differ from each other and still function within specification. Gold indicates 5 percent and silver 10 percent tolerance. **Figure 15-4** shows the significance of each band used on a resistor.

Next, we can look at two actual examples shown in **Figure 15-5**. In **Figure 15-5A** the arrangement is:

1st stripe = red = 2

2nd stripe = green = 5

3rd stripe = red = multiplier to power of 2 or 10^2, so you *add* 2 zeros = 2,500 Ω

4th stripe = gold = tolerance rated at within 5 percent

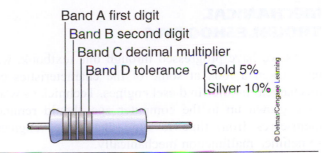

Figure 15-4 Interpreting bands on a carbon resistor.

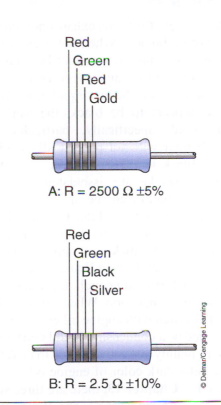

Figure 15-5 Interpreting band resistance codes.

At first glance, **Figure 15-5B** does not appear to be unlike the previous resistor:

1st stripe = red = 2

2nd stripe = green = 5

3rd stripe = black = multiplier is zero, so the value is 25 Ω

4th stripe = silver = tolerance rated at within 10 percent

CAUTION *There are few acceptable uses for a 12V DC test light on modern vehicle electrical circuits because of the potential for damage. Learn to rely on a DMM.*

MECHANICAL TROUBLESHOOTING

As we have progressed through this textbook, we have taken a look at some of the characteristics of mechanical failures in diesel engines. Technicians who have grown up in the computer age should remind themselves from time to time that diesel engines sometimes malfunction mechanically.

Smoke Analysis

An engine can emit from its exhaust noxious gases, water, and carbon dioxide. When these emissions remain in the gaseous state they cannot be seen exiting the exhaust piping. For exhaust gas to be identified as smoke, the emission must be in a liquid or a solid state. When smoke appears to be black, the state of the emission is solid; specifically, particulate solids through which light will not pass at all, making the smoke appear black to the observer's eye. The root causes of black smoke are associated with incompletely combusted fuel or engine oil.

When exhaust smoke is white, the emission is in the form of condensing liquid droplets from which light reflects or refracts, making it appear white to the observer's eye. The condensing liquid could be engine coolant, fuel, or when cold enough, the water that is a normal product of combustion. When smoke is blue, it is normally associated with engine oil emission. This is usually classified as a condensing liquid emission from which light is mainly reflected rather than refracted, due to the usually dark color of engine oil.

As outlined in **Chapter 13,** there are three states of matter—solid, liquid, and vapor. The technician analyzing the smoke emitted from a diesel engine is making an observation on the state of the emission. It is important to note that the absence of observable smoke does not mean that the engine is not producing noxious emissions—for example, NO_x emissions are gaseous and therefore cannot be seen. Gaseous emissions can only be identified using exhaust gas analysis equipment. Nevertheless, observable smoke emitted from an engine tells a story, and the technician should have some ability to interpret the causes of a smoking engine.

In summary, exhaust emissions from the stack(s) of a diesel engine can appear in one of the following four ways:

1. Clear. The exhaust gas stream is in a gaseous state.
2. White. Some liquid is condensing in the exhaust gas stream.
3. Black. Some solid matter (particulate) is contained in the exhaust gas.
4. Blue. Usually indicates presence of engine oil in exhaust gas.

Black Smoke

The term **black smoke** is used to describe anything from a grayish haze to heavily sooted exhaust gas emission. It is the result of the incomplete combustion of fuel and therefore has many causes, the most common of which are:

- Insufficient combustion air
- Restricted exhaust
- EGR failure
- DPF failure
- Excess or irregular fuel distribution
- Improper grade of fuel

When black smoke is observed from the tailpipe of any highway diesel manufactured after 2007 regardless of cause, the diesel particulate filter (DPF) is likely to be damaged to the point that it requires replacement.

Insufficient Combustion Air. The causes for insufficient combustion air are air starvation caused by a performance defect in the air intake system components, from the air filter, turbocharger, and boost air/ heat exchanger, to a plugged EGR heat exchanger or EGR control valve failure, through to intake valve problems, restricting emergency stop gate, and so on. Check for a problem that can be read electronically first in electronically managed engines. Test for intake air inlet restriction using a negative pressure gauge or manometer that reads in inches of H_2O (mm of H_2O); compare the readings to manufacturer specifications.

Exhaust System Restriction. Exhaust system restriction problems can usually be related to turbocharger failure, collapsed exhaust system piping, internal failure of engine silencer or catalytic converter, a plugged-up diesel particulate filter, or exhaust brake malfunction. Because of emissions systems electronic monitoring found in most post-2004 and all post-2007 highway diesel engines, connect to the chassis data bus to troubleshoot the condition, even if a fault code has not been logged.

Excess Fuel/Irregular Fuel Distribution. Excess fuel (overfueling) or irregular fuel distribution can be caused by out-of-spec injection timing (intentional or accidental), injector nozzle failure, variable timing control device failure, unbalanced fuel rack settings, lugging engine (operation at high loads at speeds below peak torque rpm), incorrect governor settings, inoperative or tampered with manifold boost management system, or defective barometric capsule (altitude compensator). When you are troubleshooting electronically managed engines, make sure you connect to the chassis data bus, select the engine PCM, and when possible, perform an electronic cylinder cutout test: this can pinpoint unbalanced cylinder fueling and usually identify the cause.

Improper Fuel Grade. When fuel is stored for prolonged periods, the more volatile fractions evaporate, altering the fuel's chemical properties. Additionally, fuel suppliers seasonally adjust fuel to accommodate temperature extremes, so using summer fuel in the middle of winter can cause problems. Although the practice is generally illegal, using used engine lube/fuel mixers produces smoking and can seriously damage exhaust aftertreatment devices in post-2007 vehicles. Heavier residual oils may not vaporize, and even if they do, there is insufficient time to properly combust them. This practice may also cause high acidity in the exhaust gas (high sulfur), which may rust out exhaust systems unusually quickly.

Tech Tip: It is essential to use the correct ULS fuel (**see Chapter 7**) with post-2007 engines equipped with diesel particulate filters (DPFs). Pre-2007 LS fuel can terminally destroy a DPF in as little as one hour of operation. NEVER use lube/fuel mixers with engines equipped with DPFs.

White Smoke

White smoke is caused by condensing liquid in the exhaust gas stream. Temperature usually plays a role when white smoke is observed, both ambient temperature and the engine-operating temperature. Remember that water is a natural product of the combustion of any hydrocarbon fuel, and in northern midwinter conditions it is normal for some of this to condense in the exhaust gas.

However, when white smoke is a problem, the following are some of the possible causes.

Cylinder Misfire. Check for a problem that can be read electronically first on electronically managed engines; perform an electronic cylinder cutout test using an EST. With nonelectronic engines, disable

each injector by cracking (loosening the line nut) high-pressure pipes in hydromechanical PLN systems.

CAUTION *Never attempt to mechanically disable the injectors on electronically controlled diesel engines.*

WARNING *When cracking high-pressure pipes, ensure that suitable eye protection is worn and that spilled fuel is not in danger of igniting.*

Low Cylinder Compression Pressure. There are a number of possible causes of low cylinder compression. These are usually mechanical problems that prevent the cylinder from sealing properly. The procedure for checking cylinder compression pressures and cylinder leakage is outlined later in this chapter.

Low-CN Fuel. This is more likely to be a problem experienced by operators that bulk-purchase fuel in large volumes (farms) and then store it for prolonged periods. Diesel fuel CN is defined by its most volatile fractions, and it is these that are most likely to boil off when exposed to high temperatures. Low-CN fuel may produce white smoke emission; this appears as a haze in the exhaust gas stream when fueling a warm engine. Test the fuel's specific gravity using a fuel hydrometer, and compare it to the fuel supplier's specification. Consult the fuel supplier as to the correct additive and proportions to correct a CN deficiency in a storage tank. Contaminated fuel (cut with waste oils/toxins) is seldom a major problem in North America, but when refueling in more remote parts of Mexico, make sure that the fuel is purchased from reputable suppliers.

Air Pumped through the High-Pressure Injection Pump Circuit. Check the fuel subsystem because this is usually the source of the air, depending on the type of system. This condition is normally accompanied by rough engine operation. Using a diagnostic sight glass can help you to diagnose the problem and locate it. A diagnostic sight glass is usually a section of optically clear Perspex tube coupled into dash-4, dash-6, or dash-8 hydraulic hose; the sight glass is coupled in series into the circuit to be tested.

Coolant Leakage to Cylinders. Confirm that the emission truly is coolant. Engine coolant has an acrid, bittersweet odor that is very noticeable at the tailpipe.

Locate the source. Some possibilities are injector cup failure, head gasket (fire ring) failure, or a cracked cylinder head. Coolant leakage may be difficult to locate; if the failure is not immediately evident, drop the oil pan, pressurize the cooling system, and observe. If necessary, place clean cardboard under the engine with the cooling system pressurized and leave for a while to attempt to identify the source; this may save an unnecessary engine disassembly. It is often easier to identify an internal coolant leak with the engine intact rather than disassembled, so explore all the options with the engine assembled first. Check out the exhaust aftertreatment apparatus (SCR and DPF) after completing repairs.

Low Combustion Temperatures. Low combustion temperatures may be the result of extreme low-temperature conditions or a fault in the cooling system management system, such as a defective thermostat, fan drive mechanism, or shutters. If the emission is just water it is not harmful to the environment, but it does compromise the efficacy of the SCR and DPF systems.

Blue Smoke

Blue smoke is usually caused by lube oil getting involved in the combustion process. Some possible causes are turbocharger seal failure, pullover of lube oil from an oil bath air cleaner sump, worn valve guides, ring failure, glazed cylinder liners, high oil sump level, excessive big end bearing oil throw off, low-grade fuel, fuel contaminated with automatic transmission fluid (ATF), or engine lube placed in fuel tanks as an additive. Any blue smoke condition can seriously damage aftertreatment hardware, so the condition should be resolved quickly.

Low Oil Pressure

Verify the problem:

1. Check the oil sump level and correct if necessary.
2. Install a **master gauge** (an accurate, fluid-filled gauge).
3. Warm engine to operating temperature and check oil psi.
4. Investigate oil consumption history.
5. Determine the cause.

Some possible causes and suggested solutions:

- Restricted oil filter or oil cooler bundle: change the oil and filter(s). If the problem persists, clean or replace the oil cooler bundle (core), and check or clean the filter and oil cooler bypass valves.

- Contaminated lube (fuel): detect by oil analysis or by odor. Engine lube contaminated with fuel can have a darker appearance and feel thin to the touch. If the cause is fuel, locate the source. This may be difficult, and the procedure varies with the type of fuel system and the routing of the fuel to the injector.

  Pressure testing the fuel delivery components may be required. Porosity in the cylinder head casting, failed injector O-ring seals, leaking fuel jumper pipes, and cracked cylinder head galleries are some possible causes. Perform repairs as required, then service the oil and filters.
- Excessive crankshaft bearing clearance: inspect the bearings to determine the cause. Visually check the crank journals to determine whether removal is required. Replace the bearings, ensuring that the clearance of the new bearings is checked.
- Excessive camshaft or rocker shaft bearing clearance: replace the bearings.
- Pump relief valve spring stuck open or fatigued: clean the valve and housing, replacing parts as necessary. Check the bypass/diverter valves in the oil cooler and filter mounting pad.
- Oil pump defect: recondition or replace the oil pump.
- Oil suction pipe defect: replace the oil suction pipe.
- Defective oil pressure gauge or sending unit: replace the oil pressure gauge or sending unit.
- Broken-down (chemically degraded) lube oil: change the oil and filters.

High Oil Consumption

Verify the condition by monitoring oil consumption and analyzing exhaust smoke. Some possible causes and suggested solutions:

- Excess cylinder wall lubrication: high oil sump level, excessive connecting rod and big end bearing oil throw off, plugged oil control/wiper rings, oil pressure too high, or oil diluted with fuel. This type of problem usually requires engine disassembly to diagnose and repair.
- External oil leaks: steam or pressure wash the engine, then load on a chassis dynamometer or road test to determine the source.
- High oil temperatures: malfunctioning boost air heat exchanger, lug down engine loading, overfueling, incorrect fuel injection timing, or problem with the oil cooler or engine cooling system.

- Worn piston ring fit abnormality: replace the rings.
- Piston ring failure: determine cause. Check for other damage, then replace the rings.
- Turbocharger seal failure: recondition the turbocharger (recore or replace).
- High oil sump level: this causes aeration. Determine the cause of high oil level. Check for operator or service error first, then for the presence of fuel and engine coolant in oil.
- Glazed cylinder liners or sleeves: caused by improper break-in procedure or prolonged engine idling. Replace cylinder liners/sleeves or use a glaze buster to reestablish crosshatch.
- Worn cylinder head valve guides or seals: measure to specification and recondition the cylinder head if required.
- Improper dipstick marking: drain oil, fill sump to manufacturer specification, then check/alter dipstick markings.

High Oil Temperature

Some possible causes and suggested solutions:

- Insufficient oil in circulation: check sump level, pump pressure, and lubrication circuit restrictions. Repair as required.
- High water jacket temperatures: test cooling system performance.
- Plugged/failed oil cooler: disassemble and inspect galleries, bundle, and bypass valve for restrictions and scaling. Service the cooling system afterward to prevent a recurrence.
- Oil badly contaminated: submit sample for analysis and repair cause.
- Engine lugdown: provide some driver training; the problem is more common when standard transmissions are used.

Cooling System Problems

Cooling system problems can be grouped into the following categories:

- Overheating
- Overcooling
- Loss of coolant
- Defective radiator cap
- Defective thermostat

From the troubleshooting perspective, we can study these problems by looking at the performance problems they produce:

- High coolant temperatures
- Low coolant temperatures

High Coolant Temperature. Accurately verify the condition:

1. Check coolant level in the radiator.
2. Install a master gauge to verify the problem.
3. Observe exhaust pyrometer readings and compare to specification.

Some possible causes and solutions:

- Incorrect mixture: the correct mixture for EGs is usually an equal proportion of water and antifreeze with somewhere between 3 percent and 6 percent coolant conditioner. Increasing the antifreeze proportion with both EG and PG reduces cooling efficiency, but when high concentrations of PG are used for arctic operation, you may have no choice.
- Aerated coolant: usually caused by combustion gases entering the cooling circuit through a defective cylinder head or failed fire ring/cylinder head gasket. Heat is not transferred efficiently through a gas. Verify the condition: check for bubbles at the water manifold.
- Fan clutch: test fanstat/thermal fan operation using a master gauge and loading the engine until it reaches the temperature required to cycle the fan.
- Radiator: check for internal and external flow restrictions. Clean externally. Internal restrictions usually require the removal of the radiator and either recoring or the use of specialized reconditioning equipment.
- System does not seal: this permits the coolant to boil at a lower temperature and cause boilover. Test with a cooling system pressure tester. Check and repair/replace the defective radiator cap and/or pressure relief valves.
- Improper air flow: incorrectly sized fan or missing/damaged radiator shroud may significantly reduce flow through the radiator and engine compartment.
- Loose pump and fan drive belts: this will cause a reduction in coolant and/or air flow. Check visually and with a belt tension gauge.
- Coolant hoses: rubber-based coolant hoses should be changed every couple of years, silicone hoses every five years. Hoses often fail internally while appearing sound externally. A collapsed hose may cause a significant flow restriction.
- Restricted air intake: may cause high engine temperatures. Test inlet restriction with a water manometer. Check with the manufacturer-recommended maximum specifications, usually logged in inches (mm) of H_2O restriction.

- Exhaust restriction: causes high engine temperatures. Uncouple the exhaust piping from turbo or remove the exhaust manifold to see if the condition is corrected. Check the aftertreatment canister(s) (muffler, catalytic converter[s], SCR, and DPF) for restriction and internal collapse.

- Shunt line failure: a restriction in the shunt line from the radiator top tank to the water pump inlet may cause a boil condition at the inlet, reducing coolant flow and causing overheating.

- Thermostat: test out of engine for opening value and full open position using a boiler and a thermometer.

- Water pump: remove and check for a loose or damaged impeller and check the impeller-to-housing clearance against specification. Rebuild or replace as required.

- Boost air heat exchanger: an air flow or internal restriction will cause a rise in engine temperatures. Test air flow through the engine compartment. Bug and winter grille covers impede air flow through the heat exchangers and may unevenly load the fan. The same applies to snow plows and other front bumper–mounted equipment.

- High-altitude operation: cooling system efficiency diminishes as altitude increases. Therefore larger cooling system capacity is required for high-altitude operation.

- Lugging: operating an engine at high loads and lower speeds means high temperatures and reduced coolant flow. Driver training is required.

- Overfueling: this raises the amount of rejected heat and may exceed the cooling system's ability to handle it. Recalibrate engine fueling to specification.

- Fuel injection timing: depending on the type of fuel system, both retarding and advancing fuel injection timing may result in engine overheating. Fuel injection timing is commonly tampered with; fuel injection timing should be set to manufacturer specification. Greatly diminished engine life is the consequence of minor injection timing adjustments, whereas major component failure is the consequence of more dramatic adjustments.

Tech Tip: The reason diesel engine operating temperatures may run higher than expected in extreme cold may be due to the use of high-concentration PG coolant. High-concentration PG solutions do provide increased antifreeze protection but also lower cooling efficiency, resulting in noticeably higher operating temperatures.

Low Coolant Temperature. Accurately verify the condition:

1. Check cooling system temperature management components such as shutter and fan operation.
2. Install a master gauge to verify the problem.
3. Observe exhaust pyrometer readings and compare to specification.

Some possible causes and solutions:

- Thermostat: a thermostat that is stuck in the open position will cause the engine to run cool. Remove the thermostat and check its start-to-open and fully open temperature specifications.

- Air vent valve: if stuck in the open position, this may cause low coolant temperatures when the engine is under light loads.

- Prolonged idle or light load operation: when little fuel is being used by the engine, there is less rejected heat for the cooling system to handle, and operating temperatures may be lower.

- Fan clutch: test thermal fan operation using a master gauge and loading the engine until it reaches the temperature required to cycle the fan. Ensure that this occurs at the specified temperature and not at a lower value.

- Improper air flow: incorrectly sized fan or missing/damaged radiator shroud may result in overcooling by creating irregular flow through the engine compartment.

Cylinder Compression Problems

Although current certification testing requires technicians to be able to use both a cylinder pressure and cylinder leakage test, both are effectively obsolete when it comes to testing cylinder compression problems in any automotive diesel engine built since 2001. The correct way to determine cylinder leakage and/or balance problems is to perform an **electronic cylinder balance test** using the manufacturer diagnostic software. Overall low or unbalanced cylinder compression can create problems of excessive blowby, low power, and vibration. When any of these driveability conditions are evident, cylinder balance should be checked.

© Ford Motor Company

Figure 15-6 Cylinder pressure test adapter inserted into the injector bore.

Cylinder Pressure Test. Cylinder pressure may be checked using a compression tester. In diesel engines, the compression test should be performed when the engine is close to operating temperature by first removing all of the injectors, fitting the compression test gauge to the injector bore in one of the cylinders, and then cranking through five rotations. The batteries should be fully charged during this test because cranking velocity should exceed 250 rpm; this is why all of the injectors should be removed. Test each cylinder sequentially and match the test results to specifications. The results should be within the engine manufacturer's cylinder compression parameters and normally within 10 percent of each other. Compression testing will locate which cylinder(s) is/are at fault, but not the cause of the problem. **Figure 15-6** shows the cylinder compression test adapter recommended by Ford in the Power Stroke engine; the adapter is inserted into the injector bore and a gauge attached.

Cylinder Leakage Test. A cylinder leakage or air test kit sometimes stands a better chance of identifying the cause of a problem. The **cylinder leakage test** consists of a pressure regulator and couplings that fit to the injector bores. Once again the engine should be close to its operating temperature at the beginning of the test. The test is performed with the piston in the cylinder to be tested at TDC, and regulated air pressure is delivered to the cylinder. The test may indicate valve seal leaks, head gasket leaks, leaks to the water jacket, leaking piston rings, and defective pistons. The cylinder leakage test regulator delivers air at a controlled volume and pressure, then measures the percentage of leakage.

High Exhaust Temperature Readings

When diagnosing high exhaust temperatures on highway diesel engines equipped with catalytic

converters and diesel particulate filters, always use the engine manufacturer's recommended procedure. In post-2007 diesel engines, thermocouple-type pyrometers are used in some DPFs to help manage regeneration cycles. Those not using pyrometers use **resistance temperature detectors (RTDs)**, also known as **platinum resistance thermometers (PRTs)**. In pre-2007 chassis with high exhaust temperatures, first verify the condition:

1. Install a **master pyrometer** (do not rely solely on the vehicle pyrometer).
2. Chassis **dynamometer** test using a full instrumentation readout.
3. If a dynamometer is not available, load the unit to its maximum extent off a test bed.

Tech Tip: Transient (temporary) high exhaust temperatures are normal on post-2007 vehicles equipped with oxidation catalysts, dosed reduction catalysts, and diesel particulate filters (DPFs). During DPF regenerative cycles, transient high exhaust temperatures are normal (may cause a visible glow at night) and can be ignored unless they log a fault code.

Some possible causes and solutions to high exhaust temperatures are:

- Air inlet restriction: test inlet restriction value to specification using a manometer.
- Flow restricted, boost air heat exchanger: check air flow through the hood intake grille.
- High engine load, low air flow operation: engines using air-to-air boost air cooling *require* air flow through the heat exchanger. A pickup truck with air-to-air boost cooling operating in a construction environment under extreme temperatures may produce high exhaust temperature readings even when operated skillfully.
- Fuel injection timing: check to specification.
- Overfueling: look for other indicators of an overfueling condition and check fuel settings to specification.

Sudden Engine Shutdown

This can result from a number of conditions. First, try to determine whether the check engine lamp (CEL) or stop engine lamp (SEL) is illuminated. If this was the case, connect to the chassis data bus, and access the engine PCM for active and inactive DTCs. Some possible causes and solutions:

- Electrical failure: use the appropriate electrical circuit troubleshooting to locate the problem. Check the most obvious causes first.
- No fuel: refuel tanks and use an appropriate priming method to get the engine running again. Remember that most current light duty diesels are self-priming.
- Air in fuel system: air may be pulled into the fuel subsystem from any of a number of locations; common causes are the fuel filter sealing gaskets, double gaskets (caused by failure to remove a used gasket during servicing), and a failed filter mounting pad assembly.
- Water in fuel system: most current secondary filters will plug on small quantities of water. Locate the source of the water. Replace and prime filters.
- Plugged fuel line: sequentially work through the fuel subsystem circuit to locate the failed component. An internally failed fuel hose can be difficult to locate because an internal flap may require moving fuel through the circuit to block flow. Similarly, something simple, such as a small decal floating in the fuel tank, can plug a fuel pickup line and shut down the engine, after which it floats away.
- Lubrication failure: caused by a failed lubrication circuit component. A lubrication circuit failure will rapidly develop into a seized engine. Try to manually bar the engine over. A lubrication failure often results in a complete engine disassembly.

Engine Runs Rough

This describes a condition in which the engine produces an inconsistent rpm roll (nonrhythmic), hunting (rhythmic), cylinder misfire, or surging at a specific rpm or throughout the engine operating range. Some possible causes and solutions:

- Air in fuel system: often accompanied by white smoke emission. Locate the source of the air, checking the suction side of the fuel subsystem first.
- Any problems in the engine breathing circuit including heat exchangers, EGR coolers, venturii control valves, VGT turbos, aftertreatment hardware, and DPFs.
- Leak or restriction on the charge side of the fuel subsystem: repair a leak if evident. Use a pressure gauge to check charging pressure to specification.

- Injection nozzle failure(s): this results in a cylinder misfire condition. Locate the affected cylinder by disabling each injector sequentially. Crack the high-pressure pipe nuts at the pump on hydromechanical injector systems. When checking electronically managed engines, use the electronically driven diagnostics and never mechanically short out an injector.
- CN value of fuel is too low: this occurs when fuel is stored for prolonged periods either in the vehicle or in base storage tanks, and the fuel deteriorates chemically or biochemically.
- Fuel cloud point is high: when fuel begins to wax, it can restrict filters and pumping apparatus. Analyze the fuel and add an approved cloud point depressant as required.
- Advanced injection timing: a common tampering problem in older hydromechanical engines. Check injection timing and reset to specification if required.
- Maladjusted valves: perform an overhead adjustment.
- Cylinder leakage: caused by a failed fire ring(s), a crack in the cylinder head/block, or valve leakage. Perform an electronic cylinder balance test, cylinder leakage test, or cylinder compression test.

Engine Blowby

Because rings create the seal of a piston in a cylinder bore, some blowby is inevitable, due to the end-gap requirement. Normally, blowby is a specification determined by the manufacturer. Remember, piston rings seal most efficiently when cylinder pressures are highest, and engine rpm is high when there is less time for gas blowby. Some causes and solutions of excessive engine blowby are:

- Cracked head or piston: diagnose cause and recondition engine.
- Worn, stuck, or broken piston rings: diagnose cause and recondition engine.
- Glazed liner/sleeve inside wall (due to improper break-in practice or prolonged idling): diagnose cause and recondition engine.
- Poor-quality or degraded engine lube: change the oil and filters.

High crankcase pressures can also be caused by:

- Air compressor (plugged discharge line, worn rings, air pumped through oil return line): test air compressor operation.
- Turbocharger (seal failure allows turbine housing to leak pressurized air through to the turbine

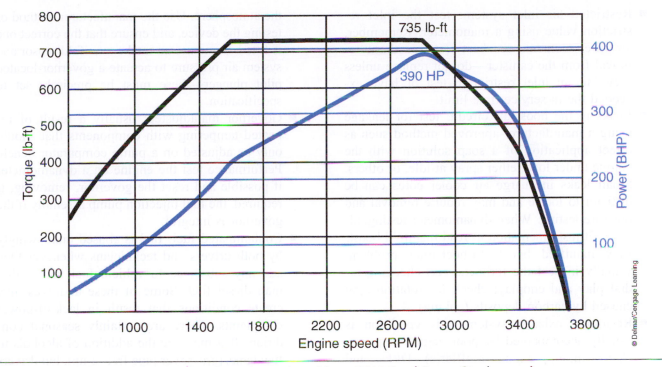

Figure 15-7 Power and torque band graph produced by a 2011 Ford PowerStroke engine.

shaft and back through the oil return piping): test turbocharger operation and recondition or replace as required.

Engine Will Not Crank

Determine whether the problem is related to the cranking circuit (electrical or pneumatic problem) or a mechanical engine condition. Focus on the cranking circuit first and eliminate any obvious causes; in most cases, the cause of a failure-to-crank condition will be found here. Try to bar the engine over by hand. If it cannot be turned over, check the engine externally and both the clutch and transmission: jack a drive wheel off the ground if necessary. If engine seizure is suspected, remove all the injectors and again attempt to bar the engine over by hand before beginning to disassemble the engine. If disassembly is required, carefully remove and label components so that the root cause of the failure can be determined.

Tech Tip: Use manufacturer software–guided troubleshooting whenever you can, even when you know the problem is not electronic. These programs are designed to prevent technicians from overlooking a critical step in a trouble-shooting procedure.

Lack of Power

''Lack of power'' (LOP) has a way of commonly showing up on shop work orders, and without running the unit on a dynamometer it can be difficult to convince a skeptical customer that the condition may lie more in his head than on the chassis. In many cases, the lack of power is more closely associated with driver expectations than a genuine engine problem. Although the technician can eliminate many causes of low power, realistically, the only proof available is a chassis dynamometer test, which usually involves taking the unit to a heavy duty auto shop.

It makes a lot of economic sense for shops that work frequently on low-power complaints to prepare a strategy check sheet that addresses the specific equipment worked on; this ensures that the work is performed sequentially (eliminating obvious/quick-to-perform tasks first) and enables each technician to pursue the steps in more or less the same manner. **Figure 15-7** shows a power and torque graph produced by a 2011 Ford Power Stroke engine. Note the wide torque band profile that characterizes the performance.

There are too many variables based on engine manufacturer and fuel system type to provide definitive lack-of-power troubleshooting strategies. The following is a general list of some causes of low power and their solutions:

- Restricted fuel filters: replace the filters. Check the vehicle fuel tanks for contamination.

- Restricted air inlet system: test the inlet restriction value using a manometer. Remember, air filter elements are best tested without removal from the canister—do not replace unless they fail an inlet restriction test or have exceeded the in-service time limit.

- Leaks in the boost air circuit: test for leakage using a manufacturer-approved method such as direct application of a soap solution with the engine under load, ether spray at idle, or others. Small leaks in charge air cooler cores can be difficult to locate and may require removal and pressure testing. When dynamometer testing, always fit instrumentation to read manifold boost.

- Low manifold boost—turbocharger problem: visually inspect the turbocharger, checking radial play and endplay; check for rotation drag caused by carbon deposits (coking).

- Restricted exhaust system: this condition is usually accompanied by poor engine response. Check piping, engine muffler(s), DPFs, and catalytic converters when equipped. Mufflers can fail internally when baffles and resonator walls collapse; test backpressure value to specification.

- Low fuel subsystem charging pressure: fit a pressure gauge and test fuel pressure at idle and high-idle speeds, comparing results to the specifications.

- Valve lash maladjustment: often accompanied by smoking and top end/valve clatter. Adjust valves to specification.

- Defective boost pressure sensing and fuel control devices: all current engines use either variable capacitance or piezo-resistive turbo-boost sensors that seldom malfunction. Older engines used different types of aneroid devices. The devices used on hydromechanical engines were not only more likely to malfunction, they could be easily tampered with. Most aneroids had a set trigger specification (predetermined manifold boost value), and their function was to limit fueling until there was enough air in the cylinders to properly combust it. Some aneroids permitted a graduated increase of fueling, proportional to increase in manifold boost, whereas others simply switch at a predetermined boost value, limiting fueling until it is achieved. Replacing a properly functioning boost fueling control device with one with a lower trigger value will not cure a low-power complaint or increase engine power, although it will cause puff smoking at shifting, waste fuel, and contaminate the atmosphere. Use the manufacturer method of testing the device, and ensure that the correct one is fitted. Some use an intake manifold sensor and system air pressure to actuate a governor-located pilot plunger—these must be precisely set to specification.

- Governor maladjusted: usually a result of repeated tampering with components that should only be adjusted on a pump comparator bench. Performance test the engine on a dynamometer if possible and reset the governor, removing, if required, the fuel injection pump assembly if the governor is integral.

- Contaminated fuel: fuel is abused unknowingly by both drivers and technicians whenever they add any substance to vehicle fuel tanks other than diesel fuel. Some of these additives may create conditions that result in lack-of-power complaints. There are certainly seasonal conditions that mandate the addition of alcohols to fuel tanks (crossover pipe freeze-up), but drivers and technicians should be educated to understand that dumping additives into fuel tanks, especially aftermarket additives of dubious chemistry, should be avoided. Purchasing number 1D fuel is much cheaper than purchasing inexpensive, summer grade number 2D fuel and paying for the problems it creates when operating equipment in extreme winter conditions.

- Defective fuel: not a common problem in North America and difficult to diagnose without the use of specialized equipment. Check with the fuel supplier. Testing specific gravity may identify a fuel in which the lighter fractions have boiled off, lowering the CN, but the exact original values must be known. Defective fuel problems are usually the consequence of storing a fuel for prolonged periods or using fuels outside the season in which they were purchased.

Engine Vibration

While driveline vibrations are not uncommon, true engine vibrations are not often a problem. When investigating a vibration complaint, the technician should eliminate all possible causes in the driveline behind the engine first. In doing this, it should be remembered that the clutch may disengage the driveline from the transmission and beyond from the engine, but that most of the mass of the clutch is rotated with the engine, engaged and disengaged.

- Cylinder misfire: see section "Engine Runs Rough."

- Loose vibration damper: check bolts for shear damage and damper fastener holes for elongation. Visually inspect damper for other damage. Retorque and test operation.
- Defective vibration damper: viscous-type dampers should be replaced at each engine overhaul but seldom are—visually inspect. The slightest external defect is reason to replace the unit. To dynamically test a vibration damper, a lathe and dynamic balance apparatus is required—a procedure beyond the scope of most shops. Visually inspect rubber drive ring dampers.
- Defective external driven component: every driven engine accessory is capable of unbalancing the engine. Some common accessories that may cause an unbalanced engine condition when they or their bearings fail are the fan assembly (broken/damaged blades, failed bearings), water pump, accessory drive bearings, idler pulley bearings, alternator, and others.

Soot in Inlet Manifold

Some soot in the intake manifold is normal in most diesel engines. It is a common indicator of an engine operated at low loads and speeds for prolonged periods if the engine was manufactured prior to 2004. Under such operating conditions, the valve overlap duration is at a maximum in real-time values and manifold boost is minimal or nonexistent. Excessive soot in the intake manifold may be an indication of imminent turbocharger failure, an injection timing problem, or a problem related to the EGR circuit in post-2004 diesels. We take a brief look at EGR circuit troubleshooting in the next section.

EGR Circuit Malfunctions

When the EGR circuit fails, in most cases, at least one fault code is logged because the circuit is so thoroughly monitored. When a fault code is logged, make sure you adhere to the manufacturer software-driven diagnostic routine. Typical diagnostic fault codes that relate to the EGR circuit:

- Short to ground
- Dead short
- EGR valve position circuit fault

A typical manufacturer diagnostic routine to verify EGR PWM-actuated valve function would appear as follows:

- Check for multiple codes: when other engine codes are logged, these should be solved first.

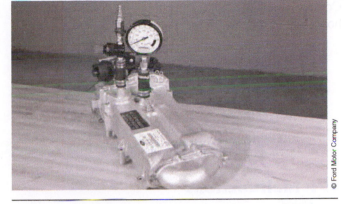

Figure 15-8 EGR pressure tester used to test EGR heat exchanger coolant leaks.

- Disconnect the EGR valve electrical connector: check for pin damage and corrosion and repair as necessary.
- Reconnect the EGR valve electrical connector. Key on, engine off (KOEF): use the manufacturer software to monitor the EGR valve's actual position.
- If it reads less than the specified KOEF position (usually expressed as a percentage): command the EGR PW to step to 50 percent. If it fails to move to within 5 percent of the input 50 percent, the EGR valve is defective and should be replaced. If it moves within 5 percent of the requested 50 percent, next command it to move to 100 percent PW. In this position, actual valve position should be at 95 percent or higher. If it is not, replace the valve. If it is:
- Remove the EGR valve and check for physical damage and carbon/sludge buildup. Either condition requires the valve to be replaced, but in the case of carbon/sludge buildup, check the EGR cooler for internal failure, which can result in coolant leakage and sludge buildup. **Figure 15-8** shows a Ford EGR pressure tester used to test EGR heat exchanger coolant leaks.

Manifold Boost Problems

When manifold boost is either too high or too low, engine performance complaints result. Common causes of lower-than-specified manifold boost are:

- Air system restriction upstream or downstream from turbocharger impeller
- Air leakage downstream from turbocharger impeller
- Low fuel delivery
- Mismatched turbocharger

Common causes of higher-than-specified manifold boost are as follows:

- Mismatched turbocharger
- Sticking vane ring on VGT turbos
- Excessively high inlet air temperatures
- Deposits on turbine volute or nozzle
- Overfueling caused by governor or by calibration programming
- Advanced fuel injection timing

DPF Malfunctions

DPF malfunctions are characterized by one or more of the following:

- Black smoke emission
- DPF collects excessive soot
- DPF is restricted
- DPF temperatures are high

Begin troubleshooting a DPF malfunction by inspecting it externally for indications of dents or cracks. Remove the DPF and check for indications of fuel or engine oil. If fuel is present, perform a cylinder balance test to identify a defective injector. Check cylinder exhaust tracts with an infrared gun; look for an exhaust tract variable of greater than 10 percent. If oil is present, check the turbocharger(s) for oil leakage. Note that any problem upstream in the engine breathing circuit can produce excessive soot loads and black smoke. Pay special attention to the venturi for the EGR system, EGR mixing valves, and the aftercooler.

DPF Restriction.

The engine management electronics are designed to respond to DPF restriction caused by excessive soot dump by logging fault codes and power derate strategies. This type of malfunction can be caused by operating the vehicle for prolonged periods when passive regeneration cycles are inhibited either by the driver or by the mode of operation. Most have a threshold that once exceeded requires an active regeneration cycle to be commanded. In some cases, when the soot load threshold has been exceeded by a considerable margin (say 200 percent), a factory password may have to be downloaded to clear the code and derate condition.

DPF Differential Pressure Problems.

DPF high differential pressure problems are usually caused by a buildup of soot or ash in the particulate trap. The appropriate repair strategy is to command an active regeneration cycle if this is still possible. If not, the unit must be either laundered on-chassis or removed and replaced, depending on manufacturer. Low differential pressure is usually caused by a failure of the tubing between the particulate trap and the pressure differential sensor. Check this circuit and replace components as necessary.

High DPF Intake Temperatures.

High DPF intake temperatures can be caused by a range of engine circuits upstream from the DPF, including:

- Restricted boost air heat exchanger
- Air inlet restriction
- Exhaust restriction
- High-altitude operation
- Overfueling
- Low coolant level

In every case, to troubleshoot high DPF intake temperatures, check for any active and historical codes: in most cases, high DPF intake temperatures are not sourced in the DPF, although an exhaust restriction downstream from the aftertreatment canister could cause this condition.

Mechanical Engine Knock

- Bottom end knock: produces an easily recognizable low-frequency thump. The cause is big end or connecting rod journal bearing failure, a condition that rapidly develops into a crankshaft failure if not attended to. Replace the bearings and thoroughly inspect and measure the throw journals.
- Failed crankshaft: diagnose the problem in-chassis by visual inspection of journals and their bearings and the critical crankshaft stress points.
- Damaged gears: often identified by a high-frequency whine. Remove the timing gear cover and replace failed components, remembering that debris from the failure may have been pumped through the entire lubrication circuit.
- Failure of feedback circuit component: check rocker trains and camshaft. Repair components as required, once again remembering that debris from the failure may have been pumped through the entire lubrication circuit.

Combustion Knock

This is sometimes known as *diesel knock*. If the noise is rhythmic, it may be that it is being produced from one engine cylinder, which may be verified by comparing cylinder temperatures with an infrared thermometer. Some causes and solutions:

- Fuel injection timing: usually produces erratic knock, amplified at higher loads and rpms. Check fuel injection timing to specification.
- Air in fuel: usually accompanied by white smoke emission. Check for air admission on the suction side of the fuel subsystem.

- Low-grade fuel: fuel with a low CN (caused by additive contamination, prolonged storage, non-highway ASTM grade, etc.) will extend the ignition lag phase. This results in excess fuel in the cylinder at the time of ignition, rapid pressure rise, and detonation. Analyze the fuel quality using the fuel supplier/refiner's recommended procedure.
- Fueling out of balance: a maladjusted rocker over an EUI (VW prior to 2010) can produce combustion knock. Check overhead adjustment. Check individual EUI rockers for binding.

Fuel Pump/Injector Scuffing

Fuel injector pump plunger scuffing has become a problem in recent years in part due to the lower lubricity of LS and the current ULS diesel fuels, and excessive addition of alcohol into fuel. Surface scuffing of the pump plungers can be difficult to diagnose, especially when the surface scuffing is minor. This type of failure is not common, but when it appears, it is usually found on PLN, EUI, HEUI, and EUP high-pressure pump plungers. Its causes are water in the fuel (usually in solution with methyl hydrate alcohols allowing it to pass through separators) or low-lubricity fuel. To overcome the problem, some manufacturers coat both the plunger and barrel assembly of high-pressure fuel injectors with tungsten carbide. The symptoms of fuel injector scuffing are:

- Low power: results from severe scuffing.
- Slight engine miss: most notable at low-load operation.

- Misfire: again, most notable under low-load operation.
- Smoking: some high-load hazing evident when multiple injectors are affected.

An electronic cylinder balance test (aka **cylinder contribution test [CCT]**) can sometimes identify an injector scuffing condition before it sets fault codes, especially when using diagnostic software capable of graphing displays. Severe cases are more likely in older electronic and hydromechanical diesel engines that lack the diagnostic software capable of identifying the condition.

Tech Tip: Caution owners against adding excessive quantities of methyl hydrate and diesel fuel conditioners to fuel tanks. Both form a solution with any water present in the fuel tank, and the solution can bypass water separators to enter the high-pressure fuel injection circuit, where it causes the scuffing damage.

QUICK REFERENCE DIAGNOSTIC CHARTS

The following sets of diagnostic charts (see **Table 15-2 to Table 15-17**) repeat some of the information we have already outlined and are intended as a guideline to diagnose some typical problems. As much as possible the approach is generic, but where it is not, the specific engine or fuel system is referenced.

TABLE 15-2: SYMPTOM: ENGINE WILL NOT CRANK	
Possible Cause	**Action**
Discharged or dead batteries	Test battery. Charge or replace as required.
Loose or corroded terminals	Identify location using voltage drop (VD) testing. Clean, tighten, or replace terminals.
High resistance in cranking circuit cables	VD test cables: (1) Switch to starter. (2) Battery to starter motor. Replace as required.
Defective starter motor or starter solenoid	Check operation using DMM. Repair as required.
Defective key switch	Test and repair as indicated by electronic troubleshooting. Replacement switch may require reprogramming.
Seized engine	Attempt to manually bar engine over through two revolutions. Diagnose cause of seizure.
Wake-up circuit not energized/activated or not communicating with powertrain data bus	Connect EST to the DDL. Verify communication into the powertrain gateway.
Driver door module circuit not communicating with powertrain bus, launching antitheft strategy	Connect EST to the DDL. Verify communication into the powertrain gateway and check antitheft logic status.

TABLE 15-3: SYMPTOM: ENGINE CRANKS, WILL NOT START

Possible Cause	Action
Slow crank	Reference Table 15-2: engine will not crank.
Check fault codes	Codes logged: use software diagnostics. No codes logged: continue using this chart.
No fuel to engine	(1) Check fuel tank level. (2) Check for restriction in fuel subsystem lines and filters. Replace as required.
Defective fuel transfer pump	(1) Check fuel-charging pressure to spec. (2) Check for air in fuel. Repair or replace.
Degraded fuel, water in fuel	Drain fuel from tank. Replace fuel and fuel filters. Flush new fuel through system before cranking.
Incorrect oil viscosity (cold weather condition)	Replace engine oil and filters when possible: 15W-40 engine lube gels in cold weather and cannot be drained.
Low compression	Perform cylinder compression test: use Table 15-15 to troubleshoot.
Defective charge pressure regulator	Remove charging pressure regulator and inspect; replace if required.
Defective electrical connections	Perform wiggle wire tests at PCM and critical engine sensors.

TABLE 15-4: SYMPTOM: ENGINE MISFIRE

Possible Cause	Action
Codes logged	Run diagnostic routines. Check the PCM audit trails.
Degraded or contaminated fuel	Drain fuel tanks and replace filters.
Low-charging pressure	Check fuel tank level. Sequentially check fuel subsystem for restrictions or air ingestion. Use Hg manometer in primary circuit, pressure gauge in secondary circuit. Check charging pressure regulator valve or restrictor fitting.
Valve lash adjustment	Perform top end adjustment.
Failed camshaft	Adjust valve lash. Run engine hard, preferably on a dyno for 30 minutes. Recheck valve lash. Check cam lift to specification. Replace camshaft if required.
Valves not seating	Remove cylinder head(s). Recondition valves and seats.
Defective fuel injector or pump	Condition should be identified by codes in all current engines: if the problem is intermittent, wiggle wires, examine audit trails, and perform an electronic cylinder balance test.
Air in fuel	Locate source of air ingestion. Check filter pads on primary circuit first. Use an inline diagnostic sight glass to locate source.
Blown cylinder head gasket	Locate source of leak by external visual inspection. Compression test each cylinder to check for blown fire rings. Check coolant and oil for contamination.
Leakage at high-pressure pipes	Check for trace leakage. Check pipe nut torque on high-pressure pipes. Identify source of leak and check all other high-pressure pipes on PLN, EUP, and CR fuel systems.

TABLE 15-5: SYMPTOM: ENGINE STALLS	
Possible Cause	**Action**
Codes logged	Run diagnostic routines. Check audit trails. If no active codes, road or dyno test using EST snapshot mode: set trigger to collect data beginning at code log event.
Idle speed set low	Reset or reprogram idle speed.
High parasitic load cut-in events	Check air compressor and alternator, especially on city and highway buses.
Fuel tank vent circuit plugged or restricted	Check vent or filter. Tank vent filters can rapidly plug in construction and agricultural operations. Clean/replace vent/filters.
Fuel starvation	Check fuel tank pickup tube(s). Check fuel tank for foreign matter: a decal or piece of paper over the pickup tube inlet can shut system down. Check fuel lines for internal restriction such as a rubber flap. Replace lines as required.
Low fuel supply operating on rough terrain	Check fuel tank level and replenish if necessary.
Defective injector nozzle(s)	Identify affected injector(s) and replace. Identify cause of nozzle failure and remedy; if caused by water, identify source.
Defective injection pump(s)	This condition should produce codes in current engines. In older engines, use the manufacturer service literature.

TABLE 15-6: SYMPTOM: IRREGULAR ENGINE RPM	
Possible Cause	**Action**
Air ingested into primary side of fuel subsystem: results in erratic engine rpm roll.	Locate air source and repair: check primary fuel filter pad and fuel tank pickup tube first.
Fractured fuel transfer pump, pulsation damper: results in erratic engine rpm roll.	Test charging pressure in secondary fuel circuit. Replace pulsation dampening disc.
Unbalanced fueling: result is rhythmic engine rpm fluctuation known as hunting.	Perform electronic cylinder balance test.
Cylinder leakage.	Perform cylinder balance test on electronically managed engines. Cylinder leakage can be caused by blown fire-ring (head gasket), cylinder valves, or piston rings.

TABLE 15-7: SYMPTOM: LACK OF POWER

Possible Cause	Action
Codes logged	Run diagnostic routines and check audit trails before doing anything else.
EGR mixing chamber problem	Inspect mixing gate door and leakage. Repair as needed.
Boost side leak	Check for leaks downstream from the turbocharger impeller housing. Test the charge air cooler for leaks.
Plugged or restricted fuel tank vent(s) or filters	Clean vents and replace filters if necessary.
Plugged air filter	Check filter restriction using a water-filled manometer. Replace if outside of specification.
Degraded fuel	Determine cause of fuel degradation, especially when biodiesel is used. Drain fuel and replace fuel filters. Refuel with fresh fuel.
Low fuel charging pressure	(1) Check fuel tank level. (2) Check fuel filter restriction. (3) Check for air ingestion. (4) Check fuel pressure regulator. Fill fuel tank and replace defective components.
Defective injector nozzle(s)	Identify affected injector(s) and replace. Identify cause of nozzle failure and remedy; if caused by water, identify source.
Valve lash adjustment	Perform a valve adjustment.
Turbocharger malfunction	Remove flow piping: inspect radial play at turbine and impeller. Check for contact or lube leakage on both sides. Replace or recore turbo.
Exhaust restriction	Check for restriction in exhaust piping and DPF canister. In vehicles with low horizontal exhausts, check exit piping from DPF/muffler canister.
Externally or internally plugged charge air cooler (CAC)	Visually check for external plugging of exchanger fins (results in insufficient cooling). Remove CAC and perform flow and leakage tests.
Restricted boost side ducts and pipes	Remove and visually inspect; replace if defective.

Note: Caused by many factors. If possible, verify the lack-of-power complaint by chassis dyno testing. Never act on a driver's word alone: over 50 percent of LOP complaints prove to be baseless.

TABLE 15-8: SYMPTOM: POOR FUEL MILEAGE

Possible Cause	Action
Codes logged	Run diagnostic routines and check audit trails before doing anything else.
Restricted air intake circuit	Use an H_2O manometer (upstream from impeller housing) and pressure gauges (downstream from impeller housing) to identify location. Repair/replace as required.
Fuel leakage	Check for evidence of external fuel leakage and repair.
Defective fuel injectors	Dribbling nozzles may not trip a fault code in some systems. Usually accompanied by more frequent DPF regen cycles. Test older hydromechanical nozzles and repair/replace.
Internal engine wear	Identify root cause. You may find an electronic cylinder balance test helps. Worn rings may also produce other symptoms, such as plugging of PCV filters and more frequent DPF regen cycles. Repair engine as required.

TABLE 15-9: SYMPTOM: EXCESSIVE OIL CONSUMPTION

Possible Cause	Action
External oil leaks	Check engine for obvious external leakage. Check all seals and gaskets. Add UV dye to oil, then pressure wash, if the source of the leakage is difficult to locate.
Turbocharger oil leakage to manifold	Check for oil in turbocharger inlet and soot in exhaust outlet. Check for high inlet restriction (plugged air filter). Perform repairs as necessary.
Leakage at EGR valve piston shaft	Inspect EGR circuit for oil loss. Replace EGR valve if necessary.
Front and rear main seal leakage	Check for oil tracking at both seals. Check seal races on crankshaft. Replace and realign main seals and install wear sleeves if necessary.
Air compressor passing oil	Check air supply circuit downstream from the air compressor. Remove compressor and check drain-back ports for plugging. Replace compressor flange gasket and/or compressor.
Plugged crankcase breather or PCV circuit	Identify cause of clogging and clean or replace components.
High exhaust backpressure	Check exhaust backpressure. Check exhaust circuit for internal restrictions such as collapsed resonator baffles.
Worn valve guides and seals	Check valve guide bore size to specification and seal integrity. Repair/replace components as necessary.
Worn piston rings	You may identify this condition by running a cylinder balance test if you are familiar with the duty cycle results produced by a healthy engine. Perform a cylinder leakage test in nonelectronic engines. Rebuild engine.
Leakage from HEUI oil actuation circuit to the fuel supply gallery	Condition is usually caused by rolled or failed center O-rings in the HEUI injector assembly.

TABLE 15-10: SYMPTOM: ENGINE OVERHEATING

Possible Cause	Action
Low coolant level	Locate cause. Visually inspect for external leaks and pressure-test circuit. Repair and replenish coolant.
Loose or worn fan/water pump drive belts	Adjust belt tension or replace drive belts.
EGR heat exchanger leak	Replace EGR cooler; do not attempt to repair.
Restricted air flow through radiator	Carefully clean bugs and road dirt from rad cooling fins using low-pressure hose. Test after cleaning.
Defective radiator cap	Dynamic pressure-test rad cap to ensure it seals to specified pressure. Replace if necessary.
Defective thermostat(s)	Test thermostat opening temperature; also check temperature sensor accuracy. Replace components as necessary.
Defective fan cycle controls	Check viscous fan operation by running engine to the specified fan temperature. With on/off fan controls, check the means used to disengage/engage the fan, electric and pneumatic.
Combustion gas in coolant	Check for evidence of cylinder head gasket and injector sleeve failure. Repair as necessary.
Defective water pump	Check water pump drive. Remove water pump and check impeller-to-shaft integrity. Repair or replace water pump and reinstall.
Plugged oil cooler	Usually produces higher-than-specified oil temperatures along with normal-range coolant temperatures. Remove oil cooler and disassemble to clean/remove restriction.
CAC restricted	Check grille for bugs, road dirt, etc.
Improper driver shifting practices	Check engine audit trail if available. Refer operator to driver trainer.

TABLE 15-11: SYMPTOM: HIGH EXHAUST TEMPERATURE

Possible Cause	Action
Codes logged	Run diagnostic routines to identify root cause. Where separate DPF module (noncatalyzed type) is used, check for DPF codes.
Improper driver shifting practices (high load and steep terrain operation)	Check engine audit trail if possible. Advise driver on how to shift.
High EGR temperatures	EGR flow rate too high or ineffective EGR heat exchanger. Check EGR mixing gate operation and EGR heat exchanger for clogging.
Restricted air intake	Use an H_2O manometer to check for inlet restriction and repair the cause.
Leak in air intake circuit	Check hose clamps, pipes, and ducts. Repair as required.
Exhaust leak upstream from turbine housing	Check for external leaks under load. Replace gaskets and manifold coupling seals as necessary.
Exhaust restrictions	Check for exhaust circuit restrictions, including the cylinder head exhaust tract insulation, loose baffles, plugged catalytic converters, and DPFs. Replace or repair as necessary.
Overfueling or advanced timing	More common in hydromechanical engines; check audit trail for fuel injection system–related events, and run a cylinder balance test. Perform a tune up on nonelectronic engines verifying base injection timing.
High-load operation with insufficient ram air	High-load operating a ram air-cooled CAC when the vehicle is moving at low velocity can produce overheat. Review the type of operation the engine has been spec'd for with the owner.

TABLE 15-12: SYMPTOM: LOW OIL PRESSURE

Possible Cause	Action
Low oil level	Add oil and check for oil leaks or an oil-burning condition.
Incorrect oil viscosity	Drain oil and replace filters. Replenish with oil meeting the manufacturer-specified viscosity for the temperature conditions.
Defective oil pressure gauge or sensor	Check the operation of the oil pressure sensor first, then the pressure gauge. Replace components as necessary.
Restricted full-flow oil filter	Plugged full-flow filters trip the bypass valve on the oil filter flange and may, or may not, cause low oil pressure. Replace oil and filters.
Engine oil diluted with fuel	Check for fuel system leaks; this varies according to the type of fuel system used. Check O-rings on fuel injectors and EUPs. Repair and replace as necessary.
Oil pump failure	Check mounting arrangement. When the oil pump is remote mounted, carefully check the pickup plumbing for air ingestion. Replace or repair components as necessary.
Oil filter and filter mounting pad problems	Remove filter(s) and visually inspect gaskets. Check for possible restrictions to the in/out ports of the oil filter. Repair, clean, or replace components as necessary.
Excessive clearance between crankshaft journals and bearings	Perform a bearing roll in. Note that this is not a common problem today as friction bearing technology has advanced significantly.
EGR valve leaking excessively	Check EGR valve operation and repair as necessary.
Degraded engine oil	Low-quality engine oils, especially those with high detergent loads, can break down and thin, resulting in low oil pressure. Replace engine oil and filters with those specified by the manufacturer.

TABLE 15-13: SYMPTOM: OIL IN COOLANT

Possible Cause	Action
Defective oil cooler core or bundle O-rings	Remove and disassemble oil cooler. Test bundle or plate stack. Replace if defective. Rolled or failed bundle O-rings can also cause this condition. Replace as necessary. Clean out entire cooling system using detergent.
Failed head gasket	Replace cylinder head gasket. In cases where nonintegral grommets are used, take special care in placing these.
Cylinder head or cylinder block porosity	Identify the location of the failure and replace as necessary.

TABLE 15-14: SYMPTOM: COOLANT IN ENGINE OIL

Possible Cause	Action
Cracked cylinder head	Disassemble and repair oil cooler. This is a less common outcome of an oil cooler failure than oil in coolant.
Head gasket failure	Pressure-test cooling system and replace the cylinder head gasket.
Injector cup seal failure	Hydrostatically test cylinder head and replace injector cups as required.
Cylinder wet liner failure	Pressure-test system and replace defective components as required.
Defective oil cooler core	Hydrostatically test cylinder head to locate the failure and replace cylinder head.

TABLE 15-15: SYMPTOM: LOW COMPRESSION

Possible Cause	Action
Incorrect valve lash adjustment	Perform a complete overhead tune up, making sure the valves are set to spec.
Blown fire ring(s) in head gasket	Replace head gasket.
Broken or fatigued valve springs	Check and replace as required. In the case of fatigued valve springs, it makes sense to replace them all.
Valves not seating	Remove cylinder head and check valve seat faces. Replace valves and valve seats.
Piston ring failure	Rings may be stuck, worn, broken, or glazed. Determine root cause of failure and replace the ring sets.
Failed camshaft and/or lifters	Determine root cause and replace failed components, being sure to check every component in the valve train.

TABLE 15-16: SYMPTOM: FUEL IN LUBE

Possible Cause	Action
Excessive idling in cold weather (cylinder washdown)	Program optimized idle, install auxiliary cold weather assists, and instruct driver how to handle the auto in cold weather.
Injector nozzle malfunction	Mostly a condition on older hydromechanical engines. Remove and test injectors. Replace nozzle valves and reassemble injectors.
Injector O-ring failure	Pull injector assembly and inspect for worn, torn, or rolled O-rings. Replace all O-rings whether observed to be defective or not.
HEUI IAP circuit failures (Ford up to 2010)	Check the HEUIs, jumper pipes if used, and the cylinder oil and fuel manifolds. Replace components as required.

TABLE 15-17: SYMPTOM: EGR SYSTEM MALFUNCTIONS

Possible Cause	Action
Excessive idling—especially in cold weather, it causes carbon coking in EGR heat exchanger that chokes down on flow	Program optimized idle and instruct driver how to handle the vehicle in cold weather. Equip the vehicle with auxiliary power for cold weather operation.
High EGR temperature	Indicates a failed EGR heat exchanger. Replace the heat exchanger or heat exchanger core.
EGR valve housing cracked	Inspect the EGR housing and replace if necessary.
EGR gate valve sticking	Check oil supply to valve. Check for oil coking or corrosion around the actuator bushing. Also check electrical connections to the actuator. Clean, repair, and replace components as required.
EGR valve not functioning properly due to low oil pressure	Troubleshoot the low oil pressure condition using Table 15-12.
Turbo VGT actuator does not respond	Determine whether problem is mechanical, pneumatic, or electrical (depending on the manufacturer). Check for codes. If none, focus on a mechanical cause at the turbo VGT actuator or within the turbocharger.

Tech Tip: When a sensor logs an active DTC, try disconnecting the sensor and observing whether the code changes. It there is no change, it suggests that the fault lies in the wiring circuit. If the code changes, it would suggest that the problem lies in the sensor.

DIAGNOSING BUS FAULTS

Before beginning to diagnose any type of bus fault, it is necessary to consult the manufacturer service literature and identify the type of bus you are working with. Manufacturer software is designed to diagnose data bus problems, but the truth is, some do it a lot better than others. A scopemeter **(see Figure 15-2)** can be a useful tool in identifying bus communication problems.

Using a Scopemeter to Verify CAN Activity

Be sure to read the information provided by the scopemeter manufacturer. The scopemeter may be equipped with accessories such as a J1962 connector, allowing a hand-held scopemeter to be directly connected to the chassis data bus. The same would apply to a scopemeter that uses a CA and PC software. If the scopemeter is not so equipped, the probes should be connected to pin cavities 6 (CAN high) and 14 (CAN low) as highlighted in **Table 15-2** and demonstrated

in **Figure 15-9,** which shows a typical CAN bus configuration.

Troubleshooting One-Line Bus Malfunctions

Because one-line bus multiplexing can be regarded as an older technology, for the troubleshooting sequence that follows we will assume that a scan tool is being used. Five general types of problems can result from a one-wire bus line:

1. Short to power side of circuit. Shuts down bus communications between all the modules with an address on the bus because no binary logic exchange can take place.
2. Short to ground side of circuit. Shuts down bus communications between all the modules with an address on the bus because no binary logic exchange can take place.
3. **Open circuit.** The result of an open varies with the type of bus. In a star network, this prevents data exchange between modules on the opposite side of the fault. When a single open occurs in a loop topography bus there may not be any communication problems because there are multiple paths for message packets to travel.
4. Interference. May be troublesome because message packets can become scrambled as binary high and low logic is pushed outside of range. Check with a scopemeter as described

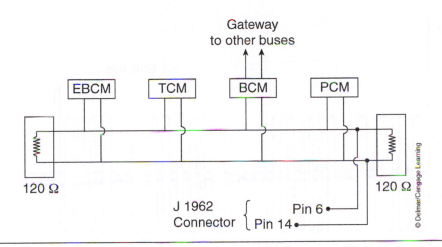

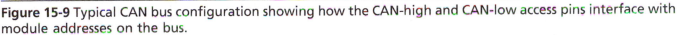

Figure 15-9 Typical CAN bus configuration showing how the CAN-high and CAN-low access pins interface with module addresses on the bus.

earlier in the chapter. Causes can be inappropriate repairs, physical rerouting/repositioning of bus lines, and missing terminating resistors.

5. High resistance. When severe enough, the effect can be similar to that of an open. Not a common problem due to the high impedance of module internal circuitry, so it has to be pretty severe before messaging is affected.

One-Wire Circuit Diagnostics

First determine whether communication with all the modules with addresses on the bus is possible. If the scan tool fails to power up or fails to recognize a handshake (illuminate communications LED), check the EST and the connection cables before proceeding to suspect an on-chassis problem. Always verify the functionality of an EST before assuming a malfunction is chassis based. If any one or more of the modules handshakes with the EST, you can assume that there is no short: this means items 1 and 2 listed previously can be immediately excluded as a cause, and you can focus on items 3 through 5.

No Handshake with Any Module. If the EST has been verified as being functional and there is no communication with any module on the bus:

1. Disconnect the EST from the diagnostic link connector (DLC).
2. If the DLC is a J1962 connector, use a DMM to verify that there is a good ground at pins 4 and 5 (see **Table 15-1**).
3. Connect the DMM probes to pin 4 and the suspect circuit on the chassis side of the DLC. Set the DMM to read Min/Max.

4. Do something that should trigger bus activity. This will vary according to model year, but depressing a key fob command (lock/unlock) is reliable.
5. Observe the DMM readings. If both Min and Max are significantly above zero volts, a short to power is indicated.

Tech Tip: Voltage fluctuations that exceed 12V DC and drop below zero indicate electromagnetic or radio frequency interference.

6. If both Min and Max are close to zero volts, a short to ground is indicated: switch the DMM to read resistance. If a resistance reading is lower than one kilohm (1 kΩ), a short to ground is likely. If an out-of-range (OL) reading is displayed, check the wire/connections from the DLC to the chassis network.

Tech Tip: Actual resistance readings will vary according to the number of modules on the bus because they are wired in parallel, but they should never exceed 1,000 Ω.

No Communication with One Module. If communication is lost between the EST and one module on a data bus:

1. Disconnect the affected module.
2. Check the power and ground circuit connections at the connection stub.

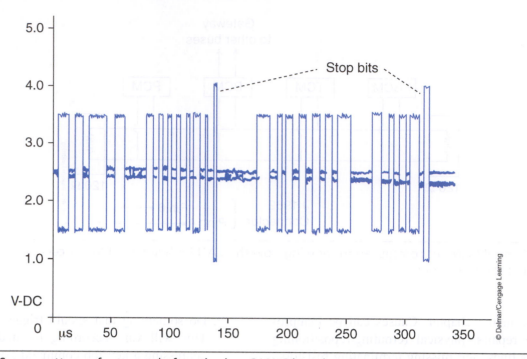

Figure 15-10 Scope pattern of a properly functioning CAN-C bus showing a 2.0V differential pattern. The two 3.0V differentials between CAN high and CAN low indicate end-of-message stop bits.

3. Create some bus activity to check for signals reaching the bus stub at the module. If these three check okay, the module is likely defective.

Replacing a Module. A module should not be replaced before verifying the loaded resistance of both the power and ground circuits connected to the module. This can be done by connecting an electrical load equivalent to the current draw of the module in question (from the power to the ground at the module connector stub). The voltage reading should be no more than 0.5V DC less than the system voltage of 12.6V DC. If for instance the voltage reading was 11.6V DC (a 1.0V DC differential), high resistance at the power or ground circuits is likely. If high resistance is indicated during this test, it should be repaired and the failed module diagnosis reevaluated.

Troubleshooting Two-Wire Bus Malfunctions

Because two-line bus multiplexing can be regarded as a current technology, for the troubleshooting sequence that follows we will assume that PC-based software and a CA is being used, although scan tools will work with some systems. **CAN-C** is a voltage differential bus, meaning that the digital bit logic is messaged by a high bit voltage not exceeding 4.0V DC and a low bit voltage that does not drop below 1.0V DC. The low-voltage line is the dominant bit.

On a properly functioning bus **(see Figure 15-10)** the voltage differential is typically 2.0 volts, the exception being a 3.0V DC differential to punctuate an end-of-message signal known as a stop bit. In addition, a 12.0V DC "wake-up" pulse is used to activate all of the modules on the bus. It is common for a body control module (BCM) to quarterback bus activity (wake-up/emergency response strategy). A major advantage of a CAN-C bus is its ability to tolerate interference.

CAN-C Fault Modes

As with one-wire buses, five general types of problems can occur in a two-wire bus line, along with one additional one:

1. Short to power side of circuit. Shuts down bus communications between all the modules with an address on the bus because no binary logic exchange can take place.
2. Short to ground side of circuit. Shuts down bus communications between all the modules with an address on the bus because no binary logic exchange can take place.
3. Short between CAN high and CAN low. Caused by pinching the two wires together during repair procedures or by a collision.
4. Open. Result of an open varies with the type of bus. In a star network, this prevents data

exchange between modules on the opposite side of the fault. When a single open occurs in a loop topography (CAN-C) bus, there may not be any communication problems because there are multiple paths for message packets to travel.

5. Interference. May be troublesome because message packets can become scrambled as binary high and low logic is pushed outside of range. Check with a scopemeter as described earlier in the chapter. Causes can be inappropriate repairs, physical rerouting/repositioning of bus lines, and missing terminating resistors.

6. High resistance. When severe enough, the effect can be similar to that of an open. Not a common problem due to the high impedance of module internal circuitry, so it has to be pretty severe before messaging is affected.

CAN Diagnostic Routines

Begin by establishing communication between the PC, the CA (hereafter referred to as the EST), and the DLC. When a failure-to-communicate condition occurs, begin by checking the EST, its connectors and cables. Use another vehicle if necessary to corroborate EST functionality. For purposes of outlining the following diagnostic routines, it will be assumed that the malfunction has occurred on the powertrain bus.

No Handshake with Any Module.
When the EST fails to communicate with any module on the bus, use the following routine:

1. Disconnect the EST from the DLC.
2. Use a DMM to verify a good ground at J1962 pins 4 and 5.
3. Attach DMM probes to pin 6 (positive probe) and pin 14 on the chassis J1962 connector.
4. Set the DMM to the Min-Max mode on the V-DC setting to measure voltage differential.
5. Trigger some activity on the bus, such as turning the ignition key fob.
6. The minimum voltage should be close to 0 and the maximum close to 2.0V.

If the correct readings are displayed, the problem is likely in the EST. If the readings are outside of specification, run the following diagnostic routines:

1. Connect the positive DMM probe to pin 6 and the negative DMM probe to pin 4. The specification should read a maximum of 3.5V and a minimum of 2.5V.

2. Connect the positive DMM probe to pin 14 and the negative DMM probe to pin 4. The specification should read a maximum of 2.5V and a minimum of 1.5V.

Based on what you have measured during the preceding sequence you should be able to pinpoint the cause to one of the following conclusions:

- If the minimum and maximum of both the preceding circuits are above their normal values by the same amount, either a module or its ground integrity are at fault.
- If both 6 (CAN-high) and 14 (CAN-low) show a maximum of 2.5V, the circuits are shorted together.
- If either circuit reads a maximum and minimum of close to zero volts, the circuit is shorted to ground. An exception to this would be in the event of an open between the DLC and the remainder of the network. Should this be the case, the vehicle will start and run normally, but there could be no EST communications.
- If either CAN high or CAN low displays a minimum and maximum above the specified pulse value, the circuit is shorted to power.

No Communication with One CAN Module.
If communication is lost between the EST and one module on a data bus:

1. Disconnect the affected module.
2. Check the power and ground circuit connections at the connection stub.
3. Create some bus activity to check for signals reaching the bus stub at the module. If these three check out okay, the module is likely defective.

Replacing a CAN Module.
The procedure is identical to replacing a one-wire bus module but it is repeated here. A CAN module should not be replaced before verifying the loaded resistance of both the power and ground circuits connected to the module. This can be done by connecting an electrical load equivalent to the current draw of the module in question (from the power to the ground at the module connector stub). The voltage reading should be no more than 0.5V DC less than the system voltage of 12.6V DC. If for instance the voltage reading is 11.6V DC (a 1.0V DC differential), high resistance at the power or ground circuits is likely. If high resistance is indicated during this test, it should be repaired and the failed module diagnosis reevaluated.

Summary

- The ESTs used to service, diagnose, and reprogram auto engine management systems are onboard diagnostic lights, DMMs, scan tools, and PCs with CAs and the appropriate software.

- Flash codes are an onboard method of accessing diagnostic codes. Most systems display active codes only, but some display active and historic (inactive) codes.

- A DMM is one of the most important ESTs available to the technician. You should understand how to spec one out and use it to perform electrical and electronic troubleshooting.

- A continuity test is a quick resistance test that distinguishes between an open and a closed circuit.

- Circuit resistance and voltage are measured with the test leads positioned in parallel with the circuit.

- Direct measurement of current flow is performed with the test leads located in series with the circuit. In other words, during the test, current flow is routed through the DMM.

- Most manufacturers are moving toward using the PC and a Windows environment loaded with proprietary software as their primary diagnostic and programming EST.

- ESTs connect with the vehicle data bus connector PCM(s) via a 16-pin data connector known as a J1962 ALDL or DLC.

- Most PC-based ESTs require the use of a CA serial link to make the connection between the chassis data bus and the EST unit. In some cases, a scan tool can act as a CA. Wireless EST to data bus communications are also enabled when the appropriate land-to-chassis hardware and software are used.

- Most current scan tools and PC-based ESTs are updated by Web-driven downloads.

- All manufacturers require the use of their online SIS to produce accurate results with their diagnostic software.

- The diesel technician should be able to analyze exhaust smoke emission and use this to help diagnose engine and fuel system malfunctions.

- White exhaust smoke indicates the presence of condensing liquid in the exhaust gas.

- Black exhaust smoke indicates the presence of particulates in the exhaust gas.

- Blue smoke emission is generally associated with an engine oil-burning condition.

- A scopemeter can be used to determine whether the communication activity on a data bus produces a normal or abnormal profile.

- A DMM connected to J1962 pins 6 and 14 can be used to diagnose a range of CAN-C bus communications malfunctions, including short to ground, short to power, short across bus lines, and opens.

Internet Exercises

Use search engines and the URLs listed here to check out the following:

1. How to update Bosch MCVI diagnostic software.

2. Surf to www.picotech.com. Download the free software and see what you can do with it.

3. If you are a member of www.iatn.net, connect with it and follow some of the threaded troubleshooting discussions. If you are not a member, check it out and determine what is required to join.

4. Identify a troubleshooting problem you have experienced with a light duty diesel engine. Use as few words as possible to broadcast it onto the Web and see what solutions you can come up with.

5. Select one of the light duty diesel engine manufacturers. Determine the costs of subscribing to their Web-based SIS and diagnostics software programs.

Shop Tasks

1. Use a scan tool or PC-based EST to connect to a J1962 DLC on a vehicle chassis. Key off, disconnect one of the modules on either a one- or two-wire bus. Use the troubleshooting profile in this chapter to establish the cause of the malfunction you have created.

2. Connect a DMM in V-DC mode to pins 6 and 14 on a vehicle J1962 DLC. Create some bus activity and check the voltage values to specification using the values outlined in the chapter or in manufacturer service literature.

3. Connect a labscope to pins 6 and 14 on a vehicle J1962 DLC. Create some bus activity and compare the waveforms produced to those outlined in this chapter.

4. Connect an EST to the vehicle J1962 DLC. Making sure that the chassis is key off, disconnect engine pressure, temperature, and shaft speed sensors sequentially and observe what effect this has on the engine and EST display. Do not attempt to disconnect sensors monitoring the exhaust aftertreatment components.

5. Use a search engine to access online chat lines on Dodge (Cummins), Ford, GM (Duramax), BMW, and VW-Audi threaded problem-solving discussions. Make a note of any complaints common to all of the light duty highway diesel manufacturers.

Review Questions

1. Which of the following is usually required to hard-wire couple a diagnostic PC with a vehicle data bus?

   A. Modem
   B. Jumper wires
   C. DLC and CA
   D. Parallel link connector

2. The output (signal) of an inductive pulse generator shaft speed sensor is measured in:

   A. V-DC
   B. Ohms
   C. V-AC
   D. Amperes

3. Which pin on a J1962 DLC represents CAN high?

   A. 4
   B. 6
   C. 12
   D. 14

4. Which of the following procedures cannot be performed using a PC loaded with the manufacturer's diagnostic software?

   A. Erasing inactive fault codes
   B. Erasing active fault codes
   C. Customer data programming
   D. Cylinder balance tests

5. Technician A says that on a J1962 DLC, pin 14 is a ground pin. Technician B says that on a J1962 DLC pin 3 is a ground pin. Who is correct?

   A. Technician A only
   B. Technician B only
   C. Both A and B
   D. Neither A nor B

6. Which of the following is the preferred source for service literature covering a repair procedure?

   A. Online SIS
   B. Hardcopy service manual
   C. Technical service bulletin
   D. Mechanic with a good memory.

7. Removing an engine coolant thermostat would most likely result in an increase of which noxious emission?

   A. HC

   B. NO_x

   C. SO_2

   D. CO_2

8. Which of the following operating modes would be most likely to produce valve float?

   A. Overfueling

   B. Lugdown

   C. Ether abuse

   D. Overspeeding engine

9. Technician A states that an excessive amount of oil in the crankcase can produce a low oil pressure problem in a diesel engine. Technician B states that the first thing to check when oil pressures fluctuate is the crankcase oil level. Who is correct?

   A. Technician A only

   B. Technician B only

   C. Both A and B

   D. Neither A nor B

10. Technician A states that white smoke emission can be caused by air starvation to the engine. Technician B states that blue smoke is associated with an engine that is burning oil in its cylinders. Who is correct?

   A. Technician A only

   B. Technician B only

   C. Both A and B

   D. Neither A nor B

Glossary

abrasive wear occurs when hard particles get between two moving surfaces, cutting or scratching the surfaces. In the case of lubed mating components, the particles have to be thicker than the lubricant film for abrasive wear to take place.

absolute maximum power the highest power an engine can develop at sea level with no limitations on speed, fuel/ air ratio, or fuel quantity.

accumulator device for storing energy. Often used to describe the high-pressure chamber or common rail used in some electronically controlled fuel injection systems.

ACERT Caterpillar technology that uses a four-phase emissions reduction strategy to meet 2004, 2007, and 2010 EPA standards. Critical ACERT components include series turbocharging, variable valve timing, MEUI and HEUI injectors, and exhaust gas aftertreatment.

A-circuit voltage regulator located on the ground side of the rotor and battery, voltage is picked up by the field circuit inside the alternator.

acronym a word formed by the initial letters of other words.

active codes an electronically monitored system circuit, condition, or component that is malfunctioning and logs an ECM code, which may be displayed or read with an EST.

active DPF system a diesel particulate filter that actively generates the heat required to regenerate (burn off soot) the device. Heat is usually derived from a heat grid or by injecting diesel fuel.

active regeneration term used to describe diesel particulate filter regeneration aided by fuel dosing, sometimes in combination with spark ignition.

actuators hardware that puts the results of computer processing into action. Examples of actuators would be the injector drivers in a diesel EUI system.

adaptive when used as a prefix, such as in *adaptive* cruise control, this means electronically managed with *soft* parameters.

adaptive trim ECM/PCM software that evaluates fuel flow performance of electrohydraulic injector and EUIs at set engine operating hour intervals and makes corrections to ensure fueling balance. Sometimes abbreviated to *A-trim*.

AdBlue an aqueous urea solution of 68 percent water and 32 percent urea used for selective catalytic reduction (SCR)–type catalytic converters. This term is used primarily in Europe: AdBlue is identical to what is more commonly known as *diesel exhaust fluid (DEF)* in North America.

adhesive wear occurs when moving surfaces make contact without adequate lubrication, producing heat through friction and elevating surface temperatures to melting point; can result in the surfaces adhering to each other. The fastest-progressing wear characteristic.

adsorption a chemical term that is best defined as *adhesion*: when a substance is adsorbed it does not undergo any significant chemical change. Adsorption is the principle used in diesel NO_x adsorber catalysts (NACs). Base metal oxides initially "store" NO_x compounds, after which a rhodium reduction catalyst in combination with *dosing* reduces them back to N_2 and O_2.

advanced combustion and emissions reduction technology (ACERT) Caterpillar combustion management approach to emissions reduction introduced in 2004. Currently used on all Caterpillar on-highway and most off-highway engines. See *ACERT*.

advanced diesel engine management (ADEM) Caterpillar acronym used to describe its management electronics and associated hardware.

afterburn term that can be used to describe the normal combustion of fuel in a diesel engine cylinder after injector nozzle closure or random ignitions of fuel pockets after primary flame quench in an engine cylinder.

afterburn injection see *dosing injection*.

after top dead center (ATDC) any engine position during piston downstroke.

air-conditioning (A/C) the cooling circuit in an HVAC climate control system.

air/fuel ratio (AFR) the mass ratio of an air-to-fuel mixture.

air-to-air aftercooling (ATAAC) heat exchanger located downstream from the turbocharger compressor housing; cools air prior to routing it to the intake manifold.

alcohol any of a group of distillate hydrocarbon liquids containing at least one hydroxyl group; sometimes referred to as *oxygenates*.

algorithm software term that describes a programmed sequence of operating events.

alloy the mixing of a molten base metal with metallic or nonmetallic elements to alter the metallurgical characteristics.

all-speed governor another term for *variable-speed governor*.

alpha data represented by letters of the alphabet.

alternating current (AC) current flow that cyclically alternates in direction, usually produced by rotating a coil within a magnetic field.

altitude compensator device used on older diesel fuel injection systems containing a barometric capsule; used to measure atmospheric pressure and derate fueling at altitude to prevent smoking.

altitude deration the engine fuel delivery cutback that is managed to occur on the basis of increase in altitude, to prevent engine overfueling as the air charge becomes less oxygen dense. Power deration is typically 4 percent per 1,000 feet (305 meters) of altitude in boosted engines.

American Petroleum Institute (API) classifies lubricants and sets standards in the petroleum-refining industry.

American Society for Testing Materials (ASTM) agency that sets industry standards and regulations, including those for fuel and lube oils. Standards are set by accord and influence both recommended practice and legislated standards.

American Standard Code for Information Interchange (ASCII) widely used data-coding system found on PCs.

American Trucking Association (ATA) organization with a broad spectrum of representation responsible for setting standards in the U.S. commercial trucking industry.

ampere unit of electrical current flow equivalent to 6.28 × 1,018 electrons passing a given point in a circuit per second.

ampere turns (At) the basic unit of measurement of magnetomotive force.

amplification term used in electronic circuits to describe what happens when very small currents are used to switch much larger ones using transistors.

amplifier piston hydraulically actuated piston that pumps fuel to injection pressure values in a Cat/International HEUI; also known as intensifier piston.

anaerobic sealant paste-like sealant that cures (hardens) without exposure to air.

analog the use of physical variables, such as voltage or length, to represent values.

analog signal a communication line signal consisting of a continuous electrical wave.

AND gate an electronic switch in which all the inputs must be in the ON state before the output is in the ON state.

aneroid a device used to sense light pressure conditions. The term is used to describe manifold boost sensors that limit fueling until there is sufficient boost air to combust it. The device usually consists of a diaphragm, spring, and fuel-limiting mechanism.

annular ring shaped.

annuli plural of *annulus*.

annulus a ring.

anode positive electrode; the electrode toward which electrons flow.

anodizing oxide coating on the surface of a metal formed by an electrolytic process; results in hardening the surface area. Many aluminum alloy pistons are *anodized* to improve service life.

antifreeze a liquid solution added to water to blend the engine coolant solution that raises the boil point and lowers the freeze point. Ethylene glycol (EG), propylene glycol (PG), and extended-life coolants (ELCs) are currently used.

antinodes portion of a wave below the zero, mean, or neutral point in a waveband.

antithrust face used to describe the minor thrust face of a piston; the outboard side of the piston as its throw rotates off the crankshaft centerline through the power stroke.

antithrust side piston term meaning minor thrust side.

API gravity measure of how the weight of a petroleum liquid compares to the weight of water; measured with a hydrometer. If API gravity is greater than 10, the petroleum-based liquid is lighter than, and therefore will float on water. If the API gravity specification is less than 10, it is heavier than water and will sink. Used to compare the relative densities of petroleum liquids and is important if you are doing performance testing on engine dynamometers.

app(s) widely used short form for (computer and phone) applications.

application software programs that direct computer processing operations.

aqueous urea the reducing agent used in SCR systems. Aqueous urea is a solution of 32 percent urea and distilled water. Known as DEF in North America and AdBlue in Europe.

arcing bearing or gear failure caused by electric arcing.

articulating piston a two-piece piston with separate crown and skirt assemblies, linked by the piston wrist pin and afforded a degree of independent movement. The wrist pin is usually full floating or bolted directly to the con rod, in which case it is known as a crosshead piston.

ash (1) the powdery/particulate residues of a combustion reaction. (2) Solid residues found in crude oils. Present in trace quantities in engine lubricating oils and diesel fuels.

assembly line data link (ALDL) the current OBD II data link. A 16-pin connector also known as an *SAE J1962 connector*.

Association of Diesel Specialists (ADS) organization to which fuel injection specialty shops belong and which monitors industry standards of practice and education.

ASTM 1D fuel fuel recommended for use in high-speed, on-highway diesel engines required to operate under variable load and variable speeds. Minimum CN must be above 40. In theory, the ideal fuel for mobile diesel engines but in practice, it is not used as often as number 2D fuel because it has less heat energy by weight, making it less economical.

ASTM 2D fuel fuel recommended for use in high-speed, on-highway diesel engines required to operate under constant loads and speeds. Like number 1D fuel, the minimum CN is required to be above 40. Widely used in highway diesel engines because it produces better fuel economy than number 1D fuel due to its higher calorific value, albeit at the expense of slightly inferior performance.

ASTM 975 standard to which all petroleum-based diesel fuels must conform.

ASTM 6751 standard to which pure biodiesel-based diesel fuels must conform.

asynchronous transfer mode (ATM) a method of digital transmission and switching that can handle vast amounts of data at high speed.

ATA connector used on heavy commercial trucks; describes a J1708 data bus connector (six-pin Deutsch) but often used to reference either the J1708 or J1939 data link.

ATA data link an SAE/ATA standard J1587/J1708, six-pin Deutsch connector required to connect to older commercial truck data buses.

atm a unit of atmospheric pressure equivalent to 14.7 psi (101.3 kPa). Used as a unit of measurement in the United States and United Kingdom, especially on fuel calibration instruments. Close, but not exactly equivalent, to a European unit of *bar*.

atomization the process of breaking liquid fuel into small droplets by pumping it at a high pressure through a minute flow area.

atomized droplets the liquid droplets emitted from an injector nozzle.

A-trim see *adaptive trim*.

audit trail a means of electronically tracking electronically monitored problems in an engine management system. May be discreet, that is, not read by some diagnostic ESTs and programs; also known as *tattletale*.

AUTOEXEC.BAT a batch file loaded into the DOS kernel that governs boot-up protocol.

automotive governor a term sometimes used to describe a *limiting speed*, mechanical governor.

Automotive Service Excellence National Institute (ASE) organization dedicated to setting certification test standards for auto and truck technicians.

axis the point about which a body rotates; the center point of a circle. Plural: *axes*.

B5 standard petroleum-based diesel fuel cut with 5 percent biodiesel.

B20 standard petroleum-based diesel fuel cut with 20 percent biodiesel.

B100 term used to describe pure, uncut biodiesel fuel meeting the ASTM standard D6751.

backbone data bus consisting of a twisted wire pair.

background computations computer operating responses of lower priority than foreground operations, that while important do not require immediate response; monitoring of engine fluid temperatures would be classified as a background computation.

back leakage test an injector bench fixture test in which nozzle valve-to-nozzle body leakage is measured.

balanced atom an atom in which the numbers of electrons and protons are equal.

bandwidth volume or capacity of a data transmission medium; the number of packets that can be pumped down a channel or multiplex backbone.

bar a metric unit of pressure 105 Newtons per square meter (equal to 14.3 psi); approximately, but not exactly, one unit of atmosphere (14.7 psi) or 1 atm.

barometric capsule a barometer device used on some hydromechanical injection pumps to limit high-altitude fueling.

barometric pressure sensor (BARO) an electronic barometric pressure–sensing device.

base circle the smallest radial dimension of an eccentric. Used to describe cam geometry; the train that the cam is responsible for actuating would be unloaded on the cam base circle. Also known as *inner base circle* or IBC.

basic input/output system (BIOS) when a computer is booted, the CPU looks to the BIOS chip for instructions on how to interface between the disk-operating system and the system hardware.

baud times per second that a data communications signal changes and permits one bit of data to be transmitted.

baud rate the speed of a data transmission.

bay a vacant location in the computer housing/system designed to accommodate system upgrades.

bearing shell a half segment of a friction bearing such as that which would be used as a crankshaft main bearing.

bedplate replaces a set of main bearing caps with a single casting plate that bolts to the engine cylinder block; enables a lighter cylinder block to withstand high resistance to torsionals.

before top dead center (BTDC) a piston location in the cycle before full piston travel, usually abbreviated BTDC.

beginning of energizing (BOE) moment that an EUI or EUP is electrically energized.

beginning of injection (BOI) in engine management, a specific point at which injection begins.

bell crank a single arm lever with its fulcrum at the apex of a shaft, often used as a mechanical relay. The word originates from medieval church bell ringing mechanisms.

benzene hydrocarbon fuel fraction obtainable from coal or petroleum; known to be a carcinogen.

big end the crankshaft throw end of a connecting rod.

binary system a two-digit arithmetic, numeric system commonly used in computer electronics.

biocide bacteria-killing agent that can be added to stored diesel fuel or biodiesel that is under attack by bacteria colonies.

biodiesel fuel derived from farm products with a vegetable and alcohol base; when used in current diesel engines should meet ASTM standard D6751.

bipolar transistor a three-terminal transistor that functions as a sort of switched diode.

bit a binary digit that can represent one of two values, on or off; presence of voltage or no voltage; the smallest piece of data a computer can manipulate. There are eight bits to a byte.

bits per second (BPS) a measure of the speed at which data can be transferred.

black smoke smoke that appears black to the observer is caused by particulate (solids) emission in the exhaust gas stream; light is blocked by the particulate, making it appear black.

blended torque transmission the CVT transmission used in HEV buses which allows drive torque input from either the diesel engine powerplant, the electric motor/generator, or both simultaneously, providing infinitely variable output torque.

B-life manufacturer-generated ratings that project the percentages of engines in a series that will fail at various mileage intervals. A B10-life correlated to a mileage value means that 10 percent of that engine family have failed at that mileage.

blink codes fault codes blinked out using diagnostic lights; also known as flash codes.

blotter test an inaccurate and generally obsolete method of testing used engine oil for viscosity and contamination.

blue smoke usually associated with engine oil combusted in the engine cylinder; caused by the mixture of condensing droplets and particulate emitted when oil is burned in an engine.

Bluetooth a wireless network suited to applications where lower volumes of data have to be transferred; used primarily for phones, headsets, and PDAs.

body control module (BCM) the gateway used to integrate external and internal (intrabus and bus-to-bus networking). The BCM is used differently by manufacturers and is usually programmed with a range of other processing responsibilities such as bus wake-up and vehicle security.

boil point the temperature at which a liquid vaporizes.

bomb calorimeter test a test used to calculate the heating value of a fuel; a known quantity of the substance is combusted and the heat that is released is calculated.

boosted engine any turbocharged engine; turbo-boosted.

boot the process of loading an operating system into RAM or main memory.

boot up to load an operating system into RAM, electronically reload a system program, or reset a computer.

bore an aperture. The internal diameter of a pump or engine cylinder or the act of machining a cylindrical aperture.

Bosch Mastertech a generic scantool whose software is well supported by manufacturers so that it can read and undertake some diagnostics on a range of vehicles.

Bosch VCI-CA the Bosch vehicle communications, communications adapter (serial link) required to connect a Windows environment PC with a vehicle to access the data bus.

bottom dead center lowest point of travel of piston in an engine cylinder during its cycle. Usually abbreviated BDC.

boundary lubrication thin-film lubrication characteristics of an oil.

Boyle's law states that for a given, confined quantity of gas, pressure is inversely related to the volume, so as one value goes up, the other goes down. In compressing gas in an engine cylinder during piston upstroke, cylinder volume is reduced, so cylinder pressure accordingly is increased.

brake horsepower standard expression for brake power commonly used in the vehicle industry. See *brake power*.

brake power power developed by an engine measured at the flywheel by a dynamometer or brake. Factored by torque and rpm.

brake specific fuel consumption (bsfc) a measure of the fuel required to perform a unit of work; used in graphs of engine data designed to show fuel efficiency at specific engine loads and rpm.

breakout box a diagnostic device fitted with coded sockets that accesses an electrical or electronic circuit by teeing into it; used in conjunction with a DMM.

breakout T term used to describe a breakout box or in some cases a diagnostic device that tees into two- or three-wire circuits to enable diagnoses by DMM of a single component such as a sensor.

bridge the software and/or hardware used to make electronic connections, such as that used to connect nodes in a network.

British thermal unit (Btu) the amount of heat required to raise the temperature of 1 pound of water 1°F at 60°F. The standard unit of heat energy measurement.

brittle describes the property of a material that is unable to sustain any plastic deformation without fracturing; you could describe a china coffee cup as being brittle.

broach a boring bit used for final, accurate bore sizing.

brushes stationary devices used to conduct current to and from a rotor in electrical components such as alternators and motors.

bubble collapse the condition caused by wet liner combustion pressure impulses acting on the coolant, resulting in vapor bubbles that collapse and cause cavitation.

buffers memory locations used to store processed data before it is sent to output devices.

bundle multiple arrangement of cooling tubes that form the core of a heat exchanger.

buret see *vial*.

bus (1) a transit vehicle. (2) An electronic connection; transit lines that connect the CPU, memory, and input/output devices; increasingly used to mean "connected."

bushing any of a number of types of friction bearings designed to support shafts.

bus systems term used to describe data highways.

buttress an additional/auxiliary support device such as a gusset.

buttress screws transverse bolts used in addition to vertical plain, main bearing cap screws with the objective of limiting cylinder block torque twist.

butt splice the joining of two pieces in a series connection.

bypass filter a filter assembly plumbed in parallel with the lubrication circuit, usually capable of high-filtering efficiencies.

bypass valve a diverter valve fitted to full-flow filter (series) mounting pads, designed to reroute lubricant around a plugged filter element to prevent a major engine failure.

byte unit of measure of computer data, comprised of eight bits; used to quantify computer data memory.

cab-over-engine (COE) truck or van chassis in which the engine compartment is located directly underneath the driver cab, eliminating the hood. Usually abbreviated to *COE*.

cache high-speed RAM located between the CPU and main memory; used to increase processing efficiency.

cage a computer system housing location accommodating two or more bays.

calibration adjusting performance specifications to a standard; fuel trimming of diesel fuel injection components is known as *calibration,* or the process of correlating a set of readings with a standard.

calibration parameters the specific values required when setting performance to specification.

California Air Resources Board (CARB) the state of California agency responsible for driving emissions legislation and enforcement. By engineering and establishing standards that exceed federal standards and effecting them earlier, CARB has led the emission control initiative throughout North America. CARB standards have now been adopted by over 20 states, with more planning to follow.

calipers comparative measuring instrument used for measuring od (outside diameter) or id (inside diameter).

calorific value the heating value of a fuel measured in Btu, calories, or joules.

cam an eccentric. An eccentric portion of a shaft, often used to convert rotary motion into reciprocating motion.

cambox the lower portion of a port-helix metering injection pump in which the actuating camshaft is mounted and the lubricating oil sump is located.

cam geometry the shaping of a cam profile and the effect it produces on the train it actuates.

cam ground trunk-type pistons that are machined slightly eccentrically. Because of the greater mass of material required at the wrist pin boss, this area will expand proportionally more when heated, allowing the piston to assume a true circular shape when heated to operating temperatures.

cam heel the point on a cam profile that is exactly opposite the toe or center point of the highest point on the cam.

cam nose the portion of the cam profile with the largest radial dimension; its center point would be the cam toe. That portion of the cam profile that is OBC.

cam plate the input shaft–driven, rotating-reciprocating member used to actuate the distributor plunger in a Bosch-type, sleeve-metering, rotary distributor injection pump such as the VE.

cam profile the cam geometry; simply, the shape of the cam.

camshaft a crankshaft-driven shaft, machined with eccentrics (cams) designed to actuate trains positioned to ride the cam profiles; the engine feedback assembly actuator responsible for timing/actuating cylinder valves and fuel injection apparatus. Driven at half engine speed on four-stroke cycle engines and at engine speed on two-stroke cycles.

camshaft position sensor (CPS) any of a number of types of engine position sensors using either an inductive pulse generator or Hall effect electrical principle.

CAN 2.0 the current high-speed data bus used as a platform for CAN-C (auto/light duty) and J1939 (heavy duty) powertrain multiplexing. See *controller area network.*

CAN-A a single wire (early generation) data line with data transfer speeds not exceeding 33.3 kb/ps. Nonhigh performance such as GM-LAN. Slow by today's standards.

CAN-B a single or two-wire bus line with data transfer speeds not exceeding 125 kb/ps. Rapidly becoming obsolete.

CAN-C the current two-wire, high-speed data bus with data transfer speeds not exceeding Mb/ps. The automotive equivalent of J1939 based on CAN 2.0 architecture. Current (2011) average maximum transaction speeds are 500 kb/ps.

canister a cylindrical container.

capacitance measure of how much electrical charge can be stored for a given voltage potential; measured in farads.

capacitor an electrical device that can store an electrical charge or block AC and pass DC. Also known as a *condenser.*

carbon (C) an element found in various forms, including diamonds, charcoal, and coal. It is the primary constituent element in hydrocarbon fuels. Atomic number 6.

carbon dioxide (CO_2) the product of combusting carbon in the oxidation reaction of a HC fuel. An odorless, tasteless gas that is nontoxic and not classified as a noxious engine emission, but that contributes to greenhouse gases that concern environmentalists.

carbon monoxide (CO) a colorless, odorless, and poisonous gas that is produced when carbon is not completely oxidized in combustion.

carcinogen a cancer-causing agent.

Carnot cycle relates to ratio of work output to heat input that should equal the difference between the temperatures of the heat source and rejected heat combined, divided by the temperature of the heat source.

cartridge a removable container; used to describe the housing that encloses a filter.

cartridge tape data storage medium of the sequential type, currently used for high data volume PC data backup.

catalyst a substance that stimulates, accelerates, or enables a chemical reaction without itself undergoing any change.

catalytic converter an exhaust system device that enables oxidation and reduction reactions.

Caterpillar Engine Company a major diesel engine manufacturer. Corporate center is in Peoria, Illinois.

Caterpillar information display (CAT-ID) the Caterpillar digital dash display that provides the driver with ECM/PCM feedback data such as fuel economy and engine parameters.

cathode negative electrode; the electrode from which electrons flow.

cathode ray tube (CRT) device used as a computer monitor.

cavitation describes metal erosion caused by the formation and subsequent collapse of vapor pockets (bubbles) produced by physical pulsing into a liquid such as that of a wet liner against the wall of coolant that surrounds it. Bubble collapse causes high unit pressures and can quickly erode wet liners when the protective properties of the coolant diminish.

CD-ROM an optically encoded data disk that is read by a laser in the same way an audio CD is read and is designed for read-only data.

central processing unit (CPU) computer subcomponent that executes program instructions and performs arithmetic and logic computations.

centrifugal filter a filter that uses a centrifuge consisting of a rotating cylinder charged with pressurized fluid and canted jets to drive it; centrifugal filters often have high efficiencies and are often of the bypass type.

centrifugal force the force acting outward on a rotating body.

centrifuge a device that uses centrifugal propulsion or a centrifugal force principle of operation.

cetane improvers see *ignition accelerators*.

cetane number (CN) the standard rating of a diesel fuel's ignition quality. It is a comparative rating method that measures the ignition quality of a diesel fuel versus that of a mixture of cetane (good ignition characteristics) and heptamethylnonane (poor ignition characteristics). A mixture of 45 percent cetane and 55 percent heptamethylnonane would have a CN of 45. Diesel fuels refined for use on North American highways are classified by the ASTM as 1D and 2D and must have a minimum CN of 40. Typical is a CN of 45 during summer months, slightly higher during the winter.

chain hoist a mechanical or power-operated ratcheting lifting device that consists of an actuating block, lift chains, and hook.

characters per second (CPS) speed rating of a printer device.

charge air cooler (CAC) the heat exchanger that effects charge air cooling; in mobile diesel engines, this is usually ram air assisted.

charge air cooling the cooling of turbo-boost air by means of ram air or coolant medium heat exchangers.

charge differential electrical pressure usually described as potential difference and measured in voltage.

charging circuit the portion of the fuel subsystem that begins with the charging or transfer pump and is responsible for delivering fuel to the injection pumping/metering apparatus. In a port-helix metering pump, this extends through the charging gallery of the injection pump.

charging pressure a term used to describe the pressure on the charge side of the transfer pump in a fuel subsystem. Charging pressure parameters are defined by the cycle speed of the charging pump, the flow area it unloads to, and the regulating valve.

charging pump the pump responsible for moving fuel through the fuel subsystem. Plunger, gear, and less commonly, vane-type pumps are used.

charging system consists of the battery, alternator, voltage regulator, associated wiring, and the electrical loads of the vehicle. The charging circuit provides the current loads required by the vehicle and recharges the batteries when necessary.

Charles law states that the volume occupied by a fixed quantity of gas is directly proportional to its temperature if the pressure remains constant.

chassis dynamometer a test bed that measures brake power delivered to the vehicle wheels by having them drive roller(s) to which torque resistance is applied and accurately measured.

chassis-mounted charge air cooling (CMAC) method of cooling turbo-boost air using a ram air heat exchanger; effective in highway applications, less so in off-road service.

chatter a nozzle bench test characteristic in which a nozzle valve rapidly opens and closes; caused by the slow rate of pressure rise when testing nozzle valves.

check engine light (CEL) a dash warning light that is often used as a first-level alert to the driver.

chemical bonding the force holding atoms together in a molecule or a crystal.

cherry picker a single boom, portable hydraulic hoist useful for such tasks as removing engines from vehicles.

chip a complete electronic circuit that has been photo-infused to a semiconductor material such as silicon; also known as *I/C* (integrated circuit), *microchip*.

chopper wheel the rotating disc that cuts a magnetic field to produce rotational speed or rotational position data to an ECM either by producing an AC voltage value or by pulse-width modulation.

Class II bus single-wire bus line that became commonly used after 1997 with the advent of OBD II, when it became the standard scan tool communication format. Class II has a default zero-volt status, and modules with an address on class II pull the voltage to 7V DC to create the high logic bit change. Max speeds of 10.2 kb/s.

clean gas induction (CGI) a Caterpillar exhaust gas recirculation strategy that sources dead gas downstream from the exhaust aftertreatment circuit, hence the term "clean."

clearance volume the volume in an engine cylinder when the piston is at top dead center.

clevis a yoke that is often used in conjunction with a clevis pin and a lever to convert rotary motion to linear or vice versa.

client anything in a computer processing cycle or multiplex data transaction that can be described as having a need.

clipboard temporary storage location for data during cut, paste, and program transfer operations.

clock speed the measure of how fast a CPU can process data measured in MHz (megahertz) or millions of cycles per second.

clockwise (CW) right-hand rotation.

closed circuit an electrical circuit through which current is flowing.

closed-circuit voltage (CCV) voltage measured in an energized circuit.

closed crankcase ventilation (CCV) an EPA requirement for diesel engine crankcases beginning in 2007 (off-highway 2008). Diesel engine manufacturers have adopted positive or centrifugal-type filtration of crankcase vapors prior to rerouting them to the intake upstream from the turbocharger impeller. Equivalent to the automotive *positive crankcase ventilation (CCV)*.

cloud point the temperature at which wax crystals present in all diesel fuels become large enough to make the fuel appear hazy. It is also the point at which plugging of fuel filters becomes a possibility. The cloud point is usually 5°F (3°C) above the fuel's pour point.

cluster the smallest data storage unit on a diskette.

CMP sensor camshaft position sensor.

coalesce to combine to form a single whole.

coaxial cable type of wiring used to transmit signals with almost unlimited bandwidth but unable to carry two-way signals.

coder/decoder (Codec) device that converts analog voice signals to digital signals and vice versa.

coefficient of friction a means of rating the aggressiveness of friction materials; alters with temperature and the presence of any kind of lubricant.

coefficient of thermal expansion the manner in which a material behaves as it is heated and cooled. For instance, aluminum has a higher coefficient of thermal expansion than steel, meaning that when a similar mass of each material is subjected to an identical amount of heat, the aluminum will expand more.

coils electromagnetic devices used as the basis of solenoids, transformers, and motors, and in electronics, to shape voltage waves.

coincidental damage in failure analysis, the term is used to refer to secondary component or system failures created by the *root cause* failure.

cold-start strategy a programmed startup sequence in an electronic management system in which the timing, fuel quantity, and engine-operating parameters are managed on the basis of ambient and engine fluid temperatures. During this process, other inputs such as throttle position may be ignored by the ECM.

collateral damage in military parlance, destruction beyond the intended target. In failure analysis, the term is used to describe *coincidental damage*.

combustion the act of burning a substance. An oxidation reaction.

combustion pressure usually refers to peak cylinder pressure during the power stroke.

come-along a ratcheting chain or cable device designed to exert linear (lift or pull) on an object.

command circuit used to describe input sensors such as the TPS (throttle position sensor) that commands (requests) an output from the ECM.

command sensors input sensors that signal a command status to the PCM or ECU: a throttle position sensor would be a good example.

common platform a term created to describe what happens when two major manufacturers merge and proceed to develop common technology but for purposes of brand identification, badge and market the technology under different names.

common rail (CR) system fuel injection system in which injection pressures are created by a pump that then supplies fuel to an accumulator or common rail connected to fuel injectors. The fuel injectors are then electrically or electrohydraulically actuated by the ECM/PCM.

communications adapter (CA) the serial communications adapter required to handshake PC software with the chassis data bus. Most have some processing and device-security capability. Generic CAs permit the required read-only connectivity to scrutinize the required OBD fields. The current automotive compliance standard for CAs is J2534.

commutator conductors connected to (starter motor) armature windings.

compact disc (CD) optically encoded digital data storage.

compacted graphite iron (CGI) a new-generation cylinder block casting composite used because of its high strength, resistance to torque twist, and light weight. One example of CGI is known as GJV-450.

companion cylinders term used to describe pistons paired by their respective crank throws to rotate together through the engine cycle, such as 1 and 4 in an inline, four-cylinder engine.

comparative measuring instruments that gauge a dimension but require another instrument to produce an actual value. Dividers would require a tape measure to convert the dimension measured to a value.

comparitor anything used to compare one value to another.

comparitor bench a fuel injection pump test fixture used to compare the performance and output values of an injection pump with a set of master specifications. Usually consists of a means of driving the injection pump (as if it were being driven by the engine it is designed to fuel), a drive turret equipped with a protractor (for phasing), and graduated vials (means of measuring fuel quantity injected). The term can also be used to describe the test fixtures used to set up mechanical and electronic unit injectors.

composite steel trunk piston a trunk-type piston in which the crown and skirt sections are manufactured separately and then screwed together using a proprietary process by Mahle; see *Monocomp piston*.

compound (1) a substance consisting of two or more elements held together by chemical force and not necessarily retaining any characteristics of the composite elements. (2) The process of increasing the force acting on a plunger or piston by using both mechanical and fluid forces.

compound turbocharging a turbocharger circuit in which the turbine housing, instead of driving an impeller, is connected to a reduction gearing coupling with an output shaft indirectly connected to the engine crankshaft. This allows the turbocharger to directly transmit torque to the engine drivetrain. An example is the 2010 DDC *axial power turbine* (APT).

compressed air the means of powering many tools and commercial truck chassis equipment in many applications. It is usually pressurized to between 90 and 150 psi, plumbed throughout the shop, and accessed by quick couplers.

compressed natural gas (CNG) pressurized natural gas used for commercial and automotive vehicles; consists largely of methane.

compression ignition (CI) an engine in which the fuel/air mixture is ignited by the heat of compression.

compression pressure cylinder pressure generated by the piston as it travels upward on the second stroke of the four-stroke cycle. In diesel engines, this pressure is usually four to six times that in a current gasoline-fueled engine.

compression ratio the ratio of piston swept volume to total cylinder volume with the piston at bottom dead center; a volumetric ratio, not a pressure ratio.

compression ring the ring(s) designed to seal cylinder gas pressure located in the upper ring belt.

compressional load a force that attempts to compress or squeeze from diametrically opposite directions to a common point in the component under load.

compressor housing the section of a turbocharger responsible for compressing the intake air and feeding it into the intake circuit; also known as *impeller housing*.

computer-assisted design (CAD) the commonly used industrial component design method.

computer-assisted machining (CAM) programmable computer-managed machining.

concentric circles having a common center.

concept gear found in some diesel engine timing gear-trains, a concept gear is a two-piece assembly that uses coaxial springs between the hub and outer toothed ring to maintain zero-lash tooth contact with the gears it is in contact with.

condensation the changing of a vapor to a liquid by cooling.

condenser see *capacitor*.

conductance the ability of a material to carry an electrical current.

conduction heat transmission through solid matter.

conductors materials that readily permit the flow of electrons from atom to atom; usually metallic elements that have fewer than four electrons in their outer shells.

connecting rod the rigid mechanical link between the piston wrist pin and the crankshaft throw.

consolidated engine controller an ECM or PCM that houses the microcomputer and output switching for such things as injector drivers.

constant geometry (CG) describes a turbocharger in which all the exhaust gas is routed through the turbine housing, which has no internal or external controls.

constant horsepower sometimes used to describe a high-torque-rise engine.

constant volume sampling (CVS) an exhaust gas analysis and measurement procedure used before certification.

contact stress fatigue occurs when two surfaces slide or roll against each other, developing high stress, surface movement, and fatigue cracks in one or both surfaces.

Contamination Control a Caterpillar program started in 1999 aimed at minimizing mechanical damage cause by wear; extends to lube quality and handling plus shop floor maintenance practices.

continuity an unbroken circuit; used to describe a continuous electrical circuit. A continuity test would determine if a circuit or circuit component was capable of current flow.

continuously open throttle valve (CTV) brake Mercedes-Benz variation on the internal engine compression brake using small valves (the CTVs) fitted into the engine cylinder head that allow some cylinder leakage to the exhaust during both the compression and exhaust strokes under braking.

continuously variable transmission (CVT) describes any transmission capable of infinite output ratios. Can be achieved by the use of drive belts/chains and tapered input/output rotors or by using planetary gearsets.

controller area network (CAN) a data bus system developed by Robert Bosch and Intel for vehicle applications. CAN is a serial data transmission network used as the basis for the current CAN-A, CAN-B, CAN-C (automotive), and SAE J1939 (truck) data backbones.

control rack the fuel control mechanism on an MUI or multicylinder port-helix metering pump that when moved linearly, rotates the pumping plunger(s) in unison.

control sleeve the component that is tooth meshed to the control rack and connects to the plungers by means of slots; used to rotate the plungers in the barrels.

control strategy the manner in which a PCM/ECM has been programmed to manage the engine, especially in the event of an electronically monitored problem.

control unit the part of a computer CPU responsible for fetching, decoding, executing, and storing.

convection heat transfer by currents of gas or liquids.

conventional memory the first data logged into RAM on boot up, used primarily to retain the operating system.

conventional theory (of current flow) asserts that current flows from a positive source to a negative source. Despite the fact that it is fundamentally incorrect, it is nevertheless widely accepted and used. See *electron theory*.

cooled exhaust gas recirculation (C-EGR) introduced in 2002 to diesel emission standards. Dead-end gas is first cooled, then rerouted into the intake system to dilute the intake charge and lower temperatures, reducing NO_x emission.

coprocessor a chip or CPU enhancement designed for specific tasks such as mathematical calculation.

cordierite ceramic material used for honeycomb flow-through substrates on diesel particulate filters and catalytic converters; not in itself a catalyst.

corrosion chemical change and deterioration of the metal surfaces; caused by acidic, alkaline, or electrolytic conditions.

corrosive alkaline or acidic substances that dissolve metals and skin tissue.

coulomb one coulomb is equal to $6.28 \times 1,018$ electrons.

counterclockwise (CCW) left-hand rotation.

counterflow radiator a double-pass radiator in which coolant is cycled through U-column tubes, usually from a bottom-located intake tank to a bottom-located output tank. This type has higher cooling efficiencies than other radiator designs.

covalent bonding the atomic condition that occurs when electrons are shared by two atoms.

covert term that means undercover, but commonly used in vehicle electronics to describe the logging of data that cannot be read using the commonly available diagnostic software. Events such as engine overspeed conditions that could affect the system warranty are often written or backed up covertly to an electronic system.

cracked rod connecting rod manufactured and machined in one piece, after which the big end is separated by a precisely defined fracture. This ensures a cap-to-rod fit of the highest precision.

crank angle a location in an engine cycle noted by rotational degrees through the cycle.

crank axis center point about which a crankshaft rotates.

crankcase the lower portion of the engine cylinder block in which the crankshaft is mounted and under which is the lubrication oil sump.

crankshaft a shaft with offset throws designed to convert the reciprocating movement of pistons into torque.

crank throw the offset journal on a crankshaft to which a connecting rod is connected.

creep describes the independent movement of two components clamped by fasteners when they have different coefficients of thermal expansion or have different mass, which means their expansion and contraction rates do not concur. A function of a gasket is to accommodate component creep while maintaining an effective seal.

crimping pliers jawed tools designed to crimp a terminal to a wire without crushing or damaging the terminal.

critical flow venturi (CFV) one type of MAF sensor sometimes known as a pressure differential flow sensor. Its operating principle is based on the relationship between inlet pressure and flow rate through a venturi, so it uses an inlet and outlet pressure sensor either side of a venturi; this enables the PCM/ECM to calculate flow mass.

critical pollutants the U.S. EPA's list of six important airborne pollutants: ozone, lead, carbon monoxide, sulfur dioxide, nitrogen dioxide, and respirable particulate matter.

cross-flow radiator a usually low-profile design of radiator (used with aerodynamic hood/nose), in which the entry and output tanks are located at either end and coolant flow is horizontal.

cross-flow valve configuration a cylinder head valve configuration in which the intake and exhaust valves are located in series in the cylinder head, meaning that gas flow from the inboard valve differs from (and may interfere with) that of the outboard valve.

crosshead piston an articulating piston with separate crown and skirt assemblies in which the connecting rod is bolted directly to the wrist pin.

crossover pipe a pipe that connects a pair of fuel tanks mounted on either side of a vehicle frame at the sump level, enabling fuel to be drawn from one tank while maintaining an equal fuel load in each tank.

crown the leading edge face of a piston, or in articulating pistons, the upper section of the piston assembly. Crown geometry (shape) plays a large role in defining the cylinder gas dynamic.

crude oil the organic fossil fuel pumped from the ground from which diesel fuel, gasoline, and many other petroleum products are refined; raw petroleum.

Cummins Engine Company a major manufacturer of diesel engines. Corporate center is Columbus, Indiana.

Cummins ISB Interact System B series, inline four- and six-cylinder engines that have progressed through several generations. Current displacement for six-cylinder engine used by Chrysler Dodge is 6.7 liters.

current the flow of electrons in a closed electrical circuit.

current transformer a DMM accessory that permits high electrical current flow values to be transduced and read.

cursor the underline character or arrow that indicates the working location on a computer screen display.

customer data programming any class of PCM or ECU programming over which the owner of the vehicle theoretically has ownership: an example is tire rolling radii which can be reprogrammed using an EST.

cybernetics the science of automated (computer) control of machines, systems, and nature. Diesel engine controllers or PCMs/ECMs are cybernetic devices.

cycle (1) a sequence of events that recurs, such as those of the *diesel cycle*. (2) One complete reversal of an alternating current from positive to negative.

cylinder balance test an electronically managed diagnostic test in which engine cylinders are sequentially cutout to measure the percentage of work performed by each. Also known by terms such as *cylinder contribution test* and *electronic cylinder balance test*.

cylinder block the main frame of any engine to which all the other components are attached.

cylinder contribution test a diagnostic test in which engine cylinders are sequentially cutout to measure cylinder balance—or percentage of work performed by each engine cylinder. Also known by terms such as *cylinder balance test* and *electronic cylinder balance test*.

cylinder gas dynamic engine cylinder gas movement during the cycle; high turbulence was an objective in many older diesel engines, while lower turbulence or quiescent dynamics are used in many newer diesels with high injection pressures.

cylinder head the components clamped to a cylinder block containing the engine breathing and fueling control mechanisms.

cylinder leakage tester device used to test cylinder leakage by applying regulated air to the cylinder at a controlled volume and pressure and producing a percentage of leakage specification.

cylinder volume total volume in an engine cylinder with the piston at BDC; the sum of swept volume and clearance volume.

Darlington pair two transistors arranged to form an amplifier that permits a very small current to switch a large one; used in ECMs as drivers.

dash display unit (DDU) any of a number of different visual digital display units used in current vehicles: can display data bus information, ECU status, running conditions, GPS data, etc.

data raw (unprocessed) information.

database a data storage location or program.

data bus multiplex backbone consisting of a twisted wire pair; the neural backbone of a network system.

data communication link (DLC) the means used to access a chassis data bus. The current OBD II data link is an *assembly line data link (ALDL)*, a 16-pin connector also known as an *SAE J1962 connector*.

data compression a means of reducing the physical storage space for data by coding it.

data frame a data tag consisting of 100 to 150 bits for transmission to the bus. Each tag codes a message for sequencing transmission to the data bus and also serves to limit the time it consumes on the bus.

data hub the hub of a network system. Used by most chassis/engine manufacturers to log data such as warranty status, repair history, and proprietary programming of on-board ECMs.

data link the connection point or path for data transmission in networked devices.

data logging the tracking of computer data for later analysis.

data processing the production and manipulation of data by a computer.

dead volume fuel fuel that is statically retained for a portion of the cycle; usually refers to the fuel retained at residual line pressure in a high-pressure injection pipe that connects injection pump elements with hydraulic injectors in PLN systems.

decoding a CPU control unit operation that translates program instructions.

default preselected option in computer processing outcomes that kicks in when a failure occurs outside the programmed algorithm. Failure strategy that permits limited functionality when a critical input is lost. Revert to basics. Limp-home mode.

delivery valve a combination check and pressure management valve that is used on many hydromechanical diesel fuel injection systems.

delta an alternator diode bridge arrangement in which three diodes are connected in a triangular arrangement for purposes of rectifying AC to DC.

desktop a computer term that either describes a non-portable, desk-based PC system or the screen display at any given moment of PC operation.

detonation combustion in an engine cylinder occurring at an explosive rate, accelerated by more than one flame front; caused by a number of different conditions, but in diesel engines often by prolonged ignition lag when ambient temperatures are low, when it is known as diesel knock.

Detroit Diesel Corporation (DDC) a major diesel engine manufacturer, part of the DaimlerChrysler/Freightliner Corporation. Corporate center is Dearborn, Michigan.

Detroit Diesel electronic controls (DDEC) DDEC I was introduced in 1985 and marketed in 1987. It was the first full-authority engine management system available on a North American engine. DDEC has evolved through a number of versions and generations of engines.

Deutsch connector a widely used, weatherproof, proprietary electrical and electronic connector.

device drivers software used to control input and output devices.

Dexcool a brand of EG-based antifreeze.

diagnostic routines term used to describe manufacturers structured, software-guided troubleshooting sequences.

diagnostic sight glass a section of optically clear Perspex tube coupled into number 6, number 8, or number 10 hydraulic hose used to troubleshoot air admission into fuel and hydraulic circuits; the sight glass is coupled in series into the circuit to be tested.

diagnostic trouble code (DTC) means of classifying logged codes either numerically or in text.

dial bore gauge an instrument designed to facilitate rapid bore comparative measurements, much used by the diesel engine rebuilder.

dial indicator an instrument designed to measure movement, travel, or precise relative dimensions. It consists of a dial face, needle, and spring-loaded plunger. Dial indicators can measure values down to one hundred thousandth of an inch or thousandths of a millimeter.

diamond dowels diamond-shaped alignment dowels used on flywheel housings that are less inclined to deform than cylindrical dowels.

Diamond Plus Navistar International's chassis data bus management software built on CAN 2.0/J1939 architecture. Accessed using Navistar *EZ-Tech* or *ServiceMaxx*

diatomic a molecule consisting of two atoms of the same element.

dielectric insulator substance such as the separation plates used between the conductor plates in a typical capacitor.

diesel coolant additives (DCA) proprietary supplemental coolant additives.

diesel cycle the four-stroke, compression ignition cycle patented by Rudolf Diesel in 1892. Though the term *diesel* can be used to describe some two-stroke cycle CI engines, the diesel cycle is necessarily a four-stroke cycle.

diesel exhaust fluid (DEF) an aqueous mixture of 32% urea and distilled water used in diesel SCR systems. Known as *AdBlue* in Europe.

diesel fuel a simple hydrocarbon fuel obtained from crude petroleum by means of fractioning and usually containing both residual and distillate fractions.

diesel fuel conditioning module (DFCM) term used by some manufacturers to describe the fuel subsystem module consisting of devices such as the charge (transfer) pump, sensors, water separator, and primary and secondary filters,

diesel knock a detonation condition caused by prolonged ignition lag.

diesel multistage filter (DMF) a combination DPF, catalytic converter, and muffler assembly used on post-2007 highway diesel engines.

diesel oxidation catalyst (DOC) single-stage oxidation catalysts that have been used on highway diesels since the mid-1990s.

diesel particulate filter (DPF) a diesel soot scrubber that physically traps particulate from the exhaust gas, then burns it off during regeneration cycles. Most are PCM/ECM managed to regenerate under passive (preferred) and active cycles that require fuel dosing and a heat source. Heat is usually derived by upstream injection of diesel fuel or from a heat grid. Required in most post-2007 highway diesel engines.

diffuser the device in a turbocharger compressor housing that converts air velocity into air pressure.

digital the representing of data in form of digits or other discrete methods.

digital audiotape (DAT) high-density data storage tape written to by a helical scan head.

digital calipers a precise id and od measuring instrument with the appearance of Vernier calipers, the accuracy of a micrometer, and the ability to convert from the standard to metric system at the push of a button.

digital computer a calculating and computing device capable of processing data using coded digital formats.

digital diagnostic reader (DDR) another term for a scan tool–type EST.

digital diagnostic tool (DDT) a reader/programmer EST.

digital display unit (DDU) any device used to display digital data.

digital micrometer a micrometer that displays dimensional readings digitally.

digital multimeter (DMM) a voltage, resistance (ohms), and current (amperes) reading instrument.

digital signals data interchange/retention signals limited to two discernable states; combinations of ones and zeros into which data, video, or human voice must be coded for transmission/storage and subsequently reconstructed.

digital video disc–read-only memory (DVD-ROM) an optically encoded data storage medium that can retain 25 times the data of a CD-ROM.

digitizing the process used to convert data to digital format.

diode a two-terminal, solid state device that allows current to flow in one direction only: a sort of electronic one-way check valve.

direct current (DC) current flow through a circuit in one direction only.

direct injection (DI) describes any engine in which fuel is injected directly into the engine cylinder and not to any kind of external prechamber. Most current diesel engines are direct injected.

discrete in computer technology, this means *coded*. For instance, coding analog values into binary values is expressing those values in a *discrete* format.

diskettes a common portable data storage medium for PCs.

disk-operating system (DOS) the set of software commands that govern computer operations and enable functional software programs to be run.

displacement on demand (DOD) term used to describe the emission reduction and fuel consumption strategy of shutting down (by not fueling) engine cylinders. Used on diesel engines since the early 1990s.

distillate any of a wide range of distilled fractions of crude petroleum, some of which would be constituents of a diesel fuel. Refers to the more volatile fractions in a fuel. Sometimes used to refer to diesel fuels.

distributor head section of a Bosch-type, sleeve-metering, rotary distributor pump in which the plunger moves; contains delivery passages, an electric fuel shutdown, screw plug with vent screw, and delivery valves.

distributor plunger center and cross-drilled plunger used in rotary distributor pumps to feed injection fuel to the hydraulic head or distributor head supply passages.

distributor rotor drive shaft–driven rotor on an inlet-metering, opposed-plunger distributor pump that connects the pump chamber with the supply passages in the hydraulic head for fuel delivery to the injectors.

dividers a comparative-type measuring compass, usually with an adjusting screw for setting precise dimensions.

doping the process of adding small quantities of impurities to semiconductor crystals to provide them with either positive or negative electrical characteristics.

dosing the process of injecting fuel, usually from the fuel subsystem, into the exhaust for purposes of regenerating a diesel particulate filter or for activating a NO_x adsorber catalyst.

dosing injection describes injection of fuel into an engine cylinder late in the power stroke that is not intended to be combusted in the cylinder; dosing fuel vaporizes and is sent unreacted into the exhaust, where it can assist DPF regeneration and NAC aftertreatment devices. Also known as *afterburn injection*. The term is also used by some manufacturers to describe DEF injection to DPFs.

double helix a port-helix plunger design with both upper and lower helix characteristics that results in a variable beginning and ending of the pump effective stroke.

double-pass radiator a counterflow radiator in which the coolant is routed to make two passes, therefore entering and exiting from separate tanks both located either at the top or the bottom of the radiator. A high-efficiency radiator.

downflow radiator a typical radiator in which hot coolant from the engine enters at the top tank, flows downward, and exits through a bottom tank.

downlink the transmission signal from a communications satellite to an Earth receiver or the receiver itself.

download data transfer from one computer system to another; often used to describe proprietary data transfer when reprogramming vehicle ECMs.

DPF pulse cleaner one type of off-vehicle DPF cleaning device; uses high-volume, low-pressure air pulses to remove contaminants.

driver another name for a power transistor. A transistor capable of switching high-current loads.

driver display unit (DDU) a term used to describe the dash mounted driver information display unit.

droop an engine governor term denoting a transient speed variation that occurs when engine loading suddenly changes.

droop curve a required hydromechanical governor characteristic in which fueling drops off in an even curve as engine speed increases from the rated power value to high idle.

dry liners liners that are fitted either with fractional looseness or fractional interference that dissipate cylinder heat to the cylinder block bore and have no direct contact with the water jacket.

dry sump an engine that uses a remotely located oil sump; not often seen on highway diesel applications but used in some bus engines to reduce the profile of the engine.

dual helices a plunger geometric design with identical helices machined diametrically opposite each other on the plunger. Commonly used, it helps prevent side loading of the plunger at high-pressure spill.

ductile describes materials (usually steels) that are malleable or flexible; capable of temporary plastic deformation when a load is applied without fracturing.

dumb node a network node with no independent processing or data retention capability.

Duramax Isuzu-built, GM diesel engines available in 6600 (6.6 liter) and 7800 (7.8 liter) versions.

dynamic RAM (DRAM) RAM with high-access speed.

dynamometer a testing device that loads an engine by applying a resistance to turning effort (torque) and factors this against *time* to produce brake power values. Often used to performance test or break in engines after reconditioning.

EEPROM electronically erasable, programmable read-only memory. Vehicle computer memory category that can be rewritten or flashed with customer or proprietary reprogramming. Includes an ECM's write-to-self capability.

effective stroke describes that portion of a constant-travel plunger or piston stroke used to actually pump fluid.

electrohydraulic injector (EHI) an ECM-switched injector used on CR and EUP injection systems; opening and closing values are soft, being controlled by the ECM.

electrolysis a chemical change produced in electrolyte by an electrical current, often resulting in decomposition.

electrolyte a solution capable of conducting electrical current.

electromagnetic interference (EMI) low-level radiation (such as that emitted from electrical power lines, vehicle radar, etc.) that can interfere with signals on data buses unless suppressed.

electromagnetism describes any magnetic field created by current flow through a conductor.

electromechanical injector (EMI) a pintle-type injector in which the pintle nozzle is integral with an armature; when the EMI is energized the nozzle opens. Used in GFI and as a dosing injector in diesel DPFs.

electromechanical switch any switch in which output status is controlled by manually or automatically switching electrical circuits on and off; differentiated from *smart* or *ladder* switches.

electromotive force (EMF) voltage or charge differential.

electron a negatively charged component of an atom.

electronically erasable, programmable read-only memory see *EEPROM*.

electronic control analyzer programmer (ECAP) Caterpillar PC-based reader/programmer instrument.

electronic cylinder balance test an electronically managed diagnostic test in which engine cylinders are sequentially cutout to measure the percentage of work performed by each. Also known by terms such as *cylinder balance test* and *cylinder contribution test*.

electronic distributor unit (EDU) term when electronic injector drivers (switching mechanisms) are housed in a different module from the PCM/ECM.

electronic diesel controls (EDC) Bosch term for its engine PCMs/ECMs.

electronic/engine control module (ECM) refers to the computer and integral switching apparatus in an electronically controlled vehicle system. The SAE-recommended term for describing the engine control electronics using the MID 128 address on the chassis data bus; most engine manufacturers adhere to this recommendation but not all; the term powertrain control module (PCM) is more commonly used in light duty.

electronic engine management computerized engine control.

electronic foot pedal assembly (EFPA) pedal mechanical travel managed by the TPS.

electronic governor any kind of governing using computer controls.

electronic management system management by computer or computers.

electronics branch of electricity concerned with the study of the movement of electrons through hard wire, semiconductor, gas, and vacuum circuits.

electronic service tool (EST) covers a range of instruments including DMMs, diagnostic lights, generic and proprietary scan tools, and PC-based communications devices.

electronic smart power (ESP) Cummins optional engine derate/fuel economy engine management package.

electronic technician (ET) Caterpillar PC-based software that enables the technician to diagnose system problems, reprogram ECMs, and access system data for analysis to produce fuel mileage figures and driver performance profiles.

electronic unit injector (EUI) the cam-actuated, electronically controlled pumping mechanism used to fuel engines such as the VW family of diesels until 2010.

electronic unit pump (EUP) a cam-actuated, ECM-controlled pumping and metering unit that supplies a hydraulic injector by means of a high-pressure line.

electron theory the theory that asserts that current flow through a circuit is by electron movement from a negatively charged point to a positively charged one. See *conventional theory*.

electrostatics the force field that surrounds an object with an electrical charge.

element (1) any of more than 100 substances (most naturally occurring, some man-made) that cannot be chemically resolved into simpler substances. (2) A component part of something such as a pump element.

emissions control device (ECD) term used to describe emission control hardware.

emulsify to disperse one liquid into another or to suspend a fine particulate in a solution.

emulsion the dispersion of one liquid into another, such as water in the form of fine droplets into diesel fuel.

end gas the gas that results from combusting fuel in engine cylinders; usually means the gases present at flame quench, that is, before any exhaust gas treatment occurs; typically a mixture of CO_2, H_2O, and whatever noxious gases are present.

ending of energizing (EOE) denotes the end of the switched duty cycle of an EUI.

ending of injection (EOI) instant that fuel injection ceases.

end of line (EOL) usually in reference to terminating a programming procedure.

energized-to-run (ETR) any of a group of solenoids that must be electrically energized to remain in On status, a non-latching–type solenoid.

energy best expressed as stored potential; its unit is kilowatt/hours (kW/h) or horsepower/hours (hp/h). In technology we more commonly use the term *potential energy*.

engine a machine that converts one form of energy to another.

engine brake any type of engine retarder. The term usually describes an internal engine compression brake but may also refer to an exhaust compression brake or an engine-mounted hydraulic retarder.

engine displacement the sum of the swept volume of all the engine cylinders.

engine dynamometer a dynamometer used for testing the engine on a test bed outside of the chassis.

engine/electronic control module (ECM) the SAE recommended term used to describe an electronic engine controller. That said, it is used to describe other vehicle control modules. The term *powertrain control module (PCM)* is preferred by automotive manufacturers and while others use *electronic control unit (ECU)*.

engine/electronic control unit (ECU) the SAE recommended term used to describe an electronic controller other than the engine. That said, it is used to describe other vehicle control modules.

engine family rating code (EFRC) engine series ratings.

engine hours a means of comparing engine service hours to highway mileage. Most engine manufacturers equate 1 engine hour to 50 highway line-haul miles (80 km), so a service interval of 10,000 miles (16,000 km) would equal 200 engine hours. The term *service hours* is also used.

engine longevity the engine life span. In highway diesel engines it is usually reckoned in miles on highway engines and hours in off-highway applications.

Engine Manufacturer's Association (EMA) association of engine manufacturers that by majority opinion work with the ASTM, SAE, and TMC to recommend industry standard practices.

engine position sensor (EPS) shaft position sensor using a reluctor pulse generator or Hall effect principle.

engine silencer a muffler that uses sound absorption and resonation principles to change the frequency of engine noise.

Environmental Protection Agency (EPA) federal regulating body that sets and monitors noxious emissions standards, among other functions.

EPS engine position sensor.

etching bearing or other component failure caused by chemical action.

ethylene glycol (EG) an antifreeze of high toxicity that the EPA hopes to phase out.

E-Trim Caterpillar EUI fuel flow specification required to be programmed to the ADEM ECM whenever an EUI is removed and replaced. E-Trim data is important because it enables the ECM to balance fueling to each cylinder.

execute effect an operation or procedure.

executive the resident portion of a computer or program operating system.

exhaust blowdown the first part of the cylinder exhaust process that occurs at the moment the exhaust valves open.

exhaust brake an external engine compression brake that operates by choking down the exhaust gas flow area; sometimes used in conjunction with an internal engine compression brake, meaning that the piston is contributing to retarding effort on both its upward strokes.

exhaust gas recirculation (EGR) a means of routing "dead"-end gas back into the intake to "dilute" the intake charge of oxygen, reducing combustion heat and therefore NO_x. In most diesel engine EGR systems, exhaust gas is cooled by a heat exchanger before rerouting to the intake.

exhaust manifold the cast iron or steel component bolted to the cylinder exhaust tracts responsible for delivering the end gases to the turbocharger and the exhaust system.

expansion board a circuit board added to a computer system to increase its capability.

explosion an oxidation reaction that takes place rapidly; high-speed combustion.

extended life coolant (ELC) coolant premix that provides extended service and a lower level of monitoring over EG mixed solutions.

external ground refers to a vehicle electrical component that receives power (positive) from an insulated wire and completes the ground path by being mated to chassis ground rather than a dedicated ground wire.

EZ-Tech International Navistar PC-to-chassis data bus software and hardware. An EZ-Tech laptop connects to the chassis data bus by means of a Navistar CA and J1939 data port (nine-pin Deutsch).

failure analysis diagnosis of a failed component, usually out of the engine.

fanstat a combination temperature sensor and switch (usually pneumatic) used to control the engine fan cycle.

farad a measure of capacitance. One farad is the ability to store $6.28 \times 1,018$ electrons at a 1V charge differential.

fault mode indicator (FMI) a J1939 (heavy duty) mode of defining a component or circuit failure numerically to an EST by ascribing to it one of up to 30 possible failure or imminent failure modes (SAE).

fax short form of facsimile; a method of reproducing an image or text digitally and transmitting it using the phone system.

feedback assembly the engine's mechanical, self-management components consisting of a geartrain, camshaft, valve trains, MUI and EUI actuating trains, fuel injection pumping apparatus and valves.

Ferrotherm™ piston a Mahle trademark for two-piece articulating piston assemblies consisting of a forged steel crown and aluminum skirt. Used by most heavy duty commercial vehicle diesel manufacturers from the early 1990s until the advent of single-piece, forged or composite steel, trunk pistons around the 2004 emissions year.

ferrous metals are composed primarily of iron and usually attracted by a magnet.

fetching CPU function that involves obtaining data from memory.

fiber optics the transmission of laser light waves through thin strands of fiber used to digitally pulse data at much higher speeds than copper wire with zero chance of electromagnetic interference. At present, used for infotainment most buses.

FIC module fuel injection control module. Usually a slave module connected by a proprietary CAN bus to the engine ECM/PCM, responsible for controlling an injection pump.

field coils used in electric motors and other devices to flow current through to establish and electromagnetic field.

field effect transistor (FET) group of transistors used to switch or amplify within a circuit.

fields specific items of (electronic) information.

field service bulletin (FSB) one manufacturer's term for *technical service bulletins (TSBs)*.

field service tips (FSTs) one manufacturer's term for *technical service bulletins (TSBs)*.

file a collection of related data.

filter monitor a pressure-sensing device that monitors DPF or aftertreatment device backpressure and alerts the vehicle data bus when a plug-up condition is imminent. Some are also capable of signaling temperature.

fire point the temperature at which a combustible produces enough flammable vapor for a continuous burn; always a higher temperature than flash point.

fire ring normally used to refer to the fixed ring that may be integral with the cylinder head gasket responsible for sealing the cylinder. Sometimes used to refer to the top compression ring, but this usage is uncommon and probably incorrect.

FireWire a digital audio/video serial bus interface standard offering high-speed communications that can be used to network with vehicle data buses.

fixed disk a data storage device used in PCs and mainframe computers consisting of a spindle and multiple stacked data retention platters.

flame front during flame propagation, the leading edge of the flame in an engine cylinder.

flame propagation the flame pattern from ignition to quench during a power stroke in an engine cylinder.

flame quench the moment that the flame ceases to propagate or extinguishes in an internal combustion engine.

flammable any substance that can be combusted.

flash term used to describe the downloading of new software to EEPROM.

flashback a highly dangerous condition that can occur in operating oxyacetylene equipment in which the flame may travel behind the mixing chamber in the torch and explode the acetylene tank using the system's oxygen. Most current oxyacetylene torches are equipped with flashback arresters.

flash codes the PCM/ECM-generated fault codes that are usually displayed by means of diagnostic lights and alert the driver or technician to the nature of an electronically monitored malfunction; also known as *fault codes* or *blink codes*.

flash memory nonvolatile computer memory that can be electrically erased and reprogrammed. It is primarily used in memory cards and USB flash drives (thumb drives, jump drives) and as data retention on vehicle ECM/PCM personality modules.

flash point the temperature at which a combustible produces enough flammable vapor for momentary ignition.

flash programming term that has come to mean any reprogramming procedure.

flash RAM nonvolatile RAM; NV-RAM.

floppy disks See *diskettes*.

floptical a diskette that combines magnetic and optical technology to achieve high-density storage capacity.

flow area the most restricted portion of a fluid circuit; for instance, a water tap sets a flow area and as the tap is opened, the flow area increases, thereby increasing the volume flow of water.

flow control refers to any device that can proportionally control flow through a circuit. A thumb over the end of a hose is a flow control device.

fluid any substance that has fluidity. Both liquids and gases are fluids. Fluid power incorporates both hydraulics and pneumatics.

fluid friction the friction of dynamic fluids, always less than solid friction.

fluidity the state of a substance that permits it to conform to the shape of the vessel in which it is contained. Both liquids and gases possess fluidity.

fluid power term used to describe both hydraulics and pneumatics.

flutes protruding lands with grooves in-between.

flywheel an energy and momentum storage device usually bolted directly to the crankshaft.

flywheel housing concentricity a critical specification that ensures that the relationship of the flywheel and anything connected to it is concentric.

follower used to describe a variety of devices that ride a cam profile and transmit the effects of the cam geometry to the train to be actuated; also known as *tappet*.

font typeface size and appearance.

force the action of one body attempting to change the state of motion of another. The application of force does not necessarily result in any work accomplished.

foreground computations computer-operating responses that are prioritized, such as the response to a critical command input such as the TPS (throttle position sensor) whose signal must be acted on immediately to generate the appropriate outcome.

forged steel trunk pistons a trunk-type piston assembly with an open skirt; manufactured by Mahle under the trade name *Monotherm*. First appeared for the 2004 model year but rapidly have become the piston of choice in medium- and large-bore diesel engines.

format (1) to alter the appearance or character of a program or document. (2) To prepare data retention media, such as diskettes, to receive data by defining tracks, cylinders, and sectors, a process that removes any previously logged data.

forward leakage an injector bench fixture test that tests the nozzle seat's sealing integrity.

fossil fuel unrenewable, organically derived fuels such as petroleum and coal.

four-terminal EUI a more recently introduced, dual-actuator EUI; operates on the same principles as a two-terminal EUI but has an ECM-controlled, electrohydraulic nozzle which allows soft (as opposed to fixed-value) NOPs—used mainly on heavy duty engines.

fractions refers to separate compounds of crude petroleums separated by distillation and other fractioning methods such as catalytic and hydrocracking and classified by their volatility.

fractured rod a con rod manufacturing process; see *cracked rod*.

fretting occurs when two parts that fit tightly are allowed to move slightly against each other, sometimes resulting in microwelding, which creates small surface irregularities. Can result from a head gasket that fails to allow thermal creep as cylinder head and block pass through heat and cool cycles.

friction the resistance an object or fluid encounters in moving over or through another.

friction bearing a shaft-supporting bearing in which the rotating member can directly contact the bearing face or race.

fuel substance that can be used as a source for heat energy.

fuel conditioner usually unknown quantities of cetane improver and pour point depressants suspended in an alcohol base.

fuel control actuator any of a number of electronically controlled devices used as fuel control mechanisms.

fuel demand command signal (FDCS) International HEUI output driver signal.

fuel filter device for filtering sediment from fuel, rated by entrapment capability.

fuel heater a heat exchanger device used in extreme cold to prevent diesel fuel from waxing in the fuel subsystem.

fueling algorithm the set of PCM-programmed rules and procedures designed to produce the desired performance and emissions from an engine at any given moment of operation.

fuel map a diagram or graph used to indicate fueling through the entire performance range of an engine; also used to describe the ECM/PCM fuel algorithm.

fuel pressure sensor (FPS) a pressure-sensing mechanism, usually of the variable capacitance type, that measures the charging pressure in the fuel subsystem and signals its value to the ECM/PCM.

fuel rate actual rate of fuel pumped through an injector to an engine cylinder; factored by cam geometry and engine rpm.

fuel subsystem the fuel circuit used to pump fuel from the vehicle fuel tank and deliver it to the fuel metering/injection apparatus. The fuel subsystem typically comprises a fuel tank, water separator, primary filter, transfer or charge pump, secondary filter, and the interconnecting plumbing.

fuel tank the fuel storage reservoir on a vehicle.

full-flow filter a filter plumbed in series on the charge side of the pump that feeds a circuit.

full-wave rectification occurs when a diode bridge is used in an alternator to convert both the node and antinode of AC voltage into a DC voltage characteristic.

function keys numerical keys prefixed by the letter F that act as program commands and shortcuts.

fuzzy logic a PCM or computer processing outcome that depends on multiple inputs, operating conditions, and operating system commands; opposite to *closed loop*.

gallons per hour (gph) the means of rating liquid flow in a hydraulic circuit.

galvanometer a meter used to measure small electrical currents.

gas analyzer a test instrument for measuring and identifying the exhaust gas content.

gas dynamics the manner in which gases behave during the compression and combustion strokes and the processes of engine breathing.

gasket yield point the moment that a malleable gasket is crushed to its desired shape to conform to the required shape between two clamped components to provide optimum sealing.

gasoline a hydrocarbon fuel composed of the volatile petroleum fractions from the aromatic and paraffin ranges.

gates routing switches with either digital or mechanical actuation.

gateway a multiplexing term used to describe a module with an address on more than one bus. In cases where the buses use different communications protocols and/or speeds, the *gateway* acts as a transducer.

gear pump a positive displacement pump consisting of intermeshing gears that uses the spaces between the teeth to move fluid through a circuit.

genset a complete electricity generating unit consisting of an internal combustion engine and an electricity generator.

geosynchronous orbit the park orbit of communications satellites 35,400 km (22,300 miles) from the Earth's equator; differentiated from low-orbit GPS satellites. Used for data exchange.

gerotor a type of gear pump that uses an internal crescent gear pumping principle, often used as an oil pump.

gigabyte a billion bytes; a measurement of digital memory capacity.

glazing friction wearing of a component to a mirror finish.

global diagnostic system (GDS) PC-based software required to troubleshoot all GM vehicles post-2010 model year.

global positioning satellite (GPS) refers to communications using low-orbit telecommunications satellites. Currently used for vehicle tracking and navigation.

GM-MDI the GM Windows environment software known as *multiple diagnostic interface (MDI)*. Used in conjunction with a communications adapter (CA) to interface with the vehicle data bus(es). Required to effectively work with post-2010 GM vehicles.

governing algorithm the processing cycle map built on input circuit sensor status signals and programmed software instructions that manage fuel injector duty cycles.

governing map see *governing algorithm*.

governor a component that manages engine fueling on the basis of fuel demand (accelerator) and engine rpm; may be hydromechanical or electronic.

governor spring the force, usually variable, that opposes centrifugal force in mechanical governors; often amplified by accelerator pedal travel.

graduates see *vial*.

graphical user interface (GUI) software such as MS Windows that is icon and menu driven.

gray scaling used by monochrome digital displays and scanners to code color to black, white, and shades of gray.

gross power expression of raw engine power with all potential parasitic loads and emissions controls removed. (There is an SAE formula for gross power—J1995). See also *net power*.

ground describes the point or region of lowest voltage potential in a circuit; the portion of a vehicle electrical circuit serving multiple system loads by providing a return path for the current drawn by the load. Used in vehicle systems using 48V or less and ideal for the commonly used 12V vehicle systems.

ground circuit the negative or noninsulated side of a chassis electrical circuit.

ground strap a conductive strap, usually braided wire, that extends a common ground electrical system.

groupware software that allows multiple users to work together by sharing information.

gumming a term used to describe unburned fuel and lubrication oil residues when they sludge in piston ring grooves and other areas of the engine.

half-wave rectification occurs when only half of the AC voltage (node or antinode) is converted to DC.

Hall effect a method of accurately sensing rotational or linear position and speed. Normally used to produce a digital signal, but may be configured to produce an analog signal using a DAC. A moving metallic shutter alternately blocks and exposes a magnetic field from a semiconductor sensor.

handshake establishing a communications connection, especially where two electronic systems are concerned. The communication protocols must be compatible.

hard in electronics, this is a value that should not or cannot be changed. The term has evolved in definition so that it references anything in computer technology that can be physically touched. An example would be service literature; this might be accessed online, in which case it would be called *soft* format, or printed out, in which case it would be converted to a *hard* format.

hard copy computer-generated data that is printed to a page rather than retained on disk.

hard disk see *fixed disk; hard drive*.

hard drive a data storage device used in PCs and mainframe computers consisting of a spindle and multiple stacked data retention platters.

hard parameter a fixed value that cannot or should not be altered or rewritten. Maximum engine rpm would be an example of a hard parameter.

hard value used to describe a rigid or unchanging value in computer terminology: its opposite is *soft value* which would mean a changing parameter or one that can be manipulated.

hardware computer equipment excluding software.

Hazard Communications Legislation federal legislation that incorporates the Right to Know clauses that pertain to workplace hazards. Administered by OSHA.

head crash occurs when a computer disk head collides with the hard disk surface, causing loss of memory.

headers the manifold deck to which the coolant tubes are attached in a heat exchanger bundle, or the term used to describe low-gas restriction, individual cylinder exhaust pipes that converge at a point calculated to maximize pulse effect.

headland the area above the uppermost compression ring and below the leading edge of the piston.

headland piston a piston design that has minimized the headland volume.

headland volume the headland gas volume in a cylinder.

heat energy an expression of the energy potential a substance possesses. It is actually the amount of kinetic energy at the molecular level in an element or compound.

heat engine a mechanism that converts thermal energy into mechanical work.

heat exchanger any of a number of devices used to transfer heat from one fluid to another where there is a temperature difference using the principles of conduction and radiation.

heating value the potential heat energy of a fuel; also known as *calorific value*.

heating, ventilating, and air-conditioning (HVAC) acronym used to describe the climate control system, usually integrated on current vehicles.

helical gear a gear with spiral-cut teeth.

helical scan technology used to write data at high density on tape helically as opposed to longitudinally.

helices plural of *helix*.

helix a spiral groove or scroll. The helical-cut recesses in some injection pumping plungers that are used to meter fuel delivery. Plural: helices.

Hg manometer a mercury (Hg)-filled manometer.

high-idle speed the highest no-load speed of an engine.

high-pressure pipes the pipes or lines that deliver fuel from an injection pump element to the injector nozzle.

high-pressure washer a high-pressure water pump used to clean equipment and components before repair and inspection that has generally replaced steam cleaners.

high spring injector a type of hydraulic injector nozzle that locates the injector spring high in the injector/nozzle holder body. NOP is usually adjusted by an adjusting screw that acts directly on the spring.

histogram a graphic display in which data is represented by rectangular columns used for comparative analysis.

historic codes fault codes that are no longer active but are retained in PCM/ECM memory (and displayed) for purposes of diagnosis until they are erased; also known as *inactive codes*.

hone any of a number of types of abrasive stones used for finishing metals. Rotary hones are electrically or pneumatically driven and are used for sizing and surface finishing cylinder liner bores.

horsepower the standard unit of power measurement used in North America, defined as a work rate of 33,000 lb.-ft. per minute; equal to 0.746 kW.

host computer a main computer that is networked to other computers or nodes.

hotspot the geographical area over which a WiFi client device can effectively network.

H₂O manometer a water-filled manometer.

hunting rhythmic fluctuation of engine rpm, usually caused by unbalanced cylinder fueling.

hunting gears an intermeshing gear relationship in which after timing, the gears may have to be turned through a large number of rotations before the timing indices realign.

hybrid bus a chassis data bus incorporating both *star* and *loop* configurations.

hydraulically actuated electronic unit injector (HEUI) Caterpillar, Navistar, or Ford oil-pressure–actuated, high-pressure fuel pumping/injecting element.

hydraulic head stationary member of an inlet-metering, opposed plunger rotary distributor pump within which the distributor rotor rotates; as the distributor rotor turns, it is brought in and out of register with each discharge port connecting the pump with the fuel injectors.

hydraulic injectors any of a group of injectors that are opened and closed hydraulically as opposed to electronic; this would include the nozzle assemblies used in many EUI and HEUI units. One manufacturer uses the term *mechanical injector* in place of hydraulic injector.

hydraulics (1) the science and practice of confining and pressurizing liquids in circuits to provide mechanical power. (2) Control and actuator circuits that use confined liquids under pressure.

hydrocarbon (HC) describes substances primarily composed of elemental carbon and hydrogen. Fossil fuels and alcohols are both hydrocarbon fuels.

hydrodynamic suspension the principle used to float a rotating shaft on a bed of constantly changing, pressurized lubricant.

hydrogen (H) the simplest, most abundant element in the known universe, occurring in water and all organic matter. Colorless, odorless, tasteless, and explosive. Atomic number 1.

hydromechanical engine management all engines managed without computers.

hydromechanical governing engine governing without the use of a computer; requires a means of sensing engine speed (centrifugal force exacted by flyweights/fuel pressure) and a means of limiting fuel.

hydrometer an instrument designed to measure the specific gravity of liquids, usually battery electrolyte and coolant mixtures. Not recommended for measuring either in diesel engine applications where a refractometer is the appropriate instrument due to its greater accuracy.

hypermedia a multimedia presentation tool that permits rapid movement between screens to display graphics, video, and sound.

hypertext link the highlighting and bolding of a text word/phrase to enable Web page or program selection by mouse click. Increases user-friendliness of Internet and many other programs.

hypothesis a reasoned supposition, not necessarily required to be true.

hysteresis (1) in hydromechanical governor terminology, a response lag. (2) Molecular friction caused by the lag that occurs between the formation of magnetic flux lines and the magnetomotive force that creates it. Explains why solenoid responses are not immediate.

icons pictorial/graphical representations of program menu options displayed on-screen.

idle speed the lowest speed that an engine is run at, usually managed by the governor.

ignition accelerators volatile fuel fractions that are added to a fuel to decrease ignition delay. They increase CN.

ignition lag the time period between the entry of the first droplets of fuel to the engine cylinder and the moment of ignition based on the fuel chemistry and the actual temperatures of the engine components and the air charge.

impeller (1) the driven member of a turbocharger, responsible for compressing the air charge. (2) The power input member of a pump, such as on a torque converter or hydraulic retarder.

inactive codes fault codes that are no longer active but are retained in ECM/PCM memory (and displayed) for purposes of diagnosis until they are erased; also known as *historic codes*.

indicated power expression of gross engine power usually determined by calculation and in the United States, expressed as indicated horsepower.

indirect injection (IDI) describes any of a number of methods of injecting fuel to an engine outside of the cylinder.

This may be to an intake tract in the intake manifold or to a cell adjacent to the cylinder such as a precombustion chamber. Little used in today's engines.

induction circuit refers to the engine air intake circuit but more appropriately describes air intake on naturally aspirated engines than on boosted engines.

inert chemically unreactive. Any substance that is unlikely to participate in a chemical reaction.

inertia in physics, it describes the tendency of a body at rest or in motion to want to continue in that state unless influenced by an external force.

InfoMax vehicle-to-land station data transfer system that uses WiFi short-range wireless technology. Permits download of vehicle and driver performance data and upload of customer data programming.

INFORM Cummins PC-based data management system.

infrared the wavelength just greater than the red end of the visible light spectrum, but below the radio wave frequency.

infrared thermometer accurate heat measuring instrument that can be used for checking cylinder fueling balance.

inhibitor something that stalls or delays a reaction or event. A rust *inhibitor* slows oxidation while an oil oxidation inhibitor protects a lube from being prematurely oxidized.

ink jet printer nonimpact printer that uses a nozzle to shoot droplets onto a page.

injection actuation pressure (IAP) Caterpillar's HEUI actuation oil pressure.

injection control pressure (ICP) actuation oil pressure in HEUI fuel systems.

injection lag a diesel fuel injection term describing the time lag between port closure in a pumping element and the actual opening of the injector.

injection pressure regulator (IPR) ECM/PCM-controlled device that manages HEUI-actuating oil pressure.

injection quantity calibration data fuel flow specification assigned to EUIs, EUPs, and electrohydraulic injectors; also known by terms such as *QR* and *E-trim* programming.

injection rate a diesel fuel injection term that is defined as the fuel quantity pumped into an engine cylinder per crank angle degree. In systems except for the HEUI and common rail (CR), injection rate is determined by the pump-actuating cam profile geometry.

injector a term broadly used to describe the holder of a hydraulic nozzle assembly. May also be used to describe metering, pumping injector assemblies such as EUI and HEUI assemblies.

injector driver module (IDM) separate injector driver unit used in International's versions of HEUI up to 1997. IDM functions are integrated in a single engine controller ECM in current applications.

injector drivers the ECM-controlled components that electrically switch EUI and HEUI assemblies. Injector drivers may be integral with the main ECM housing or contained in a separate module or housing.

injector nozzle the tip of an injector machined with an orifice or orifices through which high pressure fuel is forced.

injector response time (IRT) time in ms between injector-driven signal and EUI control valve closure.

inlet metering any injection pump that meters fuel quantity admitted to the pump chamber. For instance, in an inlet metering, opposed plunger injection pump, all of the fuel admitted to the high-pressure pump chamber is injected each time that the plungers are actuated.

inlet restriction a measure of the pressure value below atmospheric, developed on the pull side of a pumping mechanism. Air inlet restriction and fuel inlet restriction are common specifications used by the diesel technician.

inlet restriction gauge instrument that measures (usually air) inlet restriction, often on-chassis.

inner base circle (IBC) in cam geometry, the portion of the cam profile with the smallest radial dimension; also known as *base circle/IBC*. When the train riding the cam profile is on IBC, it is unloaded.

input the process of entering data into a computer system.

input devices the hardware, such as a keyboard on a PC, or sensors on a vehicle system responsible for signaling/switching data to a computer system.

inside diameter (id) diametrical measurement across a bore.

inside micrometer standard or metric micrometer consisting of a spindle and thimble but no anvil; used for making internal and bore measurements.

InSite Cummins PC software.

INSPEC Cummins Windows-driven PC vehicle ECM diagnostics and programming software.

insulated circuit the positive side of a chassis electrical system which supplies components by means of insulated cables and wires.

insulators materials that either prevent or inhibit the flow of electrons; usually nonmetallic substances that contain more than four electrons in their outer shell.

intake circuit the series of components used to route ambient air into engine cylinders. In a diesel engine, includes filter(s), piping, turbo compressor housing, charge air cooler, and intake manifold.

intake manifold the piping that is clamped to the intake tract flange faces, responsible for directing intake air into the engine cylinders.

intake module assembly of air guide components used in some post-2004, and most 2007 diesel engines consisting of intake manifold(s), EGR mixing chambers, and a heat exchanger.

integrated circuit (IC) an electronic circuit constructed on a semiconductor chip, such as silicon, that can replace many separate electrical components and circuits.

intensifier piston Caterpillar HEUI hydraulically actuated piston that pumps fuel to injection pressure values; also known as *amplifier piston*.

Interact System (IS) Cummins's term used to describe its integrated electronic engine management systems with the ability to connect with fleet management and analysis software. Acronym IS is used ahead of the engine series letter.

Interact System B (ISB) series Cummins's inline six-cylinder 5.9L to 6.7L engines.

interface the point or device where an electronic interaction occurs. Separate vehicle system ECMs will sometimes require interface hardware. In terms of EST usage, *interface* means to connect or create a *handshake*.

interference angle used to provide aggressive valve-to-valve seat bite; achieved by machining a valve seat at ½-degree less than cylinder head mating seat. Not used when valve rotators are used, therefore not common in diesel engines.

interference fit the fitting of two components so that the od of the inner component fractionally exceeds the id of the outer component. Liners are sometimes interference fit to cylinder bores. Interference fitting requires the use of a press, chilling, heating, or other forceful means.

internal cam ring the means of actuating the pumping plungers in an opposed plunger, inlet-metering rotary injection pump. Cam profiles are machined inside the cam ring, and as the plunger rollers rotate within it, they are forced inboard to effect a pump stroke.

internal compression brake any of a number of engine brakes that use the principle of making the piston perform its usual work through the compression stroke and then negate the power stroke by releasing the compression air to the exhaust system at TDC on the completion of the compression stroke.

internal ground a chassis electrical device that is supplied by a dedicated cable or wire rather than being merely clamped to chassis ground.

International service information system (ISIS) see *ISIS*.

International Truck & Engine see *Navistar*.

Internet the global computer multimedia network communications system accessed through the phone system. A network of networks.

Intranet an internal computer network designed for organizational communications using Internet protocols; also known as *LAN*.

ion an atom with either an excess or a deficiency of electrons, that is, an unbalanced atom.

iron (Fe) the primary constituent of steel.

ISIS acronym for International (online) service information system. The Navistar proprietary SIS. ISIS is designed to interact with Master Diagnostics and Diamond Plus data bus management electronics.

isochronous governor a zero-droop governor or one that accommodates no change in rpm on the engine it manages as engine load varies. In electronically managed diesel engines, the term is sometimes used to describe engine operation when in PTO mode.

J1667 SAE standards for emission testing of highway diesel engines manufactured before 1991 EPA standards. Currently used by many jurisdictions for enforcement of commercial diesel emissions.

J1850 data backbone hardware and protocols used in second-generation light duty CAN-B multiplexing systems.

J1939 data backbone hardware and protocols used in heavy-duty CAN multiplexing systems. Based on CAN 2.0, the automotive equivalent is CAN-C.

J1962 connector correct term used to describe the ALDL connector and standardize its terminal assignments.

Jacobs retarders Jacobs is known mainly for its internal engine compression brakes but also manufactures driveline retarders.

joule unit of energy that describes the work done when an electrical current of 1 ampere flows through a resistance of 1 ohm in 1 second or in mechanical terms.

kaizen Japanese word meaning "continuous improvement." It has become a catchword in industry and is often linked with the practice of TQM.

KAMPWR electrical circuit that powers KAM (Navistar).

Karman vortex flow MAF sensors see *vortex flow MAF sensors*.

keep alive memory (KAM) nonvolatile RAM.

keepers split locks fitted to a peripheral groove at the top of a cylinder valve stem; hold the spring retainer in position.

kernel the resident portion of a disk or program operating system.

kerosene a petroleum-derived fuel with a lower volatility than gasoline and fewer residual oils than diesel fuel.

keyboard the data entry device used on PC systems enabling alpha, numeric, and command switching.

keystone the trapezoidal shape that gets its name from the trapezoidal stones used in a classic Roman arch bridge.

keystone ring a trapezoidally shaped piston ring commonly used in diesel engine compression ring design.

keystone rod a connecting rod with a trapezoidal eye (small end) to increase the loaded sectional area.

kilobyte a quantitive unit of data consisting of 1,024 bytes.

kilowatt a unit of power measurement equivalent to 1,000 watts. Equal to 1.34 BHP.

kinetic energy the energy of motion.

kinetic molecular theory states that all matter consists of molecules that are constantly in motion and that the extent of motion will increase at higher temperatures.

Kirchhoff's first law states that the current flowing into a point or component in an electrical circuit must equal the current flowing out of it.

Kirchhoff's second law states that the voltage will drop in exact proportion to the resistance in a circuit component and that the sum of the voltage drops must equal the voltage applied to the circuit; also known as *Kirchhoff's law of voltage drops*.

kPa kilopascal. Metric unit of pressure measurement. Atmospheric pressure is equivalent to 101.3 kPa.

lacquering the process of baking a hard skin on engine components usually caused by high-sulfur fuels or engine oil contamination.

LAN see *local area network*.

ladder switch a "smart" switch named because it contains a ladder of resistors, usually five per switch, known as a ladder bridge. The processor that receives data from the ladder switches on the data bus has a library of resistor values that enables it to identify switch status and its commands.

lambda the Greek letter λ used as a symbol to indicate stoichiometric combustion. See definition of *stoichiometric*.

lambda sensor an exhaust gas sensor used on electronically managed, SI gasoline-fueled engines to signal to the ECM/PCM the oxygen content in the exhaust gas.

lamina a thin layer, plate, or film.

lands the raised areas between grooves, especially on the ring belt of a piston.

laptop computer a portable PC, larger than a notebook but smaller than a desktop PC.

large bore in the trucking industry describes diesel engines with displacements between 12 and 16 liters.

laser any of many devices that generate an intense light beam by emitting photons from a stimulated source. Used in computer technology to read and write optically.

laser printer a common PC printer device that aims a laser beam at a photosensitive drum to produce text or images on paper.

latching solenoid a solenoid that locks to a position when actuated and usually remains in that position until the system is shut down.

latent heat thermal energy absorbed by a substance undergoing a change of state (such as melting or vaporization) at a constant temperature.

leakoff pipes/lines the low-pressure return circuit used in most current diesel fuel injection systems.

lean NO$_x$ catalyst (LNC) a catalytic converter designed to reduce NO$_x$ under conditions of excessive oxygen; usually ECM/PCM managed and requires HC injection to enable the reduction process using a rhodium catalyst.

lever a rigid bar that pivots on a fulcrum and can be used to provide a mechanical advantage.

lifters components that ride a cam profile and convert rotary motion of the camshaft into linear motion or lift. Lifters used in most diesel engines are generally solid or roller types.

lift pump the term used to describe a low-pressure pump, often located within a fuel tank, that supplies a transfer pump in the fuel subsystem.

light-emitting diode (LED) diode that converts electrical current directly into light (photons) and therefore is highly efficient as there are no heat losses.

limiting speed governor (LSG) a standard automotive governor that defines the idle and high idle fuel quantities and leaves the intermediate fueling to be managed by the operator within the limitations of the fuel system.

limp-home see *default*.

linear magnet a proportional solenoid. This term is used by Mack Trucks to describe its rack actuator mechanism.

line-haul terminal-to-terminal operation of a truck, meaning that most of its mileage is highway mileage.

liners the normally replaceable inserts into the cylinder block bores of most diesel engines that permit easy engine overhaul service and greatly extended cylinder block longevity.

link and lever assembly Caterpillar intermediary between the governor and MUIs.

liquefied petroleum gas (LPG) another term used to describe propane, a petroleum-derived gas consisting mostly of methane.

liquid crystal display (LCD) flat panel display consisting of liquid crystal sandwiched between two layers of polarizing material. When a wire circuit below is energized, the liquid crystal medium is aligned to block light transmission from a light source, producing a low-quality screen image.

load ratio of power developed versus rated peak power at the same rpm.

load-dump a chassis electrical condition that can occur when a major electrical current consumer is opened, resulting in a high voltage spike through the electrical circuit.

local area network (LAN) also sometimes known as Intranet, a usually private computer network used for communications and data tracking within a company or institution. LAN access is usually by subscription only.

local bus an expansion bus that connects directly to the CPU.

local interconnect network (LIN) term used to describe an early generation, slow speed electrical data bus. May use a one- or two-wire circuit.

locking tang a tab on a component, such as a bearing, that may help position and lock it.

logical processing data comparison and mapping operations by the CPU.

log-on an access code or procedure used in network systems (such as vehicle manufacturer data hubs) used for security and identification.

longevity long life or life span.

loop a serial data bus configuration in which data is broadcast by one module and travels through each module in the loop, ending back at the broadcast address.

lower helix the standard helix milled into most port-helix plungers used in diesel applications. These produce a constant beginning, variable delivery characteristic when no external variable timing mechanism is used.

low-pressure pump vague term that can be used to describe pumps in the fuel subsystem: may refer to a *lift pump* or *charge pump* depending on the manufacturer that uses the term.

low spring injector an injector design that locates the spring directly over the nozzle valve, thereby reducing the mass of moving components compared to a high spring model. Injector spring tension is usually defined by shims.

LS fuel low-sulfur fuel used on highway diesel engines up until 2006 when a phase-out began (completed in 2010); contains a maximum of 0.05 percent sulfur.

lubricity literally, the oiliness of a substance.

lugging term used to describe an engine that is run at speeds lower than the base of the torque rise profile (peak torque) under high loads, that is, high cylinder pressures.

machine cycle the four steps that make up the CPU processing cycle; fetch, decode, execute, and store.

magnetic flux test magnetic flux crack detection used to identify defects in crankshafts, connecting rods, cylinder heads, and other parts. An electric current is flowed through the component being tested, and iron particles suspended in liquid are then sprayed over the surface. The particles will concentrate where the magnetic flux lines are broken up by cracks.

magnetism the phenomenon that includes the physical attraction for iron observed in lodestone and associated with electric current flow. It is characterized by fields of force, which can exert a mechanical and electrical influence on anything within the boundaries of that field.

magneto an electric generator using permanent magnets and capable of producing high voltages.

magnetomotive force (mmf) the magnetizing force created by flowing current through a coil.

mainframe large computers that can process and file vast quantities of data. In the transportation industry, the data hubs to which dealerships and depots are networked are usually mainframe computers.

main memory RAM; electronically retained data pipelined to the CPU. Data must be loaded to RAM to be processed by a computer.

major thrust side when cylinder gas pressure acts on a piston, it tends to pivot off a vertical centerline; the major thrust side is the inboard side of the piston as its throw rotates through the cycle.

malleable possessing the ability to be deformed without breaking or cracking.

manifold boost turbo-boost.

manometer a tubular, U-shaped column mounted on a calibration scale. The tube is water or mercury filled to balance at 0 on the scale, and the instrument is used to measure light pressure or vacuum conditions in fluid circuits.

manual regeneration term used to describe a DPF regeneration which is either driver (dash switch) or technician (EST) activated.

mass the quantity of matter a body contains; weight.

mass air flow (MAF) sensors used to provide the PCM with mass airflow data for purposes of calculating AFR. The most common type uses pressure differential calculation based on the relationship between inlet pressure and flow rate through a venturi.

master bar a test bar used to check the align bore in engine cylinder blocks.

Master Diagnostics (MD) Navistar International EZ-Tech–driven software designed to troubleshoot Navistar engines and chassis components.

master gauge a diagnostic gauge of higher quality used to corroborate readings from an in-vehicle gauge.

master program the resident portion of an operating system. In a vehicle ECM, the master program for system management would be retained in ROM.

master pyrometer an accurate thermocouple pyrometer used when performing dynamometer testing.

MasterTech generic access Bosch EST system that is either scan tool (slower speed/memory) or CA based using PC software.

material safety data sheets (MSDS) a data information sheet that must be displayed on any known hazardous substance; mandated by WHMIS.

matter physical substance; anything that has mass and occupies space.

MBE-900 series of Mercedes-Benz inline four- and six-liter engines ranging up to 6.4 liter displacement.

MDI See *multiple diagnostic interface.*

MDI Manager GM PC-based software required to drive the MDI CA and Web network with the GM LAN.

mean average.

mean effective pressure average pressure acting on a piston through its complete cycle, the net gain of which converts to work potential. Usually calculated by disregarding the intake and exhaust strokes, and subtracting mean compression pressure from mean combustion pressure.

mechanical advantage the ratio of applied force to the resultant work in any machine or arrangement of levers.

mechanical efficiency a measure of how effectively indicated horsepower is converted into brake power; factors in pumping and friction losses.

mechanical governor a governor in which the centrifugal force developed by the rotating flyweights used to sense rpm is the force used to move the fuel control mechanism.

mechanical injectors one manufacturer's term for hydraulic injectors.

mechanically actuated, electronically controlled, unit injector (MEUI) Caterpillar term for an EUI; see *electronic unit injector.*

mechanical unit injector (MUI) cam-actuated, governor-controlled unit injectors used by DDC and Caterpillar in engines up to the mid-1990s.

media-oriented system transfer (MOST) term used to describe the dedicated infotainment bus by some manufacturers. Commonly uses a fiber optic data bus.

medium bore in the trucking industry, describes diesel engines with displacements between 8 and 12 liters.

megabytes one million bytes. A quantitive measure of data. Often abbreviated to "meg."

megahertz a measure of frequency. One million cycles per second. The system clock is speed rated in megahertz.

megapascal (MPa) one million pascals. Metric pressure measurement unit.

memory address the location of a byte in memory.

menu a screen display of program or processing options.

message identifier (MID) a J1939 identifier logged by numeric code when read by an EST (SAE).

metallurgy the science of the production, properties, and application of metals and their alloys.

metals any of a group of chemical elements such as iron, aluminum, gold, silver, tin, and copper that are usually good conductors of heat and electricity and can usually form basic oxides.

metering the process of precisely controlling fuel quantity.

metering recesses the milled recesses in MUI and port-helix plungers that are used to vary the fuel quantity and timing during pumping.

meter resolution a measure of the power and accuracy of a DMM.

Metri-Pack connector a type of commonly used, sealed electrical/electronic connector.

Mexican hat piston a piston design in which the center of the crown peaks in the fashion of a sombrero. Commonly used in DI diesel engines.

micron one-millionth of a meter, equivalent to 0.000039 inch. The Greek letter mu is used to represent micron and is written as μ.

microorganism growth a condition that may result from water contamination in fuel storage tanks.

microprocessor a small processor. Sometimes used to describe a complete computer unit.

microwaves radio waves used to transmit voice, data, and video. Limited to line-of-sight transmission to distances not exceeding 30 km.

Miller cycle variation on the Otto cycle engine that uses a supercharger and holds open the intake valves for the first part of the compression stroke, providing higher efficiencies.

millions of instructions per second (MIPS) rating of processing speed.

minor thrust face the outboard side of the piston as its throw rotates away from the crankshaft centerline on the power stroke. See *thrust faces*.

minor thrust side when cylinder gas pressure acts on a piston, it tends to pivot off a vertical centerline; the minor thrust side is the outboard side of the piston as its throw rotates through the cycle.

mixture the random distribution of one substance with another without any chemical reaction or bonding taking place. Air is a mixture of nitrogen and oxygen.

modem a communications device that converts digital output from a computer to the analog signal required by the phone system.

modulation in electronics, the altering of the amplitude or frequency of a wave for purposes of signaling data.

module a housing that contains either a microprocessor or switching apparatus or both.

monatomic a molecule consisting of a single atom.

monitor the common output screen display used by a computer system; a CRT.

monitoring sensors any of various sensors that signal temperature, position, speed, and pressure values to the PCM or ECU.

Monocomp™ piston a Mahle trademark for a trunk-type piston in which the crown and skirt sections are manufactured separately. The two separate sections are then screwed together using a proprietary process. The piston crown section is manufactured from high-temperature steel then screwed into the steel skirt assembly.

Monosteel™ piston a Federal Mogul–manufactured steel trunk piston the equivalent of a Mahle Monotherm piston.

Monotherm™ piston a Mahle trademark for forged steel, trunk-type piston assemblies with an open skirt.

MOST see *media-oriented system transfer*.

motherboard the primary circuit board in the computer housing to which the other components are connected.

motive power automotive, transportation, marine, and aircraft.

motoring running an engine at zero throttle, with chassis momentum driving engine.

mouse input device that controls the curser location on the screen and switches program options.

muffler an engine silencer that uses sound absorption and resonation principles to alter the frequency of engine noise.

multimedia the combining of sound, graphics, and video in computer programs.

multiple diagnostic interface (MDI) the proprietary GM communications adapter (CA) used to connect a GM data bus with a PC loaded with GM MDI Manager software and via the Web, the GM LAN. Capable of hardwire or WLAN connectivity with vehicle interface.

multiple diagnostic interface manager (MDI-M) the GM Windows environment software required to troubleshoot and reprogram GM's post-2010 vehicles using the CAN-C bus.

multiple-orifice nozzle a typical hydraulic injector nozzle whose function it is to switch and atomize the fuel injected into an engine cylinder. Consists of a nozzle body machined with the orifii, a nozzle valve, and a spring. Used in diesel port-helix injection pumps, MUIs, EUIs, EUPs, and HEUIs. The commonly used nozzle until 2007; no engines meeting post-2010 emissions standards use them.

multiple splice an electrical connection that joins a number of wires at a single junction.

multiplexing used to describe the connecting of two or more electronic system controllers on a data backbone to synergize system operation and reduce the number of common components and hard wiring.

multipulse injection a feature of most current diesel fuel injection systems in which a fueling pulse can be divided into up to seven separate injection events within a single cycle.

multitasking the ability of a computer to simultaneously process multiple data streams.

Nalcool a brand of antifreeze coolant solution supplemental additive.

nanosecond one-billionth of a second.

natural gas naturally occurring subterranean organic gas (gaseous crude oil) composed largely of methane.

naturally aspirated (NA) describes any engine in which intake air is induced into the cylinder by the lower-than-atmospheric pressure created on the downstroke of the piston and which receives no assist from boost devices such as turbochargers.

Navistar a major manufacturer of truck chassis and engines. Previously International Harvester and often referred to by the slang term *binder* sourced from the strong agricultural heritage of the company. Corporate center is in Chicago, Illinois.

needle valve nozzle another way of describing a multi-orifii, hydraulic injector nozzle (DDC).

net power an SAE formula to calculate actual engine power by running it with all potential parasitic loads and emissions controls activated. SAE net power formula is specified in J1349.

network a series of connected computers designed to share data, programs, and resources.

network access point (NAP) wireless data transfer module used for data transfer from vehicle to land station.

networking the act of communicating using computers.

neutral safety switch transmission located switch that opens the cranking control circuit when the transmission is in gear preventing engine startup. Neutral safety switches are designed to close when the transmission is in park or neutral.

neutron a component part of an atom with the same mass as a proton, but with no electrical charge. Present in all atoms except the simplest form of hydrogen.

newton unit of mechanical force defined as the force required to accelerate a mass of 1 kilogram through 1 meter in 1 second.

nibble four bits of data or half a byte.

Ni-Resist insert a high-strength, nickel alloy piston ring support insert in an aluminum trunk-type piston with a similar coefficient of heat expansion to aluminum.

nitrogen (N) a colorless, tasteless, and odorless gas found elementally in air at a proportion of 76 percent by mass and 79 percent by volume. Atomic number 7.

nitrogen dioxide (NO_2) one of the oxides of nitrogen produced in vehicle engines and a significant contributor to the formation of photochemical smog.

nodes (1) dumb terminals (no processing capability) and PCs connected to a network. (2) portion of a wave signal above a zero, mean, or neutral point in the band.

noise in electronics, unwanted pulse or waveform interference that can scramble signals.

normal rated power the highest power specified for continuous operation of an engine.

NOT gate any circuit whose outcome is in the On or one state until the gate switch is in the On state, at which point the outcome is in the Off state.

notebook computer briefcase-sized PC-designed for portability.

NO_x adsorber catalyst (NAC) a two-stage exhaust after-treatment system that uses base metal oxides to initially adsorb (store) NO_x compounds, followed by the use of a rhodium reduction catalyst and *dosing* to reduce NO_x back to N_2.

noxious emissions engine end gases that are classified as harmful. Includes NO_x and HC but does not include CO_2 (a greenhouse gas) and H_2O.

NO_x sensor measures exhaust gas NO_x by essentially electrolytically reducing the compound and comparing the "reduced O_2" with the O_2 in the atmosphere.

nozzle the component of most hydraulic and electronic injector assemblies responsible for switching and atomizing fuel.

nozzle closing pressure (NCP) the specific pressure at which a hydraulic injector nozzle closes, always lower than NOP due to nozzle differential ratio. Also known as *valve closing pressure*.

nozzle differential ratio the ratio of nozzle valve seat to nozzle valve shank sectional areas. This ratio defines the pressure difference between NOP and nozzle closure values.

nozzle opening pressure (NOP) the trigger pressure value of a hydraulic injector nozzle.

nozzle seat the seat in an injector nozzle body sealed by the nozzle valve in its closed position.

nozzle valve motion sensor (NVMS) motion sensor used to detect nozzle valve movement (at NOP) by inducing a signal voltage proportional to nozzle valve speed of motion.

nucleus the center of an atom incorporating most of its mass and usually made up of neutrons and protons.

numeric data represented by number digits.

numeric keypad microprocessor-based instrument with numeric-only input keys such as those on a ProLink EST.

Occupational Safety and Health Administration (OSHA) U.S. federal agency responsible for administering safety in the workplace.

octane rating denotes the ignition and combustion behavior/rate of a fuel, usually gasoline. As the octane number increases, the fuel's antiknock characteristics increase and the burn rate slows.

ohm a unit for quantifying electrical resistance in a circuit.

Ohm's law the formula used to calculate electrical circuit performance. It asserts that it requires 1V of potential to pump 1A of current through a circuit resistance of 1. Named for Georg Ohm (1787–1854).

oil cooler a heat exchanger designed to cool oil, usually using engine coolant as its medium.

oil pan the oil sump, normally flange mounted directly under the engine cylinder block.

oil window the portion of the upper strata of the Earth's crust in which crude petroleum is formed.

opacimeter see *opacity meter*.

opacity meter a light extinction means of testing exhaust gas particulate and liquid emission that rates density of exhaust smoke based on the percentage of emitted light that does not reach the sensor, so the higher the percentage reading, the more dense the exhaust smoke.

open circuit any electrical circuit through which no current is flowing, whether intentional or not.

open circuit voltage (OCV) voltage measured in a device or circuit through which there is no current flow.

opens an electrical term referring to open circuits/no continuity in a circuit, portion of the circuit, or a component.

operand machine-language directive that channels data and its location.

operating environment defines the monitor display character and the GUI (graphical user interface) consisting of icons and other symbols to increase user-friendliness.

operating system (OS) core software programs that manage the operation of computer hardware and make it capable of running functional programs.

opposed plungers reciprocating members of an inlet-metering, opposed-plunger rotary distributor pump. Usually a pair of opposed plungers are used, but some pumps use two pairs. The plungers are forced outward as fuel is metered into the pump chamber; when the plunger-actuating rollers contact the internal cam profiles, they are driven inboard (toward each other), simultaneously pressurizing the fuel in the pump chamber.

optical character recognition (OCR) scanners that read type by shape and convert it to a corresponding computer code.

optical codes graphic codes that represent data for purposes of scanning, such as bar codes.

optical disks digital data storage media consisting of rigid plastic disks on which lasers have burned microscopic holes. The disk can then be optically scanned (read) by a low-power laser.

optical memory cards digital data storage media the size of a credit card capable of retaining the equivalent of 1,600 pages of text.

OR gate a multiple input circuit whose output is in the On or one state when any of the inputs is in the On state.

Organization of Petroleum Producing Countries (OPEC) a cartel of oil-producing countries that regulates oil supplies to maintain pricing.

orifice a hole or aperture; plural *orifii*.

orifice nozzle a hydraulic injector nozzle that uses a single orifice (unusual) or a number of orifii through which high-pressure fuel is pumped and atomized during injection.

original equipment manufacturer (OEM) term used to describe the manufacturer of original product, distinct from aftermarket manufacturer (replacement product).

oscilloscope an instrument designed to graphically display electrical waveforms on a CRT or other display medium.

Otto cycle the four-stroke, spark-ignited engine cycle patented by Nicolas Otto in 1876. The four strokes of the cycle are induction, compression, power, and exhaust.

outer base circle (OBC) the portion of a cam profile with the largest radial diameter.

output the result of any processing operation.

output devices components controlled by a computer that effect the results of processing. The CRT and printer on a PC system and the injector drivers on a diesel engine are all classified as output devices.

outside diameter (od) outside measurement of a shaft or cylindrical component, but can also be used to mean any outside dimension.

outside micrometer a standard micrometer designed to precisely measure od or thickness. Consists of an anvil, spindle, thimble, barrel, and calibration scales.

overhead adjustment term used to refer to setting cylinder head valves and timing injectors; also known as *tune up*.

overspeed a governor condition in which the engine speed, for whatever reasons, exceeds the set high-idle speed or top engine limit.

oversquare engine an engine in which the cylinder bore diameter is larger than the stroke diameter: most current spark ignited engines are *oversquare*.

oxidation the act of oxidizing a material; can mean combusting or burning a substance.

oxidation catalyst a catalyst that enables an oxidation reaction. In the oxidation stage of a catalytic converter, the catalysts platinum and palladium are used.

oxidation stability describes the resistance of a substance to be oxidized. It is a desirable characteristic of an engine lubrication oil to resist oxidation, so one of its specifications would be its oxidation stability.

oxides of nitrogen (NO_x) any of a number of oxides of nitrogen that may result from the combustion process; they are referred to collectively as NO_x. When combined with HC and sunlight, reacts to form photochemical smog.

oxyacetylene a commonly used cutting, heating, and welding process that uses pure compressed oxygen in conjunction with acetylene fuel.

oxygen colorless, tasteless, odorless gas; the most abundant element on the Earth; occurs elementally in air and in many compounds, including water.

ozone an oxygen molecule consisting of three oxygen atoms (triatomic). Exists naturally in the Earth's ozonosphere (6 to 30 miles altitude), where it absorbs ultraviolet light, but can be produced by lightning and by photochemical reactions between NO_x and HC. Explosive and toxic.

packet a data message delivered to the data bus when a ladder switch resistance changes, indicating a change in switch status.

palladium an oxidation catalyst often used in catalytic converters.

pallet the "bearing" end of a rocker that directly contacts a valve stem or yoke pad.

Palm Pilot hand-sized microprocessor unit with PC compatibility; used with proprietary software to troubleshoot some vehicle electronic systems.

parallel circuits electrical circuits that permit more than a single path for current flow.

parallel hybrid term usually used to describe a commercial vehicle powertrain in which a diesel engine drives a genset, producing electricity to drive electric motors that provide torque to the wheels.

parallel hydraulic drive (PHD) one manufacturer's way of describing a *parallel hydraulic hybrid drive (PHH)* system.

parallel hydraulic hybrid (PHH) an Eaton Corp. parallel drive system in which a conventional diesel-driven drivetrain is assisted by a hydraulic system consisting of a reversible piston pump and motor coupled to the driveshaft by a clutch, accumulators, plumbing, and a control circuit.

parallel ports peripheral connection ports for computer devices that require large-volume data transmission such as printers and scanners.

parallel port valve configuration engine cylinder valve arrangement that locates multiple valves parallel to crank centerline, permitting equal gas flow through each (assuming identical lift).

parameter a value, specification, or limit.

parameter identifier (PID) code component within an MID system.

parent bore term used to describe an engine with integral cylinder bores machined directly into the cylinder block. Not often used in diesel engines, and when used, the bore surface area may be induction hardened to provide improved longevity.

parity the even or odd quality of the number of 0s (zeros) and 1s (ones), a value that may have to be set to handshake two pieces of electronic equipment.

partial authority term widely used in vehicle technology to describe a hydromechanical system that has been adapted for management by computer; a good example is the Bosch rotary injection pump that was adapted for electronic management.

particulate matter (PM) solid matter. Often refers to minute solids formed by incompletely combusted fuel and emitted in the exhaust gas.

passive DPF a catalyzed DPF that uses latent diesel exhaust heat to burn-off collected particulate (soot) from a wall flow–type filter.

passive DPF mode the primary regeneration cycle of a DPF capable of active and passive modes; it uses latent diesel exhaust heat to burn-off collected particulate (soot).

passive matrix an LCD screen display used in older notebook-type PCs and some vehicle digital display units. They use a single transistor for each row and column, producing a low-quality image.

passive regeneration usually the primary generation mode of a diesel particulate filter in which latent exhaust heat is used to combust accumulated soot.

password an alpha, numeric, or *alpha-numeric* value that either identifies a user to a system or enables access to data fields for purposes of download or reprogramming.

peak pressure the highest pressure attained in a hydraulic system.

peak torque maximum torque. In an internal combustion engine, peak torque always occurs at peak cylinder pressure, and in most cases this will be achieved at a lower speed than rated power rpm.

Peltier effect a heat exchanger principle used in some exhaust gas analyzers to drop a test sample below the *dew point* to remove water.

pencil injector nozzle a slim, pencil-shaped hydraulic injector that often uses an internal accumulator that eliminates the need for a leak-off circuit. Nearly obsolete.

peripherals input and output devices that support the basic computer system, such as the CRT and the printer.

periphery in cam geometry, the entire outer boundary of the cam; cam profile.

peristaltic pump a pump used in some exhaust gas analyzers to remove water from a test sample while allowing the gas to have minimal contact with the water.

personal computer (PC) any of a variety of small computers designed for full function in isolation from other units but which may be used to network with other systems.

Personal Computer Card International Association (PCMCIA) cards credit card–sized data storage cards that can be inserted into PC expansion slots.

personality module Caterpillar and Navistar PROM/EEPROM component.

personality ratings term used by Caterpillar and Navistar to describe PROM and EEPROM functions.

petroleum any of a number of organic fossilized fuels found in the upper strata of the Earth's crust that can be refined into diesel fuel and gasoline among other fuels.

pH used to evaluate the acidity or alkalinity of a substance. From a logarithm of the reciprocal of the hydrogen ion concentration in a solution in moles per liter; p = power, H = hydrogen.

phasing the precise sequencing of events; often used in the context of phasing the pumping activity of individual elements in a multicylinder injection pump.

photochemical reaction a chemical reaction caused by radiant light energy acting on a substance.

photochemical smog smog formed from airborne HC and NO_x exposed to sunlight; also known as *photoelectric smog* or *photosynthetic smog*.

photoelectric smog see *photochemical smog*.

photonic semiconductors semiconductors that emit or detect photons or light.

photons a quantum of electromagnetic radiation energy; when visible, known as light.

photovoltaic the characteristic of producing a voltage from light energy.

pickup tube a suction tube or pipe in a fuel tank or oil sump.

piezoelectric actuators see *piezo injectors*.

piezoelectricity some crystals become electrified when subjected to direct pressure, producing a voltage that increases with pressure increase. In piezoelectricity, the direction of polarization reverses if the direction of applied stress changes; that is, from compression to tension. Piezo effect is therefore reversible, and this reversibility is the principle used in piezo actuators.

piezo injectors diesel fuel injectors that are switched by a piezo actuator located in a control circuit or integrated directly into the injector valve shaft. A piezo actuator is constructed of several hundred piezo crystals, and mechanical movement occurs almost instantly when the wafer stack has voltage applied to it. They have faster response times than solenoid-actuated injectors.

pilot ignition a means of igniting a fuel charge that might normally require a spark, by injecting a short pulse of diesel fuel into a cylinder to ignite a premixed charge of gaseous fuel and air.

pilot injection the injection of a short-duration pulse of diesel fuel, followed by a pause to await ignition, followed by the resumption of the fuel pulse. Used as a cold-start strategy in some systems to prevent diesel knock, and used throughout the fueling profile by others, notably, later versions of HEUI. Can also be used to achieve ignition in applications that use an alternative fuel that does not readily compression ignite. In such instances, a short pulse of diesel fuel is injected to act as the ignition means for the primary fuel.

pin boss the wrist pin support bore in a piston assembly.

pintle nozzle a type of hydraulic injector nozzle used in some IDI automobile, small-bore diesel engines until recently.

pipelining rapid sequencing of functions by the CPU to enable high-speed processing.

piston the reciprocating plug in an engine cylinder bore that seals and transmits the effects of cylinder gas pressure to the crankshaft.

piston pin a wrist pin that links the piston assembly to the connecting rod eye.

piston speed the distance traveled by one piston in an engine per unit of time.

pitting a wear pattern that results in small *pock marks* or holes.

pixels picture elements. A measure of screen display resolution—each dot that can be illuminated is called a pixel.

plain old telephone service (POTS) telecommunications using the telephone system for part or all of the transaction, still the backbone of most telecommunications.

plasma a gas of positive ions and free electrons with an approximately equal positive and negative charge.

Plastigage a shaft-to-friction bearing clearance measuring system consisting of nylon cord that deforms to conform with the clearance dimension so it can be measured against a coded scale on the packaging envelope.

platinum an oxidation catalyst often used in catalytic converters.

platinum resistance thermometer (PRT) resistive thermal devices that measure the change in the electrical resistance of gases that occurs with temperature change. PRTs can replace thermocouple-type pyrometers where temperatures do not exceed 600°C (1100°F).

PLD term used to describe the engine controller module on Mercedes-Benz diesel engines; a German acronym meaning *ECM*.

plunger the reciprocating member of a plunger pump element.

plunger geometry term used to describe the shape of the metering recesses/helices in a pumping plunger and therefore the pump timing and delivery characteristics.

plunger leading edge the point on a pumping plunger closest to the pump chamber.

plunger pump any pump that uses a reciprocating piston or plunger and, in most cases, is hydraulically classified as positive displacement.

pneumatics the science of the mechanical properties of gases, especially in confined circuits designed to provide motive power.

pocket technician Palm Pilot–based tool that plugs into a dash adapter; used to assist technicians with diagnosis, but a driver's version is also an option. Capable of analyzing fuel economy and engine operating conditions and displaying fault codes.

policy adjustment a polite way of describing a shop floor error that requires a service facility to write-off time on a repair invoice.

polymer a compound composed of one or more large molecules, formed by chains of smaller molecules.

popping pressure see *nozzle opening pressure*.

pop test the testing of NOP on a hydraulic injector nozzle using a bench or pop tester.

port (1) an aperture or opening. (2) A computer connection socket used to link a computer with input and output devices.

port closure the beginning of effective stroke in a plunger and barrel pumping element, occurring when the plunger leading edge closes off the spill/fill port(s).

port opening the ending of effective stroke in a plunger and barrel pumping element, occurring when the fill/spill ports are exposed to the chamber.

positive crankcase ventilation (PCV) an EPA requirement for diesel engine crankcases beginning in 2007 (off-highway 2008). Diesel engine manufacturers have adopted positive or centrifugal-type filtration of crankcase vapors prior to rerouting them to the intake upstream from the turbocharger impeller. Usually called *closed crankcase ventilation (CCV)* in diesels.

positive displacement describes a pumping principle in which the quantity of fuel pumped (displaced) per cycle does not vary, so the volume pumped depends on the rate of cycles per minute. When a positive displacement pump unloads to a defined flow area, pressure rise will increase in proportion to rpm or cycles per minute.

positive filtration a filter in which all of the fluid (gas or liquid) to be filtered is forced through the filtering medium. Most air, fuel, coolant, and oil filters used today employ a positive filtration principle.

POST power-on self-test. A BIOS test run at boot up to ensure that all components are operational.

potential difference electrical charge differential measured in voltage.

potentiometer a three-terminal variable resistor or voltage divider used to vary the voltage potential of a circuit. Commonly used as a throttle position sensor.

pour point a means of evaluating a fuel or lubricant's low-temperature flow characteristics. The pour point of a fuel is slightly higher in temperature than its gel point.

powdered metal technology refers to sintering production processes that may include certain types of alumina ceramics. See also *sinter*.

power the rate of accomplishing work; it is necessarily factored by time.

power line carrier communication transactions delivered through a power line. Signals are converted to radio frequencies for the transaction and subsequently decoded by the receiver ECM. An example is the power line carrier use of the auxiliary (blue) wire in a standard seven-pin trailer connector for trailer-to-tractor ABS communications.

Power Stroke Ford diesel engines. The first generations were Navistar-built, HEUI-fueled V8 engines available in 7.3 and 6.0 liter displacement versions. The current Power Stroke engines (2011) are Ford engineered and built with a displacement of 6.7 liters.

power take-off (PTO) an engine- or transmission-located device used to provide auxiliary power. Can also mean the primary coupling between the engine and powertrain, so in a diesel engine, this would be the flywheel.

powertrain the components of a system directly responsible for transmitting power to the output mechanisms. In an engine, the powertrain components are piston assemblies, connecting rods, crankshaft, and flywheel.

powertrain control module (PCM) the preferred automotive term to describe the control module that manages the engine and other powertrain components. Also known as the *engine control module (ECM)*.

power transistor a transistor used as the final switch in an electronic circuit to control a solenoid or other output; sometimes known as a *driver*.

predelivery inspection (PDI) the full chassis inspection that should be performed before a new vehicle is delivered to a customer.

prefix addition of a syllable or letter(s) or numbers at the beginning of a word or acronym.

pre-injection metering (PRIME) a Caterpillar term for the pilot injection concept used in its second generation of HEUI injectors.

prelubricator a pump used to charge the lubrication circuit on a rebuilt engine before startup.

pressurizing the process of raising the pressure in a circuit.

preventive maintenance (PM) routine scheduled maintenance on vehicles.

primary filter usually describes a filter on the suction side of a fuel subsystem, whereas the term *secondary filter* describes the filter on the charge side of the transfer pump.

prime mover an initial source of power; for instance, the prime mover of a diesel fuel subsystem is the transfer pump.

processing the procedure required to compute information in a computer system. Input data is processed according to program instructions, and outputs are plotted.

program set of detailed instructions that organize *processing* activity.

programmable logic controllers small computers that perform switching functions in much the same way as a relay.

programmable read-only memory (PROM) a chip or chips used to qualify ROM data to a specific chassis application. In early vehicle computers, this was usually the only method of reprogramming data to an ECM; this PROM function has now been superseded by the EEPROM capability found in most ECMs.

ProLink Scan Tools generic electronic service tools (ESTs) manufactured by Nexiq capable of scanning and low-level programming functions on a chassis data bus. Subscription or purchase of software cartridges, multi-protocol data cards, or downloaded software are used to read each manufacturer's system. This family of EST has progressed through several generations and the current version is known as the Nexiq iQ.

propagate to breed, transmit, or multiply. The word is often used to describe the combustion process in an engine cylinder, as in flame propagation.

propane a petroleum-derived gas consisting mostly of methane, often known as *liquefied petroleum gas* or LPG.

proportional solenoid a solenoid whose armature will be positioned according to how much current is flowed through its coil. Often an ECM-actuated output.

proprietary data programming programming to a PCM/ECM that is "owned" by the vehicle manufacturer. This includes programming fields such as the fuel map and power rating.

proprietary OS OS that is privately owned and specific to a manufacturer or operator.

propylene glycol (PG) A less toxic glycol-based antifreeze solution than EG. PG mixture strength must be tested with a refractometer with a PG scale and not mixed with EG.

protocols sets of rules and regulations. Often used to define communication language.

proton positively charged component of an atom located within its nucleus.

psi pounds per square inch. Standard unit of pressure measurement.

pulsation damper on Cummins gear-type supply pumps, pulsation dampers are used to smooth the pressure waves caused by a gear-type pump as it loads fuel into its outlet.

pulse exhaust a tuned exhaust system used to optimize the gas dynamic of exhaust gas delivered to the turbocharger.

pulse wheel the rotating disc used to produce rpm or rotational position data for an ECM. The term is most often applied to the rotating member of a Hall effect sensor, but at least one manufacturer uses the term to describe an AC reluctor wheel.

pulse width (PW) usually refers to EUI duty cycle measured in milliseconds.

pulse-width modulation (PWM) constant frequency, digital signal in which on/off time can be varied to modulate the duty cycle.

pump drive gear the gear responsible for imparting drive force to a pump.

pump-line-nozzle (PLN) the hydromechanical or electronically managed injection pump-to-line-to-nozzle fuel injection principle used in most diesel fuel systems until the introduction of EUI engines.

pushrods cylindrical solid rods located between a follower and a rocker assembly that transmit the effects of cam profile to action at the rocker arm.

push tubes hollow, cylindrical tubes located between a follower and a rocker assembly that transmit the effects of cam profile to action at the rocker arm.

pyrometer a thermocouple-type, high-temperature sensing device used to signal exhaust temperature. Consists of two dissimilar wires (pure iron and constantan) joined at the hot end with a millivoltmeter at the read end. Increase in temperature will cause a small current to flow, which is read at the voltmeter as a temperature value.

quantum a defined quantity of energy, proportional to the frequency of radiation it emits.

quick response (QR) code Denso electrohydraulic injector fuel flow calibration data specification that must be programmed to the ECM; allows the ECM to precisely balance fueling to the engine cylinders.

QuickServe OnLine (QSOL) Cummins online engine and electronics server for repair, troubleshooting, and service bulletins.

quiescent a term used to describe any low-turbulence engine cylinder dynamic. Its root is from the word *quiet*.

rack actuator the ECM-controlled output device used on electronically controlled versions of port-helix metering injection pumps to move the fuel control rack.

radial piston pump means of creating injection pressures in some current common rail diesel fuel injection systems, notably Cummins CAPS and Bosch CR. Multicam profiles actuate reciprocating plungers that unload to an accumulator or rail.

radial vector the radial angle off a reference point, say TDC, in a crankshaft that indicates the mechanical advantage of a throw in its relationship with the crankshaft centerline.

radiation the transfer of heat or energy by rays not requiring matter such as a liquid or a gas.

radiator a heat exchanger used in liquid-cooled engines designed to dissipate some of the engine's rejected heat to atmosphere.

radioactive any substance or set of physical conditions capable of emitting radioactivity. Exposure to high-level radioactivity can be life threatening, while low-level radioactivity (such as electrical or radar waves) represents debatable hazards.

rail a manifold.

rail actuator see *fuel control actuator*.

rail pressure control valve (RPCV) a linear proportioning solenoid with integral spool valve used as an ECM/PCM output in a common rail injection system; its function is to precisely manage rail pressure. In ECM/PCM processing, it is looped with the rail pressure sensor (inputs actual rail pressure) in an attempt to maintain "desired" rail pressure.

rail pressure sensor (CR) V-Ref supplied, variable capacitance–type pressure sensor that signals actual rail pressure to the ECM/PCM at any given moment of operation.

ram air air fed into engine cooling and intake circuits by the velocity of a moving vehicle; increases proportionally with vehicle speed.

ramps in cam geometry, the shaping of the cam profile between the IBC and the OBC. The ramp geometry defines the actuation/unload characteristics of the train that rides its profile.

random-access memory (RAM) electronically retained "main memory" of a computer system.

rapid start shutoff valve Cummins common rail fuel system electric (solenoid) shutoff valve that traps fuel in the rail on shutdown to enable an almost instant restart.

rated power the peak horsepower produced by a diesel engine; often expressed as *rated speed* because it is always correlated to a specific rpm.

rated speed the rpm at which an engine produces peak power.

rate shaping a fuel injection term that describes the ability of a fuel system to control fuel delivery to the cylinder independent of the hard limitations of cam geometry and engine rpm. Because HEUI injectors are actuated hydraulically and the hydraulic actuation pressure can be controlled by the ECM/PCM, this system is capable of rate shaping, as are all CR systems.

ratio quantitative relationship between two values expressed by the number of times one contains the other.

reactive substances that can become chemically reactive if they come into contact with other materials, resulting in toxic fumes, combustion, or explosion.

reader-programmer a term used in the commercial truck arena to describe generic ESTs designed to scan, reprogram, and perform a limited level of diagnostics on an electronic system.

read-only memory (ROM) data that is retained either magnetically or by optical coding and designed to be both permanent and read-only.

ream the machining process of accurately enlarging an orifice using a steel boring bit with straight or spiral fluted cutting edges.

recording density number of bits that can be written to an inch of track on a disk; measured in bpi or bits per inch.

rectifier device used to convert AC into DC.

reference coil Mack Trucks rack position sensor magnetic field temperature reference; validates input from the rack position sensor.

reference voltage (V-Ref) the ECM-controlled output to onboard sensors.

refraction the extent to which a light ray is deflected (bent) when it passes through media such as water, coolant, or fog.

refractive index truly the ratio of the speed of light in a vacuum versus the speed of light through a specified medium, but in practice, used to express natural light refractometer readings in coolant or battery electrolyte.

refractometer an instrument that directs light through a liquid to measure its refractive index: used in commercial vehicle service facilities to measure the health of battery acid and antifreeze mixtures. More accurate than hydrometers.

regeneration cycle term used to describe the burn-off cycle of a diesel particulate filter in which accumulated soot is combusted.

regeneration system (RS) the *diesel particulate filter (DPF)* self-cleaning management system; the means used to manage and burn off accumulated DPF soot.

regenerative braking vehicle retarding effort achieved in a parallel hybrid drive unit when the drive electric motor magnetic field is reversed, both applying retarding torque and generating electricity that can be used to charge the batteries.

register alignment or track point of two components.

registers temporary storage locations in a CPU.

rejected heat that portion of the potential heat energy of a fuel not converted into useful kinetic energy.

relief valve a commonly used valve in hydraulic circuits (such as fuel subsystem and lubrication circuits) that defines maximum circuit pressure. The simplest type would consist of a ball check, loaded by a spring to seal a return line. When circuit pressure was sufficient to unseat the ball check, circuit fluid would be diverted from the main circuit to the return.

reluctance resistance to the movement of magnetic lines of force.

reluctor a term used to describe a number of devices that use magnetism and motion to produce an AC voltage.

remote data interface (RDI) DDEC communications link between the vehicle electronics and a fleet's PC or PC network.

remote sensing describes any kind of telemetric signaling. An example is the means some alternators use to measure potential directly from the battery rather than at the field coils.

reprogram general term used to cover a range of rewrite and overwrite procedures in computer technology.

residual line pressure the pressure that dead volume fuel is retained at in a high-pressure pipe in a PLN fuel system that uses delivery valves at the injection pump; usually around two-thirds of the NOP value.

resistance opposition to electrical current flow in a circuit.

resistance temperature detector (RTD) resistive thermal devices that, measure the change in the electrical resistance of gases that occur with temperature change. They are made of platinum, so these devices are commonly known as *platinum resistance thermometers (PRTs)*.

resolution the smallest interval measurable by an instrument. In computer terminology, it usually describes the image clarity of a CRT display in pixels. It also defines range in a DMM.

resonation principle a noise-reducing principle used in engine silencers that scrambles sonic nodes and antinodes by reflecting sound back toward its source, thereby altering the frequency.

Resource Conservation and Recovery Act (RCRA) U.S. federal legislation that regulates the disposal of hazardous materials.

restriction gauge used to display inches of vacuum pull through an air filter: when vacuum pull exceeds a threshold, the air filter can be assumed to be plugged. Measured in inches of H_2O vacuum.

retarder generally refers to braking action, that is, the retarding of vehicle movement.

retraction collar/piston the component on a delivery valve core that is designed to seal before it seats and therefore helps define the residual line pressure value.

retraction spring any spring in any component that causes an assembly to mechanically withdraw or retract.

rhodium a hard white metal occurring naturally in platinum ores and used as an NO_x-reduction catalyst.

Right to Know legislation a provision of the U.S. federal Hazard Communications legislation that imposes on employers the duty of fully revealing the potential dangers of hazardous materials to which their employees may be exposed.

ring belt the area of the piston in which the piston ring grooves are machined.

ring groove the recess or groove machined into a piston in which piston rings are placed.

road speed governing (RSG) the managing of engine output on the basis of a specific road speed. Can also be used to mean the maximum programmed road speed limit.

road speed limit (RSL) usually the maximum programmed road speed value programmed to a vehicle management system, meaning that the vehicle should not travel faster than this speed.

road speed sensor a sensor, usually of the pulse generator type, located at the transmission tailshaft or a wheel assembly that signals road speed data.

rocker arm see *rockers*.

rocker assemblies the entire rocker assembly, consisting of rockers, rocker shaft, and pedestals.

rocker pallet the end of a rocker that contacts the injection pumping tappet, the valve stem, or the valve bridge.

rockers shaft-mounted, pivoting levers that transmit the effects of cam profile to valves and injection pumping apparatus.

rod eye the upper portion of a connecting rod that connects to the piston wrist pin; also known as *small end*.

root cause in failure analysis, the original source of a failure; should be differentiated from *coincidental damage* that can often mislead in the troubleshooting process.

root mean square (RMS) a term used to describe averaging when performing machining operations and also used to describe AC voltage measurement by ascribing to it an equivalent DC value.

Roots blower a positive displacement air pump consisting of two gear-driven, intermeshing spiral fluted rotors in a housing; used to scavenge (not to boost) two-stroke cycle diesel engines a generation ago.

rotor any rotating member or shaft in a motor or pump.

router connection between two networked computers or two vehicle PCMs/ECMs.

RS-232 port a serial communications port (often Com 1) in a computer system that accepts a standard phone jack. Some scan tools are equipped with an RS-232 port, which can drive a printer or PC data display.

run-in usually describes the engine break-in procedure following a rebuild outlined by the manufacturer.

sac a spherical cavity. Refers to the chamber in some multiorifii injector nozzles beyond the seat and from which the exit orifii extend.

SAE horsepower a structured formula used to calculate brake power data that can be used for comparison purposes.

SAE J-standards standards developed by SAE industry committees and generally agreed to, usually without any statutory obligation.

SAE J1587 electronic data exchange protocols used in data exchange between heavy-duty, electronically managed systems.

SAE J1667 standards for heavy-duty, diesel stack emissions testing.

SAE J1708 serial communications and hardware compatibility protocols between microcomputer systems on a J1587 data bus. Its data link is a six-pin Deutsch connector.

SAE J1850 midgeneration light duty (automotive) data bus based on CAN 2.0 architecture. Superseded by faster CAN-C.

SAE J1939 the set of heavy-duty multiplexing standards that replaced both J1587 and J1708. Both software and hardware protocols and compatibilities are covered by J1939, which is accessed by a standard SAE nine-pin Deutsch connector. A CAN 2.0 data bus.

SAE J1962 connector the *data communication link (DLC)* is used to access a chassis data bus. The current OBD II data link is known as an *assembly line data link (ALDL)*, a 16-pin connector.

SAE viscosity grades the industry standard for grading lubricating oil viscosity.

sampling the process a computer system uses to monitor noncommand-type input data from the sensors in a system. Inputs from oil pressure, ambient temperature, coolant temperature, and so on, would be monitored by sampling.

saturation condition of an electromagnet in which a current increase results in no increase in the magnetic flux field.

scan tool term used to described the handheld *electronic service tool (EST)* used to access the data bus on light duty vehicles. The term originates from an era when these ESTs could only read (scan) the bus: more recent scan tools can read, program, and navigate troubleshooting on the data bus.

scavenge term used generally to describe the process used to expel end gases from an engine cylinder and specifically to describe: (1) the final stage of the exhaust process in a four-stroke cycle engine that occurs at valve overlap; (2) cylinder breathing on a two-stroke cycle diesel engine.

scissor jack an air-actuated floor jack with a pair of clevises that lock to the vehicle frame rails and can lift one end of the chassis well clear of the floor.

scopemeter either a hand-held meter with an integral video display or a CA that functions with software loaded to a PC; used to visually display both analog and digital waveforms for analysis.

scraper ring piston rings below the top compression ring that play a role in sealing cylinder gas as well as managing the oil film on the cylinder wall.

screen any computer output display from LCDs through CRTs.

scrolling the moving of lines of data up or down on a computer display screen.

SCSI controller controller that can support multiple disk drives and enable high data transfer rates.

SCSI port a type of parallel port that can support multiple devices to a single port.

scuffing a superficial scraping of metal against metal damage mode.

search engine software that, when provided with a key word or phrase, scans and retrieves data from memory.

secondary filter usually refers to a filter downstream from the transfer or charge pump in a typical fuel subsystem. It is in most cases, under pressure and capable of much finer filtration than a primary filter, which is usually under suction.

section modulus relates the shape of a beam, cylinder, or sphere to section and stiffness; the greater the section modulus, the higher the rigidity and resistance to deflection. A factor of RBM or resist bending moment.

sector in computer terminology, a pie-shaped section of a disk or section of track.

selective catalytic reduction (SCR) NO_x reduction converters for diesel engine applications that use *aqueous urea* injection. Upon injection, the urea vaporizes to gaseous ammonia that reacts with NO_x compounds to reduce them back to elemental nitrogen and water.

self-regeneration onboard regeneration (cleaning) cycles of a *diesel particulate filter (DPF)*; essentially, it refers to the combustion of accumulated soot during normal operation while on the vehicle.

self-test input (STI) diagnostic scan initiated by depressing a dash STI button.

semiconductor materials that neither conduct well nor insulate; they have four electrons in their outermost shell.

sending unit a variable resistor and float assembly that signals a gauge and/or ECM the liquid level in a tank.

sensing voltage the voltage signal an alternator voltage regulator uses to switch field current on and off.

sensor a term that covers a wide range of command and monitoring input signal devices.

sequential storage storage of data on media such as magnetic tape where data is read and written sequentially.

sequential troubleshooting chart commonly used to structure electronic troubleshooting. The technician navigates the troubleshooting path through the chart on the basis of test results in each step.

serial port port connection that transfers data one bit at a time and therefore more slowly than a parallel port. A mouse is connected to a serial port.

series circuit a circuit with a single path for electrical current flow.

server in the computer processing cycle or multiplex transaction, the fulfilling of a client need is provided by a server.

service hours a means of comparing engine service hours to highway mileage. Most engine manufacturers equate one engine hour to 50 highway linehaul miles (80 km), so a service interval of 10,000 miles (16,000 km) would equal 200 engine hours. The term *engine hours* is also used.

service information systems (SIS) term manufacturers use to describe their service instructions, procedures, and specifications; in most cases, SIS is online because of the ease of updating, correcting, and tagging information with TSBs.

service literature general term used to cover manufacturer service information regardless of whether it is hard (paper) copy, disk-based, or online sourced.

shear the stress produced in a substance or fluid when its layers are laterally shifted in relation to one another. Viscosity describes a fluid's resistance to shear.

shell term used to describe the concentric orbital paths of electrons in atomic structure.

short block a slang term used to describe a (reconditioned) engine minus cylinder head(s) and peripherals.

shutdown solenoid an ETR or latching solenoid that functions to no-fuel the engine to shut it down.

shutterstat a temperature-sensing, pneumatic switch used to manage air shutter operation.

sight glass see *diagnostic sight glass*.

signals codes, signs, or symbols used to convey information.

silicon a nonmetallic element found naturally in silica, silicon dioxide in the form of quartz.

silicon carbide (SiC) conductive material used to make wall flow DPFs. SiC has some advantages over cordierite because it conducts heat away from hot spots and should have longer service life.

silicone any of a number of polymeric organic compounds of the element silicon associated with good insulating and sealing characteristics.

silicone-controlled rectifier (SCR) similar to a bipolar transistor with a fourth semiconductor layer added. Extensively used to switch DC in vehicle electronic and ignition systems.

single-actuator EUI an electronic unit injector (EUI) with a single control cartridge: the VW diesel EUI used until 2010 had a single actuator. Twin actuator EUIs are not used in light duty diesel engines.

single-pass radiator any radiator through which flow is unidirectional.

single-phase mains standard two wire AC mains at 100 to 125V AC.

single speed control (SSC) used to describe isochronous PTO governing.

sinter a means of alloying metals in which the constituent materials are mixed in powdered form and then coalesced by subjecting them to heat and pressure. Produces more uniform metallurgical characteristics than alloying.

sintered steel a steel produced by a sintering process; used in certain engine and fuel system components to produce especially tough and durable material characteristics.

Six Sigma a production improvement methodology developed by Motorola Corporation as a corporate means of eliminating defects and improving customer service.

sleeve metering a means of varying the effective stroke in injection pumps by using a movable (by the governor) control sleeve. Used in a number of older injection pumps but more recently, only in the Bosch sleeve-metering, rotary injection pumps such as the VE.

sleeves see *liners*.

slip rings stationary devices used to conduct current to an alternator rotor.

small bore in the trucking industry, describes diesel engines with displacements between 5.9 and 8 liters displacement.

small end the connecting rod eye.

smart general and some would say slang, term used to describe computer controls; see also *smart logic*.

smart injector term used to describe an *electrohydraulic injector*.

smart logic used to describe computed outcomes that use a broad range of input and memory factors to produce "soft" outcomes rather than adhere to hard values. The term is also used to describe computer peripherals that possess some processing capability.

smart switch so named because it contains a ladder of resistors, usually five per switch, known as a ladder bridge; switch status can be determined by the processor mastering the multiplex using a programmed resistance library that identifies the switch and its status.

smog a word formed by combining the words fog and smoke. A haze produced by suspended airborne particulates. Two major types exist; sulfurous smog produced by combusting sulfur-laden fuels such as coal and heavy oils, and photochemical smog, a primary cause of which is vehicle emissions.

snapshot test a diagnostic test performed on an EST that captures frames of running data before and after an event that can be identified by either an automatic (such as a fault code) or manual trigger.

soak tank a usually heated tank that is filled with a detergent or alkaline solution; used to clean engine components.

Society of Automotive Engineers (SAE) organization responsible for setting many of the manufacturing standards and protocols of the motive power industries and dedicated to educating and informing its members.

soft term used in electronics to indicate a flexible value. It has evolved to refer to anything in electronic/computer technology that cannot be physically touched. An example would be service literature; this might be accessed online, in which case it would be called *soft* format, or printed out, in which case it would be converted to a *hard* format.

soft copy data that is retained electronically or on disk as opposed to being on paper.

soft cruise a cruise control mode programmed into some vehicle/engine management electronics in which the road speed is managed within a window extending both above and below the set speed. Soft cruise can increase fuel economy and is often used in conjunction with vehicle maximum speed programming below maximum cruise speed. May include accident avoidance logic (braking) where radar or video is used on the vehicle.

soft parameter a value that varies and depends on input and processing variables (see *fuzzy logic*). The term is often used to describe current cruise control systems that permit a cushion both above and below the set value. See also *soft cruise*.

software the programmed instructions that a computer requires to organize its activity to produce outcomes.

solar cells PN or NP silicon semiconductor junctions capable of producing up to 0.5V when exposed to direct sunlight.

solenoid an electromagnet with a movable armature.

solid state components that use the electronic properties of solids such as semiconductors to replace the electrical functions of valves.

solid-state storage volatile storage of data in RAM chips.

sound absorption principle a means of converting sound waves to friction then heat, which is then dissipated to atmosphere; used in engine silencers/mufflers.

sound card multimedia card capable of capturing and reproducing sound.

spalling surface fatigue that occurs when chips, scales, or flakes of two surfaces in contact with each other separate because of fatigue rather than wear. Also known as *contact stress fatigue*.

spark ignited (SI) any gasoline-fueled, spark-ignited engine, usually using an Otto cycle principle.

special field notification (SFN) a type of TSB that usually outlines a mandatory fix.

specific fuel consumption (SFC) fuel consumed per unit of work performed.

specific gravity the weight of a liquid or solid versus that of the same volume of water.

spectrographic analysis a low-level radiation test that can accurately identify trace quantities of matter in a fluid; used to analyze engine oils.

spike an electrical (voltage) or hydraulic pressure surge.

spill timing term used to describe port closure timing of a port-helix metering injection pump to the engine it fuels, and when establishing injection pump bench phasing.

spindle an intermediary, responsible for transmitting force. In a hydraulic injector, it relays spring force to the nozzle valve.

splice to join.

split locks the keepers fitted to a peripheral groove at the top of a cylinder valve stem that hold the spring retainer in position.

split shot injection pilot injection used on the first generation of multipulse injectors.

spreader bar a rigid lifting aid that permits an engine to be raised on a single-point chain hoist using two, three, or four lift points on the engine.

spreadsheet software that enables numeric data organization and calculation.

spur gear a gear with radial teeth.

square engine an engine in which the bore and stroke dimensions are the same or nearly the same.

stanchion a vertical bracket bolted to the chassis used to mount exhaust aftertreatment canisters, stack(s), and other equipment; usually on commercial trucks.

star bus see *star network*.

star module the hub module of a star network.

star network multiplexing topology in which multiple modules network through a single central module, sometimes called a *gateway*.

starter circuit the chassis electrical circuit that supplies the high current load required to crank an engine.

starter motor a vehicle cranking motor that converts electrical energy into cranking torque.

starter relay the magnetic switch that when energized by the low-current starter control circuit, closes the high-current cranking circuit to energize the starter motor.

starter system the electrical circuit that manages the starter motor consisting of a low-current control circuit and a high-current cranking circuit.

start of injection (SOI) describes the moment atomized fuel exits nozzle orifii, an event that always occurs after pump port closure (PLN systems) or the injector pulse trigger (electronic systems).

state of health (SOH) a common control module test of an actuator often performed before system startup and during operation.

static charge capacitive charge accumulation.

static discharge occurs when accumulated capacitive charge rapidly unloads usually creating an arc such as during a thunderstorm or when a negatively charged hand contacts a computer motherboard.

static electricity accumulated electrical charge not flowing in a circuit.

static friction the characteristic of a body at rest to attempt to stay that way; see *inertia*.

static RAM a RAM chip with a medium-to-large volume memory retention and high access speed.

static timing term used to describe port closure timing of a fuel injection pump to an engine.

stator the stationary element in a rotating component such as the stationary member of an alternator into which electrical current is induced by the rotor.

steel an alloy of iron and a small quantity of carbon; there are many different alloys of steel with a broad range of performance characteristics.

stiction stationary friction; an example would be thread contact friction during fastener torquing.

stoichiometric ratio an air-fuel ratio (AFR) term meaning that at ignition, the engine cylinder has the exact quantity of air (oxygen) present to combust the fuel charge; if more air is present, the AFR mixture is lean, if less air is present, the AFR is rich.

stoichiometry the science of determining the ratio of reactants required to complete a chemical or physical reaction.

stop engine light (SEL) a level III driver alert indicating that the driver should shut down the engine or if programmed to do so, the management system will shut down the engine. Usually illuminated 30 to 60 seconds before shutdown.

storage media any nonvolatile data retention devices; floppy disks, data chips, CD-ROM, and PCMCIA cards are examples.

strategy action plan. In computer technology it relates to how a series of processing outcomes is put together; for instance, startup strategy deals with how the sequence of events are switched by the ECM/PCM that results in the engine being cranked and started.

stress raiser a feature in either the shape (section modulus) or composition (metallurgical) of a component that causes a localized increase in stress; this can result in initiating a failure.

stroke linear travel of a piston or plunger from BDC to TDC. In an engine, piston stroke is defined by the crank throw dimension.

sublimation the process of converting a solid directly to a vapor by heating it; may also mean refine.

substrate (1) the supporting material on which an electric/electronic circuit is constructed/infused. (2) Thermally stable, inert material on which active catalysts are embedded on a vehicle catalytic converter.

subsystem identifier (SID) branch circuit within an MID off the data bus used for diagnostic reporting.

suction circuit the portion of a lubrication or fuel subsystem that is on the pull side of the transfer pump.

suffix addition of a syllable or letter(s) or numbers at the end of a word or acronym.

sulfur an element present in most crude petroleums, but refined out of most current highway fuels. During combustion, it is oxidized to sulfur dioxide. Classified as a noxious emission.

sulfur dioxide the compound formed when sulfur is oxidized; the primary contributor to sulfurous-type smog. Vehicles contribute little to sulfurous smog problems due to the use of low-sulfur fuels.

sump the lubricating oil storage device on an engine more commonly referred to as an oil pan.

supercharger technically any device capable of providing manifold boost, but in practice used to refer to gear-driven blowers such as the Roots blower.

supplemental cooling (system) additives (SCAs) conditioning chemicals added to antifreeze mixtures.

swash plate pump a pump that uses a rotating, circular plate (the swash plate) set obliquely on a shaft to act as a cam and convert rotary motion into reciprocating movement to actuate pistons or plungers; also known as a *wobble plate*.

swept volume volume displaced by a piston as it travels from BDC to TDC.

synchronizing position the Caterpillar MUI (3116) fuel-balancing procedure.

synergize the act of creating synergy.

synergy any process in which the combined effect of a group of subcomponents exceeds the sum of their individual effects. Used to describe somewhat independent vehicle systems that share hardware and software to cut down on duplicated components and to increase overall effectiveness or processing speeds.

synthetic oil petroleum-based and other elemental oils that have been chemically compounded by polymerization and other laboratory processes.

system board see *motherboard*.

system clock device that generates pulses at a fixed rate to time/synchronize computer operations.

system pressure regulator a usually hydromechanical device responsible for maintaining a consistent line pressure; located downstream from a pump.

system unit the main computer housing and its internal components.

tang release tool a lock release tool required to service the sealed connector blocks on electronic engines.

tappets used to describe a variety of devices that ride a cam profile and transmit the effects of the cam geometry to the train to be actuated; also known as *followers*.

tattletale an audit trail that may be discreetly written to ECM/PCM data retention; the recording of electronic events for subsequent analysis.

Tech II the GM proprietary scan tool. Slow transaction speeds (max 18 kb/s) and minimal memory limit its functionality on current systems.

Tech Central International Trucks central data hub.

technical service bulletin (TSB) the more common term used to describe up-to-date amendments and corrections to a service procedure or product recall.

Technology and Maintenance Council (TMC) division of the ATA that sets and recommends safety and operating standards for commercial vehicles.

telecommunications any distance communication regardless of transmission medium.

teleconferencing audio/video communication using computers, a camera, and modem linkages.

telematics information transmittal of computer data from a mobile vehicle to data hubs: may include voice, text, and location data.

telemetry the processes of obtaining and transmitting data from sensors for processing or display. For instance, the telemetry of a typical drive-by-wire, analog-type TPS requires a V-Ref supply, potentiometer, and chassis ground, plus the connection wiring to connect the device to the ECM/PCM that requires the signal.

teleprocessing networked station-to-station data exchange and processing.

Tempilstick™ a heat-sensing crayon used for precise determination of high temperatures.

template torquing procedure used on torque-to-yield fasteners that usually involves torquing to a specified value with a torque wrench followed by turning the fastener through an arc of a specified number of degrees measured by a protractor or template. Produces more consistent clamping pressures than torque-only methods. See *torque-to-yield*.

tensile strength unit force required to physically separate a material; in steels, tensile strength exceeds yield strength by around 10 percent.

terabyte a trillion bytes.

terminal (1) a computer station or network node. (2) An electrical connection point.

thermal efficiency measure of how efficiently an engine converts the potential heat energy of a fuel into usable mechanical energy, usually expressed as a percentage.

thermatic fan a fan with an integral temperature-sensing mechanism that controls its effective cycle.

thermistor a commonly used temperature sensor that is supplied with a reference voltage and by using a temperature-sensitive variable resistor, signals back to the ECM/PCM a portion of it.

thermocouple a device made of two dissimilar metals, joined at the ''hot'' end and capable of producing a small voltage when heated. The principle used in pyrometers that monitor DPF and engine exhaust temperatures.

thermostat a self-contained, temperature-sensing/coolant flow modulating device used to manage coolant flow within the engine cooling system.

thick film lubrication lubrication of components where clearance factors tend to be large and unit pressures low.

thin film lubrication see *boundary lubrication*.

three-phase mains high voltage electrical mains circuits used in shops to run such things as welding and compressor equipment. Runs at pressures ranging from 220 to 600V AC.

three-way splice the uniting of three wires at a junction.

threshold values outside limits or parameters.

throttle mechanism that controls air flow to the intake manifold in SI gasoline and diesel engines with pneumatic governors. The term is commonly used to describe the speed control/accelerator/fuel control mechanism in a diesel engine.

throttle delay a mechanical device used to create a lag between accelerator demand and fuel delivered, usually to cut down on smoke emission.

throttle position sensor (TPS) device for signaling accelerator pedal angle to the PCM. A TPS usually receives a V-Ref input and uses a potentiometer (contact) or Hall effect (noncontact)–type operating principle.

thrust linear force.

thrust bearing a bearing that defines the longitudinal or endplay of a shaft.

thrust collar in a mechanical governor, the intermediary between the centrifugal force exacted by the flyweights and the spring forces that oppose it. The thrust collar in a governor is usually connected to the fuel control mechanism.

thrust faces a term used to describe loading of surface area generally but most often of pistons. When the piston is subject to cylinder gas pressure there is a tendency for it to cock (pivot off a vertical centerline) and load the contact faces off its axis on the pin.

thrust washer see *thrust bearing*.

thyristor three-terminal solid-state switch.

timing the manner in which events or actions are sequenced. The term is used in many applications in engines relating to valves, injection, ignition, and others.

timing advance unit a hydromechanical or electronically controlled timing advance mechanism used with port-helix metering pumps.

timing bolt means of locking a component or engine to position for purposes of timing.

timing dimension tool a tune-up tool used to set the tappet height on EUIs.

tone wheel the rotating disc used to produce rpm and rotational position data for an ECM/PCM. The term is most often applied to the rotating member of a Hall effect sensor, but at least one manufacturer uses the term to describe an AC reluctor wheel.

top dead center outboard extremity of travel of a piston or plunger in a cylinder. Usually abbreviated to TDC.

topology the configuration of a computer network or architecture of a digital packet; a network schematic.

torched piston a piston that has been overheated to the extent that meaningful analysis of cause is not possible.

torque twisting effort or force. Torque does not necessarily result in accomplishing work.

torque rise the increase in torque potential designed to occur in a diesel engine as it is lugged down from the rated power rpm to the peak torque rpm, during which the power curve remains relatively flat. High-torque rise engines are sometimes described as constant-horsepower engines.

torque rise profile diagrammatic representation of torque rise on a graph or fuel map.

torque-to-yield bolts with metallurgical properties that allow them to stretch to their yield point as they are tightened; used where more precise clamping loads are required, such as on cylinder heads and connecting rods. Torqued using a *template-torque* method.

torque twist effect one way of describing the twist forces a cylinder block must sustain as engine torque is transferred to the drivetrain; increases as engine torque increases.

torsion twisting force.

torsional stress twisting stresses. A crankshaft is subject to torsional stress because a throw through its compression stroke will travel at a speed fractionally lower than mean crank speed, whereas a throw through its power stroke will accelerate to a speed fractionally higher than mean crank speed. These occur at high frequencies.

total base number (TBN) measure of lube oil acidity reported in lab lube oil analysis; will increase as low-ash oils such as CJ-4 become commonplace.

total dissolved solids (TDS) dissolved minerals measured in a coolant by testing the conductivity with a current probe (TDS tester). High TDS counts can damage moving components in the cooling system such as water pumps.

total indicated runout (TIR) a measure of eccentricity of a shaft or bore usually measured by a dial indicator.

total quality management (TQM) a customer service philosophy that maintains that customer service is the core of all business activity and that every employee in an organization should be made to feel part of a team with a common objective.

tower computer a PC housed in an upright case, usually capable of greater system expansion than the horizontal desktop style.

toxic materials that may cause death or illness if consumed, inhaled, or absorbed through the skin.

trackball the cursor control device often used in some notebook computers consisting of a rotating ball integral with the keypad.

train a sequence of components with a common actuator. See also *valve train*.

transducer an input circuit device that converts temperature, pressure, linear, and other mechanical signals into electrical signals to be sent to an ECM/PCM. A transducer may produce either analog or digital signals to be sent to control modules.

transducer module component responsible for transducing (converting) all engine pressure values (such as oil/turbo-boost, etc.) to electrical signals for ECM/PCM input.

transfer pump describes the fuel subsystem pump used to pull fuel from the fuel tank and deliver it to the injection pumping/metering apparatus.

transformer an electrical device consisting of electromagnetic coils used to increase/decrease voltage/current values or isolate subcircuits.

transient short-lived, temporary; often refers to an electrical spike or hydraulic pressure surge.

transistors any of a large group of semiconductor devices capable of amplifying or switching circuits.

transmission control protocol/Internet protocol (TCP/ IP) a standard for data communications that permits seamless connections between dissimilar networks.

transorb diode used to protect sensitive electronic circuits (such as an ECM) from the inductive kick that can be created by solenoids when their magnetic field collapses.

transponder a ground-based satellite uplink—can be either mobile or stationary.

transposition error a data entry error.

trapezoid a quadrilateral with one pair of parallel sides.

trapezoidal ring see *keystone ring*.

trapezoidal rod a connecting rod with a trapezoidal small end or rod eye designed to maximize the sectional area of the rod subject to compressive pressures; more commonly referred to as a *keystone rod*.

triangulation see *trilateration*.

triatomic a molecule consisting of three atoms of the same element.

tribology the study of friction, wear, and lubrication.

trilateration the locating of a common intersection by using three circles each with different centers; the mathematical basis of GPS technology.

troubleshooting the procedures involved in diagnosing problems.

turbo-boost sensor (TBS) a device used to signal manifold boost values to the ECM/PCM. An aneroid is used in hydromechanical engines and either a piezo-resistive or variable-capacitance device is used in electronically controlled engines.

trunk-type piston a single-piece piston assembly usually machined from aluminum alloys in diesel engine applications.

truth table a table constructed to represent the output of a multiswitch circuit, based on the switch status within the circuit.

tune-up term used to refer to the setting of valves and timing injectors in diesel engines; more commonly known as *overhead adjustment*.

turbine a rotary motor driven by fluid flow such as water, oil, or gas.

turbo commonly used short form of *turbocharger*.

turbocharger an exhaust gas–driven, centrifugal air pump used on most diesel engines to provide manifold boost. Consists of a turbine housing within which a turbine is driven by exhaust gas, and a compressor housing within which an impeller charges the air supply to the intake manifold.

twisted wire pair used as the data backbone in two-wire multiplex systems such as CAN-C. The wires are twisted to minimize EMI.

two-stage filtering any filtering process that takes place in separate stages.

two-terminal EUI an older style EUI (pre-2007) with a single actuator or control cartridge; this generation of EUIs were equipped with hydraulic injector nozzles that were limited to a fixed NOP.

typeface the design appearance of alpha characters.

ultracapacitors function like two capacitors in series. Used as storage-assist devices in HEV applications.

ultralow sulfur fuel (ULS) required for on-highway use in all jurisdictions throughout the United States and Canada. At this time, ULS is required to have a maximum sulfur content of 0.005 percent, the equivalent of 15 ppm.

ultraviolet (UV) radiation having a wavelength just beyond the violet end of the visible light spectrum. Emitted by the sun, but much of it is filtered out by the ozone shield in the Earth's stratosphere.

ultrawide SCSI an SCSI unit that can support up to 15 devices in any combination of internal and external devices. The last device in the chain must be terminated both internally and externally on shutdown.

undercrown the reverse side of a piston crown; in most current diesels, a lube oil cooling jet is targeted at a specific location on the undercrown.

undersquare engine an engine in which the bore dimension is smaller than its stroke dimension. Refers to high-compression engines. Most diesel engines are undersquare.

Underwriters Laboratories (UL) federal product safety standards committee that approve equipment performance to protect consumers.

unit injector a combined pumping, metering, and atomizing device.

universal asynchronous transmit and receive (UART) a single-wire bus, used into the late 1990s, driven by V-Ref; modules with addresses on UART communicate by pulling down the V-Ref value to create a recessive bit (i.e., the "low") binary mode.

universal product code (UPC) the commonly used commercial bar code designed for optical scanners.

universal serial bus (USB) a means of connecting peripherals (external devices) to a PC or network system.

uplink signal transmission from a stationary or mobile ground station to a telecommunications satellite.

upload the act of transferring data from one medium or computer system to another.

upper helix a helix milled into the upper portion of a pumping plunger and giving the characteristic of a variable beginning of pump effective stroke.

urea crystallized nitrogen-based compounds in solution with distilled water; aqueous urea is injected into SCR-type reduction catalytic converters.

USB flash drives memory data storage devices integrated with a USB (universal serial bus) interface; they are small, portable, very lightweight, and rewritable by flashing.

user ID a password used for security and identification on multiuser computer systems.

vacuum restriction used to describe a restriction on the suction side of a fluid circuit; for instance, a plugged filter.

valence number of shared electron bonds an element can make when it combines chemically to form compounds. A valence electron is one in the outer shell of an atom.

validation (1) data confirmation/corroboration. (2) The third and final stage of a vehicle ECM/PCM reprogramming sequence in which a successful data transfer is confirmed to a central data hub.

valve any device that controls fluid flow through a circuit.

valve bridges a means of actuating a pair of cylinder valves with a single rocker; also known as *valve yoke*.

valve closes orifice (VCO) (nozzle) a sacless hydraulic injector nozzle.

valve closing pressure (VCP) the specific pressure at which hydraulic injector nozzle closes, always lower than NOP due to nozzle differential ratio. Also known as *nozzle closing pressure*.

valve float a condition caused by running an engine at higher-than-specified rpms in which valve spring tension becomes insufficient, causing asynchronous (out of time) valve closing.

valve margin dimension between the valve seat and the flat face of the valve mushroom; critical valve machining specification.

valve pockets recesses machined into the crown of a piston designed to accommodate cylinder valve protrusion when the piston is at TDC.

valve polar diagram a valve mapping exercise that makes use of circles or a spiral configuration to map the valve closing and opening event during the engine cycle.

valve train all the components between the cam and the valve; typically would include followers/tappets, push tubes/rods, rocker assemblies, and valve bridges/yokes.

valve yoke See *valve bridges*.

vaporization changing the state of a liquid to a gas.

variable capacitance a common linear sensing device consisting of a coiled spring supplied with a reference voltage: the output signal consists of a proportion of the V-Ref. Commonly used as pressure-sensing devices on engines.

variable geometry (VG) term usually applied to turbochargers that have either external (wastegate) or internal means of managing the way in which exhaust gas acts on the turbine.

variable nozzle (VN) one type of VG turbocharger that uses an ECM/PCM-controlled actuator to vary the turbine volute flow to determine turbine efficiencies.

variable-speed governor (VSG) a governor in which the speed control/throttle mechanism inputs an engine speed value and the governor attempts to maintain that speed as the engine load changes.

variable speed limit (VSL) rpm limiting on a moving vehicle for PTO operation.

variable valve actuator (VVA) electronically controlled and hydraulically (engine lube) actuated, variable valve timing; can delay intake valve closure, reducing the compression charge and ratio.

variable valve timing (VVT) used to optimize power while minimizing emissions; ECM/PCM controlled and hydraulically actuated (engine lube) when used.

vector a straight line between two points in space; a line that extends from the axis of a circle to a point in its periphery.

VECTRO Volvo electronic engine management.

VED-12 Volvo VECTRO managed, inline 12-liter, six-cylinder engine.

vehicle control center (VCC) vehicle-based communications and control center that organizes telematics, communications, and navigation. VCC is usually networked to the powertrain data bus (CAN-C or J1939) and manages connections to the outside world using technologies such as GPS, Bluetooth, and two-way satellite communications.

vehicle control unit (VCU) a vehicle management module that acts as a gateway on the powertrain data bus that is used by some manufacturers to optimize powertrain and chassis operations.

vehicle electronic programming system (VEPS) Caterpillar and International initial ECM proprietary programming, usually executed on the assembly line.

vehicle interface connector (VIC) a *gateway*, the physical connector that networks the engine electronics to the chassis electronics.

vehicle personality module (VPM) Caterpillar/Navistar term used to describe the PROM and EEPROM memory components in a chassis management module.

vehicle speed sensor (VSS) venting the act of breathing an enclosed vessel or circuit to atmosphere to moderate or equalize pressure.

venting the means used to aspirate or breathe a chamber to atmosphere.

vial calibrated cylindrical glass vessel (test tube) used for precise measuring of fuel delivery on a fuel injection pump calibration bench. Also known as *graduate* or *buret*.

video display terminal (VDT) a CRT, LED, or plasma display screen.

videographics array (VGA) the quality category of a video display monitor.

viscosity often used to describe the fluidity of lubricant, but correctly defined, it is a fluid's resistance to shear.

visible image area (VIA) the means of assessing the actual size of a video display monitor by measuring diagonally across the screen. A nominal 15-inch monitor may actually measure only 13.5 inches.

vocational automotive diesels engines that are designed to work rather than power leisure vehicles such as automobiles.

volatile memory RAM data that is only retained when a circuit is switched on.

volatile organic compounds the boiled-off, more volatile fractions of hydrocarbon fuels. The evaporation to atmosphere occurs during production, pumping, and refueling procedures; also known as *VOCs*.

volatility the ability of a liquid to evaporate. Gasoline has greater volatility than diesel fuel.

Volkswagen 507.00 specification an extended-service lubricant recommended for VW and Audi diesel engines. The 507.00 lube is close to the API CJ-4 lube standard, but it has slightly lower ash and a lower general level of wear-enhancing additives.

volt a unit of electrical potential; named after Alessandro Volta (1745–1827).

voltage electrical pressure. A measure of charge differential.

voltage drop voltage drops in exact proportion to the resistance in a component or circuit. The voltage drop calculation is made to analyze component and circuit conditions.

voltage-drop testing closed electrical circuit testing in which voltage loss is measured in fractions of a volt to locate high resistance wiring, terminals, or components.

voltage regulator the solid state device used to manage field current through an alternator.

volumetric efficiency a measure of the breathing efficiency of an engine cylinder: the ratio of gas at atmospheric pressure versus the actual amount of air in the cylinder at the completion of the intake stroke. Usually better than 100 percent in turbo-boosted engines.

volute a snail-shaped, diminishing sectional area such as those used in turbocharger geometry; may possess a fixed or variable flow area.

vortex flow MAF sensors MAF sensor with an obstruction in the air flow that causes air passing around it to generate vortices (like mini-tornadoes); these increase with air flow velocity, so by locating an ultrasonic speaker and pickup (microphone) across the stream of vortices that spin off in opposing directions, a frequency-modulated shift can be signaled to the ECM; the frequency increases in proportion to air velocity.

VSS vehicle speed sensor.

wall flow filter an extruded ceramic or SiC DPF filter with hundreds of square channels per square inch through which exhaust gas is routed; the objective is to entrap soot.

water-in-fuel (WIF) sensor usually located in the filter/water separator sump, it signals an alert when covered with water.

waterless engine coolant (WEC) an EG-based premixed coolant that can last the life of the engine without the need for testing or replacement so long as it is never contaminated with water.

water separator a canister located in a fuel subsystem used to separate water from fuel and prevent it from being pumped through the injection circuit.

watt a unit of power commonly used to measure mechanical and electrical power. Named after James Watt (1736–1819).

wavelength frequency or the distance between the nodes and antinodes of a radiated or otherwise transmitted wave.

weatherproof connectors commonly used, sealed electric and electronic circuit connectors used in many electronically managed circuits.

wet liners cylinder block liners that have direct contact with the water jacket and therefore must support cylinder combustion pressures and seal the coolant to which they are exposed.

white smoke caused by liquid condensing into droplets in the exhaust gas stream. Light reflects or refracts from the droplets, making them appear white to the observer.

wide open throttle (WOT) a term usually used in the context of SI gasoline-fueled engines to mean full fuel request. Used by one diesel manufacturer to describe *high-idle* speed.

windings any of a number of different types of coils used in electromagnetic devices such as solenoids, starter motors, and alternators.

Windows the Microsoft Corporation graphical user interface (GUI) program manager widely used in PC systems including all those used in the automotive industry.

wireless the use of radio frequencies to transmit analog or digital signals.

wireless fidelity (WiFi) wireless communications that conform to Institute of Electrical and Electronics Engineers (IEEE) protocol 802.11; the WiFi communications standard used in North America.

wobble plate pump slang term for a swash plate pump.

word processing using a computer system to produce mainly text in documents and files.

word size the number of bits that a CPU can process simultaneously, a measure of processing speed. The larger the word size, the faster the CPU. A 286 CPM is a 16-bit processor while a Pentium is a 64-bit processor.

work when force produces a measurable result, work is accomplished.

Workplace Hazardous Materials Information System (WHMIS) the section of the Hazard Communications legislation that deals with tracking and labeling of hazardous workplace materials.

wrist pin the pin that links the connecting rod eye to the piston pin boss; also known as *piston pin*.

write once, read many (WORM) optical disk that can be written to once (permanently) and read many times.

wye a diode bridge arrangement in which the diodes are connected in a "Y" arrangement, with each diode extending from a neutral junction. Used to rectify AC to DC.

yield point the point at which a material succumbs to stress to permanently deform.

yield strength unit force required to permanently deform a material; in steels, yield strength is approximately 10 percent less than tensile strength.

zener diode a diode that will block reverse bias current until a specific breakdown voltage is achieved.

Acronyms

A ampere

ABS antilock brake system

AC alternating current

A/C air-conditioning

ACC adaptive cruise control

ACT air cleaner temperature

ADC analog-to-digital converter

ADEM advanced diesel engine management (system)

ADL automatic door locks

ADS Association of Diesel Specialists

AFC air/fuel control

AFR air/fuel ratio

ALCL assembly line communications link

ALDL assembly line diagnostic link

ALU arithmetic and logic unit

AMU air manifold unit

ANSI American National Standards Institute

AOS automatic occupant sensor

API American Petroleum Institute

APT American pipe thread

AQI air quality index

ASCII American Standard Code for Information Interchange

ASE (National Institute for) Automotive Service Excellence

ASME American Society of Mechanical Engineers

ASTM American Society for Testing Materials

At ampere turns

ATC automatic traction control

ATDC after top dead center

ATF automatic transmission fluid

atm unit of pressure equivalent to one unit of atmospheric pressure

ATM asynchronous transfer mode

AWG American wire gauge

BARO barometric pressure sensor

BBM body builder module

BCM body control module

BDC bottom dead center

BEC bussed electrical center

BHM bulkhead module

BHP brake horsepower

BIOS basic input/output system

BMEP brake mean effective pressure

BOE beginning of energizing

BOI beginning of injection

BP barometric pressure

BP brake power

BPS bits per second

bsfc brake-specific fuel consumption

BTDC before top dead center

BTM brushless torque motor

Btu British thermal unit

C carbon

CA communications adapter (serial link)

CAC charge air cooling

CAD computer-assisted design

CAFE corporate average fuel economy

CAM computer-assisted machining/manufacturing

CAN Controller Area Network

CARB California Air Resources Board

CCA cold cranking amps

CCP climate control panel

CCV closed-circuit voltage

CCV closed crankcase ventilation

CCW counterclockwise or left-hand rotation

CD compact disc

CD-ROM compact disc–read-only memory

C-EGR cooled exhaust gas recirculation

CEL check engine light

CEO chief executive officer

cfm cubic feet (per) minute

CFR ceramic fiber reinforced

CG constant geometry

CGI compacted graphite iron

CGT constant geometry turbocharger

CI compression ignition

cid cubic inch displacement

CLM coolant level module

CMAC chassis-mounted charge air cooling

CMP camshaft position

CN cetane number

CNG compressed natural gas

CO carbon monoxide

Codec coder/decoder

CPL control parts list (Cummins parts number)

CPS camshaft position sensor

CPS characters per second

CPU central processing unit

CRC cyclic redundancy check

CRT cathode ray tube

CSA Canadian Safety Association

CTS coolant temperature sensor

CTV continuously open throttle valve

CVRSS continuously variable road sensing suspension

CVSA Commercial Vehicle Safety Alliance

CW clockwise

DAC digital-to-analog converter

DC direct current

DCA diesel coolant additives

DCL data communication link

DCS dealer communication system

DDL diagnostic data link

DDR digital diagnostic reader

DDT digital diagnostic tool

DDU digital (or dash) unit

DEF diesel exhaust fluid

DI direct injection

DIC driver information center

DID driver information display

DIMM double inline memory module

DIN Deutsch industrial norm (German standards: some OEM wiring schematics)

DIP dual inline package (chip)

DLC data link connector

DMM digital multimeter

DOC diesel oxidation catalyst

DOHC double overhead cam

DOS disk operating system

DOT Department of Transportation

DMF diesel multistage filter

DPF diesel particulate filter

DRAM dynamic RAM

DRL daytime running lights

DSP digital signal processing

DTC diagnostic trouble code

DVD digital video disc

DVD-ROM digital video disk–read-only memory

DVOM digital volt ohmmeter

DZM door zone module

EBCM electronic brake control module

ECB electronic circuit breaker

ECD emission control device

ECI electronically controlled injection

ECM electronic/engine control module

ECS evaporative (emission) control system

ECT engine coolant temperature

ECU electronic/engine control unit

EEC electronic engine control

EEPROM electronically erasable programmable read-only memory

EFPA electronic foot pedal assembly

EFRC engine family rating code

EG ethylene glycol

EGR exhaust gas recirculation

EHI electrohydraulic injector

EIA Electronics Industries Association

ELC extended life coolant

EMA Engine Manufacturer's Association

EMF electromotive force

EMI electromagnetic injector

EMI electromagnetic interference

EMM extended memory managers (ECMs)

EOE ending of energizing

EOF end of frame (multiplexing packet)

EOI ending of injection

EOL end of line (programming)

EPA Environmental Protection Agency

EPS engine position sensor

ESI electronic service information

ESS engine speed sensor

EST electronic service tool

ET Electronic Technician (Caterpillar diagnostic software)

ETR energized to run

EUI electronic unit injector

EUP electronic unit pump

E&C entertainment and comfort

FE iron

FET field effect transistor

FIC fuel injection control module

FM frequency modulation

FMI failure mode indicator (SAE)

FMVSS Federal Motor Vehicle Safety Standard

FPDM front passenger door module

FPS fuel pressure sensor

FRC fuel ratio control

FST field service tips

GDS global diagnostic system (GM)

GM General Motors (Corporation)

gnd ground

gph gallons per hour

GPS global positioning satellite

GUI graphical user interface

H hydrogen

HC hydrocarbon

HDEO heavy-duty engine oil

HE hydro-erosive (machining technology)

HEUI hydraulically actuated electronic unit injector

HEV hybrid electric vehicle

Hg mercury

HLA hydraulic launch assist

H₂O water

HUD head-up display

HVAC heating, ventilating, and air-conditioning

IAP injection actuation pressure

IBC inner base circle

I/C integrated circuit

ICP injection control pressure (Navistar)

ICU instrument control unit

id inside diameter

ID identify

IDI indirect injection

IDM injector driver module (Navistar)

I-EGR internal exhaust gas recirculation

IHP indicated horsepower

IMEP indicated mean effective pressure

INSITE Cummins PC software

IP instrument panel

IPC instrument panel cluster

IPR injection pressure regulator (Navistar)

IRT injector response time

ISB Interact System B series (Cummins)

ISC Interact System C series (Cummins)

ISIS International (trucks) (online) service information system

ISO International Standards Organization

IVS idle verification switch

KAM keep-alive memory

KAMPWR electrical circuit that powers KAM (Navistar)

KDD keyboard display driver

km kilometers

KPa kilopascals

LAN local area network

LCD liquid crystal display

LCM lighting control module

LED light-emitting diode

LGM liftgate module

LNC lean NO$_x$ catalyst

LNG liquefied natural gas

LPG liquid petroleum gas

LS limiting speed

LS low sulfur

LSG limiting speed governor

m meter

MAP manifold actual pressure

MB-906 Mercedes-Benz 6.4L engine

MCI Mastertech communications interface

MCI-CA Mastertech communications interface—communications adapter

MCT manifold charge temperature

MD Master Diagnostics (Navistar troubleshooting software)

MDI multiple diagnostic interface (GM CA)

MEP mean effective pressure

MEUI mechanically actuated, electronically controlled, electronic unit injector

MHz megahertz

MID message identifier (SAE)

MIL malfunction indicator lamp

MIPS millions of instructions per second

mm millimeter

mmf magnetomotive force

MON motor octane number

MOSFET metal oxide semiconductor field effect transformer

MOST media-oriented system transfer (infotainment bus)

MPa megapascals

MPC multiprotocol cartridge/card (ProLink)

MPP massively parallel processors

MSDS material safety data sheets

MSM memory seat module

MTS Mastertech scan tool (Bosch)

MUI mechanical unit injector

MUX multiplexed

MVCI Mastertech vehicle communications interface (Bosch CA)

MY model year

N nitrogen

NAC NO_x adsorber catalyst

NAFTA North American Free Trade Agreement (Mexico, U.S.A., and Canada).

NAP network access point (wireless data transfer module)

NBF German acronym for needle motion sensor (Bosch)

Nm Newton-meter

NO₂ nitrogen dioxide

NOP nozzle opening pressure

NOₓ oxides of nitrogen

NPN negative-positive-negative (semiconductor)

NTC negative temperature coefficient

NVMS nozzle valve motion sensor

NV-RAM nonvolatile random access memory

O oxygen

OBC outer base circle

OBD onboard diagnostics

OC occurrence count

OCR optical character recognition

OCV open circuit voltage

od outside diameter

OEM original equipment manufacturer

OOS out-of-service

OPEC Organization of Petroleum Exporting Countries

OS operating system

OSHA Occupational Safety and Health Administration

Pa Pascal

Pb lead

PC personal computer

PC port closure (spill timing)

PCM powertrain control module

PCM pulse code modulation

PCMCIA personal computer card International Association

PCU powertrain control unit

PCV positive crankcase ventilation

PDI predelivery inspection

PDM power distribution module

PE pump (injection) enclosed (actuation—integral camshaft)

PF pump (injection) foreign (actuation—external camshaft)

PG propylene glycol

pH power hydrogen (measure of acidity/alkalinity)

PHC partially burned hydrocarbons

PHH parallel hydraulic hybrid

PID parameter identifier (SAE)

PLC power line carrier (multiplexing)

PLC programmable logic controller (smart relay)

PLD German acronym meaning ECM (MB engines)

PLD engine control unit (Mercedes-Benz)

PLN pump-line-nozzle (diesel fuel injection)

PM particulate matter

PM preventive maintenance

PN positive-negative (junction semiconductor)

PNP positive-negative-positive (semiconductor)

POST power-on self-test

POTS plain old telephone service

ppb parts per billion

PPID proprietary parameter identifier (specific to one OEM)

ppm parts per million

PRIME pre-injection metering (Caterpillar HEUI)

PRT platinum resistance thermometer

PSID proprietary subsystem identifier (specific to one OEM)

PTC positive temperature coefficient

PTO power take-off

PW pulse width

PWM pulse-width modulation

QSOL QuickServe OnLine (Cummins)

R resistance

RAM random access memory

RCRA Resource Conservation and Recovery Act

R&D research and development

RDI remote data interface

RE Bosch rack actuator

RFA remote function actuator

RKE remote keyless entry

rms root mean square

ROM read-only memory

RON research octane number

RP recommended practice (ATA-TMC)

RQV Bosch VS governor

RS regeneration system (DPF operation)

RS-232 port standard telephone jack

RSG road speed governing

RSL road speed limit

RSV Bosch VS governor

RTD resistance temperature detectors

RVC rear vision camera

S sulfur

SAE Society of Automotive Engineers

SAE J1587 data bus software protocols

SAE J1667 HD emission testing standards

SAE J1708 data bus hardware protocols

SAE J1850 generation 2, light duty data bus

SAE J1930 recommended acronyms/terminology

SAE J1939 data bus hardware/software protocols (CAN 2.0 HD)

SAE J1962 current light duty, 16-pin data connector

SBD strategy-based diagnostics

SCA supplemental cooling (system) additive

SCFR squeeze cast, fiber-reinforced

SCR selective catalytic reduction

SCR silicone-controlled rectifier

SCSI small computer system interface

SEL stop engine light

SEO stop engine override

SFC specific fuel consumption

SFN special field notification (type of TSB)

SHH series hydraulic hybrid

s.i. Système international (metric system)

SI spark ignited

SiC silicon carbide

SID subsystem identifier (SAE)

SIMM single inline memory module

SIR supplemental inflatable restraint

SIS service information systems (generic and Caterpillar specific)

SMCC stepper motor cruise control

SOH state of health

SOHC single overhead cam

SOI start of injection

SPI serial parallel interface

SPL smoke puff limiter

SRAM static random access memory

SRS supplemental restraint system (generic)

SRS synchronous reference sensor (DDEC)

STEO stop engine overide

STI self-test input

STID station ID (GM OnStar)

STOP stop engine light

STS Service Technicians Society

SVGA super video graphics array

SWM steering wheel module

SWPS steering wheel position sensor

TBN total base number (lubes)

TBS turbo-boost sensor

TCC torque converter clutch

TCP throttle charge pressure

TCP transmission control protocol (multiplexing)

TCS traction control system

TCU transmission control unit

TDC top dead center

TDS total dissolved solids

TEL tetraethyl lead (gasoline)

TEL top engine limit (Cat: high idle)

TEM timing event marker

TIG tungsten inert gas (welding)

TIM tire inflation monitor

TIP throttle inlet pressure

TIR total indicated runout

TMC Technology and Maintenance Council (heavy trucks)

TML tetramethyl lead (gasoline)

TP throttle position

TPM tire pressure monitor

TPS throttle position sensor

TQM total quality management

TRS timing reference sensor

TSB technical service bulletin

TTS transmission tailshaft speed

UART universal asynchronous receive and transmit

UBEC underhood bussed electrical center

UGDO universal garage door opener

UHC unburned hydrocarbons

UIS unit injection system (Bosch EUI fuel system)

ULEV ultralow emissions vehicle

ULS ultralow sulfur (fuel)

ULSD ultralow sulfur diesel

UNC unified (thread) coarse

UNF unified (thread) fine

UPC universal product code

USB universal serial bus

V volt

V-Bat battery system voltage

VCO valve closes orifice (nozzle)

VCP valve closing pressure

VCU vehicle communications unit

VCU vehicle control unit

VDT video display terminal

VEPS vehicle electronics programming system

VES variable effort steering

VF vacuum fluorescent

VGA video graphics array

VGT variable geometry turbocharger

VI viscosity index

VIA visible image area

VIC vehicle interface connector

VIN vehicle identification number

VIP vehicle interface program

VIU vehicle interface unit

VOC volatile organic compound

VPM vehicle personality module (Caterpillar/ Navistar)

VR voltage regulator

V-Ref reference voltage (almost always ±5V DC)

VS variable speed

VSC variable speed control

VSG variable speed governor

VSL variable speed limit

VSL vehicle speed limit

VSS vehicle speed sensor

VVT variable valve timing

W watt

WEC waterless engine coolant

WHMIS Workplace Hazardous Materials Information System

WIF water in fuel (sensor)

WiFi wireless fidelity

WORM write once, read many

WOT wide open throttle

WSS wheel speed sensor

WT World Transmission (Allison)

Index

Note: Page numbers referencing figures are italicized and followed by an "*f.*" Page numbers referencing tables are italicized and followed by a "*t.*"